774 CAU

Handbook
of Optical Holography

Contributors

Gilbert April

Henri H. Arsenault

Paul L. Bachman

N. Balasubramanian

Stephen A. Benton

Gerald B. Brandt

David Casasent

W. Thomas Cathey

H. J. Caulfield

Burton R. Clay

Donald H. Close

Mary E. Cox

John B. DeVelis

Kalyan Dutta

Thomas K. Gaylord

James W. Gladden

Walter Koechner

Robert L. Kurtz

Sing H. Lee

Matt Lehmann

Robert D. Leighty

E. N. Leith

Hua-Kuang Liu

Robert B. Owen

George O. Reynolds

William T. Rhodes

Anthony Tai

Brian J. Thompson

Juris Upatnieks

J. R. Varner

F. T. S. Yu

Handbook of Optical Holography

H. J. CAULFIELD

Aerodyne Research, Inc.
Bedford, Massachusetts

1979

ACADEMIC PRESS

A Subsidiary of Harcourt Brace Jovanovich, Publishers

New York   London   Toronto   Sydney   San Francisco

ACADEMIC PRESS, INC.
111 Fifth Avenue, New York, New York 10003

United Kingdom Edition published by
ACADEMIC PRESS, INC. (LONDON) LTD.
24/28 Oval Road, London NW1 7DX

Library of Congress Cataloging in Publication Data

Main entry under title:

Handbook of optical holography.

Includes bibliographies and index.
1. Holography––Handbooks, manuals, etc.
I. Caulfield, Henry John, Date
TA1540.H36   774   79–51672
ISBN 0–12–165350–1

PRINTED IN THE UNITED STATES OF AMERICA

79 80 81 82   9 8 7 6 5 4 3 2 1

To my mother and father

Contents

Contents

4 Major Hologram Types

5 Variations

6 Image Formation/*Juris Upatnieks*

7 Cardinal Points and Principal Rays for Holography/*Henri H. Arsenault*

8 Equipment and Procedures

9 Special Problems

10 Application Areas

Contents

List of Contributors

Numbers in parentheses indicate the pages on which the authors' contributions begin.

Gilbert April (165), Laboratoire de Recherches en Optique et Laser, Université Laval, Quebec G1K 7P4, P.Q., Canada

Henri H. Arsenault (165, 239), Laboratoire de Recherches en Optique et Laser, Université Laval, Quebec G1K 7P4, P.Q., Canada

Paul L. Bachman (89), Aerodyne Research, Inc., Bedford Research Park, Bedford, Massachusetts 01730

N. Balasubramanian (269, 621), Optics Consultant, Cupertino, California 95014

Stephen A. Benton (349), Research Laboratories, Polaroid Corporation, Cambridge, Massachusetts 02139

Gerald B. Brandt (463), Westinghouse Research and Development Center, Pittsburgh, Pennsylvania 15235

David Casasent (503), Department of Electrical Engineering, Carnegie-Mellon University, Pittsburgh, Pennsylvania 15213

W. Thomas Cathey (127, 191, 199, 205, 219), Department of Electrical and Computer Engineering, University of Colorado, Denver, Colorado 80202

H. J. Caulfield (181, 209, 367, 587, 601), Aerodyne Research, Inc., Bedford Research Park, Bedford, Massachusetts 01730

Burton R. Clay (415), RCA Corporation, Burlington, Massachusetts 01803

Donald H. Close (573), Hughes Research Laboratories, Malibu, California 90265

Mary E. Cox (561), Department of Physics and Astronomy, University of Michigan-Flint, Flint, Michigan 48503

John B. DeVelis (69, 139), Department of Physics, Merrimack College, North Andover, Massachusetts 01845

Kalyan Dutta (13), Block Engineering, Inc., Cambridge, Massachusetts 02139

Thomas K. Gaylord (379), School of Electrical Engineering, Georgia Institute of Technology, Atlanta, Georgia 30332

James W. Gladden (277), Center for Coherent Optics, U.S. Army Engineer Topographic Laboratories, Fort Belvoir, Virginia 22060

Walter Koechner (257, 613), Science Applications, Inc., McLean, Virginia 22102

Robert L. Kurtz (299), TAI Corporation, Huntsville, Alabama 35803

List of Contributors

Sing H. Lee (537), Department of Applied Physics and Information Science, University of California, San Diego, La Jolla, California 92093

Matt Lehmann* (447), Systems Techniques Laboratory, SRI International, Menlo Park, California 94025

Robert D. Leighty (277), Center for Coherent Optics, U.S. Army Engineer Topographic Laboratories, Fort Belvoir, Virginia 22060

E. N. Leith (1), Environmental Research Institute of Michigan, Ann Arbor, Michigan 48107

Hua-Kuang Liu (299), Department of Electrical Engineering, University of Alabama, Tuscaloosa, Alabama 35401

Robert B. Owen (299), Space Sciences Laboratory, Marshall Space Flight Center, Alabama 35812

George O. Reynolds (69, 139), Arthur D. Little, Inc., Cambridge, Massachusetts 02140

William T. Rhodes (373), School of Electrical Engineering, Georgia Institute of Technology, Atlanta, Georgia 30332

Anthony Tai (51), Department of Electrical and Computer Engineering, Wayne State University, Detroit, Michigan 48202

Brian J. Thompson (29, 43, 157, 609), College of Engineering and Applied Science, The University of Rochester, Rochester, New York 14627

Juris Upatnieks (225), Environmental Research Institute of Michigan, Ann Arbor, Michigan 48107

J. R. Varner (595), Physics Division, Research Laboratories, Eastman Kodak Company, Rochester, New York 14650

F. T. S. Yu (51), Department of Electrical and Computer Engineering, Wayne State University, Detroit, Michigan 48202

* Present address: Information Systems Laboratory, Electrical Engineering Department, Stanford University, Stanford, California 94305.

Preface

Holography is in its third historical cycle. The first, in the late 1940s, derived from Gabor's first papers in *Nature*. The field attracted brilliant researchers (Lohmann, Rogers, etc.) but little general interest. The second, in the mid 1960s, derived from the Leith and Upatnieks papers in the *Journal of the Optical Society of America* and from the almost simultaneous availability of continuous wave visible lasers. That time the enthusiasm of holography's proponents was so great that the reality appeared to fall far short of the promise. Major holography efforts were started but soon dissolved. Many "holographers" were forced into other fields. Government support dried up. The third and present phase has no clear birthdate and no clear seminal paper. It began in the mid 1970s with a slow but steady rebirth of interest and support. This is a phase in which enthusiasm is great but is tempered by realism. In the midst of this phase, a number of us who have worked in holography for many years thought it wise to gather together what we knew so far, in the hope that such a "handbook" would help the field we enjoy and love to progress in a rapid and orderly way.

This, then, is a book with a mission. The success of that mission requires that readers not seek the wrong things from this book. It is not intended as either a self-study book or a college textbook (although it might supplement other books for those purposes). This is a book for people who want to use holography—whether for industry, government, health services, education, or research. Here you can go to answer such questions as

Is holography of any potential value in solving my particular technical problem?

How good is a holographic lens?

What is the formula for vibrational sensitivity for the kind of hologram I am making?

How do I choose components for my holography setup?

What is this undefined jargon in the technical paper I am reading?

What recording medium should I use?

Preface

It is not intended that this book be read through as one might read a novel or even a textbook. Rather, it should be the book to which the reader turns when he has a specific question.

The list of potential users includes research workers and students, teachers, application engineers, government technical administrators, contract monitors, and policy makers, and users of holographic equipment.

Deliberately omitted to keep the size of the book within reasonable bounds are many important areas of nonoptical holography such as acoustic, microwave, γ- and x-ray, electron, and computer holography.

I have enjoyed editing this book, or at least it seems so now that the inevitable browbeating and clerical problems are behind me. The authors took their assignments seriously and deserve much credit for their good manuscripts. Beyond thanking them, I want to thank some patient employers, J. S. Draper and E. R. Schildkraut, a marvelous secretary, Shirley Fedukowski, and the editorial staff of Academic Press.

Handbook
of Optical Holography

1

Introduction

E. N. Leith

The basic process of photography consists of forming an image of an object (either two or three dimensional) and projecting this image onto a light sensitive surface. Each object point is converted into a corresponding image point, and one is concerned only with the brightness, or irradiance, distribution of the image.

Holography, although also a photographic process, is radically different in concept. Here, the goal is not to record merely the irradiance distribution of an image, but in effect to record the complete wave field as it intercepts the recording plane, which in general is not even an image plane. Recording of the complete wave field means recording the phase as well as the amplitude. The problem lies of course in recording phase. The amplitude (or its square, the irradiance) is easily recorded; any photographic recording material can do that. All detectors are totally insensitive to the phase differences among the various parts of the field. Yet, information about the object is carried in the phase structure, as well as in the amplitude structure, of the field, and both must be sensed if the wave field is to be wholly recorded.

Gabor (1948, 1949, 1951), in his invention of holography, solved the basic problem by means of a background wave, which converts phase differences into intensity differences; thus, phase becomes encoded into a quantity that photographic film can recognize. To this record Gabor applied the name *hologram,* meaning whole record. The pattern of the wave is in effect imprinted into the hologram in such a way that at any desired later time the wave field can be exactly regenerated simply by illuminating the hologram with an appropriate beam of light. This beam, upon passing through the hologram, acquires the phase and amplitude modulation characteristics of the original wave field. It is as though the original wave were captured by the plate and later released. The reconstructed wave then propagates as if it had never been interrupted. An observer in the path of the beam will find it indistinguishable from the original wave. He will seem to see the original object, just as he would have seen it if it were still there. He will see it with all the optical properties one expects from viewing the real world; there will be full three dimensionality and all the normal parallax relations of real life. This striking realism is certainly what has made holography a subject of enormous fascination for scientist and laymen alike. Indeed, holography is a most radical departure from conventional photography.

Holography had an important precursor in the Bragg x-ray microscope (Bragg, 1929, 1939, 1942) and in the even earlier work of Wolfke (1920). Bragg, too, had been concerned with obtaining a complete record of the scattered wave field from an object, in his case, a crystal illuminated with x rays. Like holography, Bragg's method was a two-step diffraction process. The scattered x rays from the crystal were photographically recorded, then used to create an analogous field with visible light. In Bragg's case (as well as in Wolfke's) the crystal was a three-dimensional periodic structure, hence under plane wave illumination, only one diffracted wave component (spatial frequency) was produced at a time, in accordance with the rules of Bragg diffraction. This difference is not fundamental to the theory. In any event, one must record the phase and the amplitude, and of course detectors record only the amplitude. Bragg's method was to choose a particular kind of crystal with a symmetry such that the far field diffraction pattern (Fourier transform) of the object distribution is purely real, having no phase. Further, the crystals under consideration were those with a heavy atom at the center, thus providing a bias background which made the Fourier transform positive, as well as real. Thus it sufficed to measure only the magnitudes of the plane waves representing the Fourier components. Bragg, after recording the wave amplitude, would construct a mask consisting of openings whose positions and size represented the values of the Fourier components. The mask, when illuminated with coherent light, would form a far field diffraction pattern that was an image of the atomic structure of the crystal. This work was extended by Buerger (1950), and Boersch (1967) carried out similar experiments in Germany.

This work had been in part anticipated in 1920 by Wolfke, whose work in the meantime had been forgotten. Wolfke also considered the possibility of using the recorded x-ray diffraction pattern from a crystal to obtain an optical image of the crystal lattice and then illuminating the diffraction pattern transparency with a beam of monochromatic light to produce the lattice image, noting that the object must be symmetrical and "without a phase structure."

Gabor's process of holography was suggested by the Bragg microscope. His aim was to improve the image quality of the electron microscope, which suffered from spherical aberration that could not be corrected to the high degree that optical lenses are aberration corrected. The electron lenses are magnetic fields, and their properties cannot be controlled with the precision that can be achieved with optical lenses. Gabor's solution was ingenious and a sharp departure from traditional electron microscopy. He would record the scattered field of the illuminated object, then regenerate the field with optical waves. The spherical aberration would carry over to the optical domain, where it could be corrected by the well known techniques of the lens designer. Prior to undertaking the electron microscope project, he demonstrated the feasibility of the technique, using optical waves for both the making and the reconstruction processes.

Aside from the wavelengths involved (and the use of electron waves instead of electromagnetics), Gabor's proposed method differed from Bragg's in a number of ways. Gabor's process did not produce Bragg diffraction, and the entire field was available at one instant for recording. Also, Gabor's process dealt with Fresnel rather than Fraunhofer diffraction; this distinction is not fundamental, but it did facilitate carrying out the process. The principal distinction is that Gabor's process did not depend on a special class of objects that produced a positive real Fourier transform. Gabor's method required a coherent background wave, analogous to the strong scattering center of the Bragg method, but he was able to produce his coherent background as he wished. In this method, a transparency $s_0 + s$ is illuminated with a coherent light beam, where s_0 is the uniform part of the transparency (the portion of zero spatial frequency) and s is the nonzero spatial frequency part. The Fresnel diffraction pattern can be written

$$u_0 + u, \quad \text{where} \quad s_0 = u_0$$

(i.e., the coherent background is unaltered by the diffraction process), and the irradiance is

$$|\,u_0 + u\,|^2 = |\,u_0\,|^2 + |\,u\,|^2 + u_0 u^* + u_0^*\,u;$$

this is the basic equation of the Gabor technique. If this irradiance distribution is recorded and the record illuminated with a coherent beam, a portion of the resulting field will represent the term $u_0^* u$, which is a regeneration of the nonzero spatial frequency part of the nondiffracted field. Combining this with the background term $|\,u_0\,|^2$ produces a wave which seems to emanate from a virtual object $s_0 + s$ located at the position of the original object.

The process lends itself to two basic interpretations, depending on whether or not we choose to regard s_0 as a part of the object. If s_0 is part of the object, then a photographic recording of the object field results in a complete loss of phase of the object field. But by choosing the object so that the uniform part predominates, the phase of the diffraction pattern is nearly constant, and the loss of the phase is relatively unimportant. This view stresses the similarity to the Bragg process, where because of symmetry and the strong background scatterer, there is no phase to be lost, and the reconstruction can be exact. With the strong background but without the symmetry, as in Gabor's case, this loss of phase, although not catastrophic, does lead to the difficulty of the twin image term $u_0 u^*$.

By an alternative view we think of the object as being only the portion s, with the uniform part being added so as to produce a strong background wave. Again, recording the intensity results in a loss of phase of the total wave $u_0 +$ u, but the phase of the signal part u is preserved, although imperfectly, because of the presence of the other term $u_0^* u$.

Following Gabor's invention of holography, many researchers began working in this new area. Haine, Dyson, and Mulvey continued the effort to make successful holograms with the electron microscope (Haine and Dyson, 1950; Haine and Mulvey, 1952). As with Gabor, the results were less than had been desired. Numerous practical difficulties barred success, including object instability and voltage instabilities in the electron lens power supply. Others pursued purely optical holography, including Rogers (1952), El-Sum and Kirkpatrick (1952), El-Sum (1952), Baez (1952), and Lohmann (1956). The imaging obtained with holography, however, was poor, and interest in this technique subsided until by the 1950s there was little remaining activity in this once promising area. The primary reason for the poor imagery was the twin image. There were other difficulties; the term $|u|^2$ (i.e., self-interference among the scattered waves from the various object points), extraneous terms due to the inevitable nonlinearities of the recording process, and the scattered light from various scattering centers, such as dust and scratches on the various optical elements, all produced noise which overlay the reconstructed image, giving a displeasing appearance. The scatterer noise is not a defect of holography per se but is intrinsic to the coherent light used for holography. Any scatterer in the system produces a wake of scattered light which propagates downstream with the background beam, interfering with it, and producing extraneous patterns that are recorded on the hologram and ultimately overlie the final image.

It has been said that the lack of a bright coherent source (e.g., the laser) caused the early failure of holography. We doubt that this is the case; our own experience in holography and coherent optical processing during our prelaser period, 1955–1962, in general indicated that the brightness and coherence levels obtainable with the mercury arc source were adequate for a wide range of applications, not only for laboratory experiments but even for operational equipment. In short, we had quite phenomenal success.

It was during the ebb of holography that our work, which led to the revival of holography, began. This revival process was a complicated one, with some rather unusual aspects; in particular, it was not just one wave, but several, with each reaching successively further.

The first, which is perhaps best regarded as a precursor, resulted in a minirevival of holography. In 1955, while working in the area of radar, we rediscovered Gabor's process of holography. Our theory was that if radar returns were recorded on photographic film, or a similar optical transparency,

4

and then illuminated with a beam of coherent light, the resulting diffracted light waves could be replicas in miniature of the original radar waves that impinged on the receiving aperture of the radar system. The theory, as it was originally developed, considered both the cases of conventional, real antenna systems and the synthetic aperture system. From the standpoint of holography, it is of course unimportant whether the sample wavefronts are recorded simultaneously (the real aperture) or sequentially (the synthetic aperture). We developed an extensive theory of holography that in many ways paralleled Gabor's original work, which at that time was not known to us.

Despite the prior work of Gabor, our work had some original aspects. First, it introduced into holography the concept of the carrier frequency (i.e., the off-axis technique), which has so effectively disposed of the twin image problem. Second, it addressed the problem of lateral dispersion, which has to do with the tendency of carrier frequency holograms, because of their gratinglike properties, to spectrally disperse the reconstructed waves, thus leading to greater monochromaticity (i.e., temporal coherence) requirements for the off-axis hologram. It proposed the use of a grating that matches the spatial carrier of the hologram in order to compensate for the lateral chromatic dispersion of the hologram. Third, it proposed the use of a Fresnel zone plate to compensate for the longitudinal chromatic dispersion of the hologram, which results in the image plane forming at a distance proportional to the wavelength. This is, of course, the reason that Gabor's holography process requires monochromatic light for the reconstruction process, with an equivalent reason applying for the monochromaticity requirement in the hologram recording process. Thus, when we also consider that the coherence requirements for recording radar data (or indeed, any electrical data) are inherently identical whether the recording process is done in the in-line or off-axis (carrier) mode, it follows that carrier frequency holography, as originally conceived, had considerably less requirement for monochromaticity than had Gabor's original in-line method. This situation may seem surprising to many since it is often, but incorrectly, assumed that off-axis holography intrinsically has a greater monochromaticity requirement than in-line holography.

Finally, our work in a sense turned Gabor's original work around; instead of going from very short wavelengths to optical wavelengths, we went from long wavelengths to optical wavelengths. The technology for performing this alternative operation was much better in hand. It was easy to make holograms at radar wavelengths; the problems that plagued Gabor in the electron domain were not problems at all in the microwave domain. Furthermore, the basic accomplishments of holography, the preservation of the phase of a wave and the subsequent use of the phase, as well as amplitude, to create either a second wave or an image of the original object distribution, was not at all the problem here; the recording of phase and its recovery on readout, which had been Gabor's goal, had in fact been routine at radio wavelengths for many years.

1 Introduction

Indeed the theory of holography we developed was essentially a new way of interpreting old established processes. What had originally been described as an optical computation system was now described in terms of holography. This new method of describing old processes seemed to offer many new insights into the optical processing of synthetic aperture radar data. Although slow in gaining acceptance by the radar community, it eventually became firmly established by about 1960. Thus the first wave of holographic revival was hardly earthshaking, although its ultimate effects were considerable.

It is interesting to note that Rogers (1956, 1957), working at about the same time in New Zealand, also applied holography to radio waves, by recognizing that radio waves scattered from the ionsphere, if photographically recorded, could be treated as holograms.

In 1960 we experimented with optical holography, first of all duplicating Gabor's original experiment. Although the quality of the imagery was at that time hardly impressive by the standards of conventional photography, the results were nevertheless startling, inasmuch as this process seemed to create something (the image) from what appeared to be nothing. There in the optical system was an image, produced by rays of light which could be traced upstream in the optical system, toward the source, but only as far as that unintelligible piece of film called the hologram. It contained no discernible object corresponding to the image, yet the image forming rays ended abruptly there. The process, to one unversed in holography, seemed mysterious and inexplicable. Our reaction to this holographic experiment was one of fascination. How much more fascinating it must have been to Gabor and his colleagues when they observed these same effects for the first time!

Our enthusiasm prompted us to seek means for improving the imagery (Leith and Upatnieks, 1962, 1963, 1964). We reasoned that the twin image was basically an aliasing problem and the solution was to place the holographic signal on a spatial carrier. The mechanism for so doing was to introduce a separate coherent background wave, which we called the reference beam. It was to impinge on the recording plate at some nonzero angle with respect to the object wave. This resulted in the Fresnel diffraction pattern of Gabor's holographic process being overlaid with a fine fringe pattern. The photographic record of this two beam overlay became the carrier frequency, or off-axis, hologram, with its fine-line structure. Such a hologram looked like and behaved like a diffraction grating.

When we illuminated this new type of hologram, we produced, as expected, a zero-order wave which behaved like the reconstructed wave of the traditional Gabor hologram, producing the usual inseparable twin images and containing all of the other defects of the in-line case, including the intermodulation term and terms due to nonlinearities in the hologram recording process.

However, also emanating from the hologram was a pair of side orders not before seen from a hologram. These waves separated from the zero order,

6

revealing a new set of images of a quality heretofore unseen in a hologram. One side order formed a virtual image that was completely free from its twin image term and all of the other undesirable terms that had previously plagued holography. The other first-order formed a real image of similar quality. Furthermore, the images were positive, rather than the negative images formed in the zero order, as well as in the conventional Gabor technique.

This last point is worth noting in order to dispel a commonly held myth that holography as opposed to conventional photography forms a positive image from a negatively formed recording. The positiveness of the images formed from holographic negatives has nothing to do with the basic process of holography. Just as conventional photographs on photographic film form negatives, so conventional in-line holograms form negative images. Placing an image on a spatial carrier results in an image that is insensitive to the polarity of the recording process, and the image produced in this way will always be positive. Various imaging processes exist wherein images are routinely placed on a carrier and reimaged using a diffracted order. Exactly the same physical considerations lead to the expectation that a carrier-type hologram will always produce a positive image.

Our initial off-axis holography work was carried out in the prelaser area, with only the conventional mercury arc source; even with this source we had considerably more coherence than we needed.

When the laser became available, we experimented with it while continuing to use the conventional mercury arc source. Each had its own special advantages, and it was not entirely clear which was better. The laser method resulted in shorter exposures (seconds instead of minutes), made careful equalization of object and reference beam paths unnecessary, and did not require special techniques to prevent large path differences between the beams from accumulating at positions in the beam displaced from the equalization parts. On the other hand, the coherence of the laser was hundreds or thousands of times greater than that required by the process of holography as it was then practiced, and the noise, which was the principal problem in holography, became an even greater problem. We eventually chose the laser, but we found that good quality holograms could be made with either source; the decision was a matter of which set of advantages to exploit and which set of problems to attack.

Next, we introduced the concept of diffuse illumination holography, accomplished by placing a diffuser, such as ground glass, between the light source and the object. Thus the vast increase of spatial frequency bandwidth and the consequent redundancy effectively obliterated the artifact noise that had until then plagued holography. Now not only was the extremely severe requirement on clean technique considerably mitigated, but indeed, large portions of the hologram could be damaged without noticeably affecting the image. It was at this stage that holography acquired the well-known property that each portion produces the entire image. Finally, we extended the process to record the radiation scattered from real world reflecting 3D objects. This involved little new theory but considerable new experimental technique. Now for the first time, holography moved from the conventional table to the granite bench, since the introduction of reflecting objects (as well as greater separations between object and reference beam paths) led to greatly increased stability requirements.

The achievement of these experimental results had been difficult, first, because of the great stability requirements and second, because the coherence requirements were severe. This type of object superimposed upon a formerly modest coherence requirement a new requirement many orders of magnitude greater, viz., all parts of the object had to be simultaneously coherent with the reference beam. Thus the coherence length had to be of the order of twice the object depth. Now for the first time, we required the great coherence of the laser. Often the laser was not coherent enough, as for example when it oscillated in nonaxial modes, or when the frequency drifted because of cavity instabilities.

Of the various objects we chose for our initial work, one proved particularly troublesome. This object was a paper sheet torn from a calendar and pasted to an aluminum block. As many times as we holographed it, the result was always the same: a bright reconstruction occurred everywhere except at one position, where the image was persistently obliterated, the reconstruction there producing only a dark spot. Examination revealed that a hole had been drilled in the aluminum block, and the calendar sheet stretched across it had become a vibrating membrane.

Certainly the most frustrating experience, however, was with the holography of our earliest 3D reflection object, a collection of junk retrieved from odd corners of the laboratory. We knew that the image from the resulting hologram should be unlike any that had ever before been produced, having full natural 3D with complete parallax, an exact recreation of the original object. Yet observations failed to confirm these marvelous expectations. The problem was that the hologram, having been produced on a small plate, measured only about 2 cm square, and only one eye at a time could look into this holographic window.

As holography advanced through these steps the imaging improved in quality and dramatic impact, but the means for producing it became more sophisticated and more difficult. For example, stability requirements for in-line holography are exactly those of conventional photography (assuming equal exposure time in the two cases), whereas in going to off-axis holography, diffuse-illumination holography, and 3D object holography, the requirements became progressively greater, with the final jump to 3D objects being a step much greater than all the other steps together. Similarly, the coherence requirements increased along the way. For in-line holography, the coherence requirements are modest. For off-axis holography, they are basically no greater, contrary to what is often supposed. For diffuse illumination holography, there is a jump, but not so great as to require the laser. Finally, with 3D objects, the requirements make a jump that is many times greater than all of the previous ones, and one that indeed requires the laser.

About the same time that we were pursuing our research in holography, Denisyuk (1962, 1963, 1965) of the Soviet Union reported a major advance in which the process of holography was combined with a form of color photography invented in 1891 by the French physicist Gabriel Lippmann. The Denisyuk hologram can produce either monochromatic or color images when it is viewed in white light from a point source. This result is achieved by having the object and reference beams travel in opposite directions, resulting in fine fringes which are surfaces aligned nearly parallel to the film surface and with spacing of the order of a half wavelength of light. Thus ordinary emulsions with thicknesses of about 15 nm will contain about 30 or so such fringes. Denisyuk's holograms are therefore termed volume holograms, since they require the third, or depth, dimension of the emulsion as well as the lateral dimension for their operation. The consequence is that such holograms when illuminated with white light from a point source reflect a narrow wavelength band of light to form the holographic image, whereas the remaining wavelengths pass through the hologram as if through a sieve, producing no effect.

Denisyuk's work is one of the cornerstones of holography and has led to some of the best holographic images ever produced; yet it required a few years for the significance of Denisyuk's great advance to be fully appreciated.

Although the sum total of the research results in holography has become enormous, there are some that have special significance. Certainly the use of Fourier-transform holograms as complex spatial filters, such as matched filters, is highly significant and stands by itself as a major advance in the field of spatial filtering. Such filters were developed in the early 1960s in various forms for various applications, particularly for the optical processing of radar data. However, the form which now dominates is one introduced by Vander Lugt (1963) for image processing.

Hologram interferometry is an equally significant advance, and one that stands on its own as a major advance in the field of interferometry. Stemming from the 1964–1965 period, its invention has various curious, perhaps enigmatic, aspects. Holography was already 17 years old by 1964, and although dozens had worked in holography, no one had discovered hologram interferometry. And then suddenly it was independently discovered by perhaps as many as a half-dozen groups. The earliest report was by Powell and Stetson (1965), who described the time average form. Then within a few months the other forms (double exposure and real time) were reported by several groups, all working independently. Viewed as an advance in interferometry, hologram interferometry is astonishing; it permitted interferometric comparison of arbitrary waves that existed at different times, an attainment unthinkable in the context of traditional interferometry.

But why the plethora of independent, nearly simultaneous discoveries at such a late period in holographic history? I offer this explanation: Hologram interferometry arises as a consequence of failure to meet the stability requirements in holography; but those requirements are rather slight as long as the object is of the transmission rather than the reflecting type, and experiments with reflecting objects were reported only in late 1963. Thus only at this time was the stage set for hologram interferometry; prior to this time such a discovery had low probability, but after the discovery became inevitable. This explanation is perhaps somewhat oversimplified, since as we have noted, stability requirements for holography had increased all along as the degree of sophistication increased. But clearly holography was not ready for hologram interferometry before the 1960s.

It seemed apparent from the beginning of 3D arbitrary object holography that the display area was the natural one for holography, and by the mid-1960s holographers were busily exploiting this possibility. Some of the most marvelous imagery the world has ever seen was created in those years. But the science of such image formation had greatly outpaced the economics. Such holograms were expensive to make and view. And so the technology did not reach a fruition much beyond the laboratory stage.

In the 1970s this picture began to change as a consequence of several important developments. First, Benton (1969) introduced the rainbow hologram, a thin or planar hologram viewable in white light. Since this hologram utilizes the entire white light spectrum instead of just a narrow wavelength band, this hologram can be extremely bright, even when the source is one of only moderate brightness, say a 100-W bulb. Such holograms therefore can be viewed conveniently and inexpensively.

A second major development was the composite or multiplex hologram as produced by Cross (1977). This hologram is a clever engineering synthesis of many technologies (Pole, 1967; De Bitetto, 1968, 1969; King, 1968; McCrickerd and George, 1968; George et al., 1968; Redman, 1968; King et al., 1970). The hologram is formed from a great many ordinary photographs made in a conventional way. The pictures, made from different positions, constitute many views of the object, and they have in their totality all the essential information contained in the hologram. The composite hologram in the form developed by Cross can be viewed in white light just as can the rainbow hologram.

Multiplex holograms can be made from any arbitrary object, they can be made relatively inexpensively, they can be mass produced by replication, their viewing system is inexpensive, and a basic problem of holography, the inability to magnify a three-dimensional image equally in the lateral and longitudinal dimensions, is overcome. These factors give the multiplex hologram a viability in the commercial display area unmatched by other hologram types.

10

The perfection of the rainbow and multiplex holograms, along with vast improvements in the technology of Denisyuk (or volume) holograms (especially in the Soviet Union) has placed holography more than ever in the public eye.

The major advances in holography, including the new white light method, were surprising and essentially unanticipated developments. From past experience, we fully expect that there are more such surprising advances to come, and we can only speculate about their nature and their impact. With white light readout now well in hand, perhaps some really effective method will be found to produce the hologram in white light.

REFERENCES

Baez, V. A. (1952). Focusing by diffraction, *Amer. J. Phys.* **20**, 311.

Benton, S. A. (1969). Hologram reconstruction with incoherent extender sources, *J. Opt. Soc. Amer.* **59**, 1545.

Boersch, H. (1967). Holographie und Elektronenoptik, *Phys. Bl.* **23**, 393.

Bragg, W. L. (1929). An Optical method of representing the results of x-ray analyses, *Z. Kristallogr. Kristallgeometrie Kristallphys. Kristallchem.* **70**, 475.

Bragg, W. L. (1939). A new type of "x-ray microscope," *Nature* **143**, 678.

Bragg, W. L. (1942), The x-ray microscope, *Nature* **149**, 470.

Buerger, M. J. (1950). The photography of atoms in crystals, *Proc. Nat. Acad. Sci. USA* **36**, 330.

Cross, L. (1977). Multiplex holograms, *Proc. SPIE Seminar 3D Imaging*.

De Bitetto, D. J. (1968). Bandwidth reduction of hologram transmission systems by elimination of vertical parallax, *Appl. Phys. Lett.* **12**, 176.

De Bitetto, D. J. (1969). Holographic panoramic stereograms synthesized from white light recordings, *Appl. Opt.* **8**, 1740.

Denisyuk, Yu. N. (1962). Photographic reconstruction of the optical properties of an object in its own scattered radiation field, *Sov. Phys.—Dokl.* **7**, 543.

Denisyuk, Yu. N. (1963). On the reproduction of the optical properties of an object by the wave field of its scattered radiation, Pt. I, *Opt. Spectrosc. (USSR)* **15**, 279.

Denisyuk, Yu. N. (1965). On the reproduction of the optical properties of an object by the wave field of its scattered radiation, Pt. II, *Opt. Spectrosc. (USSR)* **18**, 152.

El-Sum, H. M. A. (1952). Reconstructed wavefront microscopy, Ph.D. thesis, Stanford Univ., Stanford, California (available from Univ. Microfilm Inc., Ann Arbor, Michigan).

El-Sum, H. M. A., and Kirkpatrick, P. (1952). Microscopy by reconstructed wavefronts, *Phys. Rev.* **85**, 763.

Gabor, J. D. (1948). A new microscopic principle, *Nature* **161**, 777.

Gabor, J. D. (1949). Microscopy by reconstructed wavefronts, *Proc. Roy. Soc.* **A197**, 454.

Gabor, J. D. (1951). Microscopy by reconstructed wavefronts: II, *Proc. Phy. Soc.* **B64**, 449.

George, N., McCrickerd, J. T., and Chang, M. M. T. (1968). Scaling and resolution of scenic holographic stereograms, *Proc. SPIE Seminar-in-Depth Holography,* p. 117.

Haine, M. E., and Dyson, J. (1950). A modification to Gabor's proposed diffraction microscope, *Nature* **166**, 315.

Haine, M. E., and Mulvey, T. (1952). The formation of the diffraction image with electrons in the Gabor diffraction microscope, *J. Opt. Soc. Amer.* **42**, 763.

King, M. C. (1970). Multiple exposure hologram recording of a 3D image with a 360° view, *Appl. Opt.* **7**, 1641.

King, M. C., Noll, A. M., and Berry, D. H. (1970). A new approach to computer-generated holography, *Appl. Opt.* **9**, 471.

1 Introduction

Leith, E. N., and Upatnieks, J. (1962). Reconstructed wavefronts and communication theory, *J. Opt. Soc. Amer.* **52**, 1123.

Leith, E. N., and Upatnieks, J. (1963). Wavefront reconstruction with continuous-tone objects, *J. Opt. Soc. Amer.* **53**, 1377.

Leith, E. N., and Upatnieks, J. (1964). Wavefront reconstruction with diffused illumination and three-dimensional objects, *J. Opt. Soc. Amer.* **54**, 1295.

Lohmann, A. (1956). Optische Einseitenbandübertragung angewandt auf das Gabor-Mikroskop, *Opt. Acta* **3**, 97.

McCrickerd, J. T., and George, N. (1968). Holographic stereogram from sequential component photographs, *Appl. Phys. Lett.* **12**, 10.

Pole, R. V. (1967). 3D imagery and holograms of objects illuminated in white light, *Appl. Phys. Lett.* **10**, 20.

Powell, R. L., and Stetson, K. A. (1965). Interferometric vibration analysis of three-dimensional objects by wavefront reconstruction, *J. Opt. Soc. Amer.* **55**, 612.

Redman, J. D. (1968). The three-dimensional reconstruction of people and outdoor scenes using holographic multiplexing, *Proc. SPIE Seminar-in-Depth Holography*, p. 161.

Rogers, G. L. (1952). Experiments in diffraction microscopy, *Proc. Roy. Soc. Edinburgh* **63A**, 193.

Rogers, G. L. (1956). A new method of analysing ionospheric movement records, *Nature* **177**, 613.

Rogers, G. L. (1957). Diffraction microscopy and the ionosphere, *J. Atmos. Terr. Phys.* **10**, 332.

Vander Lugt, A. (1963). Signal detection by complex spatial filtering, *J. Opt. Soc. Amer.* **53**, 1341.

Wolfke, M. (1920). Über der Möglichkeit der optischen Abbildung vom Molekulargittern, *Phys. Z.* **21**, 495.

$$\mathbb{2}$$

Background

2.1 INTEGRAL TRANSFORMS

Kalyan Dutta

An integral transform of a function $f(x)$ is another function $F(s)$ of the form

$$F(s) = \int_a^b f(x)K(x, s)\, dx, \qquad (1)$$

where $K(x, s)$, a specified function of x and s, is called the kernel of the transform. Introduction of the transform $F(s)$ in place of $f(x)$ is a device commonly used in applications to physical problems where the manipulation of $F(s)$ is simpler than that of $f(x)$. In optics, the use of the Fourier transform method [with a kernel of the form $\exp(-j2\pi sx)$] is widespread in the analysis of holographic and imaging systems.

In recent years, the use of a number of such transforms in the analysis of optical systems has increased. Each of these transforms is useful in dealing with some particular aspect of behavior of a system that makes it difficult to treat using direct or Fourier transform methods. Some of these transforms allow a simplification or an extra neatness of treatment even when direct or Fourier methods may be adequate.

In this section, descriptions and definitions of several of these transforms have been compiled, together with collections of theorems and results concerning these transforms which are useful in their manipulation. In addition, a number of transform pairs of frequently used functions are listed for each transform. Other useful transform pairs may be generated from these lists by the use of one or more of the theorems.

Many of these transforms are very closely related to the Fourier transform and thus to one another. Several of these relationships are indicated in the descriptions. These relationships are sometimes of help in solving a problem

2. Background

in one transform domain using results which are known to be true in another. Not mentioned here are the discrete counterparts of some of these transforms, as well as certain other discrete transforms whose applications are primarily in the digital processing of discretely sampled data (Andrews, 1970).

We may note that, as defined above, all integral transforms may be regarded as linear operators acting on $f(x)$ to produce $F(s)$. Therefore the transforms listed here are all linear transforms. Also, as a matter of practical applicability, we are concerned only with transforms possessing an inverse, that is, ones for which solutions exist of the form

$$f(x) = \int_c^d F(s)H(x, s)\, ds,\qquad(2)$$

this formula then being referred to as the inverse transform yielding $f(x)$ from $F(s)$. In special cases the kernels for the forward and the inverse transforms may be identical, giving a symmetrical relationship between a function and its transform.

2.1.1 The Fourier Transform

Fourier transformation is by far the most widely used coherent optical data processing operation, of use wherever the frequency analysis, filtering, correlation, or classification of signals is called for. Under certain conditions (Goodman, 1968, Chapter 4) the behavior of a coherent optical system is naturally described as that of a Fourier transform operator, most commonly performing a two-dimensional Fourier transformation.

The complex Fourier transform of a (possibly complex-valued) one-dimensional function $f(x)$ can be defined (Bracewell, 1965) as

$$F(s) = \int_{-\infty}^{\infty} f(x) \exp(-j2\pi sx)\, dx.\qquad(3)$$

The customary definition for the inverse Fourier transform is then

$$f(x) = \int_{-\infty}^{\infty} F(s) \exp(j2\pi sx)\, ds.\qquad(4)$$

Various other definitions of the Fourier transform and its inverse are possible and in common use (Bracewell, 1965, Chapter 2); in the above forms, application of the forward and then the inverse transformation to a function yields the original function. $F(s)$ is usually called the Fourier spectrum of $f(x)$; alternatively, $f(x)$ may also be regarded as the spectrum of $F(s)$.

The Fourier transform of a two-dimensional function $f(x, y)$ can be defined as

$$F(u, v) = \int\int_{-\infty}^{\infty} f(x, y) \exp[-j2\pi(ux + vy)] \, dx \, dy, \tag{5}$$

with the inversion relation being

$$f(x, y) = \int\int_{-\infty}^{\infty} F(u, v) \exp[j2\pi(ux + vy)] \, du \, dv. \tag{6}$$

Higher dimensional Fourier transforms can be defined in a similar way (Bracewell, 1965, Chapter 12; Sneddon, 1951, Chapter 1).

2.1.1.1 Some Properties of the Fourier Transform

A number of Fourier transform theorems and other results are summarized here; results are given for two-dimensional functions wherever possible. In the following, $f(x, y)$ and $F(u, v)$ [and $g(x, y)$, $G(u, v)$] are assumed to be basic transform pairs.

Separability If $f(x, y)$ can be written as $f_1(x) \cdot f_2(y)$, then the transform $F(u, v)$ is expressible as $F_1(u) \cdot F_2(v)$, where F_1 and F_2 are the one-dimensional transforms of f_1 and f_2, respectively.

Similarity and Shift Theorems A combined form of these two results is provided. $f(\alpha x - a, \beta y - b)$ transforms to

$$\frac{1}{|\alpha\beta|} F\left(\frac{u}{\alpha}, \frac{v}{\beta}\right) \exp\left[-j2\pi\left(u\frac{a}{\alpha} + v\frac{b}{\beta}\right)\right]$$

Convolution Theorem The convolution of $f(x, y)$ and $g(x, y)$, defined as

$$f(x, y) ** g(x, y) = \int\int_{-\infty}^{\infty} f(x - \xi, y - \eta) g(\xi, \eta) \, d\xi \, d\eta, \tag{7}$$

transforms to $F(u, v)G(u, v)$. Similarly, $f(x, y)g(x, y)$ transforms to $F(u, v) ** G(u, v)$.

Autocorrelation Theorem The autocorrelation of $f(x, y)$, defined as $f(x, y) ** f^*(-x, -y)$, or

$$\int\int_{-\infty}^{\infty} f(x + \xi, y + \eta) f^*(\xi, \eta) \, d\xi \, d\eta,$$

transforms to $|F(u, v)|^2$.

2. Background

Rayleigh's Theorem

$$\int_{-\infty}^{\infty} |f(x, y)|^2 \, dx \, dy = \int_{-\infty}^{\infty} |F(u, v)|^2 \, du \, dv. \tag{8}$$

Derivative Theorem $\partial[f(x, y)]/\partial x$ transforms to $j2\pi u F(u, v)$, and similarly, $\partial[f(x, y)]/\partial y$ transforms to $j2\pi v F(u, v)$.

Differentiation under Convolution

$$\frac{\partial}{\partial x}[f(x, y) ** g(x, y)] = \frac{\partial}{\partial x} f(x, y) ** g(x, y)$$

$$= f(x, y) ** \frac{\partial}{\partial x} g(x, y), \tag{9}$$

and similarly for differentiation with respect to y.

The (One-Dimensional) Transform   of $\int_{-\infty}^{\infty} f(x, y) \, dx$ is $F(0, v)$; similarly, the transform of $\int_{-\infty}^{\infty} f(x, y) \, dy$ is $F(u, 0)$. In two and higher dimensions this result has been termed the projection-slice theorem: the projection of $f(x, y)$ on an axis is the transform of a slice of $F(u, v)$ along another axis. The result is more general than is stated here; a projection of f on *any* line in the xy plane has as its transform a corresponding slice of F and vice versa.

2.1.1.2 Some Commonly Used Fourier Transform Pairs

The functions $\delta(x)$, $rect(x)$, $sinc(x)$, and $\Lambda(x)$ are defined as

$$\delta(x) = 0, \qquad x \neq 0; \qquad \int_{-\infty}^{\infty} \delta(x) = 1, \tag{10}$$

$$rect(x) = \begin{cases} 1, & |x| < \tfrac{1}{2}; \\ 0, & |x| > \tfrac{1}{2}; \end{cases} \qquad sinc(x) = \sin(\pi x)/(\pi x), \tag{11}$$

and

$$\Lambda(x) = \begin{cases} 1 - |x|, & |x| \leq 1, \\ 0 & , & |x| \geq 1. \end{cases} \tag{12}$$

With these definitions, the following Fourier transform pairs can be listed:

$\delta(x, y)$	1
$rect(x, y)$	$sinc(u) \, sinc(v)$
$\Lambda(x)\Lambda(y)$	$sinc^2(u) \, sinc^2(v)$
$\exp[-j\pi(x + y)]$	$\delta(u - \tfrac{1}{2}, v - \tfrac{1}{2})$

16

$$\exp[-\pi(x^2 + y^2)] \qquad\qquad \exp[-\pi(u^2 + v^2)]$$

$$\sum_{m,n=-\infty}^{\infty} \delta(x - m, y - n) \qquad \sum_{m,n=-\infty}^{\infty} \delta(u - m, v - n), \, m, n \text{ integer}.$$

Extensive tables of Fourier transforms are given in Campbell and Foster (1948) and Erdelyi (1954); the Fourier transform is discussed in detail by Sneddon (1951), Champeney (1973), and Bracewell (1965) and in other references therein.

2.1.2   The Laplace Transform

Though not of direct usefulness in optics, the Laplace transform is stated here for completeness. Defined with a generalized exponential kernel, it represents an extension of the concept of Fourier transformation to functions for which the Fourier transform may not exist. If for a function $f(x)$

$$\int_{-\infty}^{\infty} |f(x)| \, dx$$

is not bounded but

$$\int_{-\infty}^{\infty} |f(x)| \exp(-\sigma x) \, dx$$

is (for some real number σ), then the (two-sided) Laplace transform of $f(x)$ with respect to the complex variable p is (Carslaw and Jaeger, 1941)

$$L(p) = \int_{-\infty}^{\infty} f(x) \exp(-px) \, dx, \tag{13}$$

with the real part of p greater than σ. The inversion formula for the transform is

$$f(x) = \frac{1}{2\pi j} \int_{c-j\infty}^{c+j\infty} L(p) \exp(px) \, dp, \tag{14}$$

with $c > \sigma$.

A statement of the one-sided Laplace transform is obtained by setting to zero the lower limit of integration in the definition of $L(p)$. It will be seen that the two-sided Laplace transform contains the one-sided transform and the Fourier transform as special cases. For p imaginary, the Fourier transform is obtained while, in general, the Laplace transform of $f(x)$ is equivalent to the Fourier transform of $\exp(-\alpha x) f(x)$, where α is the real part of p.

The Laplace transform can be defined for two- and higher-dimensional functions in a manner similar to that for the Fourier transform (Sneddon, 1951, Chapter 1).

17

2. Background

2.1.2.1 Some Properties of the Laplace Transform

Results for the Laplace transform are, in general, very similar to those for the Fourier transform and are stated here briefly. If $f(x)$, $F(p)$ are Laplace transform pairs, then so are the following:

$$f(ax) \qquad \frac{1}{|a|} F\left(\frac{p}{a}\right) \qquad \text{(similarity)}$$

$$f(x-a) \qquad \exp(-ap)F(p) \qquad \text{(shift)}$$

$$f(x)*g(x) \qquad F(p){\cdot}G(p) \qquad \text{(convolution)}$$

$$f(x){\cdot}f(-x) \qquad F(p){\cdot}F(-p) \qquad \text{(autocorrelation)}$$

$$\frac{\partial}{\partial x} f(x) \qquad pF(p) \qquad \text{(derivative)}$$

$$\int_{-\infty}^{x} f(u)\,du \qquad \frac{1}{p} F(p) \qquad \text{(integral)}.$$

2.1.2.2 Some Laplace Transform Pairs

$$\delta(x) \qquad 1$$

$$\text{rect}(x) \qquad \frac{1}{p}\left[\exp\left(\frac{p}{2}\right) - \exp\left(-\frac{p}{2}\right)\right]$$

$$\Lambda(x) \qquad \left\{\frac{1}{p}\left[\exp\left(\frac{p}{2}\right) - \exp\left(\frac{-p}{2}\right)\right]\right\}^{2}$$

$$\exp(-\alpha|x|) \qquad \frac{2\alpha}{(\alpha^2 - p^2)}$$

$$H(x) \qquad \frac{1}{p}.$$

[$H(x)$ is defined as 0 for $x < 0$, 1 for $x > 0$.]

$$xH(x) \qquad \frac{1}{p^2}$$

$$\exp(-\alpha x)\,H(x) \qquad \frac{1}{p+\alpha}$$

$$\cos(\omega x)\,H(x) \qquad \frac{p}{p^2 + \omega^2}$$

$$\sin(\omega x)\,H(x) \qquad \frac{\omega}{p^2 + \omega^2}.$$

18

Tables of Laplace transform pairs as well as detailed discussion of the transform are given by Van der Pol and Bremmer (1935) and Bracewell (1965).

2.1.3   The Fourier–Bessel Transform

This transform arises from considering the two-dimensional Fourier transform to be applied to the class of circularly symmetric functions. Most optical systems and many optical signals have just this kind of symmetry. For such two-dimensional distributions, which are functions of a radius r only, it can be shown (Goodman, 1968, Chapter 7) that the transformed functions are also circularly symmetric (and thus functions of a radial frequency ρ alone), and that a function and its transform may each be obtained from the other by applying the same symmetrical one-dimensional transformation. This operation is called the Fourier–Bessel transform and can be defined as

$$G(\rho) = 2\pi \int_0^\infty f(r)J_0(2\pi r\rho)r \, dr, \tag{15}$$

having as an inverse the identical transformation

$$f(r) = 2\pi \int_0^\infty G(\rho)J_0(2\pi r\rho)\rho \, d\rho. \tag{16}$$

In these definitions J_0 is a Bessel function of the first kind and of order zero (McLachlan, 1955).

The Fourier–Bessel transform is also known as a Hankel transform of zero order and is frequently referred to simply as the Hankel transform. An entire family of such transforms is obtainable by using instead the kernels J_ν, the Bessel functions of order ν, ν being not necessarily integer-valued. The Fourier transforms of two-dimensional radially symmetric functions that have harmonic angular variation [i.e., of the specialized form $f(r) \exp(jn\theta)$] can be shown to reduce to the Hankel transforms of higher integer order, while the transforms of radial functions of higher than two dimensions can be described using various half-order Hankel transforms (Sneddon, 1951, Chapter 2).

2.1.3.1   Theorems for the Fourier–Bessel Transform

From the Fourier–Bessel transform pair $f(r)$ and $G(\rho)$, the following pairs may be derived:

$$f(ar) \qquad\qquad (1/a^2)G(\rho/a) \qquad\qquad \text{(similarity)}$$

$$\int_0^\infty r' f_1(r') \int_0^{2\pi} f_2(r^2 + r'^2 - 2rr' \cos \theta)^{1/2} \, d\theta \, dr'$$

$$G_1(\rho) \cdot G_2(\rho) \qquad\qquad \text{(convolution)}.$$

19

2. Background

In addition, the following relations are true:

$$\int_0^\infty r f(r) \, dr = G(0)/(2\pi), \tag{17}$$

$$\int_0^\infty r f_1(r) f_2^*(r) \, dr = \int_0^\infty \rho G_1(\rho) G_2^*(\rho) \, d\rho \quad \text{(Parseval)}, \tag{18}$$

$$\int_0^\infty r |f(r)|^2 \, dr = \int_0^\infty \rho |G(\rho)|^2 \, d\rho \quad \text{(Rayleigh)}. \tag{19}$$

2.1.3.2 Some Fourier–Bessel Transform Pairs

$$\delta(r - a) \qquad 2\pi a J_0(2\pi a \rho)$$
$$\text{rect}(r/2) \qquad J_1(2\pi\rho)/\rho$$
$$\exp(-\pi r^2) \qquad \exp(-\pi\rho^2)$$
$$1/r \qquad 1/\rho.$$

Some references for the Fourier–Bessel transform are Titchmarsh (1948), Sneddon (1951), and Bracewell (1965). Bessel functions are discussed by McLachlan (1955).

2.1.4 The Fresnel Transform

In describing the free propagation of coherent optical fields, and in the analysis of diffraction under conditions less restrictive than those required for Fourier transformation, the Fresnel transform (Goodman, 1977; Papoulis, 1968) plays an important role. In its basic form (Mertz, 1965) this transformation can be defined as

$$g(x) = \int_{-\infty}^\infty f(u) \exp[j\pi s(x - u)^2/\lambda] \, du, \tag{20}$$

in which form it will be recognized simply as a convolution of $f(x)$ with an exponential chirp function $\exp(j\pi s x^2/\lambda)$. The inverse transformation is similarly expressed as a convolution:

$$f(x) = \int_{-\infty}^\infty g(u) \exp[-j\pi s(x - u)^2/\lambda] \, du. \tag{21}$$

It is possible to define the two-dimensional Fresnel transformation and its inverse in the same way:

$$g(x, y) = \int\int\limits_{-\infty}^{\infty} f(u, v) \exp\{j\pi s[(x - u)^2 + (y - v)^2]/\lambda\} \, du \, dv, \qquad (22)$$

$$f(x, y) = \int\int\limits_{-\infty}^{\infty} g(u, v) \exp\{-j\pi s[(x - u)^2 + (y - v)^2]/\lambda\} \, du \, dv. \qquad (23)$$

The Fresnel transform is closely allied with the Fourier transform. By decomposing the Fresnel transform kernel it can be shown that $g(x) \exp(-j\pi sx^2/\lambda)$ and $f(y) \exp(j\pi sy^2/\lambda)$ are related by a Fourier transform. Conversely, if $f(y)$ and $g(x)$ are a Fourier transform pair, then by manipulating the transform expression a Fresnel transform relationship can be shown between $g(x) \exp(j\pi sx^2/\lambda)$ and $f(y) \exp(-j\pi sy^2/\lambda)$. The multiplication by a chirp function in these expressions is analogous to the kind of transformation performed by a thin lens on the complex amplitude of a field incident on it (Goodman, 1968, Chapter 5). This fact, coupled with a description of field propagation between lenses as a Fresnel transformation (or convolution with a chirp function), makes it possible to develop an algebra to describe coherent optical processors in which such convolutions and multiplications form the basic operations (Carlson and Francois, 1977). Other applications of the Fresnel transform include the analysis of Fresnel holograms and the analysis of imaging systems using zone-plate coded apertures.

Since the Fresnel transform operation consists of convolution with the chirp function $\exp(j\pi sx^2/\lambda)$, the properties of the chirp function are of interest in any analysis involving Fresnel transforms. The parameter s in the above expressions is most commonly interpreted as denoting a curvature for spherical wavefronts. By a generalization of this concept, complex values for s can denote complex wavefront curvature (i.e., a spherical wavefront combined with a Gaussian intensity profile).

Properties of the two-dimensional chirp function $\exp[j\pi s(x^2 + y^2)/\lambda]$ are derived by Cathey (1974, Appendix 2) and are listed also by Carlson and Francois (1977).

2.1.5 The Hilbert Transform

The pair of equations

$$g(x) = -\frac{1}{\pi} \int_{-\infty}^{\infty} \frac{f(y)}{x - y} \, dy \quad \text{and} \quad f(x) = \frac{1}{\pi} \int_{-\infty}^{\infty} \frac{g(y)}{x - y} \, dy \qquad (24)$$

define respectively the Hilbert transform and its inverse (Titchmarsh, 1937). Unlike other functions and their transforms which are defined on conjugate

2. Background

domains, here f and g are both functions of x. A skew-reciprocal relation (reciprocal except for a minus sign) exists between f and g; $g(x)$ is said to be conjugate to $f(x)$, $-f(x)$ being conjugate to $g(x)$. Also, $g(x)$ is sometimes called the quadrature function corresponding to $f(x)$.

The forward and inverse Hilbert transforms are both seen to be convolution operations, respectively, with $-1/(\pi x)$ and $1/(\pi x)$. This fact gives a particularly simple relationship between $f(x)$ and $g(x)$ in the Fourier domain; if $F(s)$ and $G(s)$ are their Fourier transforms, then F is related to G by

$$G(s) = j\, \text{sgn}(s) \cdot F(s)$$

where the value of the function $\text{sgn}(s)$ is defined to be 1 or -1 accordingly as s is greater or less than zero.

The Hilbert transform is a useful analytical tool that associates with every real function $f(x)$ a complex function $\frac{1}{2}[f(x) - jg(x)]$. This latter function can be shown to have a Fourier spectrum which is one-sided (zero for $s \leq 0$) but otherwise identical to the spectrum of $f(x)$, in much the same way as $\frac{1}{2}\exp(j2\pi sx)$ is related to $\cos(2\pi sx)$. The concept of phasor analysis is thus extended to nonmonochromatic signals through the aid of the Hilbert transform.

This transformation also plays an important role in the description of the spectra of temporal networks and filters, whose impulse responses are required to be causal (one-sided). As there is no such restriction on the impulse response of optical systems, this description unfortunately does not have a counterpart in optical analysis. Certain particular optical techniques, however, such as the Schlieren method of imaging phase objects by introducing a knife edge to block a part of the optical spectrum, can be analyzed by the application of Hilbert transform methods (Eu and Lohmann, 1973).

2.1.5.1 Relationships for the Hilbert Transform

If $f(x)$, $g(x)$ are related by means of the Hilbert transform, then so are the following pairs:

$f(ax)$	$g(ax)$	(similarity)
$f(x - a)$	$g(x - a)$	(shift)
$f_1(x)*f_2(x)$	$-g_1(x)*g_2(x)$	(convolution).

Also, the Hilbert transform of $f_1(x)*f_2(x)$ is $g_1(x)*f_2(x)$ or $f_1(x)*g_2(x)$. In addition, the following are true:

$$\int_{-\infty}^{\infty} |f^2(x)|\, dx = \int_{-\infty}^{\infty} |g^2(x)|\, dx \qquad \text{(Rayleigh)}, \qquad (25)$$

$$\int_{-\infty}^{\infty} f(x+u)\, f^*(u)\, du = \int_{-\infty}^{\infty} g(x+u)g^*(u)\, du \qquad \text{(autocorrelation).} \quad (26)$$

2.1.5.2   Some Hilbert Transform Pairs

$$\delta(x) \qquad\qquad -1/(\pi x)$$
$$\sin(x) \qquad\qquad \cos(x)$$
$$\cos(x) \qquad\qquad -\sin(x)$$
$$1/(1 + x^2) \qquad -x/(1 + x^2).$$

Other Hilbert transform pairs are listed by Bracewell (1965), with discussion of the transform. More mathematical detail is contained in Titchmarsh (1948, Chapter 5).

2.1.6   The Mellin Transform

A definition for the Mellin transform is (Sneddon, 1951)

$$M(s) = \int_0^\infty f(x)x^{s-1}\, dx, \tag{27}$$

where s is a complex variable; and if for some $k > 0$,

$$\int_0^\infty x^{k-1}|f(x)|\, dx < \infty, \tag{28}$$

then

$$f(x) = \frac{1}{2\pi j}\int_{c-j\infty}^{c+j\infty} M(s)x^{-s}\, ds \tag{29}$$

is the Mellin inversion formula for any $c > k$. The Mellin transform can be similarly defined in two dimensions, the formula for imaginary arguments ju and jv being

$$M(u, v) = \int\!\!\int_0^\infty f(x, y)x^{ju-1}y^{jv-1}\, dx\, dy. \tag{30}$$

The Mellin transform has been used in the analysis of linear optical systems that are not space-invariant (Robbins and Huang, 1972) and in the restoration of images degraded by space-variant blur (Sawchuk, 1972). Its usefulness in these situations arises out of the following property: The modulus of the Mellin transform of a function is invariant with respect to a magnification (or linear stretch) of the input function (Baudelaire, 1974) in the same way as the modulus of the Fourier transform of a function is invariant with respect to a shift of origin. The effect in both cases is to introduce a progressive linear phase change, or phase tilt, in the transform.

23

2. Background

We note that a logarithmic change of variable is just the kind of transformation that converts a linear stretch into a shift of origin. By means of the change of variable $x = \exp(-\xi)$, the Mellin transform of $f(x)$ can be shown to be equivalent to the two-sided Laplace transform of the resulting function of ξ (Bracewell, 1965, Chapter 12). With s purely imaginary, a similar relation thus exists also between the Mellin and the Fourier transforms.

In optics the scale-invariance property of the Mellin transform is exploited in various ways. For optical systems which are characterized by spread functions that do not change shape but change only in size in the appropriate way, application of the Mellin transform yields systems that can then be analyzed using linear, shift-invariant techniques. Using a combination of the Fourier and Mellin transform operations, optical correlators can be devised which are insensitive not only to shifts but also to changes in scale between object and reference signals (Casasent and Psaltis, 1977).

2.1.6.1   Some Mellin Transform Theorems

Given the transform pair $f(x)$, $M(s)$, the following Mellin transform pairs can be derived:

$$f(ax) \qquad a^{-s}M(s) \qquad\qquad \text{(similarity)}$$

$$x^a f(x) \qquad M(a + s)$$

$$f(1/x) \qquad M(-s)$$

$$\frac{d}{dx} f(x) \qquad -(s - 1)M(s - 1).$$

Also,

$$\int_0^\infty \frac{1}{x} |f(x)|^2 \, dx = \frac{1}{2\pi j} \int_{-j\infty}^{j\infty} |M(s)|^2 \, ds \tag{31}$$

and, more generally, if $N(s)$ is the Mellin transform of $g(x)$, then

$$\int_{-\infty}^\infty \frac{1}{x} f(x) g^*(x) \, dx = \frac{1}{2\pi j} \int_{-j\infty}^{j\infty} M(s) N^*(s) \, ds. \tag{32}$$

Convolution: Defining Mellin convolution as

$$f(x) * g(x) = \int_0^\infty \frac{1}{y} f(y) g\left(\frac{x}{y}\right) dy, \tag{33}$$

$f(x) * g(x)$ transforms to $M(s) \cdot N(s)$.

24

2.1.6.2   Some Mellin Transform Pairs

$f(x)$	$M(s)$
$\delta(x - a)$	a^{s-1}
$x^n H(x - a)$	$-a^{s+n}/(s + n)$
$1/(1 + x)$	$\pi \operatorname{cosec}(\pi s)$
$1/(1 + x^2)$	$(\pi/2) \operatorname{cosec}(\pi s/2)$
$\exp(-x^2)$	$\tfrac{1}{2}\Gamma(s/2)$.

For other Mellin transform pairs see Bracewell (1965). The Mellin transform is discussed in some detail by Sneddon (1951).

2.1.7   The Abel Transform

For two-dimensional systems that are rotationally symmetric, we have seen (Section 2.1.3) that a description can be given in terms of the one-dimensional Fourier–Bessel transform. A second way in which such systems can be completely characterized is through a description of their response to a one-dimensional input such as a line or an edge. For such systems it can be shown (Jones, 1958) that the one-dimensional point spread function $f(r)$, a function of a radius r, and the line spread function $A(x)$, a function of an ordinate x, are related by means of the Abel transform, which can be defined as

$$A(x) = 2 \int_x^\infty f(r) \frac{r}{(r^2 - x^2)^{1/2}} \, dr. \tag{34}$$

The Abel inversion formula is then given by

$$f(r) = -\frac{1}{\pi} \frac{\partial}{\partial r} \int_r^\infty \frac{rA(x) \, dx}{x(x^2 - r^2)^{1/2}}. \tag{35}$$

By means of a change of variable the Abel transform equation can be put into the form of a convolution integral (Bracewell, 1965, Chapter 12). This form has been termed the modified Abel transform, and because it is space-invariant, allows the use of Fourier methods of analysis, and is useful for computational purposes as well.

From the relationship between the point spread function and the line spread function, it may be deduced that the Fourier–Bessel transform and the Abel transform are closely related. There is in fact a close relationship among the Abel transform, the Fourier–Bessel transform, and the Fourier transform. Successive application of the Abel, Fourier, and Fourier–Bessel transforms to a function yields the original function (Bracewell, 1956). In optics this result is embodied in the Abel transform relationship between the point spread

2. Background

function and the line spread function, the Fourier transform relationship between the line spread function and the (one-dimensional) optical transfer function, and the Fourier–Bessel transform relationship between the optical transfer function and the point spread function (Jones, 1958).

The Abel transform and its inversion is a particular solution of the general problem of the reconstruction of a multidimensional object from a knowledge of its projections. For an arbitrary object, the inversion operation has been termed the (inverse) Radon transform, and algorithms for performing this operation are of current interest because of their applications to tomographic image synthesis (Barrett and Swindell, 1977).

2.1.7.1 Some Abel Transform Pairs

$f(r)$	$A(x)$
$\delta(r - a)$	$2a/(a^2 - x^2)^{1/2}$
$\text{rect}(r/2a)$	$2/(a^2 - x^2)^{1/2}$
$\exp(-r^2/a^2)$	$a\pi^{1/2}\exp(-x^2/a^2)$
$J_0(ar)$	$(2/a)\cos(ax).$

2.1.7.2 Sources for Abel Transform Theory

More Abel transform pairs are listed by Bracewell (1965); the Abel integral is treated theoretically by Whittaker and Watson (1940). The relation between the line spread function and the point spread function has been treated by Marchand for both the symmetric case (Marchand, 1964) and the general case (Marchand, 1965).

REFERENCES

Andrews, H. C. (1970). "Computer Techniques in Image Processing." Academic Press, New York.
Barrett, H. H., and Swindell, W. (1977). *Proc. IEEE* **65,** 89.
Baudelaire, P. (1973). *Proc. IEEE* **61,** 467.
Bracewell, R. N. (1956). *Austral. J. Phys.* **9,** 198.
Bracewell, R. N. (1965). "The Fourier Transform and Its Applications." McGraw-Hill, New York.
Campbell, G. A., and Foster, R. N. (1948). "Fourier Integrals for Practical Applications." Van Nostrand-Reinhold, Princeton, New Jersey.
Carlson, F. P., and Francois, R. E. (1977). *Proc. IEEE* **65,** 10.
Carslaw, H. S., and Jaeger, J. C. (1941). "Operational Methods in Applied Mathematics." Oxford Univ. Press, London and New York.
Casasent, D., and Psaltis, D., (1977). *Proc. IEEE* **65,** 77.

Cathey, W. T. (1974). "Optical Information Processing and Holography." Wiley, New York.

Champeney, D. C. (1973). "Fourier Transforms and Their Physical Applications." Academic Press, New York.

Erdelyi, A. (ed.) (1954). "Tables of Integral Transforms." McGraw-Hill, New York.

Eu, J. K. T., and Lohmann, A. W. (1973). *Opt. Comm.* **9,** 257 (1973).

Goodman, J. W. (1968). "Introduction to Fourier Optics." McGraw-Hill, New York.

Goodman, J. W. (1977). *Proc. IEEE* **65,** 29.

Jones, R. C. (1958). *J. Opt. Soc. Amer.* **48,** 934.

Marchand, E. W. (1964). *J. Opt. Soc. Amer.* **54,** 915.

Marchand, E. W. (1965). *J. Opt. Soc. Amer.* **55,** 352.

McLachlan, N. W. (1955). "Bessel Functions for Engineers," 2nd ed. Oxford Univ. (Clarendon) Press, London and New York.

Mertz, L. (1965). "Transformations in Optics." Wiley, New York.

Papoulis, A. V. (1968). "Systems and Transforms with Applications in Optics." McGraw-Hill, New York.

Robbins, G. M., and Huang, T. S. (1972). *Proc. IEEE* **60,** 862.

Sawchuk, A. A. (1972). *Proc. IEEE* **60,** 854.

Sneddon, I. N. (1951). "Fourier Transforms." McGraw-Hill, New York.

Titchmarsh, E. C. (1948). "Introduction to the Theory of Fourier Integrals." Oxford Univ. (Clarendon) Press, London and New York.

Van der Pol, B., and Bremmer, H. (1955). "Operational Calculus Based on the Two-Sided Laplace Integral." Cambridge Univ. Press, London and New York.

Whittaker, E. T., and Watson, G. N. (1940). "Modern Analysis." Cambridge Univ. Press, London and New York.

2.2 INTERFERENCE AND DIFFRACTION

Brian J. Thompson

2.2.1 Properties of Coherent Fields

An optical field can, in general, be written as a function that depends upon both the spatial coordinate x and the time t. For the present discussion, we will consider a single cartesian coordinate of the electric field vector and assume that the light has a narrow spectral width; hence the optical field will be written as $V(x, t)$. The field is, in general, a complex function, and the natural fluctuations of the light beam produce variations at a rate of approximately 10^{14} times per second. Normally we are interested in detecting that field and use a detector that integrates over a time interval very long compared to 10^{-14} sec. The detected quantity is the intensity $I(x)$, defined by

$$I(x) = \langle V(x, t)V^*(x, t)\rangle, \tag{1}$$

where the angle brackets denote a time average and the star denotes a complex conjugate. Equation (1) is valid whether the field is incoherent, partially coherent, or coherent.

The field is considered to be incoherent if the light at any one point in the field is completely unrelated, in a time-averaged sense, to every other point in the field. We shall write $V(x_1, t)$ as the field at point x_1 at time t, and $V(x_2, t)$ as the field at x_2 at the same instant in time. The two points x_1 and x_2 are then considered to be incoherent if the time-averaged cross correlation of the fields at these two points is zero.

$$\langle V(x_1, t)V^*(x_2, t)\rangle = 0. \tag{2}$$

The complete field is incoherent if this statement is true for all points x_1 and x_2 in the field. Note that an incoherent source can be defined to meet these conditions, but strictly speaking, an incoherent field cannot be achieved (see Section 2.3.3).

By comparison, a completely coherent field can also be defined. The fields at x_1 and x_2 are coherent if

$$\langle V(x_1, t)V^*(x_2, t)\rangle = \text{max value}. \tag{3}$$

That is, the fluctuations with time at x_1 are exactly matched by the fluctuations

HANDBOOK OF OPTICAL HOLOGRAPHY
Copyright © 1979 by Academic Press, Inc.
All rights of reproduction in any form reserved.
ISBN-0-12-165350-1

2. Background

at x_2. The field is coherent if the maximum value is obtained for all points x_1 and x_2. It is probably obvious that the value of the cross correlation is not now dependent on the averaging process even though average measures are still made. Thus the time- and space-dependent parts of the function describing the optical field can be separated. Hence

$$V(x, t) = \psi(x) \exp(-2\pi i \nu t), \tag{4}$$

where $\psi(x)$ is the complex amplitude of the field and ν is the frequency. Thus Eq. (3) becomes

$$\langle V(x_1, t) V^*(x_2, t) \rangle = \psi(x_1)\psi^*(x_2). \tag{5}$$

The intensities at x_1 and x_2 now become

$$I(x_1) = \langle V(x_1, t) V^*(x_1, t) \rangle = \psi(x_1)\psi^*(x_1), \tag{6a}$$

$$I(x_2) = \langle V(x_2, t) V^*(x_2, t) \rangle = \psi(x_2)\psi^*(x_2). \tag{6b}$$

The intensities are still time-averaged measures even though the performance of the time-average does not change the values of the function involved.

It is instructive to consider normalizing Eq. (5) by dividing by the square root of the product of the individual intensities. We then write

$$\frac{\langle V(x_1, t) V^*(x_2, t) \rangle}{[I(x_1)I(x_2)]^{1/2}} = \frac{\psi(x_1)\psi^*(x_2)}{[I(x_1)I(x_2)]^{1/2}} . \tag{7}$$

The importance of this normalization will be immediately recognized since

$$\left| \frac{\psi(x_1)\psi^*(x_2)}{[I(x_1)I(x_2)]^{1/2}} \right| = 1. \tag{8}$$

The same normalization carried out on Eq. (2) still produces a zero value for the magnitude of the normalized cross-correlation term.

Since $\psi(x)$ is a complex quantity, it is convenient to express it in terms of a pair of functions—real and imaginary functions or amplitude and phase functions. It is normal in optics to consider the amplitude and phase of the complex amplitude. Thus

$$\psi(x) = a(x) \exp[i\phi(x)], \tag{9}$$

where $a(x)$ is the amplitude and is a real and positive function and $\phi(x)$ is the phase. We then conclude that

$$I(x) = \psi(x)\psi^*(x) = a^2(x), \tag{10}$$

$$\psi(x_1)\psi^*(x_2) = a(x_1)a(x_2) \exp\{i[\phi(x_1) - \phi(x_2)]\}, \tag{11}$$

$$\frac{\psi(x_1)\psi^*(x_2)}{[I(x_1)I(x_2)]^{1/2}} = \exp\{i[\phi(x_1) - \phi(x_2)]\}. \tag{12}$$

Equation (12) is important because it shows that the normalized cross correlation has a magnitude of unity but has a resultant phase depending on the difference of the phase of the fields at the two points. That is, the light at x_1 is coherent with the light at x_2 and with some fixed phase relationship. Coherence implies a fixed (in time) phase relationship, but not necessarily one that is in phase.

2.2.1.1 Addition of Two Coherent Fields

In many situations with coherent light it is necessary to consider the addition of two beams of light. This is fundamentally true for holography as well as interferometry, image formation, optical processing, etc. Let $\psi_1(x)$ and $\psi_2(x)$ be the complex amplitude functions of the two fields of interest; then the resultant complex amplitude function is given by

$$\psi_R(x) = \psi_1(x) + \psi_2(x), \tag{13}$$

and

$$a_R(x)\exp[i\phi_R(x)] = a_1(x)\exp[i\phi_1(x)] + a_2(x)\exp[i\phi_2(x)], \tag{14}$$

where $a_R(x)$, $a_1(x)$, and $a_2(x)$ are the appropriate amplitude functions and $\phi_R(x)$, $\phi_1(x)$, and $\phi_2(x)$ are the appropriate phase functions.

Again we are interested in the detected intensity $I_R(x)$ associated with the resultant field

$$
\begin{aligned}
I_R(x) &= \psi_R(x)\psi_R^*(x) = [\psi_1(x) + \psi_2(x)][\psi_1^*(x) + \psi_2^*(x)] \\
&= a_1^2(x) + a_2^2(x) + a_1(x)a_2(x)\exp\{i[\phi_1(x) - \phi_2(x)]\} \\
&\quad + a_1(x)a_2(x)\exp\{-i[\phi_1(x) - \phi_2(x)]\}.
\end{aligned} \tag{15}
$$

In discussions of holography it is instructive to leave the equation in the form of Eq. (15); however, it has been traditional to write it as the interference law

$$I_R(x) = I_1(x) + I_2(x) + 2[I_1(x)I_2(x)]^{1/2}\cos\{\phi_1(x) - \phi_2(x)\}, \tag{16}$$

where $I_1(x)$ and $I_2(x)$ are the intensities associated with fields 1 and 2, respectively.

2.2.2 Addition of Two Coherent Waves

It is now necessary to translate the previous discussion into a more realistic (experimentally realizable) situation. To do this it is necessary to say something about how light propagates. The optical field $V(x, t)$ propagates according to a wave equation

$$\nabla^2 V(x, t) = \frac{1}{c^2}\frac{\partial^2}{\partial t^2} V(x, t), \tag{17a}$$

2. Background

where ∇^2 is the second partial derivative with respect to the space coordinates and c is the velocity of light. When the light is coherent, we can recall Eq. (4) and make this substitution into Eq. (17a) to give the wave equation (Helmholtz equation) for the propagation of the complex amplitude,

$$\nabla^2\psi(x) + k^2\psi(x) = 0, \tag{17b}$$

where $k = 2\pi/\lambda$, λ being the wavelength of the light. Important solutions of this wave equation are

(1) a plane wave propagating along the z axis

$$\psi(z) = A \exp(ikz), \tag{18}$$

where A is a constant;

(2) the converging (negative exponent) and diverging (positive exponent) spherical wave solutions

$$\psi(r) = A \exp(\pm ikr)/r, \tag{19}$$

where r is the radius of the spherical wave.

An idealized point source will give rise to a diverging spherical wave and a point source at infinity will produce a plane wave. As a first example, we shall consider the addition of two plane waves.

2.2.2.1 Addition of Two Plane Waves

We shall consider two idealized equally intense point sources at infinity that produce two plane waves at an angle 2θ to each other. That is, the two plane wavefronts make angles $\pm\theta$ with respect to the plane in which we will record the intensity produced by their interaction (see Fig. 1). We shall assume that the two waves have the same phase at the point 0. The resultant complex amplitude is then [from Eq. (14)]

$$a_R \exp[i\phi_R(x)] = a \exp(ikx \sin \theta) + a \exp(-ikx \sin \theta), \tag{20}$$

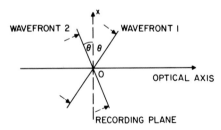

Fig. 1 Addition of two plane waves symmetrically disposed with respect to the optical axis.

and the resultant intensity $I_R(x)$ if the angle θ is small $(\sin\theta = \theta)$ is given by

$$I_R(x) = 2I(1 + \cos kx2\theta), \tag{21}$$

where I is the constant intensity associated with each individual plane wave. Finally, we note that

$$I_R(x) = 4I \cos^2(kx\theta). \tag{22}$$

From a holographic point of view it is worthwhile to write the result of Eq. (21) in the form of Eq. (15):

$$I_R(x) = 2I + I \exp(ikx2\theta) + I \exp(-ikx2\theta). \tag{23}$$

If a photographic record of $I_R(x)$ is made and then illuminated coherently with a wave $e^{ik\theta}$, the second and third terms recreate the original wave and its conjugate.

The resultant intensity from Eq. (22) is a set of cosine squared interference fringes. Fig. 2b illustrates this result. Naturally, when the two waves are incoherent, they merely add in intensity to produce a resultant intensity of $2I$. This result is illustrated for comparison in Fig. 2a. Finally, the result shown in Fig. 2c is for the partially coherent addition of two beams (see Section 2.3.2 for a discussion of this result).

2.2.2.2   Addition of a Cylindrical (or a Spherical) and a Plane Wave

We will assume that the plane wave is propagating along the optical axis of the system (Fig. 3), and the plane and cylindrical waves have zero path difference (and hence zero phase difference) at the point 0. The path difference between these two waves is then given by $x^2/2r$ when r is the radius of the spherical wave and small angles have been assumed. The phase difference is

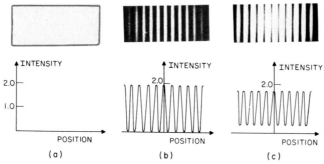

Fig. 2 An illustration of the resultant normalized intensity formed when two waves are added (a) incoherently, (b) coherently, and (c) partially coherently.

33

2. Background

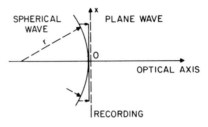

SPHERICAL WAVE

PLANE WAVE

OPTICAL AXIS

RECORDING

Fig. 3 Addition of a plane wave and a cylindrical wave.

then $kx^2/2r$. We may then write the resultant amplitude in the x plane as

$$a_R \exp[i\phi_R(x)] = a + a \exp[ikx^2/2r], \qquad (24)$$

and

$$I_R(x) = I\{2 + \exp[ikx^2/2r] + \exp[-ikx^2/2r]\} \qquad (25)$$

$$= 2I[1 + \cos(kx^2/2r)]$$

$$= 4I \cos^2(kx^2/4r). \qquad (26)$$

The resultant intensity profile is a $\cos^2$ set of interference fringes that vary with the square of the space coordinate x. This result is illustrated in Fig. 4. If the problem had been carried out with a spherical wave and a plane wave, the result would have been similar to that given by Eq. (26) and illustrated in Fig. 4a, except that the linear coordinate x would become a radial coordinate and the pattern would be radially symmetric.

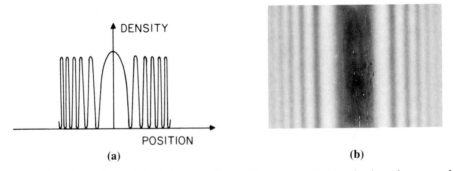

DENSITY

POSITION

(a)

(b)

Fig. 4 An illustration of the intensity distribution produced by the interference of a plane wave and a cylindrical wave. (a) Intensity plot of the profile and (b) photograph.

34

2.2.2.3 Addition of Two Cylindrical (or Spherical Waves)

The analysis of this problem is quite similar to the two previous ones. Let r_1 and r_2 be the radii of the two cylindrical waves (see Fig. 5a); then the resultant intensity is given by

$$I_R(x) = 4I \cos^2 \frac{kx^2}{4} \left(\frac{1}{r_1} - \frac{1}{r_2} \right), \tag{27}$$

where we have taken the two cylindrical waves to have zero phase difference at the point on the optical axis which defines the origin of the coordinate x of the plane in which the intensity is recorded.

If the two cylindrical or spherical waves are not propagating with their normals along the optical axis (see Fig. 5b), then there exists a linear term and a quadratic term in x and

$$a_R \exp[i\phi_R(x)] = a \exp\left[ik \left(\frac{x^2}{2r_1} + x\theta \right) \right] + a \exp\left[ik \left(\frac{x^2}{2r_2} - x\theta \right) \right] \tag{28}$$

and

$$I_R = 4I \cos^2 k \left[\frac{x^2}{4} \left(\frac{1}{r_1} - \frac{1}{r_2} \right) + x\theta \right]. \tag{29}$$

Equation (29) contains the results of Eqs. (22) and (27) in a combined form.

2.2.3 Two-Beam Interference

We can now move a step further toward a real situation by considering a hypothetical experiment in which light from two idealized point sources of equal intensity at P_1 and P_2 (Fig. 6) illuminates a plane x at distance z from the plane containing the point sources. We shall assume that z is large compared to the separation $2d$ and the maximum value of x. The path difference

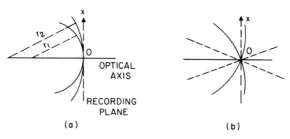

Fig. 5 Addition of two cylindrical waves propagating in the same direction (a) and at an angle to each other (b).

2. Background

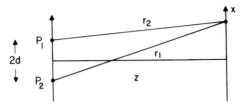

Fig. 6 Diagram to define the coordinates for the interference of light from two coherent point sources.

to some general point in the x plane from the two points P_1 and P_2 is given by $2\,dx/z$ if θ is small. Thus

$$I_R(x) = 4I\cos^2(k\,dx/z). \tag{30}$$

Equation (30) is essentially the same as Eq. (26) for the addition of two plane waves. Actually, it is really an approximate form of Eq. (29) with the quadratic term being much much smaller than the linear term.

Equation (30) is one form of the interference law and the fringe pattern in intensity that it describes is often called Young's fringes. If the two intensities of the interfering beams are not the same, then

$$I_R(x) = I_1 + I_2 + 2(I_1 I_2)^{1/2}\cos(2k\,dx/z). \tag{31}$$

2.2.4 Propagation of Coherent Light—Fresnel and Fraunhofer Diffraction

The results discussed in the previous section represent a situation in which light is emanating from two idealized points. In practice, of course, the points are usually finite apertures. The question now to be asked is, what is the effect of the finite apertures, i.e., how does coherent light propagate through an aperture?

The wave equation governing the propagation of the complex amplitude has been given in Eq. (17b) and the plane and spherical wave defined. We now wish to apply these solutions to the propagation of light through an aperture. Figure 7 shows the system we are interested in with the appropriate coordinates. Light from a very small source (effectively a point source) at P_0 in the $x_0 y_0$ plane illuminates the aperture plane $(\xi,\,\eta)$ and then propagates to the receiving plane xy. We now wish to be able to predict the complex amplitude at a point P in the xy plane. If angles are limited to being small, then the diffraction integral can be written

$$\psi(P) = \frac{-iA}{2\lambda}\iint\limits_{\text{apert}} \frac{\exp[ik(r+s)]}{rs}\,dQ, \tag{32}$$

36

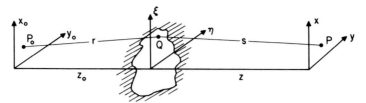

Fig. 7 Diagram to define the coordinates for the propagation of light through an aperture.

where A is a constant, r and s are the distances indicated in Fig. 7, and dQ is the elemental area in the aperture at the general point Q. Equation (32) is the Kirchhoff diffraction formula for a scalar wave and is discussed in many standard texts [e.g., Born and Wolf (1964)]. It is readily appreciated, however, by the following argument: the point source at P_0 generates a spherical wave $(e^{ikr})/r$ and fills the aperture, which in turn truncates the spherical wave. Each point on the wavefront in the aperture is a new source of diverging spherical waves $(e^{iks})/s$; the integral is then over all such points Q in the aperture plane. Equation (32) can now be written in the coordinates of the situation described in Fig. 7,

$$\psi(x, y) = \frac{-iA}{2\lambda} \frac{\exp[ik(z_0 + z)]}{z_0 z} \exp\left[\frac{ik}{2z_0}(x_0^2 + y_0^2)\right] \exp\left[\frac{ik}{2z}(x^2 + y^2)\right]$$

$$\cdot \int_{-\infty}^{\infty}\int_{\infty}^{\infty} \psi(\xi, \eta) \exp\left[\frac{ik}{2}(\xi^2 + \eta^2)\left(\frac{1}{z_0} + \frac{1}{z}\right)\right] \exp\left[-ik\xi\left(\frac{x_0}{z_0} + \frac{x}{z}\right)\right]$$

$$\cdot \exp\left[-ik\eta\left(\frac{y_0}{z_0} + \frac{y}{z}\right)\right] d\xi\, d\eta. \tag{33}$$

In classical terminology, the distribution $\psi(x, y)$ is the Fresnel diffraction pattern of the aperture function $\psi(\xi, \eta)$. Figure 8 shows a number of different Fresnel diffraction patterns associated with a circular aperture. Since these are photographic records,† they represent the intensity associated with the complex amplitude $\psi(x, y)$. Clearly, the actual intensity distribution changes rapidly with the value of z_0 and z if the other parameters are fixed.

Under certain circumstances, the quadratic term in ξ and η can be eliminated to produce a very important result. This condition is the far-field or far-zone

---

† The illustrative photographs shown in Figs. 8 and 9 are from a collection of such illustrations made over the years and used for a variety of purposes. Many of them were used for the "Physical Optics Notebook" (Parrent and Thompson, 1970).

37

2. Background

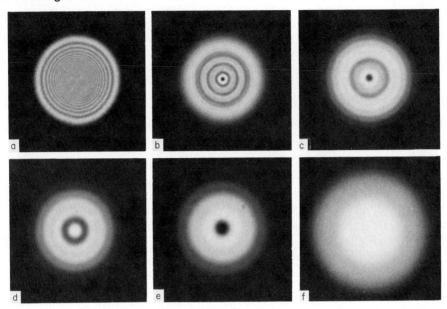

Fig. 8 Fresnel diffraction patterns associated with a circular aperture. In this series, z_0 is fixed and z increases from (a) to (f).

condition defined by the pair of inequalities

$$z_0 \gg (\xi^2 + \eta^2)_{max}/\lambda, \tag{34a}$$

$$z \gg (\xi^2 + \eta^2)_{max}/\lambda. \tag{34b}$$

Under these conditions, the distances z_0 and z are considered to be essentially infinite. Thus

$$\psi(x, y) = C \int_{-\infty}^{\infty} \int_{-\infty}^{\infty} \psi(\xi, \eta) \exp[-ik(p\xi + q\eta)] \, d\xi \, d\eta, \tag{35}$$

where $p = (x_0/z_0) + (x/z)$, $q = (y_0/z_0) + (y/z)$ and C contains the numerous constant and phase terms outside the integral. These phase terms have some importance in holographic recording and should not be forgotten.

Equation (35) represents Fraunhofer diffraction in which there is a fixed functional relationship between the aperture function $\psi(\xi, \eta)$ and the field $\psi(x, y)$. Specifically, $\psi(x, y)$ and $\psi(\xi, \eta)$ are a Fourier transform pair. Thus the complex amplitude in the classical Fraunhofer diffraction is the Fourier transform of the aperture function.

It is often of experimental value to arrange for the inequality of Eq. (34) to be met by illuminating the aperture with light from a point source at infinity

38

(or use a collimated beam produced by a lens). Under these circumstances, Eq. (35) for Fraunhofer diffraction becomes

$$\psi(x, y) = \frac{-iA}{2\lambda} \exp\left[\frac{ikz}{z}\right] \exp\left[\frac{ik}{2z}(x^2 + y^2)\right] \int\limits_{-\infty}^{\infty}\int\limits_{-\infty}^{\infty} \psi(\xi, \eta)$$

$$\cdot \exp\left[\frac{-ik}{z}(x\xi + y\eta)\right] d\xi\, d\eta, \qquad (36)$$

where the source point has been put on the optical axis so that the quadratic terms in x_0 and y_0 are also removed.

The Fraunhofer diffraction pattern can also be displayed by using a lens to image the far field which, since it is an infinity, results in the Fraunhofer diffraction pattern being located in the focal plane of the lens. Furthermore, if the aperture function is placed in the front focal plane of the lens, the remaining quadratic terms in x and y are eliminated. Thus

$$\psi(x, y) = D \int\limits_{-\infty}^{\infty}\int\limits_{-\infty}^{\infty} \psi(\xi, \eta) \exp\left[\frac{-ik}{f}(x\xi + y\eta)\right] d\xi\, d\eta, \qquad (37)$$

where f is the focal length of the lens and D is a constant. Clearly, if the receiving plane is moved from the focal plane, then Fresnel diffraction results.

It is worthwhile to look at Eq. (37) for some specific functional forms of $\psi(\xi, \eta)$. Sometimes closed-form solutions can be obtained; at other times the integrations have to be carried out numerically. Well-known and important relationships include diffraction by a rectangular aperture of height $2b$ and width $2a$. The complex amplitude in the Fraunhofer diffraction pattern is given by

$$\psi(x, y) = D4ab \operatorname{sinc}(kax/f) \operatorname{sinc}(kyb/f), \qquad (38)$$

where $\operatorname{sinc} = (\sin x)/x$. Equation (38) is often normalized to make $\psi(0, 0) = 1$, but that can disguise the fact that the constant term involves the area of the diffracting aperture. The sinc functions have equally spaced zeros along the x and y directions at intervals of $f\lambda/2a$ and $f\lambda/2b$, respectively. A photographic record of the intensity associated with this result is shown in Fig. 9a, together with the aperture. The reciprocal relationship between aperture and diffraction pattern is evident. The particular diffraction pattern is symmetric about two axes and has continuous lines of zero intensity crossing the pattern in two orthogonal directions. Thus isolated areas of intensity exist, each of which is of constant phase, and the phase changes by 180° when the intensity goes to zero. Only two phase values are present in this field; this occurs because of the symmetry of the aperture function (it is centrosymmetric). By contrast, noncentrosymmetric apertures such as the triangular aperture (Fig. 9c) have

2. Background

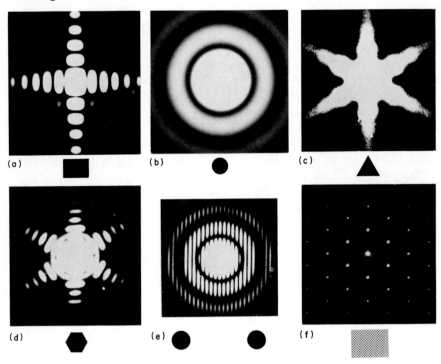

Fig. 9 Fraunhofer diffraction pattern of a few selected aperture functions.

Fourier transforms that have continuously varying phase from 0 to 2π rad. Incidentally, the diffraction pattern of the triangular aperture cannot be written down in closed form.

The diffraction pattern of a circular aperture of diameter $2a$ is circularly symmetric, and if r is the radial coordinate, then

$$\psi(r) = D\pi a^2\{[2J_1(kar/f)]/(kar/f)\}, \tag{39}$$

where J_1 is a Bessel function of the first kind and of order unity. Since a circular aperture also has a center of symmetry, the central disk is of uniform phase 0 and the rings alternate π, 0, π, etc.

We may now return to the two-beam interference result, described by Eq. (30), in which the two interfering beams are created by two circular apertures of radius a and separation $2d$. I is no longer a constant in this equation but is the distribution produced by diffraction through the circular aperture. Thus

$$I_R(x, y) = 4\left\{2J_1\left[\frac{ka(x^2 + y^2)^{1/2}}{2f}\right]\bigg/\left[\frac{ka(x^2 + y^2)^{1/2}}{2f}\right]\right\}^2 \cos^2\frac{kdx}{f}. \tag{40}$$

Figure 9e illustrates this result. The same result could be obtained in the far field and f would be replaced by z. It will be noted that the resultant transform is a product. [Since Eq. (40) is a product of two squared functions, the complex amplitude is the product of those two functions.] Thus it will be recalled that the Fourier transform of a product is the convolution of the individual transforms, and so the aperture function is expressible as a convolution. This convolution is between a function, $\delta(\xi - d) + \delta(\xi + d)$, which represents the location of the two circular apertures and the circular aperture function itself. Expressing the aperture function in this form is instructive, since it will allow the result of Eq. (40) to be written down directly.

REFERENCES

Born, M., and Wolf, E. (1964). "Principles of Optics," 2nd ed., Chapter VIII. Pergamon, New York.

Parrent, G. B., and Thompson, B. J. (1970). "Physical Optics Notebook," 2nd ed. SPIE Publ., Bellingham, Washington.

2.3 PARTIALLY COHERENT LIGHT

Brian J. Thompson

The discussions in the early part of Section 2.2 defined the meaning of coherent and incoherent light at two points in the field and hence the definition of coherent and incoherent fields. These results are two extremes of a more general situation that describes partially coherent fields.

First, we shall ask the question about two points in a field at x_1 and x_2. The time-averaged cross correlation of the field at these two points is neither a maximum value nor zero. This time-averaged cross-correlation function $\Gamma(x_1, x_2)$ is the mutual intensity between x_1 and x_2 and is defined by

$$\Gamma(x_1, x_2) = \langle V(x_1, t)V^*(x_2, t)\rangle. \tag{1}$$

The mutual intensity between the point x_1 and all other points in the field is called the mutual intensity function $\Gamma(x_1, x_2)$. This function is a measurable quantity and reduces to the normal measurable quantity, the intensity, by setting $x_1 = x_2$ in the mutual intensity function.

The mutual intensity function can be written in a normalized form by using the same method as that described in Section 2.2 for the coherent case. Thus

$$\frac{\langle V(x_1, t)V^*(x_2, t)\rangle}{[I(x_1)I(x_2)]^{1/2}} = \frac{\Gamma(x_1, x_2)}{[I(x_1)I(x_2)]^{1/2}} = \gamma(x_1, x_2), \tag{2}$$

where $\gamma(x_1, x_2)$ is the complex degree of coherence of the field and

$$0 \le |\gamma(x_1, x_2)| \le 1. \tag{3}$$

2.3.1 Addition of Two Partially Coherent Quasi-Monochromatic Fields

We can now add two fields together to find the resultant field. For this discussion we will consider adding two fields together which have a fixed degree of coherence between them; that is, the degree of coherence between the two fields at any given point in the field is constant independent of the choice of that point. For convenience, we will define that constant value of the mutual intensity as Γ_{12} where 1 and 2 refer to the two fields. If we return to the function $V(x, t)$ to define the optical field, then the result of adding two

HANDBOOK OF OPTICAL HOLOGRAPHY
Copyright © 1979 by Academic Press, Inc.
All rights of reproduction in any form reserved.
ISBN-0-12-165350-1

2. Background

fields together which have the same narrow spectral width (i.e., they are quasi-monochromatic) is

$$V_R(x, t) = V_1(x, t) + V_2(x, t), \tag{4}$$

and

$$I_R(x) = \langle V_R(x, t) V_R^*(x, t) \rangle = I_1(x) + I_2(x) + \Gamma_{12} + \Gamma_{12}^*. \tag{5}$$

Thus

$$I_R(x) = I_1(x) + I_2(x) + 2[I_1(x)I_2(x)]^{1/2} \operatorname{Re}\{\gamma_{12}\}, \tag{6}$$

where Re denotes a real part. The complex degree of coherence can be written in terms of an amplitude and phase

$$\gamma_{12} = |\gamma_{12}| e^{i\beta_{12}}. \tag{7}$$

Then Eq. (6) becomes

$$I_R(x) = I_1(x) + I_2(x) + 2[I_1(x)I_2(x)]^{1/2} |\gamma_{12}| \cos(\beta_{12}). \tag{8}$$

Equation (8) reduces to the incoherent addition of two fields when $|\gamma_{12}| = 0$ and to the in-phase coherent addition of two fields if $|\gamma_{12}| = 0$ and $\beta_{12} = 0$.

2.3.2 Interference of Two Partially Coherent Quasi-Monochromatic Beams

The interference of two partially coherent beams of uniform intensity will be a generalized form of Eq. (31) of Section 2.2. Consider the arrangement discussed in Fig. 6 of Section 2.2. The light from x_1 and x_2 is now partially coherent and hence Eq. (31) of Section 2.2 becomes

$$I_R(x) = I_1 + I_2 + 2[I_1 I_2]^{1/2} |\gamma_{12}| \cos\left(\frac{2kdx}{z} + \beta_{12}\right), \tag{9}$$

which, if $I_1 = I_2 = I$, reduces to

$$I_R(x) = 2I\left[1 + |\gamma_{12}| \cos\left(\frac{2kdx}{z} + \beta_{12}\right)\right]. \tag{10}$$

Note that in Eq. (10), I, $|\gamma_{12}|$, and β_{12} are all constant for this particular situation. Figure 2, Section 2.2, shows this result in comparison with the coherent and incoherent results.

An important way of describing the quality of the interference fringes formed in these examples is by their visibility, defined by

$$V = (I_{\max} - I_{\min})/(I_{\max} + I_{\min}) = |\gamma_{12}|, \tag{11}$$

where

$$I_{\max} = 2I[1 + |\gamma_{12}|]$$ (12)

and

$$I_{\min} = 2I[1 - |\gamma_{12}|].$$ (13)

The visibility is then a measure of the modulus of the complex degree of coherence between the two interfering beams, provided that the intensities of the two beams are identical. Thus the degree of coherence of a field can be measured by selecting pairs of points and measuring the fringe visibility as a function of position of those two points.

2.3.3 Interference of Two Partially Coherent Polychromatic Beams

We will now remove the quasi-monochromatic limitation and examine the effects of the finite spectral width. The finite spectral width can be included in the coherence function by including a time delay coordinate τ where τ is the time difference associated with the path difference to the point x from P_1 and P_2 in Fig. 6 of Section 2.2. Thus the mutual coherence function $\Gamma(x_1, x_2, \tau)$ is defined by

$$\Gamma(x_1, x_2, \tau) = \langle V(x_1, t)V^*(x_2, t + \tau)\rangle.$$ (14)

The corresponding complex degree of coherence is then given by

$$\gamma(x_1, x_2, \tau) = \frac{\langle V(x_1, t)V^*(x_2, t + \tau)\rangle}{[I_1(x)I_2(x)]^{1/2}}.$$ (15)

The two-beam interference law of Eq. (9) now becomes

$$I_R(x) = I_1 + I_2 + 2(I_1 I_2)^{1/2}|\gamma_{12}(\tau)|\cos\left[\frac{2kdx}{z} + \beta_{12}(\tau)\right],$$ (16)

which, if $I_1 = I_2 = I$, reduces to

$$I_R(x) = 2I\left\{1 + |\gamma_{12}(\tau)|\cos\left[\frac{2kdx}{z} + \beta_{12}(\tau)\right]\right\}.$$ (17)

The resultant fringe visibility is now dependent upon the value of τ, that is, upon the actual position in the fringe field. The phase (relative position) of those fringes may also change dependent upon the value of $\beta_{12}(\tau)$. Figure 1 shows the profile of a set of fringes of constant visibility and variable visibility for comparison. The visibility is the same at the center of the fringe field. Figure 1c shows a plot of the visibility as a function of position in the fringe fields. The visibility is by definition a positive quantity; however, it is clear

45

2. Background

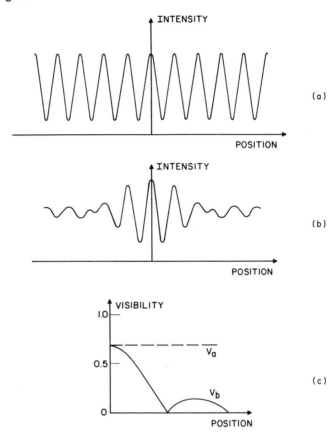

Fig. 1 Two-beam interference fringes of (a) constant visibility, (b) varying visibility, and (c) the visibility plotted as a function of position in the fringe field.

that the $\beta_{12}(\tau)$ has changed from 0 to π, after V_b goes to zero, resulting in a reversal of the position of the fringe maxima and minima.

2.3.4 Spatial and Temporal Coherence

The mutual coherence function and the complex degree of coherence are functions of both the spatial and temporal coordinates. If the light is quasi-monochromatic, i.e., $\Delta\nu \ll \bar{\nu}$ (the frequency width is much less than the mean frequency), then only the spatial dependence is of importance. Experimentally, the required condition is that maximum values of τ should be less than $1/\Delta\nu$, and hence the maximum path difference should be less than $\bar{\lambda}^2/\Delta\lambda$, where $\bar{\lambda}$ is the mean wavelength and $\Delta\lambda$ is the spectral width. Thus, experimentally, even white light can sometimes be considered quasi-monochromatic! These considerations lead to a definition of the temporal coherence of the light which is characterized by a coherence length $= \bar{\lambda}^2/\Delta\lambda$. The visibility curve plotted in Fig. 1c labeled V_b is actually a plot of the temporal coherence of the light passing through the two apertures to produce the interfering beams. The temporal part of the complex degree of coherence can be determined by setting $x_1 = x_2$ in the function and is thus designated $\gamma(x_1, x_1, \tau)$. This function is the normalized Fourier transform of the spectral distribution function of the light $G(\nu)$, i.e.,

$$\gamma(x_1, x_1, \tau) = \int_{-\infty}^{\infty} G(\nu) \exp(-2\pi i\nu\tau) \, d\nu \Big/ \int_{-\infty}^{\infty} G(\nu) \, d\nu. \tag{18}$$

A simple if not very realistic example is to consider a spectral profile that is rectangular. The temporal coherence function is then $\mathrm{sinc}(\Delta\nu\tau/2)$ where $\Delta\nu$ is the width of the rectangular spectral profile.

If the light is truly monochromatic, then we can consider the spatial coherence separately, i.e., $\gamma(x_1, x_2, 0)$. As an example, we can consider the degree of coherence in the field produced by light from an incoherent source. If the distance z from the source to the field is greater than the source dimensions and the field dimensions of interest, then the complex degree of coherence is given by the Fourier transform of the source intensity distribution, i.e.,

$$\gamma(x_1, x_2, 0) = \exp\left[\frac{ik(x_1{}^2 - x_2{}^2)}{2z}\right] \frac{\int_{-\infty}^{\infty} I_s(\alpha) \exp[-ik(x_1 - x_2)\alpha/z] \, d\alpha}{\int_{-\infty}^{\infty} I_s(\alpha) \, d\alpha},$$

$$\tag{19}$$

where $I_s(\alpha)$ is the source intensity distribution. Thus for illustrative purposes, a long thin uniform incoherent line source of width $2a$ produces a field whose coherence in the direction perpendicular to the width of the line source is given by

$$\gamma(x_1, x_2, 0) = \mathrm{sinc}[(ka(x_1 - x_2))/z]. \tag{20}$$

The degree of coherence is real but not everywhere positive; a negative value means a shift in phase of π of the coherence function $[\beta(x_1, x_2) = \pi]$.

2.3.4 Diffraction and Interference with Partially Coherent Light

In Section 2.2.4, the Fraunhofer diffraction pattern formed by passing a coherent beam of light through two circular apertures of width $2a$ and separated by a distance $2d$ was discussed [see eq. (40), Section 2.2]. If the light is

2. Background

effectively coherent over each aperture but not completely coherent between the two apertures, then Eq. (40) of Section 2.2 would become

$$I_R(x, y) = 2\left\{2J_1\left[\frac{ka(x^2 + y^2)^{1/2}}{2f}\right]\middle/\frac{ka(x^2 + y^2)^{1/2}}{2f}\right\}$$

$$\times \left\{1 + |\gamma_{12}(\tau)| \cos\left[\frac{2kdx}{f} + \beta_{12}(\tau)\right]\right\}. \qquad (21)$$

Figure 2 shows a typical set of such fringes contained within the central maximum of the envelope function which is the diffraction pattern of the circular holes. The fringes crossing this pattern are readily apparent and have a visibility that changes with position in the fringe field.

This Fraunhofer pattern can also be thought of as a convolution of the result expressed with the fully coherent answer with the image of the incoherent source function and the spectral profile function.

The coherence function can be propagated according to similar wave equations to those that apply for the propagation of $V(x, t)$ and $\psi(x)$. In particular, the mutual coherence function propagates according to a pair of wave equations,

$$\nabla_s^2 \Gamma(x_1, x_2, \tau) = \frac{1}{c^2}\frac{\partial^2}{\partial\tau^2}(\Gamma(x_1, x_1, \tau)), \qquad s = 1, 2. \qquad (22)$$

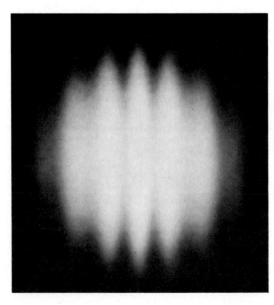

Fig. 2 Diffraction pattern of two circular holes formed with partially coherent light.

In the quasi-monochromatic limit, Eq. (22) reduces to a pair of Helmholtz equations.

$$\nabla_s^2 \Gamma(x_1, x_1) + k^2 \Gamma(x_1, x_2) = 0, \qquad s = 1, 2. \tag{23}$$

This last equation is quite important since it is really just a generalization of the propagation of the complex amplitude. For example, if we have an optical field characterized by a mutual intensity $\Gamma(\xi_1, \xi_2)$, the mutual intensity in a plane x a distance z from the incident plane is given by

$$\Gamma(x_1, x_2) = \int_{-\infty}^{\infty} \int_{\infty}^{\infty} \Gamma(\xi_1, \xi_2) \frac{\exp(ikr_1)}{r_1} \frac{\exp(-ikr_2)}{r_2} d\xi_1 \, d\xi_2, \tag{24}$$

where r_1 is the distance from a general point ξ_1 to a general point x_1 with a similar definition for r_2. Under a small angle approximation, Eq. (24) can be simplified and an analysis carried out similar to that discussed in Section 2.2.4. Furthermore, if the input field is incoherent, then Eq. (19) can be derived.

2.4 IMAGE EVALUATION

F. T. S. Yu

Anthony Tai

The formation of holographic images will be discussed in detail in Chapters 6 and 7. Before we discuss the evaluation of holographic images, it may be beneficial for us to describe briefly the image formation with conventional optical systems using spherical lenses and some of the parameters used in describing the images.

2.4.1 Image Formation

Let us assume we have two planes separated by a distance l and that the complex light distribution at the plane P_1 is $f(x, y)$ as shown in Fig. 1. This light field propagates from P_1 to P_2 and the complex light distribution $g(\alpha, \beta)$ at P_2 can be determined by means of Huygens' principle (Yu, 1973; Goodman, 1968; Born and Wolf, 1964; Caulfield and Lu, 1970, Shulman, 1970; Stroke, 1969; Kock, 1971; Collier *et al.*, 1971), and expressed as

$$g(\alpha, \beta) = C \iint_S f(x, y) \exp\{ik[l^2 + (\alpha - x)^2 + (\beta - y)^2]^{1/2}\} \, dx \, dy, \quad (1)$$

where S denotes the surface integral, C is an arbitrary constant, and k is the wavenumber. If l is large compared with the spatial dimension of P_1 and P_2, the equation can be approximated as

$$g(\alpha, \beta) = C \iint_S f(x, y) \exp(ikl)$$

$$\exp\left\{\frac{ik}{2l}[(\alpha - x)^2 + (\beta - y)^2]\right\} \, dx \, dy. \quad (2)$$

Equivalently, it can be written in the convolution form

$$g(\alpha, \beta) = C f(x, y) * h_l(x, y), \quad (3)$$

where $h_l = \exp[(ik/2l)(x^2 + y^2)]$ is the spatial impulse response, and P_1, P_2 are viewed as the input and output of the linear system. Since a converging

HANDBOOK OF OPTICAL HOLOGRAPHY
Copyright © 1979 by Academic Press, Inc.
All rights of reproduction in any form reserved.
ISBN-0-12-165350-1

51

2. Background

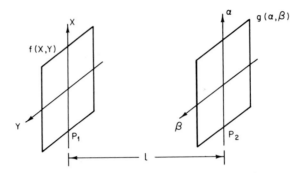

Fig. 1 The propagation of a complex light field.

(positive) lens converts a plane wave into an converging spherical wavefront, it can be shown that the lens performs a phase transformation (Yu, 1973)

$$T(\alpha, \beta) = \exp\left[\frac{-ik}{2f}(\alpha^2 + \beta^2)\right], \tag{4}$$

where f is the focal length of the lens. Now let us consider a simple imaging system as shown in Fig. 2. The light field at the output plane P_2 can be written as

$$g(\alpha, \beta) = C\{[\, f(x, y) * h_{d_1}(x, y)]T(\xi, \eta)\} * h_{d_2}(\xi, \eta), \tag{5}$$

where h_{d_1} and h_{d_2} are the corresponding impulse responses. The light fields between the input and output plane can also be related directly as

$$g(\alpha, \beta) = Cf(x, y) * h_{d_{12}}(x, y), \tag{6}$$

where

$$
\begin{aligned}
h_{d_{12}} = {} & \frac{1}{\lambda^2 d_1 d_2} \exp\left[i\frac{k}{2d_2}(\alpha^2 + \beta^2)\right] \exp\left[i\frac{k}{2d_1}(x^2 + y^2)\right] \\
& \times \iint_\epsilon \exp\left[i\frac{k}{2}\left(\frac{1}{d_1} + \frac{1}{d_2} - \frac{1}{f}\right)(\xi^2 + \eta^2)\right] \\
& \times \exp\left\{-ik\left[\left(\frac{x}{d_1} + \frac{\alpha}{d_2}\right)\xi + \left(\frac{y}{d_1} + \frac{\beta}{d_2}\right)\eta\right]\right\} d\xi\, d\eta.
\end{aligned}
\tag{7}
$$

We see that the terms containing the phase factors,

$$\exp\left[i\frac{k}{2d_1}(\alpha^2 + \beta^2)\right] \quad \text{and} \quad \exp\left[i\frac{k}{2d_1}(x^2 + y^2)\right], \tag{8}$$

are independent of the coordinate ξ, η. It can be shown that these two phase

52

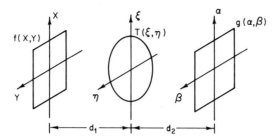

Fig. 2 Arrangement for a simple imaging system.

terms can generally be neglected since images are usually observed or recorded by their intensities. Thus the system inpulse response can be simplified to (Goodman, 1968)

$$h_{d_{12}}(x, y) \simeq \frac{1}{\lambda^2 d_1 d_2} \iint_\epsilon \exp\left[i\frac{k}{2}\left(\frac{1}{d_1} + \frac{1}{d_2} - \frac{1}{f}\right)(\xi^2 + \eta^2)\right]$$

$$\times \exp\left[-ik\left(\frac{x}{d_1} + \frac{\alpha}{d_2}\right)\xi + \left(\frac{y}{d_1} + \frac{\beta}{d_2}\right)\eta\right] d\xi\, d\eta. \qquad (9)$$

Let us assume the following condition is satisfied,

$$1/d_1 + 1/d_2 = 1/f, \qquad (10)$$

where f is the focal length of the lens. Then the impulse response can be further simplified to

$$h_{d_{12}}(x, y) \simeq \frac{1}{\lambda^2 d_1 d_2} \iint_\epsilon \exp\left\{ -ik\left[\left(\alpha + \frac{d_2}{d_1}x\right)\xi\right.\right.$$

$$\left.\left. + \left(\beta + \frac{d_2}{d_1}y\right)\eta\right]\right\} d\xi\, d\eta. \qquad (11)$$

Equation (10) is the well-known lens equation. If we look at the object light distribution as composed of many point radiators, the lens equation describes the condition for which the light field will converge and reproduce the point objects at the output plane. Equation (11) therefore describes the light distribution at this image plane, and d_2/d_1 is the lateral magnification factor.

2.4.2 Coherent and Incoherent Imaging†

In the preceding discussion, $f(x, y)$ is taken as the complex light distribution at the input plane. Let us examine the same system again, but this time let

† See Section 2.3 and Chapter 3.

2. Background

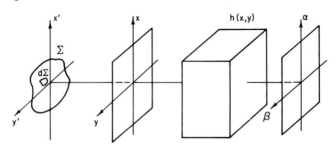

Fig. 3 An optical imaging system; the black box denotes the system with impulse response $h(x, y)$.

$f(x, y)$ be the amplitude transmittance of a signal transparency and add a monochromatic light source Σ some distance r in front of the transparency, as shown in Fig. 3. If we let the complex light field at the input plane (x, y) due to an incremental light source $d\Sigma$ in Σ be $u(x, y)$, then the light just to the right of the input transparency would be $u(x, y) f(x, y)$. The light distribution at the output plane can be written as

$$g(\alpha, \beta) = u(x, y) f(x, y) * h(x, y), \tag{12}$$

and the irradiance in the image plane due to $d\Sigma$ is

$$dI(\alpha, \beta) = g(\alpha, \beta) g * (\alpha, \beta) d\Sigma. \tag{13}$$

Therefore the total irradiance of the image due to the whole light source is

$$I(\alpha, \beta) = \iint_{\Sigma} |g(\alpha, \beta)|^2 d\Sigma \tag{14}$$

which can be written out as the convolution integral

$$I(\alpha, \beta) = \int_{-\infty}^{\infty} \!\!\!\! \iiint \Gamma(x, y; \xi, \eta) h(\alpha - x, \beta - y) h * (\alpha - \xi, \beta - \eta)$$

$$\times f(x, y) f * (\xi, \eta) \, dx \, dy \, d\xi \, d\eta, \tag{15}$$

where

$$\Gamma(x, y; \xi, \eta) = \iint_{\Sigma} u(x, y) u * (\xi, \eta) \, d\Sigma. \tag{16}$$

For the paraxial case (i.e., restricting to the wavefronts that lie close to the

54

lens axis), $\Gamma(x, y)$ can be approximated as

$$\Gamma(x, y) = \frac{1}{r^2} \int\!\!\int_{\Sigma} I(x', y') \exp\left[i\frac{k}{r}(x'x + y'y)\right] dx'\, dy'. \qquad (17)$$

Now one of the two extreme cases of the hypothetical optical imaging system can be seen by letting the light source become infinitely large. If the irradiance of the source is relatively uniform, that is, $I(\xi, \eta) \simeq K$, Eq. (17) becomes

$$\Gamma(x, y) = K_1\, \delta(x, y), \qquad (18)$$

where K_1 is an appropriate positive constant. This equation describes a completely incoherent optical imaging system.

On the other hand, if the light source is vanishingly small, Eq. (17) becomes

$$\Gamma(x, y) = K_2, \qquad (19)$$

where K_2 is a positive constant, and then the equation describes a completely coherent optical imaging system.

Referring to the completely incoherent case ($\Gamma(x, y) = K_1\, \delta(x, y)$), the irradiance at the output is

$$I(\alpha, \beta) = \int\!\!\int\!\!\int\!\!\int_{-\infty}^{\infty} \delta(\xi - x, \eta - y)h(\alpha - x, \beta - y)$$

$$\times h * (\alpha - \xi, \beta - \eta)\, f(x, y)\, f * (\xi, \eta)\, dx\, dy\, d\xi\, d\eta, \qquad (20)$$

which can be reduced to

$$I(\alpha, \beta) = \int\!\!\int_{\infty}^{\infty} |h(\alpha - x, \beta - y)|^2\, |f(x, y)|^2\, dx\, dy. \qquad (21)$$

From Eq. (21) we find that for the incoherent case the image irradiance is the convolution of the signal irradiance with respect to the impulse response irradiance. In other words, for the completely incoherent case, the optical system is linear in irradiance, i.e.,

$$I(\alpha, \beta) = |h(x, y)|^2 * |f(x, y)|^2. \qquad (22)$$

By Fourier transformation, Eq. (22) can be expressed in the spatial frequency domain:

$$I(p, q) = |H(p, q)|^2\, |F(p, q)|^2, \qquad (23)$$

where $I(p, q)$, $H(p, q)$, and $F(p, q)$ are the Fourier transforms of $I(\alpha, \beta)$, $h(x, y)$, and $f(x, y)$, respectively, and p and q are the spatial frequency

2. Background

coordinates. On the other hand, for the completely coherent case,

$$I(\alpha, \beta) = g(\alpha, \beta)g * (\alpha, \beta) = \int\int_{-\infty}^{\infty} h(\alpha - x, \beta - y) f(x, y) \, dx \, dy$$

$$\times \int\int_{\infty}^{\infty} h * (\alpha - \xi, \beta - \eta) f * (x, y) \, d\xi \, d\eta. \qquad (24)$$

From Eq. (24) it is obvious that the optical system is linear in complex amplitude, i.e.,

$$g(\alpha, \beta) = \int\int_{-\infty}^{\infty} h(\alpha - x, \beta - y) f(x, y) \, dx \, dy. \qquad (25)$$

Again, by Fourier transformation Eq. (25) becomes

$$G(p, q) = H(p, q)F(p, q). \qquad (26)$$

2.4.3 Resolution Limit

In geometric optics, with an ideal aberration-free imaging system, light rays emanating from a point would be imaged back into a point. However, this is only true when the wavelength of the light beam is infinitely small, where no diffraction takes place. Thus in practice, because of the presence of diffraction, the point image cannot be arbitrarily small, and infinite image resolution cannot be obtained with a physically realizable system. The resolution limit of an optical system depends on many factors: the wavelength of the light, size and geometry of the lenses, and the arrangement of the imaging system. The Rayleigh criterion is generally used to define the resolution limit for most imaging systems; it states that the image is resolved if the central maximum in the diffraction pattern of a point image coincides with the first dark fringe of its adjacent point. For example, if a lens with a circular aperture were used in the imaging as shown in Fig. 4, the point object would be focused into a diffraction pattern in the form of a first-order Bessel function. The distance between the central peak and the first zero would be

$$h' = 1.22\lambda l_2 / D. \qquad (27)$$

And by the Rayleigh criterion (Stone, 1963; Sommerfield, 1954; Sears, 1949; Rossei, 1957), the minimum distance between two resolvable object points would therefore be

$$h = 1.22\lambda l_1 / D. \qquad (28)$$

In applying Eq. 28 to a particular optical instrument, another relation known

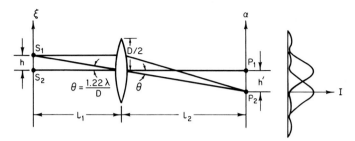

Fig. 4 Rayleigh criterion for the resolution of two-point images.

as Abbe's sine condition is often used. This is demonstrated in Fig. 5. For the general case that the refractive indexes at both sides of the lens are not the same (e.g., oil immersion microscope), it can be shown that if h and h' are small compared to l_1 and l_2, then

$$(h \sin \theta)/\lambda \simeq (h' \sin \theta')/\lambda', \tag{29}$$

where λ and λ' are wavelengths at the source and image sides of the imaging lens. For the more common case where the refractive indexes at both sides of the lens are the same (e.g., both air), then

$$h \sin \theta = h' \sin \theta'. \tag{30}$$

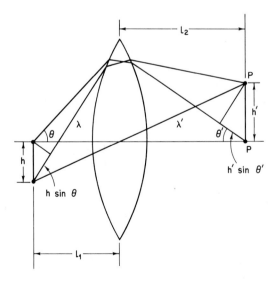

Fig. 5 Diagram demonstrating the Abbe's sine condition.

2. Background

And when θ' is small, we can make the approximation

$$\sin \theta \simeq \theta = D/2l_2 . \tag{31}$$

Substituting this into Eq. 28, we have

$$h = 1.22\lambda/(2 \sin \theta) = 0.61\lambda/\text{NA}, \tag{32}$$

where $\text{NA} = \sin \theta$ is generally known as the numerical aperture of the imaging system. The numerical aperture is also related to the focal number N by $\text{NA} = \frac{1}{2}N$.

2.4.4 Aberrations

Ideally, a converging lens should perform the phase transformation

$$T(x, y) = \exp[-i(k/2f)(x^2 + y^2)] \tag{33}$$

perfectly for the entire aperture utilized; that is, the lens should be able to transform a light field emanating from a point source at the front focal plane into an ideal plane wave, limited only by the restricting aperture of the lens. Such a lens is referred to as diffraction-limited. A simple spherical lens is unable to do this, and when it is used for imaging, defects in the image called aberrations will result. All of these aberrations can be corrected, at least partially, by using a combination of lenses of different curvatures and refractive indexes. We now describe some of the major types of aberration.

2.4.4.1 Chromatic Aberration

This form of aberration is caused by the differences in the refractive indexes of the lens for the different frequencies of light. This is usually not a problem with coherent imaging systems, including holography, since monochromatic light is used for illumination. The primary exceptions are holographic optical elements (Section 10.8) and holographic diffraction gratings.

2.4.4.2 Spherical Aberration

When the light rays passing near the edge of the lens (peripheral rays) do not focus at the same plane as those passing near the center of the lens (central rays), then spherical aberration is present. That is to say, the phase transformation by the lens is no longer a linear function of $(x^2 + y^2)$.

2.4.4.3 Curvature of Field

This form of aberration is said to be present when light rays passing through the lens at different angles to the optical axis do not focus in the same plane but form a curved field.

2.4.4.4   Coma

When the imaging system suffers from this form of an aberration, the image field is not only curved, but the peripheral rays also focus away from the central rays causing the point image to appear pear shaped.

2.4.4.5   Astigmatism

When the light rays in a plane containing the optical axis do not focus in the same plane as the rays in a plane off the optical axis perpendicular to this tangential plane, then we have astigmatism.

2.4.4.6   Incident Barrel Distortion

This form of distortion causes the image of a square to bow outward. This generally occurs when the aperture is placed in front of the lens. The magnification increases with distance from the optical axis.

2.4.4.7   Pincushion Distortion

This form of distortion makes the image of a square bow inward. This generally occurs when the aperture is placed behind the lens. The magnification decreases with distance from the optical axis.

2.4.5   Holographic Images

We have discussed the image formation with conventional imaging systems and the definition of resolution and various forms of aberration. In the following section, we discuss the resolution and aberrations in holographic reconstruction as well as magnification and signal-to-noise ratio.

2.4.5.1   Magnification

When Gabor first conceived the idea of holography, one of the first applications proposed was in the field of microscopy. By varying the wavelength or geometry of illumination in the construction and reconstruction processes, a magnification of the holographic image can be obtained. To examine the condition for magnification let us make use of the simple arrangement shown in Fig. 6 for the construction of the hologram. An off-axis point source is used as the reference to construct a hologram of two point objects a distance h apart. If the size of the aperture is assumed to be much smaller than its

2. Background

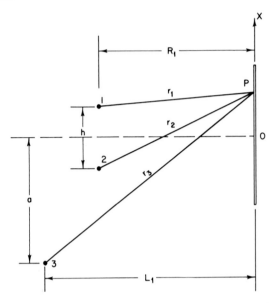

Fig. 6 Recording geometry for determining the lateral magnifications. 1 and 2 are point objects and 3 is the reference source; *P* is the photographic plate.

distance from the light source, we can use the paraxial approximation

$$r = L^2 + (x^2 + y^2) = L + \frac{(x^2 + y^2)}{2L} .$$

Using this approximation, the complex light distribution at the hologram due to these point radiators can be written as

$$u_1(\rho; k_1) \simeq A_1 \exp\left(ik_1\left\{R_1 + \frac{1}{2R_1}\left[\left(x - \frac{h}{2}\right)^2 + y^2\right]\right\}\right),$$

$$u_2(\rho; k_1) \simeq A_2 \exp\left(ik_1\left\{R_1 + \frac{1}{2R_1}\left[\left(x + \frac{h}{2}\right)^2 + y^2\right]\right\}\right), \qquad (34)$$

$$u_2(\rho; k_1) \simeq A_3 \exp\left(ik_1\left\{L_1 + \frac{1}{2L_1}[(x + a)^2 + y^2]\right\}\right).$$

Assuming that the recording is linear, the amplitude transmittance of the hologram would be

$$T(\rho; k_1) = I(\rho; k_1) = (u_1 + u_2 + u_3)(u_1 + u_2 + u_3)^*. \qquad (35)$$

The hologram is then illuminated by a divergent light of wavelength λ_2 as

60

shown in Fig. 7,

$$u_4(\rho; k_2) = A_4 \exp\left(ik\left\{L_2 + \frac{1}{2L_2}[(x - b)^2 + y^2]\right\}\right).\tag{36}$$

After some long but straightforward calculations, it can be shown that the lateral magnification of the real image can be written as (Meier, 1965; Leith *et al.*, 1965; Champagne, 1967; Diamond, 1967)

$$M_{\text{lat}}^{\text{r}} = \frac{h_r}{h} = \left(1 - \frac{\lambda_1 R_1}{\lambda_2 L_1} - \frac{R_1}{L_1}\right)^{-1},\tag{37}$$

and the magnification of the virtual image expressed as

$$M_{\text{lat}}^{\text{v}} = \frac{h_v}{h} = \left(1 + \frac{\lambda_1 R_1}{\lambda_2 L_2} - \frac{R_1}{L_1}\right)^{-1}.\tag{38}$$

We see that when a longer wavelength is used in the reconstruction, lateral magnification occurs for the virtual image. This leads to one interesting application. We indicated in the discussion of image resolution that the resolution limit is determined by the wavelength of the illuminating light. That is, the smaller the wavelength, the higher the resolution limit. We cannot, however, arbitrarily increase the frequency of illumination since as it extends beyond the ultraviolet region, the image becomes invisible. We can however, construct a hologram using invisible high-frequency light waves and reconstruct the image with lower frequency visible light. And as we can see in Eq. (38),

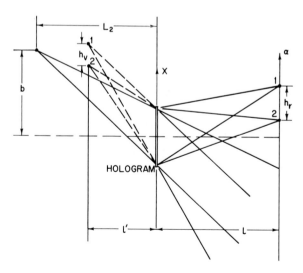

Fig. 7 Reconstruction geometry for the determination of lateral magnification.

2. Background

besides visualizing the image made with high frequency invisible light, a magnification of the image is also realized.

While we are only concerned with lateral magnification in discussing conventional imaging systems, holographic images are inherently three dimensional. Therefore, we should also look at the magnification along the longitudinal axis. Using a similar arrangement to that in Fig. 6, but with the two point objects lying on the optical axis a distance d apart, the longitudinal magnification of the real image is found to be

$$M_{long}^r = \frac{d_r}{r} \simeq \frac{\lambda_1}{\lambda_2} (M_{lat}^r)^2$$

$$= \frac{\lambda_1 \lambda_2 (L_1 L_2)^2}{[\lambda_2 L_2 R_1 - \lambda_2 L_1 L_2 - \lambda_1 L_1 R_1]^2} \qquad \text{for} \quad d \ll R_1. \qquad (39)$$

and the longitudinal magnification of the virtual image can be written as

$$M_{long}^v = \frac{d_v}{v} \simeq \frac{\lambda_1}{\lambda_2} (M_{lat}^v)^2$$

$$= \frac{\lambda_1 \lambda_2 (L_1 L_2)^2}{[\lambda_2 L_2 R_1 - \lambda_2 L_1 L_2 - \lambda_1 L_1 R_1]^2} \qquad \text{for} \quad d \ll R_1. \qquad (40)$$

Comparing the lateral and longitudinal magnifications, we find that they are not equal. Thus we expect the reconstructed three-dimensional image to be distorted if we try to obtain magnification by using different wavelengths for the illumination in the reconstruction process. Other forms of aberration would also occur, as we shall see in a later section.

2.4.5.2 Resolution

To define the resolution limit of a holographic image, we can once again make use of the Rayleigh criterion. Let us consider the same holographic system that was used in the preceding section on magnification. In addition, we assume that the hologram has a circular aperture of diameter D. It can be shown that the minimum resolvable distance between two image points in the real image reconstruction will be (Leith *et al.*, 1965; Champagne, 1967; Diamond, 1967)

$$h_{r,min} = 1.22 \lambda_1 R_1 / D. \qquad (41)$$

Similarly, for the virtual image,

$$h_{v,min} = 1.22 \lambda_1 R_1 / D. \qquad (42)$$

It should be no surprise that the lateral resolution limit of the holographic image is very similar to that obtained for the imaging system with spherical lenses. The hologram of a point object acts very much like a spherical lens; that its resolution limit comes under similar constraint is to be expected.

The longitudinal resolution limit is an important parameter in applications such as holographic contouring. Unlike the lateral resolution limit, the longitudinal resolution limit is determined by the finite frequency bandwidth of the illuminating beam. Using once again the hologram of the two-point objects illuminated by a quasi-monochromatic divergent light source in the reconstruction, the minimum resolvable longitudinal distance of the real image can be shown to be

$$d_{r,min} \simeq \Delta l_r \quad \text{for} \quad d \ll R_1, \tag{43}$$

where

$$\Delta l_r = l_r' - l_r'',$$

$$l_r' = \frac{\lambda_1 R_1 L_1 L_2}{\lambda' L_1 L_2 - \lambda' R_1 L_2 - \lambda_1 R_1 L_1}, \tag{44}$$

$$l_r'' = \frac{\lambda_1 R_1 L_1 L_2}{\lambda'' L_1 L_2 - \lambda'' R_1 L_2 - \lambda_1 R_1 L_1},$$

and λ' and λ'' are the respective low and high cutoff wavelengths of the source. We can conclude that

$$d \geq \Delta l_r (M_{long}^r)^{-1}, \tag{45}$$

where

$$M_{long}^r = \frac{\lambda_1 \lambda_2 (L_1 L_2)^2}{[\lambda_2 L_1 L_2 - \lambda_2 R_1 L_2 - \lambda_1 R_1 L_1]^2},$$

as obtained in the previous section, and $\lambda_2 = (\lambda' \lambda'')^{1/2}$ is the mean wavelength of the source. Therefore the minimum resolvable longitudinal distance can be shown to be

$$d_{r,min} = \Delta l_r (M_{long}^r)^{-1}. \tag{46}$$

Similarly, for virtual image reconstruction, we can show that the minimum resolvable longitudinal distance is

$$d_{v,min} = \Delta l_v (M_{long}^v)^{-1}, \tag{47}$$

where

$$\Delta l_v = l_v' - l_v'',$$

$$l_v' = \frac{\lambda_1 L_1 R_1 L_2}{\lambda' R_1 L_2 - \lambda' L_1 L_2 - \lambda_1 L_1 R_1}, \tag{48}$$

$$l_r'' = \frac{\lambda_1 L_1 R_1 L_2}{\lambda'' R_1 L_2 - \lambda'' L_1 L_2 - \lambda_1 L_1 R_1}.$$

2. Background

2.4.6 Holographic Aberrations

It will be pointed out in Chapter 7 that if the same plane wave is used as the reference beam in both the recording and reconstruction of the holographic image, the exact original wavefront is reproduced and the image is without any form of aberration. However, if either the wavelength or the geometry of the reference beam is changed in the reconstruction, intentionally (e.g., to produce magnification) or unintentionally, aberrations occur. In the formulation utilized for the calculation of magnifications, the paraxial approximation was made. Except for the distortion of the three-dimensional image due to the difference in longitudinal and lateral magnifications, with the paraxial approximation, the reconstructed image should exhibit no other form of aberrations. However, if a more exact formulation is used, aberrations can be shown to occur if the reconstruction beam varies from the reference used in the recording. These aberrations can be classified by the same parameters used for conventional imaging systems, namely, spherical aberration, coma, curve of field, astigmatism, and distortion (Meier, 1965; Leith *et al.*, 1965; Champagne, 1967; Diamond, 1967; Armstrong, 1965).

Let us consider the case in which a plane wave is used as the reference in the recording and a spherical wave is used for the reconstruction of the hologram as shown in Fig. 8. Then the light field at the image plane of a two-dimensional object can be written as

$$E(\sigma; k_2) = C \iint_{S_2} \left\{ \iint_{S_1} O(\xi, \eta) \exp[ik_1(x \sin \theta_1 - r_1)] \, d\xi \, d\eta \right\}$$
$$\times \exp[ik_2((\rho^2/2R) - x \sin \theta_2)] \exp(ik_2 r_2) \, dx \, dy, \tag{49}$$

where C is a complex constant, $O(\xi, \eta)$ is the two-dimensional object function, $k_1 = 2\pi/\lambda_1$, with λ_1 the recording wavelength, $k_2 = 2\pi/\lambda_2$, with λ_2 the reconstructing wavelength, $\rho^2 = x^2 + y^2$, and S_1 and S_2 denote the surface integral of the object function and the transmission function of the hologram, respectively. It is also clear that the first surface integral represents the wavefront construction, the second exponential represents the hologram illumination, and the last exponential represents the diffraction from the hologram.

Referring to Fig. 8, we can write the distances r_1 and r_2 as

$$r_1 = l_1 \left[1 + \frac{(x - \xi)^2 + (y - \eta)^2}{l_1^2} \right]^{1/2} \tag{50}$$

and

$$r_2 = l_2 \left[1 + \frac{(\alpha - x)^2 + (\beta - y)^2}{l_2^2} \right]^{1/2} \tag{51}$$

64

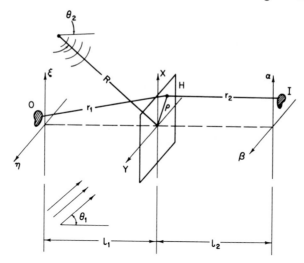

Fig. 8 Geometry for the wavefront recording and reconstruction. O, object; H, hologram, and I, image. Plane reference wave incident at θ_1 is used for recording and divergent wavefront indicent at θ_2 is used in the reconstruction.

Then, by binomial expansion, Eqs. (50) and (51) can be written as

$$r_1 = l_1 + \frac{1}{2l_1}[(x - \xi)^2 + (y - \eta)^2] - \frac{1}{8l_1{}^3}[(x - \xi)^2 + (y - \nu)^2]^2 + \cdots, \quad (52)$$

$$r_2 = l_2 + \frac{1}{2l_2}[(\alpha - x)^2 + (\beta - y)^2] - \frac{1}{8l_2{}^3}[(\alpha - x)^2 + (\beta - y)^2]^2 + \cdots. \quad (53)$$

If only the first two terms of Eqs. (52) and (53) are retained, we have the paraxial approximations, and the explicit form of Eq. (47) will be

$$E(\sigma; k_2) = C' \iint_{S_2} \left\{ \iint_{S_1} O(\xi, \eta) \right.$$

$$\exp\left[ik_1\left(x \sin \theta_1 \frac{(x - \xi)^2 + (y - \eta)^2}{2l_1} \right) \right] d\xi \, d\eta \Bigg\}$$

$$\times \exp\left[ik_2\left(\frac{\rho^2}{2R} - x \sin \theta_2 \right) \right] \exp\left[ik \frac{(\alpha - x)^2 + (\beta - y)^2}{2l_2} \right] dx \, dy,$$

$$(54)$$

where C is an appropriate complex constant. The third-order aberrations may be calculated by retaining the first three terms of Eqs. (52) and (53). The contributions of the fourth and higher terms are so insignificant that they can generally be neglected. It may be emphasized that aberrations of the wavefront

2. Background

recording and reconstruction processes depend on the exponential argument containing r_1 and r_2. With the usual paraxial approximation, the quadratix exponential factor ρ^2 is eliminated by imposing the lens condition

$$\frac{1}{R} + \frac{1}{l_2} = \frac{\lambda_2}{\lambda_1}\frac{1}{l_1},\tag{55}$$

similar to the case we discussed with the conventional imaging system. However, in the nonparaxial case it is no longer sufficient to impose the condition expressed by Eq. (55) in order to eliminate the higher order exponential term. These nonvanishing terms in the exponent constitute the aberrations in hologram images. To investigate the aberrations, we can begin with the evaluation of the phase factor $\Delta\phi = k_2 r_2 - k_1 r_1$ of the construction–reconstruction process. After a long but straightforward calculation, $\Delta\phi$ is seen to be

$$\Delta\phi = -\frac{1}{8}\left(\frac{k_2}{l_2{}^3} - \frac{k_1}{l_1{}^3}\right)\rho^4 + \frac{1}{2}\left(\frac{M^{\mathrm{r}}_{\mathrm{lat}}k_2}{l_2{}^3} - \frac{k_1}{l_1{}^3}\right)\rho^2 K^2 - \frac{1}{2}\left(\frac{(M^{\mathrm{r}}_{\mathrm{lat}})^2 k_2}{l_2{}^3} - \frac{k_1}{l_1{}^3}\right)K^4$$

$$-\frac{1}{4}\left(\frac{(M^{\mathrm{r}}_{\mathrm{lat}})^2 k_2}{l_2{}^3} - \frac{k_1}{l_1{}^3}\right)\rho^2\tau^2 + \frac{1}{2}\left(\frac{(M^{\mathrm{r}}_{\mathrm{lat}})^3 k_2}{l_2{}^3} - \frac{k_1}{l_1{}^3}\right)\tau^2 K^2,\tag{56}$$

where $\rho^2 = x^2 + y^2$, $\tau^2 = \xi^2 + \eta^2$, $K^2 = \xi x + \eta y$, and $M^{\mathrm{r}}_{\mathrm{lat}} = \lambda_2 l_2/\lambda_1 l_1$. By comparing Eq. (56) with the general treatment of lens aberrations (Born and Wolf, 1964), we can see that the first term of Eq. (56) is the spherical aberration, the second term the coma, the third term the astigmatism, the fourth term the curvature of field, and the last term the distortion.

Now let us consider the conditions under which the reconstructed image will be free of these aberrations by setting each term in Eq. (56) equal to zero. The results are tabulated in Table I.

From Table I, it can be seen that the aberrations generally cannot be corrected together except for astigmatism and curvature of field. Also, all the aberrations will vanish when lateral magnification is unity ($M = 1$).

2.4.7 Signal-to-Noise Ratio

The main concern with conventional imaging systems utilizing glass lenses is aberration. The signal-to-noise ratio (SNR) is generally so high that it is seldom a problem. The opposite is true with holographic systems. As we have mentioned earlier, it is a simple matter to produce an aberration-free holographic image. However, due to the many imperfections in the recording materials, holographic images are much noisier, and the SNR is the main concern in many applications.

SNR is generally defined as the spatial average of the light power forming

TABLE I

Aberrations and Conditions for Their Correction

Aberration	Condition
Spherical aberration	$\lambda_2/\lambda_1(l_2/l_1)^3 = 1$
Coma	$\lambda_2/\lambda_1(l_2/l_1)^3 = M;\ l_1 = l_2$
Astigmatism and curvature of field	$\lambda_2/\lambda_1(l_2/l_1)^3 = M^2;\ \lambda_2/\lambda_1 = l_2/l_1$
Distortion	$\lambda_2/\lambda_1(l_2/l_1)^3 = M^3;\ \lambda_1 = \lambda_2$

the image signal divided by the average light power of the noise. All holographic recording materials are not perfectly linear, and this nonlinearity produces higher order images which by the strict definition of SNR should be measured as part of the noise. In off-axis holography, however, these higher order images are constructed away from the desired first-order image, and they are generally not included in the SNR measurement.

To determine the SNR of the holographic reconstruction for different parameters (e.g., exposures, recording materials), there is no single standard method that can be used. Since the actual value of the SNR is dependent on the measurement method or arrangement utilized, they are usually specified. One simple measurement that can be used is to construct a hologram of a point source and measure the ratio of the light powers between the reconstructed point image and the scattered noise around it. The SNR can be plotted against the reference to object beam angle θ or spatial frequency $\sin\theta/\lambda$. To measure the SNR of the image of a diffuse object, one common geometry is shown in Fig. 9. The diffuse object is made of a back illuminated diffuse plate and the SNR is defined as the ratio of the average light intensity in the reconstructed square ring object to the average intensity of the scattered light in the center.

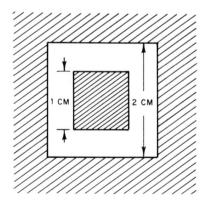

Fig. 9 Geometry of diffuse object used in the determination of SNR.

2. Background

The object distance from the hologram should also be specified since it affects the measured value.

2.4.8 Conclusion

We have presented a brief discussion on image formation and the various parameters and properties that described images for both conventional and holographic imaging systems. Strictly speaking, the holographic process is not an imaging process in the conventional sense, especially when describing the reconstruction of virtual images. Imaging means the projection or mapping of the light field from one plane to another. If this mapping is perfectly linear, then the imaging system is free of any aberrations. In holography, the original wavefront is reproduced and the mapping is therefore onto itself. In spite of this fundamental difference, the images obtained from conventional imaging systems and holographic reconstruction can be described and evaluated in a similar manner with the same parameters.

REFERENCES

Armstrong, J. A. (1965). *IBM J. Develop.* **9**, 171.
Born, M., and Wolf, E. (1964). "Principles of Optics," 2nd ed. Pergamon Press, New York.
Caufield, H. J. and Lu, S. (1970). "The Application of Holography." Wiley, New York.
Champagne, E. B. (1967). *J. Opt. Soc. Amer.* **57**, 51.
Collier, R. J., Burckhardt, C. B., and Lin, L. H. (1971). "Optical Holography." Academic Press, New York.
Diamond, F. I. (1967). *J. Opt. Soc. Amer.,* **57**, 503.
Goodman, J. W. (1968). "Introduction to Fourier Optics." McGraw-Hill, New York.
Kock, W. E. (1971). "Laser and Holography: An Introduction to Coherent Optics." Doubleday, New York.
Leith, E. N., Upatnieks, J., and Haines, K. A. (1965). *J. Opt. Soc. Amer.* **55**, 981.
Meier, R. W. (1965). *J. Opt. Soc. Amer.* **55**, 987.
Rossei, B. (1957). "Optics." Addison-Wesley, Reading, Massachusetts.
Sears, F. W. (1949). "Optics." Addison-Wesley, Reading, Massachusetts.
Shulman, A. R. (1970). "Optical Data Processing." Wiley, New York.
Sommerfield, A. (1954). "Optics." Academic Press, New York.
Stone, J. M. (1963). "Radiation and Optics." McGraw-Hill, New York.
Stroke, G. W. (1969). "An Introduction to Coherent Optics and Holography," 2nd ed. Academic Press, New York.
Yu, F. T. S. (1973). "Introduction to Diffraction, Information Processing and Holography." MIT Press, Cambridge, Massachusetts.

2.5 COMMUNICATION THEORY

John B. DeVelis
George O. Reynolds

2.5.1 Introduction

In this section we will discuss some further methods which are useful in the analysis of linear optical systems. We will assume that the optical system is a linear black box whose input–output relationship is described by the convolution process in the space domain. The linear optical system has the property that it is completely characterized by either its point spread function in the space domain or by the Fourier transform of its point spread function, the optical transfer function, in the frequency domain.

We first discuss the sampling process whereby a given band-limited function is represented by a series of discrete sample points. The technique is described in both the space and spatial frequency domains. The space–bandwidth product is also discussed in terms of the sampled function, and a number of examples are given.

Methods for describing statistical processes are introduced. These include ensemble and spatial averages, correlation functions, and the concept of spectral density. The use of these methods in linear system analysis is demonstrated by examples.

2.5.2 Sampling Theorem
2.5.2.1 Space Domain

The sampling theorem (O'Neill, 1963, Appendix A) is a curve-fitting device for representing functions with a finite spectrum centered around zero frequency. If the function to be sampled is truly band-limited, then the sampling theorem will introduce no error into the functional representation.

Theorem If a one-dimensional function $a(x)$ contains no frequencies higher than ξ_0 cycles/mm, the function is completely determined by giving its ordinates at a series of points extending throughout the space domain and spaced $(1/2\xi_0)$mm apart. For example, in optics, the function $a(x)$ could be an object scene limited by the lens transfer function; or if it is a photographic transparency, it will be limited by the film transfer function which has a finite cutoff frequency.

HANDBOOK OF OPTICAL HOLOGRAPHY

2. Background

In one dimension, the sampling theorem is given by

$$a(x) = \sum_{n=-\infty}^{\infty} a\left(\frac{n}{2\xi_0}\right) \operatorname{sinc} 2\pi\xi_0\left(x - \frac{n}{2\xi_0}\right), \tag{1}$$

where the interpolation function is the sinc function. Equation (1) shows that in order to reconstruct the function $a(x)$ we find its value at the sampled points $(n/2\xi_0)$ and at each of these points multiply the value of $a(x)$ by the interpolation function $\operatorname{sinc} 2\pi\xi_0[x - (n/2\xi_0)]$. Graphically, this corresponds to plotting sinc functions at each of the sampled ordinates such that the magnitude of the sinc function at the specific ordinate is the value of the function at that point as shown in Fig. 1. Due to the periodicity of the sinc function, it does not contribute to the function at the other sampled ordinates.

2.5.2.2 Frequency Domain

The function in Eq. (1) is a summation, over all sampled points, of the sampled function times an interpolation function (Goodman, 1968, Chapter 2). The sampled function, $a(n/2\xi_0)$, consists of the original function, $a(x)$, times an array of delta functions centered at the sampled points [comb function; $\operatorname{comb}(x) = \sum_{n=-\infty}^{\infty} \delta(x - n/2\xi_0)$], and the interpolation function is the sinc function. It is often more convenient to interpret the sampling theorem in the frequency domain rather than in the space domain. The frequency spectrum of $a(x)$ is given by the Fourier transform of Eq. (1) using the delta function interpretation of the sampled function. This Fourier spectrum $A(\xi)$ of the function $a(x)$, is a product of the replicated spectrum of the sampled function with a rect function whose size equals the bandwidth $2\xi_0$ of the function $a(x)$.

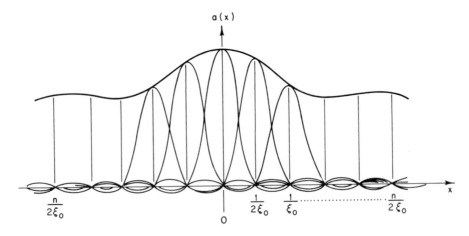

Fig. 1 Graphical representation of the sampling theorem in the space domain.

70

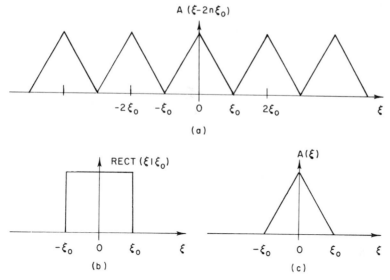

Fig. 2 Graphical representation of the sampling theorem in the frequency domain. (a) The product of the sampled function spectrum $A(\xi - 2n\xi_0)$ with (b) the spectrum of the interpolation function yields (c) the spectrum $A(\xi)$.

The rect function behaves as a linear filter which sifts the object spectrum from the sampled spectrum guaranteeing the desired representation of the object scene. This is shown graphically in Fig. 2.

2.5.2.3   Sampling Criteria and Space–Bandwidth Product

In order to successfully recover the function $a(x)$ in Eq. (1), the sampled ordinates must satisfy the condition that

$$1/x_0 \le 2\xi_0, \tag{2}$$

where x_0 is the separation of the sampled ordinates in the space domain. This ensures that the replicated spectra in Fig. 2a do not overlap. If the sampling interval x_0 is greater than $\frac{1}{2}\xi_0$, the spectra are more widely separated in Fig. 2a and the function is still recovered even though more sampling points have been used. If the sampling interval x_0 is less than $\frac{1}{2}\xi_0$, the spectra in Fig. 2a overlap and the function is not completely recovered. This degradation of the recovered function, due to undersampling, is known as aliasing.

In any practical application, the summation in Eq. (1) will extend from $-N$ to N rather than from $-\infty$ to ∞, resulting in a total of $2N$ sample points. Further, the condition given in Eq. (2) applies to each sampled point.

2. Background

The product $(2N)(2\xi_0)$, resulting from the total number of sample points and the bandwidth requirement of the function (which determines the sampling interval), is called the one-dimensional space–bandwidth product of that portion of the function being considered.

2.5.2.4 Two-Dimensional Sampling Theorem

The sampling theorem can be readily extended to two dimensions. For the two-dimensional rectangular case, we obtain the extension of Eq. (1) in the form

$$a(x, y) = \sum_{n=-\infty}^{\infty} \sum_{m=-\infty}^{\infty} a\left(\frac{n}{2\xi_0}, \frac{m}{2\eta_0}\right) \operatorname{sinc} 2\pi\xi_0\left(x - \frac{n}{2\xi_0}\right)$$

$$\times \operatorname{sinc} 2\pi\eta_0\left(y - \frac{m}{2\eta_0}\right),$$

$$(3)$$

where ξ_0 and η_0 are the cutoff frequencies associated with the ξ and η axes for the Fourier spectrum of $a(x, y)$. For the rotationally symmetric two-dimensional case, one uses the Bessel function of the first kind for sampling, and the result is given by (Goodman, 1968)

$$a(x, y) = \sum_{n=-\infty}^{\infty} \sum_{m=-\infty}^{\infty} a\left(\frac{n}{2\nu_0}, \frac{m}{2\nu_0}\right)$$

$$\times \frac{2\pi\nu_0^2 J_1(2\pi\nu_0\{[x - (n/2\nu_0)]^2 + [y - (m/2\nu_0)]^2\}^{1/2})}{2\pi\nu_0\{[x - (n/2\nu_0)]^2 + [y - (m/2\nu_0)]^2\}^{1/2}},$$

$$(4)$$

where $\xi_0 = \eta_0 = \nu_0$ for this case.

2.5.2.5 Examples

Three well-known processes which utilize the sampling principle are halftone reproduction of photography, facsimile transmission and display, and the display of television images. These applications combine electronic and optical principles to define and implement the bandwidth limitation necessary to obtain an optimum display.

Other applications which utilize sampling as a fundamental process are electrooptical scanning systems (Beiser, 1974; Dainty and Shaw, 1974). In this case the input information on the film is further band-limited by the f/number of a diffraction-limited optical scanning system. The sampling interval in the space domain, which is given by the reciprocal of the system bandwidth (as determined from a square lens aperture), is

$$x_0 = \lambda(f/\text{number})/2. \tag{5}$$

In such systems, aperture sizes are usually variable, while the scanning speed is constant; thus, the rate of information is set by the aperture size. The electronic sampling rate is determined by the sampling interval of Eq. (5) and the system speed.

Another optical example of the use of the sampling theorem is modulated imagery (Mueller, 1969). In this process the image is band-limited by a lens and sampled with a diffraction grating having the desired fundamental frequency for sampling.

The principle of angular modulation can be used to store more than one image on the film. The grating must be rotated between exposures by the amount

$$\Delta \theta_0 = 2 \sin^{-1}[\xi_{BW}/(2\xi_0)], \tag{6}$$

where $\Delta \theta_0$ is the amount of rotation, ξ_{BW} is the bandwidth of the individual images, and ξ_0 is the sampling frequency. The angle defined in Eq. (6) ensures that the image spectra are spatially separated as shown in Fig. 3a. The multiply stored images on the developed film can be individually retrieved in a coherent optical processing system by placing an aperture over the appropriate image spectrum in the transform plane of the processing system.

Aliasing results in this process when the sampling interval is too large. This results in overlapping spectra as shown in Fig. 3b.

In the retrieved image plane, the filtered image is degraded by the presence of high-frequency components from adjacent image spectra.

2.5.3   Statistical Description of Random Samples

2.5.3.1   Ensemble and Coordinate Averaging Descriptions of Random Processes

The basic entity of any communication channel (system) is "information," and the fundamental entity which most characterizes such systems is the "information capacity." Whether the system is electrical, optical, or electrooptical, it processes signal information which can best be classified as completely deterministic or statistical. In the deterministic case, the signal is usually given a Fourier series or integral representation, i.e., it is a periodic or transient waveform whose value is completely determined for all values of the independent variable (time or space). On the other hand, statistical signals take on values for any particular value of the independent variable (time or space) which are not completely determined, i.e., they are only known in a probabilistic sense. These statistical signals, usually called "random signals," are treated by introducing statistical or probabilistic methods for analyzing and synthesizing the information content of such signals. In essence, for random signals over an infinite limit, a Fourier representation does not exist, and one is forced to consider a statistical analysis. These resulting statistical

2. Background

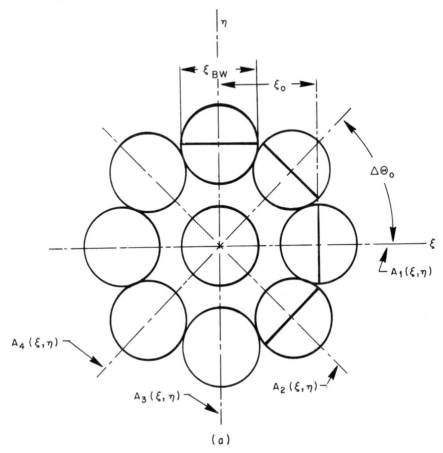

Fig. 3 (a) Angular separation of four properly sampled image spectra in frequency space. (After Mueller, 1969.) (b) Example of under sampling in frequency space showing spectra overlap which results in aliasing $\xi_0 = f_1 \lambda/p$, ξ_{BW} = bandwidth (after Mueller, 1969).

methods can be applied to the deterministic case; however, they have found wider application and interest in the analysis of random processes. In the optical case, such methods are used as the basic tool in the formulation of the classical theory of partial coherence, the analysis of film grain noise, and the analysis of coherent optical noise usually called "speckle."

A random signal (or process), $F(\bar{x}, t)$, may be defined as one which does not depend on the independent variable (either space, time, or both) in a completely deterministic manner. Generally, we work with random signals which obey the simplifying constraint, stationarity. If we have a physical process which gives rise to a random signal, such a signal will be considered

74

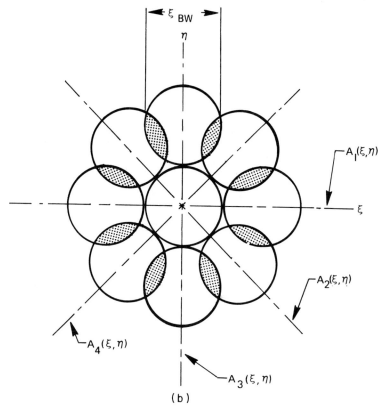

(b)

stationary with respect to the time (space) coordinate if the process depends only on the coordinate difference, i.e., $F(t_2 - t_1)$, as opposed to $F(t_1, t_2)$ or $F(x_2 - x_1)$ as opposed to $F(x_1, x_2)$. The underlying physical process which gives rise to stationary random signals is described by statistics which are not time-dependent.

In principle, we have two ways of handling these problems. In the first case, we can assume we know the function over a long period of time (space) from which we determine the probability distribution functions used to determine both time and space averages. In the second case, we have an ensemble of similar functions from which we determine the probability distribution functions by an examination of all members of the ensemble. These distribution functions are then used in the determination of ensemble averages. The ergodic assumption, in principle, tells us that coordinate and ensemble averages should yield the same results. Hence, as we now define our correlation functions, we shall assume we have ergodic stationary signals and only define the averages over spatial coordinates.

2. Background

2.5.3.2 Correlation Functions

The cross-correlation of the complex functions $a(x_1)$ and $s(x_1)$ is defined to be

$$c(x) = \langle a^*(x_1)s(x_1 + x) \rangle \equiv \lim_{L \to \infty} \frac{1}{2L} \int_{-L}^{L} a^*(x_1)s(x_1 + x)\, dx_1. \qquad (7)$$

The autocorrelation function of a complex function $a(x_1)$ is defined by

$$c(x) = \langle a^*(x_1)a(x_1 + x) \rangle. \qquad (8)$$

The following properties of the autocorrelation function are useful:

(a) it is an even function of the delay variable,

$$c(x) = c(-x); \qquad (9)$$

(b) it is a maximum at the origin,

$$c(o) > |c(x)| \qquad \text{for} \quad x \neq 0; \qquad (10)$$

(c) for x equal to zero, we get the average of the square of the function, which in many physical cases is the energy of the system,

$$c(o) = \langle |a(x_1)|^2 \rangle. \qquad (11)$$

Example *Convolution versus Correlation* The form of the integrals in Eqs. (7) and (8) should not be confused with the form of the convolution integral. The convolution process involves folding, shifting, and summing procedures, whereas the correlation procedure involves a shifting and summing without folding. This is not merely a semantic distinction. Unless the functions involved are of even symmetry, the two results are dramatically different.

SYMMETRIC FUNCTIONS If we consider the function $a(x_1)$ to be a rectangular function of width $2a$, the folding process yields the same function since $a(x_1) = a(-x_1)$. Thus, correlation of the rectangular function with itself (autocorrelation) and convolution of the rectangular function with itself both yield the same triangular function as shown in Fig. 4.

NONSYMMETRIC FUNCTIONS As an example of the difference between convolution and correlation, consider the two functions shown in Fig. 5a and given by

$$a(x_1) = \begin{cases} 1 & \text{for} \quad 0 \leq x_1 \leq 1, \\ 0 & \text{for} \quad 0 > x_1 > 1, \end{cases}$$

76

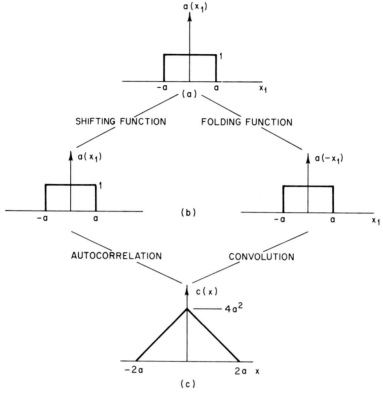

Fig. 4 Comparison of convolution and autocorrelation of symmetric functions: (a) the symmetric rectangular function, (b) the symmetric shifting and folding functions, and (c) the resulting triangular function.

and

$$s(x_1) = \begin{cases} \delta(x_1) - e^{-x_1} & \text{for } x_1 \geq 0, \\ 0 & \text{for } x_1 < 0. \end{cases}$$

Use of the convolution integral, shown schematically in Fig. 5b, yields

$$b(x) = \begin{cases} \displaystyle\int_0^x [\delta(x - x_1) - e^{(x_1 - x)}] \, dx_1 = e^{-x} & \text{for } x < 1, \\ \\ -\displaystyle\int_0^1 e^{(x_1 - x)} \, dx_1 = e^{-x}(1 - e) & \text{for } x \geq 1. \end{cases}$$

The resulting function $b(x)$ is plotted in Fig. 5c.

The autocorrelation integral given by Eq. (8) can be used to autocorrelate the same two functions, $a(x_1)$ and $s(x_1)$, shown in Fig. 6a. This autocorrela-

2. Background

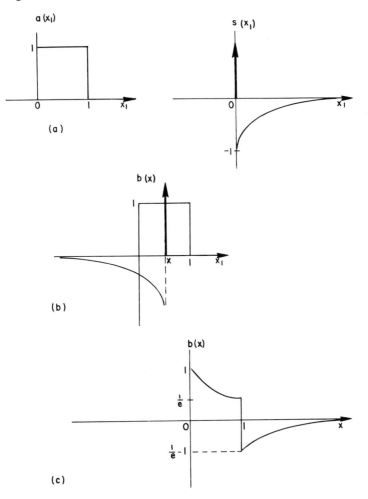

(a)

(b)

(c)

Fig. 5 Plot of convolution process for functions $a(x_1)$ and $s(x_1)$; (a) original functions, (b) convolution at the point x showing the folded function $s(x - x_1)$, and (c) result of the convolution process.

tion shown schematically in Fig. 6b yields

$$c(x) = \begin{cases} \displaystyle\int_x^1 [\delta(x_1 + x) - e^{-(x_1+x)}] \, dx_1 = 1 - e^{-2x} + e^{-(1+x)} & \text{for} \quad 0 < x < 1, \\[2ex] \displaystyle -\int_0^1 e^{-(x_1+x)} \, dx_1 = e^{-x} \, (1/e - 1) & \text{for} \quad x \le 0. \end{cases}$$

78

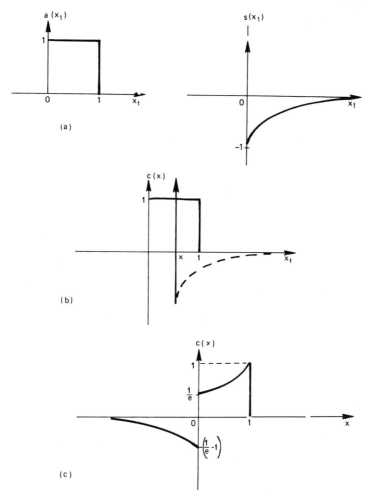

Fig. 6 Plot of the correlation process for the functions $a(x_1)$ and $s(x_1)$; (a) original functions, (b) correlation at the point x showing the sliding function $s(x_1 + x)$, and (c) result of the correlation process.

The resulting function is plotted in Fig. 6c. Comparison of Figs. 5c and 6c shows that the processes of convolution and correlation yield quite different results when the functions are nonsymmetric.

2.5.3.3 Spectral Density

The spectral density of a stationary random signal $a(x_1)$ is extremely useful in the analysis of random signals because of its measurability and its relationship to the autocorrelation function. The spectral density is sometimes referred to as the power spectral density or the power spectrum. It is defined by

$$C(\xi) = \lim_{L \to \infty} \frac{1}{L} |A^T(\xi)|^2, \tag{12}$$

where $A^T(\xi)$ is the Fourier transform of a truncated form of $a(x_1)$.

A very important relationship exists between the spectral density defined by Eq. (12) and the autocorrelation defined by Eq. (8). This relationship is the Wiener–Khinchine theorem and states that the spectral density and autocorrelation functions are Fourier transform pairs, i.e.,

$$C(\xi) = \int_{-\infty}^{\infty} c(x) \exp(-2\pi i \xi x) \, dx. \tag{13}$$

If the random signals are the input to a linear system, the statistical description of the system output is

$$C_{oo}(\xi) = |S(\xi)|^2 C_{ii}(\xi), \tag{14}$$

where $C_{oo}(\xi)$ is the spectral density of the output of the linear system, $C_{ii}(\xi)$ is the spectral density of the input to the linear system, and $|S(\xi)|$ is the modulus of the system transfer function.

2.5.3.4   Examples of Statistical Techniques

Example 1   *Linear Photographic Film*   As an example illustrating these concepts, consider an unrealistic model of a photographic film as a linear system with a Gaussian mathematical spread function given by

$$s(x) = \exp(-\pi x^2/\alpha^2), \tag{15}$$

where α is a real number defining the width of the Gaussian. Equation (15) is plotted in normalized form in Fig. 7a. The transfer function or frequency response of the film is obtained by taking the Fourier transform of Eq. (15) to obtain

$$S(\xi) = \alpha \exp(-\pi \alpha^2 \xi^2), \tag{16}$$

which is plotted in normalized form in Fig. 7b. If the input to the linear system is white noise, its autocorrelation function is given by

$$c_{ii}(x) = \overline{I^2} \, \delta(x), \tag{17}$$

where $\overline{I^2}$ is the mean square brightness of the object (input). The spectral density (power spectrum) is given by

$$C_{ii}(\xi) = \overline{I^2}. \tag{18}$$

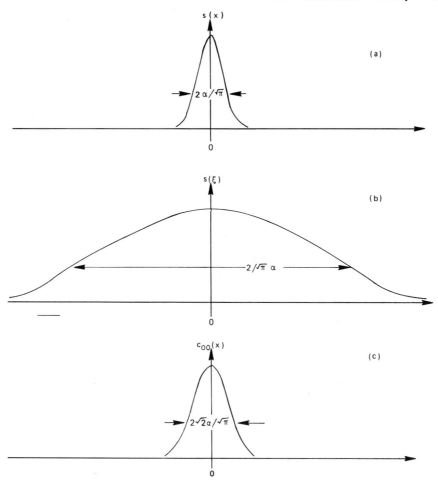

Fig. 7 Response of linear photographic film model to white noise input. (a) Plot of normalized Gaussian spread function given by Eq. (15). (b) Plot of the normalized Gaussian transfer function of the linear photographic film given by Eq. (16). (c) Plot of the normalized output correlation function given by Eq. (20) resulting from the white-noise input to the linear system.

The output power spectrum can be determined from Eq. (14) to be

$$C_{oo}(\xi) = \overline{I^2}\alpha^2 \exp(-2\pi\alpha^2\xi^2). \tag{19}$$

Taking the Fourier transform of this equation, we obtain the output autocorrelation function to be

$$c_{oo}(x) = (\overline{I^2}\alpha/\sqrt{2}) \exp(-\pi x^2/2\alpha^2). \tag{20}$$

Equation (20) is plotted in Fig. 7c in normalized form.

2. Background

This example shows that when a random function (white noise in this example) is the input to a linear system (photographic film), the output correlation function [Eq. (20)] is broader than the input correlation function [Eq. (17)]. The degree of the spread is determined by the width of the spread function α which is determined from the system bandwidth. A comparison of Fig. 7c with Fig. 7a shows that the width of the output correlation function is greater than that of the film spread function. This results from the fact that Eq. (14) involves the square of the transfer function.

In order to determine the output correlation function when the correlation function of the input differs from white noise, one must convolve the correlation function of the input with Eq. (20).

Example 2 *Filtering Signals in the Presence of Additive Noise*

WIENER FILTER   One of the fundamental problems in the application of optical spatial filtering techniques (Dainty and Shaw, 1974; Goodman, 1968; O'Neill, 1963; Yu, 1973) to actual photographs is the presence of photographic grain noise which appears as a spatially irregular structure deteriorating the image of interest. Since this irregular structure is representative of a random process, we will require statistical methods for minimizing its effect. An approach for filtering signals in the presence of additive noise has been developed and widely applied in both electrical and optical systems (Becherer and Geller, 1969; Brown, 1963; Davenport and Root, 1958; Goldman, 1969; Helstrom, 1967; Horner, 1970; Lee, 1960; Slepian, 1967).

In this approach, the filter minimizes the mean square error between the desired signal and the input signal to the filtering system. The general form of the transfer function of this optimum filtering system is (Becherer and Geller, 1969; Lee, 1960)

$$H(\xi) = C_{\mathrm{id}}(\xi)/C_{\mathrm{ii}}(\xi), \qquad (21)$$

where $C_{\mathrm{id}}(\xi)$ is the cross power spectrum of the input signal (object) and the desired signal, and $C_{\mathrm{ii}}(\xi)$ is the power spectrum of the input signal. In order to apply this technique to the filtering of photographs having granular structure, we refer to the system diagram shown in Fig. 8. The purpose of the filtering system is to extract from the degraded noisy image $I_2 + N$ the original object

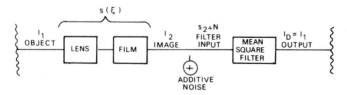

Fig. 8 Block diagram of imaging and mean square filtering system.

I_1 as exactly as possible. The transfer function of the optical spatial filtering system which performs this task in the least mean square sense is (Becherer and Geller, 1969; O'Neill, 1963)

$$H(\xi) = [C_{I_2I_1}(\xi) + C_{I_1N}(\xi)]/[C_{I_2I_2}(\xi) + C_{NN}(\xi) + 2C_{I_2N}(\xi)], \qquad (22)$$

where $C_{I_2I_1}(\xi) = C_{I_1I_1}(\xi)S^*(\xi)$ is the cross power spectrum of the object and the image, $C_{I_1N}(\xi)$ the cross power spectrum of the object and the noise, $C_{I_2I_2(\xi)} = C_{I_1I_1}(\xi)|S(\xi)|^2$ the power spectrum of the image, $C_{I_1I_1(\xi)}$ the power spectrum of the object, $C_{NN}(\xi)$ the noise power spectrum, $C_{I_2N}(\xi)$ the cross power spectrum of the image and the noise, and $S(\xi)$ the lens/film system transfer function. In general, the cross power spectrum of the object and the noise requires knowledge of the Fourier spectrum of the noise rather than its power spectrum. Therefore, in optical applications involving noise, it is assumed that the cross power spectral terms in Eq. (22) are negligibly small, i.e., the object and the noise are not correlated. In general photographic situations, the noise is directly dependent on the object; however, the lack of correlation is a reasonable approximation when the image contrast is low, such as in aerial photography or grain limited imagery. Thus, making the approximation that the object and the noise are not correlated, Eq. (22) becomes

$$H(\xi) = C_{I_1I_1}(\xi)S^*(\xi)/[C_{I_1I_1}(\xi)|S(\xi)|^2 + C_{NN}(\xi)]. \qquad (23)$$

Examination of Eq. (23) leads to two cases of interest which lend some physical insight into the filtering processes.

The first case occurs when the noise is vanishingly small so that $C_{NN}(\xi) \cong 0$. In this case, Eq. (23) reduces to

$$H(\xi) \cong 1/S(\xi). \qquad (24)$$

This is the inverse filter for the system, and we notice that it is the noise-free least-mean-square filter.

The second case of interest occurs when the noise is small but not negligible, i.e.,

$$\frac{C_{NN}(\xi)}{C_{I_1I_1}(\xi)|S(\xi)|^2} \ll 1 \quad \text{or} \quad \frac{C_{I_1I_1}(\xi)}{C_{NN}(\xi)} \gg \frac{1}{|S(\xi)|^2}.$$

For this approximation, Eq. (23) reduces to

$$H(\xi) \cong \frac{1}{S(\xi)} \left[1 - \frac{C_{NN}(\xi)}{C_{I_1I_1}(\xi)|S(\xi)|^2} \right]. \qquad (25)$$

In this approximation, where the noise is not negligible, the value of the optimum filter is reduced from the corresponding value of the inverse filter by a factor dependent upon the signal-to-noise ratio at that frequency.

For the case of a Gaussian transfer function reaching its $1/e$ value at 22.5

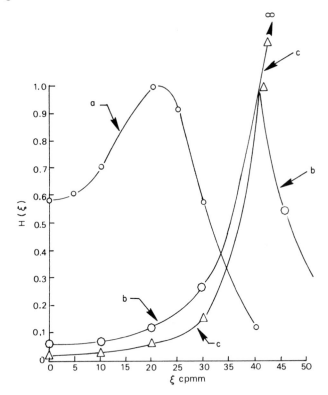

Fig. 9 Calculated optimum least-mean-square filter for assumed signal-to-noise ratio $[C_{I_1I_1}(\xi)/C_{NN}(\xi)]$ of (a) 10, (b) 100, and (c) infinity. (After Horner, 1970.)

line pairs/mm, Eq. (23) is plotted in Fig. 9 for various constant values of the signal-to-noise ratio. For decreasing values of signal-to-noise ratios in Fig. 9, we observe that the cutoff frequency of the optimum filter, as determined by the maximum ordinate of the curve, decreases. This implies that frequencies higher than the filter cutoff frequency only contribute to noise and hence are rejected by the filter. Experimental evidence showing the superior performance of the optimum filter compared to the inverse filter for the case of imaging through a turbulent medium has appeared in the literature (Horner, 1970).

MATCHED FILTER Another linear filter useful for detecting known signals from additive random background noise is the matched filter (Brown, 1963). This filter, which maximizes the ratio of peak signal to rms noise, is given by

$$H(\xi) = (\text{const})O^*(\xi)/C_{NN}(\xi), \tag{26}$$

where $O^*(\xi)$ is the complex conjugate of the signal (object) Fourier spectrum and $C_{NN}(\xi)$ is the noise power spectrum. These filters have been realized by holographic techniques and are discussed in greater detail in Section 10.5.

Example 3 *Speckle Photography (an example of ensemble averaging)*   When imaging through the atmosphere with a lens, the principal effect of the turbulent medium is to distort the transmitted wavefront from each point in the object so that the wavefront reaching the optical system contains random structure in both amplitude and phase. The thermal variations present in the atmosphere cause density inhomogeneities which, in turn, impress phase distortions on the wavefront. After sufficient propagation, the phase variations produce random amplitude variations in the entrance pupil of the lens. Phase shifts can originate near to or far from the entrance pupil of the imaging system which cause the following types of image distortions:

(1) scintillation, intensity fluctuations,
(2) distortion, shifting of all or parts of the image, and
(3) blur, broadening of the instantaneous point-spread function.

A technique for obtaining increased resolution for turbulence-degraded photographic images has been described in the literature (Labeyrie, 1970). The underlying principle is that on any fast exposure through the atmosphere, many points of the image are distorted, but at some points in the image the full-aperture resolution is present in one or more directions. Adding a series of short exposures can build up the resolution in all directions over the image format. Thus, an ensemble of very-short-exposure images of an object having a center of symmetry (double star) are recorded on film. The exposures must be fast enough to "stop" the atmospheric motion. The Fourier transforms of the individual ensemble members are added sequentially in a coherent system, such as that shown in Fig. 10. The shift theorem of Fourier analysis ensures

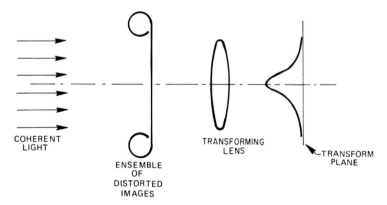

COHERENT
LIGHT

ENSEMBLE
OF
DISTORTED
IMAGES

TRANSFORMING
LENS

TRANSFORM
PLANE

Fig. 10   Optical system for adding Fourier transforms of an ensemble of distorted images.

that the Fourier transform of the object is centered on the optical axis for each ensemble member, whereas the Fourier transform of the noise is randomly distributed. A sequential recording of the intensities in the Fourier transform plane results in an addition of the signal power spectra from the ensemble members and the simultaneous averaging of their noise contributions. This addition (ensemble averaging) increases the signal-to-noise ratio of the signal by retaining all signal information, in register, from each member of the ensemble. Since the technique records the intensity of the Fourier transform, the object phase is lost and only objects with a center of symmetry can be treated successfully by this technique.

Upon development of the film in the transform plane of Fig. 10 to a gamma of one half, the enhanced image is obtained by Fourier transforming the developed film. The concept has been extended to include objects without a center of symmetry by determining the relative phase of the autocorrelation function by the use of a computer algorithm (Knox and Thompson, 1974).

2.5.4 Summary

We have discussed the sampling theorem in both coordinate space and frequency space, and the concept of the space–bandwidth product was used to relate the total number of sample points and the bandwidth requirement of the function. Optical examples illustrating the use of the sampling theorem in various applications are given. The statistical description of random signals, assuming stationarity and ergodicity, emphasizing both ensemble and coordinate averaging is presented. Correlation functions and their Fourier transforms, spectral density functions, are defined. A general comparison is made between the correlation and convolution processes for both symmetric and nonsymmetric functions. Examples illustrating the use of statistical techniques for linear optical systems subjected to random inputs are examined and interpreted. These examples include an ideal linear photographic film model, Wiener filtering, inverse filtering, and matched filtering. Finally, ensemble averaging methods are illustrated by the signal-to-noise enhancement achieved in speckle photography.

REFERENCES

Becherer, R. J., and Geller, J. D. (1969). *Soc. Photo-Optical Instrumentation Engineers—Seminar Proc.* **16**, 89.

Beiser, L. (1974). *Laser Appl.* **2**, 53.

Brown, W. M. (1963). "Analysis of Time Invariant Systems." McGraw-Hill, New York.

Dainty, J. C., and Shaw, R. (1974). "Image Science." Academic Press, London and New York.

Davenport, W. B., and Root, W. L. (1958). "Introduction to Theory of Random Signals and Noise," Chapter 11. McGraw-Hill, New York.

Goldman, S. (1953). "Information Theory." Prentice-Hall, Englewood Cliffs, New Jersey.
Goodman, J. W. (1968). "Introduction to Fourier Optics." McGraw-Hill, New York.
Helstrom, C. W. (1967). *J. Opt. Soc. Amer.* **57**, 297.
Horner, J. L. (1970). *Appl. Opt.* **9**, 167.
Knox, K. T., and Thompson, B. J. (1974). *Astrophys. J.* **193**, L45–L48.
Labeyrie, A. (1970). *Astronom. and Astrophys.* **6**, 85–87.
Lee, Y. W. (1960). "Statistical Theory of Communication." Wiley, New York.
Mueller, P. F. (1969). *Appl. Opt.* **8**, 267.
O'Neill, E. L. (1963). "Introduction to Statistical Optics." Addison-Wesley, Reading, Massachusetts.
Slepian, D. W. (1967). *J. Opt. Soc. Amer.* **57**, 918.
Yu, F. T. S. (1973). "Introduction to Diffraction, Information Processing and Holography." MIT Press, Cambridge, Massachusetts.

2.6 SILVER HALIDE PHOTOGRAPHY

Paul L. Bachman

2.6.1 Introduction

Early in the 1820s Joseph Nicephore Niepce succeeded in producing the first permanent photographic image through the insolubilization of asphalt. This accomplishment combined with the recognized light sensitivity of silver salts spurred the development of what today is commonly known as photography or silver halide photography to the imaging scientist.

Many other materials and chemistries are known which provide the means of inventing and developing practical, albeit apparently limited, imaging recording and retrieval systems. Practically everyone has heard of blueprinting, which is founded on the light sensitivity and chemistry of iron salts. The light sensitivity of diazonium salts, combined with their ability to form highly colored dyes, has provided a large industry for reprographic imaging materials used for a variety of applications from color proofing to the fabrication of printed circuit boards. Electrostatic imaging in one form or other, usually as xerography, is known to virtually everyone in the industrialized world today. The practicing holographer is undoubtedly aware of many types of image recording media, such as dichromated gelatin, photoresists, electrodeformable thermoplastics, ferroelectric crystals, various organic and inorganic photochromic materials, photoconductors in devices with other materials, magnetooptic films, and even very thin metal films (Smith, 1977). Yet among all the chemical and physical phenomena which have been investigated thus far, not one even approaches silver halide photography's unique combination of properties which combine sensitivity and stability with variety and versatility. Silver halide photographic materials consequently remain the most widely used media for image recording and retrieval for countless applications including holography.

The advent of the daguerreotype in the late 1830s marked the real beginning of silver halide photography, and its instant popularity stimulated the development of what today is a vast and highly complex science and technology. The more pertinent aspects of this science and technology for the holographer will be reviewed here. The restrictions imposed by space and by the purpose of this handbook necessarily limit the subject material and the detail with

HANDBOOK OF OPTICAL HOLOGRAPHY
Copyright © 1979 by Academic Press, Inc.
All rights of reproduction in any form reserved.
ISBN-0-12-165350-1

which it can be covered. Consequently, this chapter has been organized and written specifically for those holographers who are not versed in photographic science to provide a basic understanding and overall view of the photographic process, to provide an understanding of standard photographic jargon, to aid in translating this jargon into terms and numbers useful in holography, and finally, to provide a foundation for reading and understanding more detailed, jargon-laden articles on photographic materials. Emphasis is consequently placed on the subject of sensitometry and certain related topics, but brief attendant discussions include the physics and chemistry of the silver halide photographic process, image characteristics, and storage and handling characteristics. Section 2.6.9 provides a short compilation of additional terminology which is likely to be encountered and which does not appear in preceding sections.

For those who wish to delve more deeply into this subject, James and Higgins (1960) offer an excellent beginning. For answers to practical problems and for methods and procedures, Neblette (1962) or Thomas (1973) should be consulted. Those who desire more detailed theory with extensive references should consult James (1977). Both Collier *et al.* (1971) and Cathey (1974) discuss silver emulsions with regard to holography, and more recently, Smith has edited a book (1977) on recording materials for holography with a chapter on silver halide materials by K. Biedermann.

2.6.2 The Silver Halide Photographic Process

A brief review of the essential elements of the silver halide photographic process will provide a basic understanding and perspective for the various topics which are covered.

The photographic process can be summarized as

(1) having a photosensitive medium (the emulsion),
(2) exposing with photons to produce an invisible (latent) image,
(3) developing to render a visible image (amplification),
(4) fixing to render the image permanent.

2.6.2.1 The Emulsion

The photosensitive emulsion is in fact not an emulsion at all but rather a thin film dispersion of silver halide microcrystals in a protective colloid, typically gelatin. The term "emulsion" is used to describe both this dispersion and the coating on a support (base) such as paper, glass, metal, or polymeric film which provides the mechanical strength required for practical utilization.

Silver chloride, bromide, iodide, or their combinations are used, depending on the sensitivity (emulsion speed) and other characteristics desired. Silver chloride is used for emulsions of lowest sensitivity; chloride/bromide and bromide for somewhat greater sensitivity; and bromide/iodide for the greatest sensitivity (fastest emulsions). The iodide seldom exceeds 5%, and alone has no practical value as a photographic emulsion.

The silver halide microcrystals range in size and shape for various emulsions and within a given emulsion, although within any particular emulsion the distribution in sizes is comparatively narrow. The average size in a very fine grain Lippmann emulsion may be 0.05 μm with a range of 0.03 to 0.08 μm, while in a very fast negative-type emulsion the grain size may well be on the order of several microns.

The silver halides have a cubic crystal structure in which each silver ion Ag^+ is surrounded by six nearest neighbor halide ions X^- and vice versa. The crystal has excess halide ions (originating from the emulsion-making process) strongly adsorbed to its surfaces along with gelatin, sensitizing dyes, and other species, all of which play critical roles in stabilizing the emulsion and the latent image and in directing the results of development.

2.6.2.2   Latent Image Formation

The silver halide crystal is an n-type photoconductor with a valence band of localized electrons and with a conduction band in which injected electrons are free to migrate throughout the crystal until trapped by a lattice defect. When a photon of sufficient energy is absorbed by the crystal, an electron is promoted to the conduction band, leaving behind a positive hole represented by a free halogen atom:

$$Ag^+X^-(\text{crystal}) + h\nu \rightarrow Ag^+X^0(\text{crystal}) + e^-. \tag{1}$$

The free electron migrates until trapped at a lattice defect, which among other things can be a silver atom Ag^0. This trapped electron can then reduce a neighboring silver ion Ag^+ to produce a silver atom:

$$Ag^+X^-(\text{crystal}) + e^- \rightleftarrows Ag^0X^-(\text{crystal}). \tag{2}$$

The single silver atom is not stable, however, and has a lifetime estimated to be about one second. If during its lifetime it traps another electron, a two-atom aggregate is formed and it is stable:

$$Ag^0, X^-(\text{unstable}) + e^- \rightarrow Ag^0(e^-), X^-, \tag{3}$$

$$Ag^0(e^-), X^- + Ag^+X^- \rightarrow Ag_2^0, 2X^-. \tag{4}$$

This two-atom aggregate although stable does not constitute a latent image, that is, it will not render the crystal developable. Current theory contends that at least a four-atom aggregate (Ag_x^0, $x \geq 4$) is required to render a crystal developable. The two-atom aggregate, being stable, can trap additional mi-

2. Background

grating electrons to produce a latent image, and it therefore constitutes a stable nucleus for latent image formation:

$$Ag_2{}^0(Ag^+X^-)_n + x(e^-) \rightarrow Ag^0_{2+x}(Ag^+X^-)_{n-x}(X^-)_x. \tag{5}$$
$$\text{potential latent image} \qquad \text{latent image for } x \geq 2$$

Because of the short lifetime of the single silver atom, thermal events and photo events of very low probability, such as occur during dark storage and during very low irradiance exposures (see Section 2.6.5) may leave little or no history of their having occurred. This "relaxation" or reversibility accounts for the excellent storage life of silver halide emulsions and for the excellent stability of the latent image, particularly in comparison to other photoimaging systems, such as those based on organic dyes and photopolymers.

But we must still consider the free halogen atom (positive hole) which was formed along with the free electron. This free halogen is not totally immobile but can also migrate. Whether stationary or migrating, this halogen atom X^0 can retrap the electron,

$$X^0 + e^- \rightarrow X^-, \tag{6}$$

or oxidize silver,

$$X^0 + Ag^0 \rightarrow Ag^+X^-, \tag{7}$$

back to silver halide. To improve the efficiency of latent image formation this halogen atom must be trapped and this can be accomplished by chemical reducing agents which are added during emulsion preparation. Sulfur compounds are particularly effective, which explains the unique role played by gelatin, which is a natural protein-containing sulfur-bearing amino acid. The latent image then is at least a four-atom aggregate of metallic silver imbedded somewhere within or on the surface of the silver halide crystal.

2.6.2.3 Development

Silver halides can be chemically reduced to metallic silver by a large number of reducing agents. The net reaction for this reduction is summarized as

$$2Ag^+X^- + HOC_6H_4OH \rightleftarrows 2Ag^0 + O{=}C_6H_4{=}O + 2HX \tag{8}$$

for the common well-known hydroquinone developer. The products of the reduction are metallic silver, the oxidized form of the developer, in this case quinone, and halogen acid. The presence of alkali during development accelerates the reduction and consumes the coproduced acid. The silver halide grains in an *unexposed* photographic emulsion can also be reduced by the developer to produce unimaged density called "fog", but this reduction is comparatively slow in relation to latent-imaged grains. The presence of metallic silver ($Ag_x{}^0$, $x \geq 4$) catalyzes the chemical reduction, accelerating the rate of reaction to such a degree that a latent-imaged crystal can be converted to metallic silver before an unexposed crystal can react.

Since only a four-atom aggregate of silver is sufficient to convert the entire crystal of silver halide to metallic silver, a tremendous amplification is realized. A slow-speed, fine-grain emulsion containing 0.05-μm cubes of silver bromide contains about 2.6×10^6 silver ions per crystal, while a coarse-grain, high-speed emulsion containing 1-μm cubes contains about 2×10^{10} silver ions per crystal. Consequently, only a relatively few photons, sufficient to produce an $Ag_4{}^0$ aggregate, can produce 10^6–10^{10} atoms of metallic silver—an amplification of one million to nearly 10 trillion!

Since photographic sensitivity (speed) is a function of the developed silver (image density) produced per given exposure, we can easily appreciate the well-known reciprocal relationship between emulsion speed and resolution. A coarse-grain emulsion affords more silver per developable grain than a fine-grain emulsion, and so if high resolution is desired, one must sacrifice photographic speed, settle for a slower, fine-grain emulsion, and use longer or higher irradiance exposures.

2.6.2.4   Fixing

The unexposed, undeveloped silver halide crystals which remain after development are still photosensitive and unless removed, they limit the lifetime of the developed emulsion. Silver halides are very insoluble in water at all pH's and so must be solubilized by chemical conversion in the process of fixing. Sodium thiosulfate is a common fixing agent, well known as "hypo," and forms water-soluble complexes of silver, one of which is illustrated by

$$AgX \text{ (insol)} + 2Na_2S_2O_3 + 2H_2O \rightarrow Na_3[Ag(S_2O_3)_2] \quad \text{(sol)}. \tag{9}$$

The actual chemistry is more complicated since there are at least four complexes known and only two are readily soluble, including the one illustrated.

2.6.2.5.   Summary

As a consequence of silver halide's unique combination of solid state and chemical properties, silver halide photography remains unparalleled as an optical recording medium, especially in view of the many practical considerations attendant with numerous applications. Thus, silver halide emulsions

(1)   have excellent shelf life,
(2)   have excellent latent image stability,
(3)   can be spectrally sensitized,
(4)   offer great flexibility in being tailored for specific applications, and
(5)   offer flexibility in the quality and type of information which can be retrieved by processing.

2. Background

2.6.3 Sensitometry

Sensitometry is that aspect of photographic science which describes how the image (output) of a photographic emulsion varies as a function of the exposure (input). This basic function is described by the emulsion's characteristic curve, by its spectral sensitivity curve, and by its reciprocity failure curve, all three of which will be discussed. Another important function is the emulsion's modulation transfer function (MTF) curve which describes its resolving power, and which will be discussed in Section 2.6.6.8.

2.6.3.1. Definitions and Terminology

Exposure (E or H)† $I \times t$ where I is the radiant flux per unit area (irradiance or illuminance) incident on the film, and t is exposure time:

$$E = I \times t. \tag{10}$$

Sensitivity (S) the reciprocal of exposure E required to produce some predefined output level:

$$S = 1/E. \tag{11}$$

Transmittance (T) I/I_0 where I_0 is incident flux (influx) and I is the transmitted flux (efflux):

$$T = I/I_0. \tag{12}$$

Percent transmission $(\%\,T)$

$$100\ I/I_0. \tag{13}$$

Amplitude transmittance (T_A) the square root of transmittance:

$$T_A = T^{1/2}, \tag{14a}$$

$$T = T_A{}^2. \tag{14b}$$

Opacity (O) the reciprocal of transmission:

$$O = I_0/I. \tag{15}$$

Absorbance (A) the logarithm (base 10) of the reciprocal of transmission.

---

† The International Organization for Standardization (IOS) recommends using E for irradiance and H for exposure in contradistinction to the traditionally used I and E, respectively. Most existing references employ the traditional symbology or have deliberately defied acceptance of the new symbology, and consequently the traditional symbology is used solely in this chapter, albeit the author applauds all efforts toward universal standardization. So beware of the potential confusion which can arise from the ambiguous use of E.

Absorbance is typically used in spectroscopy:

$$A = \log(1/T), \tag{16a}$$

$$A = \log(I_0/I), \tag{16b}$$

$$A = \log(O). \tag{16c}$$

Density (*D*)   the photosensitometric equivalent of absorbance:

$$D = \log(1/T), \tag{17a}$$

$$D = \log(I_0/I). \tag{17b}$$

Irradiance (*I* or *E*)†   the radiant flux incident on the emulsion when measured in radiometric (energy) units, typically ergs per centimeter squared per second or microwatts per centimeter squared. Exposure is then typically expressed in ergs per centimeter squared or joules per meter squared.

Illuminance (*I* or *E*)†   the radiant flux incident on the emulsion in terms of white light as it affects human visual response. It is measured in photometric units as foot-candles or meter-candles (1 fc = 10.764 meter-candles). Exposures are then expressed as foot-candle-seconds or as meter-candle-seconds.

Intensity   commonly used term to describe the incident flux (*I* or *E*) regardless of units. Intensity is not a preferred term, being less specific than irradiance or illuminance. Nevertheless, it still is occasionally used and is accepted for describing the failures associated with the law of reciprocity.

2.6.3.2   Photometry versus Radiometry

Photometry (illuminance) has tended to dominate the presentation of photosensitometric data and specifications, which is not surprising since the dominant use of the photographic emulsion is representing pictorial scenes as interpreted by the human eye. Photometric units of foot-candles or meter-candles relate only to visible light, the range of which is generally accepted to be 400–700 nm. The human eye is not equally responsive to all wavelengths within this range, so its color sensitivity is described by the visual response curve of the human eye illustrated in Fig. 1, which shows the relative visibility at each wavelength as a fraction of the maximum visibility of 1.000 at 555 nm.

Since holographic work typically employs radiometric units and a well-defined wavelength, we are usually confronted with the problem of how to use the photosensitive data expressed in photometric units, which may often be the case when using a characteristic curve or a reciprocity failure curve. In later sections we shall see several examples in which this very problem is encountered, so let us take a moment here to describe the procedure for converting photometric to radiometric units.

2. Background

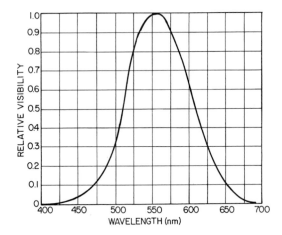

Fig. 1 Relative visual response of the human eye. Plotted from the $\bar{y}$ distribution coefficient for equal energy for the standard observer. (Data from Weast 1976–1977.)

The method is straightforward when a monochromatic light source is being considered. One watt of 555-nm radiation equals 680 lm, or conversely, 1 lm (555 nm) equals 1.47×10^{-3} W (Kodak, 1965). Since an illuminance of 1 fc equals 1 lm ft^{-2}, we find by manipulating units that an illuminance of 1.0 fc (555 nm) equals 1.47×10^{-3} W ft^{-2} which equals an irradiance of 15.8 erg cm^{-2} sec^{-1}.

Now suppose we wish to find the equivalent irradiance at 488 nm. The relative visibility at 488 nm from the curve is about 0.2 or, more exactly, is 0.192, as found in the Colorimetry Table.‡ Consequently, 1 fc (488 nm) equals 0.192×680 lm (555 nm) or 131 lm. Therefore 1 lm (488 nm) equals

$$\frac{680 \text{ lm (555 nm)}}{131 \text{ lm (488 nm)}} \times 1.47 \times 10^{-3} \quad \text{W ft}^{-2} \text{ (555 nm)}$$

or 82.3 erg cm^{-2} sec^{-1}. This protracted calculation is tantamount to dividing the irradiance equivalent for 555 nm (15.83 erg cm^{-2}) by the relative visibility corresponding to the desired wavelength, i.e. 15.8 erg cm^{-2}/0.192 = 82.3 erg cm^{-2} sec^{-1} (488 nm). Table I lists examples for some typical laser lines.

When dealing with other than a monochromatic light source, it is possible to convert irradiance to illuminance by integrating over the desired range of wavelengths. This is most easily performed by a numerical summation with each wavelength interval weighted by its proper relative visibility factor. The

‡ For a complete tabulation of distribution coefficients consult a handbook of physics under "Colorimetry," e.g., Weast (1976–1977).

TABLE I

Photometric–Radiometric Equivalences for Some Common Laser Lines

λ (nm)	Relative visibility[a]	Lm W^{-1}	W Lm^{-1} ($\times$ 10^3)	Irradiance erg cm^{-2} sec^{-1} fc^{-1}	Illuminance fc erg^{-1} cm^{-2} sec^{-1}
555	1.000	680	1.47	15.8	6.33 $\times$ 10^{-2}
488	0.192	131	7.66	82.3	1.22 $\times$ 10^{-2}
514	0.587	399	2.50	26.9	3.72 $\times$ 10^{-2}
532	0.883	600	1.66	17.9	5.59 $\times$ 10^{-2}
633	0.236	161	6.23	66.9	1.49 $\times$ 10^{-2}
694	0.0052	3.54	283	3.04 $\times$ 10^3	3.29 $\times$ 10^{-4}

[a] Data from Weast (1976–1977).

degree of accuracy achieved in arriving at a corresponding illuminance will depend on the wavelength interval chosen.

2.6.3.3   The Characteristic Curve

The image from a photographic emulsion (output) is a function of exposure (input) which is described by the characteristic curve, also referred to as the $D_{\log E}$ Curve or H & D curve after Hurter and Driffield who laid the foundations of sensitometry. Fig. 2 depicts a typical characteristic curve for a negative emulsion. Since response of the human eye to light is approximately logarithmic, density and log exposure are the most convenient and logical units of choice, rather than transmission and exposure. The curve is conveniently described by defining three segments: toe, straight-line portion, and shoulder,

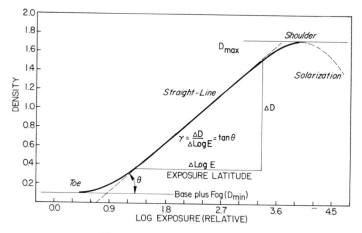

Fig. 2   The characteristic curve.

but some emulsions also show a fourth segment described as solarization (see Section 2.6.9). The toe and shoulder portions encompass the exposure ranges for nonlinear responses and are generally less useful than the straight-line range, although they can be used to great value for soliciting certain types of information or for producing desired photographic effects. The straight-line portion encompasses the preferred exposure range for most scientific applications because the output is a linear function of input.

(a) Base plus Fog Density   Every developed emulsion exhibits a background density which is usually called "base plus fog." The emulsion support (base) slightly attenuates the readout as a function of its thickness, clarity, and spectral characteristics. For typical base materials such as glass, polyester, and cellulose triacetate, the attenuation is normally spectrally flat over the visible portion of the spectrum, but it can vary significantly in the ultraviolet and infrared regions.

An unexposed emulsion has an inherent tendency to develop density to some degree, depending on many factors, such as developer composition, temperature, time, type of emulsion, development technique and emulsion age, and conditions of aging. Base plus fog combines these two effects as a single measurement, which is also referred to as D_{min} (minimum density).

(b) Gamma   Gamma (γ) is defined as the slope of the straight-line segment of the characteristic curve:

$$\gamma = \Delta D/\Delta E = \tan \Theta. \tag{18}$$

When gamma is 1, a perfect tonal reproduction is produced.

(c) Contrast   The visual difference between two levels of light leads to the concept of contrast C. The contrast ratio is defined as I_1/I_2 ($I_1 > I_2$). If, for example, this page under a given lighting had a reflectance illuminance of 2000 fc and the printing had a reflectance of 100 fc, then the contrast or contrast ratio would be 20. In sensitometry, illuminance or irradiance is attenuated according to the emulsion density, a logarithm, and consequently contrast is expressed as

$$C = D_1 - D_2. \tag{19}$$

If a photographic negative were to have a base plus fog of 0.6 and an image density of 2.0, the contrast would be 1.4 (2.0 − 0.6). You can see that when gamma increases, the image contrast increases, i.e., a given change in irradiance will produce a greater change in image (output) density. Conversely, as gamma decreases, contrast decreases. Figure 3 illustrates typical high (or hard), medium and low (or soft) contrast films.

(d) Gradient   The slope of the tangent at any point on the characteristic curve is called the gradient at that point or the point-gradient. In Fig. 4 the

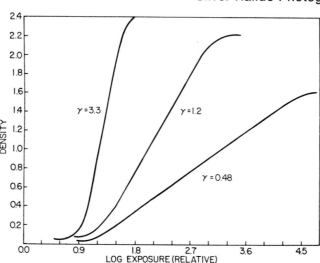

Fig. 3 Typical high, medium, and low contrast emulsions.

point-gradient at *a* is 0.34. Average gradient between two points is the slope of the straight line connecting those two points of the characteristic curve and represents the mean of all gradients in that interval. In Fig. 4 the average gradient between *b* and *c* is 0.46, and we can readily see that the maximum average gradient is gamma (γ).

(e) Maximum Density (D_{max}) The maximum density which can be produced by a particular emulsion with specified developer and development conditions is called D_{max} and corresponds to the maximum in the shoulder. D_{max} does

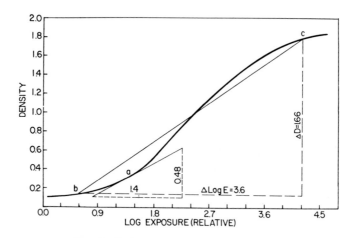

Fig. 4 Point and average gradients.

not necessarily correspond to the maximum density achievable if all the grains in the emulsion are fully developed under perhaps different conditions. That density is termed saturation density D_s, but it is not usually referred to.

(f) Exposure Latitude The exposure range $\Delta \log E$ encompassed by the straight-line segment is defined as the latitude. Since the usual recording mode employs fixed time for exposure, the latitude describes the range of irradiances to which the emulsion responds linearly. For example, if a scene to be photographed offers illumances ranging from 10 to 10,000 fc, the scene offers an exposure range or latitude of 3.0 log E units. To obtain a reasonably faithful tonal reproduction, one would not choose a high gamma emulsion with a latitude of only 2.0 since either the highs or lows or both would be lost.

2.6.3.4 Holographic Linear Recording

"Linear recording" has a different meaning to the holographer than to the photographic scientist. In photographic science, "linear recording" describes the straight-line segment of the characteristic curve where, according to Fig. 2, a difference in densities is proportional to the difference in corresponding log exposures, with the constant of proportionality being gamma:

$$D_2 - D_1 = \gamma(\log E_2 - \log E_1). \tag{20}$$

In holography, "linear recording" describes the condition in which a difference in output amplitude transmittances is proportional to the corresponding input exposures:

$$T_{A2} - T_{A1} \propto E_2 - E_1. \tag{21}$$

Only when gamma equals 2 is the condition for holographic linearity satisfied by the photographic emulsion, and let us now see why this is so.

Substituting transmittance for density [Eq. (17a)], we obtain

$$\log(1/T_2) - \log(1/T_1) = \gamma(\log E_2 - \log E_1) \tag{22}$$

which rearranges to

$$\log T_1 - \log T_2 = \gamma(\log E_2 - \log E_1). \tag{23}$$

Substituting amplitude transmittance for transmittance [Eq. (14b)], we obtain

$$\log(T_{A1})^2 - \log(T_{A2})^2 = \gamma(\log E_2 - \log E_1) \tag{24}$$

which reduces to

$$\log(T_{A1}/T_{A2}) = \log(E_2/E_1)^{\gamma/2}. \tag{25}$$

Equation (25) shows that linearity can be satisfied only when gamma is +2 or

−2. When gamma is +2 (for a negative-working emulsion), we obtain

$$T_{A1}/T_{A2} = E_2/E_1 \tag{26}$$

so that

$$T_{A1} - T_{A2} = c(E_2 - E_1), \tag{27}$$

where c is a constant. When gamma is −2 (for a positive-working emulsion), we obtain

$$T_{A1}/T_{A2} = E_1/E_2 \tag{28}$$

so that

$$T_{A1} - T_{A2} = c(E_1 - E_2). \tag{29}$$

Both Eqs. (27) and (29) verify the proportionality condition defined by Eq. (21).

In practice, an emulsion need not have a gamma of 2 for holographic linearity to be achieved. A limited exposure range corresponding to any point on the characteristic curve where the instantaneous or point-gradient is 2 will satisfy the requirement. Referring again to Fig. 2 we can define the general equation for the straight-line segment as

$$D = \gamma \log E + c, \tag{30}$$

where c is a pseudoconstant (depending on γ) defining the log E intercept of the extended straight-line segment. Rewriting Eq. (30), we obtain

$$D = \gamma \log CE, \tag{31}$$

where $c = \gamma \log C$. Substituting T_A for T as before, we ultimately arrive at

$$T_A = C^{-\gamma/2} E^{-\gamma/2}. \tag{32}$$

Equation (32) establishes $\gamma = -2$ as the condition for linearity, but $\gamma = +2$ will also satisfy the requirements for holographic linearity since only a phase shift is introduced. Note that although gamma is used in the derivation, it need not be the gamma defined as the slope of the straight-line segment. Any slope equaling 2 will satisfy the condition, but of course the required exposure is confined to a narrow exposure range or possibly two ranges (toe and shoulder) rather than to the much larger range of exposures offered by the straight-line segment (exposure latitude) when gamma equals 2.

2.6.3.5   Photographic Speed

The speed of a photographic emulsion is described by a speed number which relates the sensitivity of the emulsion to a particular type of application, such as daylight photography, CRT recording, or graphic arts copying. The speed

2. Background

number has no inherent meaning but is merely an index which provides the photographer with a convenient means of determining the minimum shutter speed or smallest aperture required to produce a desired photographic result with a given amount of light. Some typical examples of speed indexes are American Standards Association (ASA), Deutsche Industrie Normen (DIN), Aerial Film Speed (AFS), Copying Index, Printing Index, and CRT Exposure Index. Any particular speed index is usually defined as a function of E_s which, according to Eq. (11), is the sensitivity S. The function is arbitrarily designed to provide easily manageable whole numbers such as ASA 40 or ASA 120. The E_s used in these expressions is called the speed point, which is explicitly defined for each index in terms of the characteristic curve and sometimes the specific conditions by which the characteristic curve is generated.

For emulsions whose speed is defined for white-light exposures, it is very important to define the spectral composition of the "white light" used to produce the characteristic curve, because most emulsions do not exhibit equal sensitivity over the entire visible spectrum. The spectral composition of "white light" is therefore defined in terms of its color temperature which is approximately 5500°K for daylight and which typically ranges from 3200 to 3400°K for tungsten bulbs employed in sensitometers. Daylight color temperature is then simulated from a tungsten source by means of a daylight correction filter which attenuates the longer wavelengths to provide the proper blue-to-red color balance.

At this point let us take the specific example of ASA for negative working black-and-white emulsions. In this instance the ASA speed is defined as 0.8/E_s where E_s is the exposure, expressed in meter-candle-seconds, required to produce a net density of 0.1 when the characteristic curve is processed to a gamma of 0.62. Figure 5 illustrates this definition graphically and will be used to illustrate how ASA ratings for two black-and-white emulsions, A and B, are

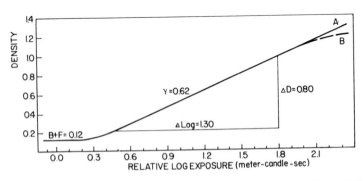

Fig. 5 ASA speed for negative black and white emulsions. ASA = $0.8/E_s$; $\gamma = 0.80/1.30$; E_s at $D = 0.10$ above base plus *fog*. Abs. $\log E_s$ emulsion $A = 2.30$, ASA = 40; abs. $\log E_s$ emulsion $B = 37.0$, ASA = 160.

determined. But first we should note that the log exposure axis is expressed in relative units, which conveniently permits the plotting of any characteristic curve. The relative log exposure is then simply converted to absolute log exposure for each curve, depending on the actual exposure required to produce a given density point on that curve. The curve in Fig. 5 defines two emulsions, A and B, both processed to the conditions required by the definition of ASA. Suppose for emulsion A the speed point E_s corresponds to a log exposure $\bar{2}.30$, which equals an exposure of 0.02 meter-candle-sec; then its ASA rating is 0.8/0.02 or 40. By comparison, suppose the speed point E_s for emulsion B is $\bar{3}.70$ which equals an exposure of 0.005 meter-candle-sec. Emulsion B is obviously faster than emulsion A, requiring only one-fourth the exposure to produce a D_{net} of 0.10, and this is indicated by its ASA rating of 160 (0.8/0.005). In other words, emulsion B is four times faster than emulsion A.

By way of contrast to ASA let us consider one other speed index, namely, Copying Index (CI), which is used in microfilming. Microfilming is primarily concerned with providing effective contrast of line copy documents rather than providing faithful reproduction of continuous tonal values. Consequently, CI is defined quite differently to be equal to $45/E_s(1.20)$ where $E_s(1.20)$ is the speed point exposure, expressed in meter-candle-sec, required to produce a density of 1.20 with the particular emulsion and processing being used and with an exposure meter reading obtained from a gray card having a reflectance of 18% at the copy plane.

By knowing the speed rating of an emulsion, its definition, and the criteria upon which the definition is predicated, you can calculate the approximate exposure requirements for a desired result, but do not overlook the potential complications arising from reciprocity effects, differences in spectral distribution, and processing conditions.

(a) **Effective Speed**   The preceding two examples should leave no doubt that speed numbers have little or only limited value when the emulsion is used and/or processed in a manner other than specified by the definition of speed number. Also, speed numbers have little or no value when comparing significantly different types of emulsions, and so these shortcomings lead to recognition of the effective speed of the emulsion. You can see from Fig. 6 that if the criterion for speed is the exposure required to produce a D_{net} of 0.10, then the effective speed of emulsion B is greater than that of emulsion A. If, on the other hand, the criterion for speed is the exposure required to produce a D_{net} of 1.20, then the effective speeds are obviously reversed and emulsion A is faster than emulsion B.

(b) **Radiometric Sensitivity**   For many photographic applications including holography there is no question that the arbitrary nature of speed index numbers leaves much to be desired for describing the speed or sensitivity of an emulsion. The obvious and logical approach is to define the speed of an

2. Background

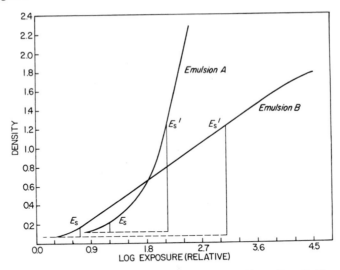

Fig. 6 Comparison of effective speeds. Speed point E_s at D_{net} 0.10; speed point E_s' at D_{net} 1.20.

emulsion as sensitivity or log sensitivity, expressed in terms of the energy required to produce a given density above base plug fog (D_{net}). The spectral sensitivity curve of an emulsion does just this and will now be discussed.

2.6.3.6 Spectral Sensitivity

Silver halide emulsions are inherently sensitive to high-energy radiation, including x rays, gamma rays, ultraviolet and blue light, and with the use of appropriate dyes this inherent sensitivity can be extended to include the green, red, and near-infrared regions of the spectrum. Consequently, five classifications of sensitized emulsions are recognized:

Blue sensitive These emulsions have only the normal ultraviolet and blue sensitivity inherent to silver halides.

Extended blue sensitive These emulsions have their inherent ultraviolet and blue sensitivity enhanced but not spectrally extended.

Orthochromatic These emulsions have their inherent sensitivity extended to include the green portion of the spectrum.

Panchromatic These emulsions have their inherent sensitivity extended to include both the green and red portions of the spectrum, typically to about 650 nm, although the sensitivity already begins dropping around 610 nm. For certain applications there are extended red emulsions which are responsive to about 700 nm.

104

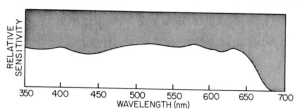

Fig. 7 Typical wedge spectrogram.

Infrared These emulsions are sensitive to ultraviolet, the entire visible portion of the spectrum, and a portion of the invisible near-infrared. Sensitivity usually extends to about 900 nm but about 1300 nm can be achieved.

The relative spectral sensitivity of an emulsion can be illustrated by a wedge spectrogram as depicted in Fig. 7, but the spectral sensitivity curve provides the quantitative data required. Fig. 8 illustrates a typical spectral sensitivity curve relating wavelength to log sensitivity expressed as centimeters squared per erg. Since sensitivity is the reciprocal of exposure [Eq. (11)], or

$$\log S = \log 1 - \log E, \tag{33}$$

the curve indirectly provides the exposure at a given wavelength required to produce the indicated net density.

Spectral sensitivity is determined with a spectral sensitometer which provides monochromatic exposures for generating a series of characteristic curves, one for each wavelength. This series of curves is then used to generate spectral sensitivity curves for desired densities included within the capability of the particular emulsion. Figure 9 illustrates how the four points (488 and 633 nm, 0.3 and 1.0 net densities) in Fig. 8 were obtained from two characteristic curves.

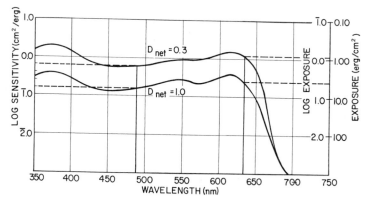

Fig. 8 Typical spectral sensitivity curve.

2. Background

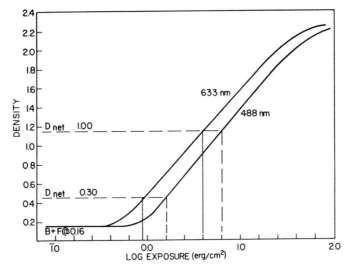

Fig. 9 Characteristic curves for monochromatic exposures of a typical emulsion.

Now let us see how the spectral sensitivity curve in Fig. 8 can be used to determine the exposure required to produce a net density of 1.0 at 488 nm. We see that the D_{net} 1.0 curve at this wavelength intersects log sensitivity at $\bar{1}.20$. Using

$$\log E = \log 1 - \log S, \tag{34}$$

we find

$$\log E = 0 - (9.20 - 10.00) \text{ or } 0.80$$

and

$$E = 10^{0.8} \quad \text{or} \quad 6.31 \quad \text{erg cm}^{-2}.$$

Now suppose we have a 488-nm monochromatic light source with a power output of 3.0 μW cm^{-2} sec^{-1}. We find, from Eq. (10), the exposure time to be 0.21 sec (6.31 erg cm^{-2}/30 erg cm^{-2} sec^{-1}). Calculations taken from a published curve cannot be relied upon to provide accuracy, so in practice an exposure series should be made using the calculated exposure time as a midpoint for the series. Spectral sensitivity curves for color films are exhibited in the same manner but with three curves, one for each primary color.

2.6.3.7 Hypersensitization

Methods are available for enhancing the sensitivity (speed) of an emulsion, but only at the expense of emulsion stability. Consequently, these methods are useful only when the emulsion is exposed and processed shortly after being hypersensitized, otherwise fogging may be encountered. The simplest method is to bathe the emulsion with water, dry, and expose. Hypersensitizing baths employing ammonia or amines such as triethanolamine are also described. Enhancement can also be achieved by degassing the emulsion under vacuum and by gassing the emulsion with hydrogen. Enhancement with a blanket preexposure is also effective and is discussed in relation to low-intensity reciprocity failure (Section 2.6.5.2).

2.6.3.8 Processing§

The characteristic curve is an inherent property of an emulsion but is dependent upon processing conditions, specifically the developer and the time and temperature of development.

(a) Developer Developer formulations differ in their ability to effect different degrees of contrast, effective speed, granularity, fog, and the time of development. All these characteristics are interrelated, and a given developer formulation is designed to provide the desired combination of characteristics as defined by the specifications of the emulsion. A fine-grain developer, for example, will usually reduce the effective speed while a high contrast developer will usually increase graininess.

(b) Development Time Development time affects gamma and D_{max}, and data are usually published to show a series of characteristic curves as a function of development time. In the early stages of development gamma and D_{max} increase with time, but beyond a given time, fog will begin developing at an ever increasing rate. A typical set of curves is shown in Fig. 10.

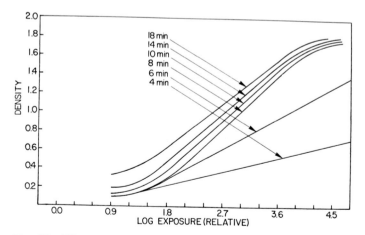

Fig. 10 Effect of development time on characteristic curve.

§ See Section 9.1.

2. Background

(c) Development Temperature As you would intuitively expect, development time decreases with increasing temperature of the developing bath. Too high a temperature will generally increase fog, produce a coarser grain, and shorten the working life of the developer solution, while too low a temperature will significantly reduce contrast. The temperature–time relationship is critical to producing specific image quality, so procedures and specifications should be adhered to rigidly. When it is not possible to employ the specified temperature, the time–temperature graph, if available, can be used to determine the appropriate adjustment in development time. Figure 11 shows a typical time–temperature graph from which you can determine the corrections required for achieving the specified characteristic such as constant gamma or constant speed.

2.6.3.9 Bleaching

Bleaching is the chemical process of dissolving the metallic silver image to eliminate the visual appearance of blackness, i.e., to lower image density. The products are soluble and/or insoluble silver salts which can be removed in a subsequent fixing step. In earlier days bleaches were used to alter contrast or to convert the black silver image to a compound of different color (a process calling toning). Their present day value lies in removing silver in color photography and in reverse-image processing (the process of generating a positive image rather than the usual negative image).

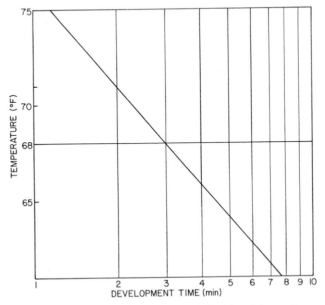

Fig. 11 Time–temperature graph for development.

The silver image of a properly processed and washed emulsion is stable for many years and is resistant to corrosion by water and nonreacting salts at all pH levels, but it can be oxidized,

$$Ag^0 \rightarrow Ag^+ + e^-, \tag{35}$$

by a large number of suitable oxidizers such as hydrogen peroxide, ferricyanide, ferric, dichromate, persulfate, permanganate, bromate, cupric, and mercuric, to mention a few. The mechanism of bleaching is complex and the rates and relative degrees of bleaching high and low density areas will vary among systems. Ammonium persulfate, for example, can be used to preferrentially bleach the higher densities without affecting the lower densities. Some systems are autocatalytic so that the rate of bleaching accelerates as the process proceeds and other systems combine bleaching and fixing in a single process step. Dry bleaching has been accomplished using bromine and chlorine vapors (Graube, 1974) with the result that the distortions and thickness variations arising from wet bleaching and subsequent drying are avoided (see Section 2.6.6.2).

2.6.4 Densitometry

The image in silver halide photography is composed of finely divided particles of metallic silver, called grains, and their effective concentration per unit volume (or unit surface area since the emulsion thickness is essentially constant) is related to the exposure by the characteristic curve. An incident flux I_0 (influx) on passing through the developed emulsion is scattered by the silver grains so that the emerging flux I (efflux) is attenuated. We have already seen that the degree of attenuation is described by the density [Eq. (17)] so that its evaluation, and hence the characteristic curve, depend on the manner by which the efflux I is measured. This topic with its associated considerations is called densitometry.

Let us consider what happens when an influx I_0 enters the emulsion nearly perpendicular to its surface. A portion of this influx will emerge more or less on axis as though it had not been scattered or only very slightly scattered. Another portion will be scattered to the extent that it never emerges. The remainder is scattered so that it emerges from the emulsion at all angles over 180°. That portion which emerges on or nearly on axis is called specular (I_s) and the total efflux over 180° is called diffuse (I_D). Since $I_s < I_D$, specular density will measure greater than diffuse density. Now consider an influx I_0 which enters the emulsion over the full angular range of 180° (diffuse influx). That portion which enters at large angles to the axis is more likely to be scattered widely and never emerge from the emulsion. Consequently the efflux I will be less, whether specular or diffuse, than when the influx is specular. The result of all these possibilities reduces to four extreme cases for measuring density: specular/specular, specular/diffuse, diffuse/specular, and doubly diffuse. The typical densitometer likely to be encountered today gives doubly diffuse densities as the agreed-to standard. All four cases and intermediate conditions can be encountered in various photographic applications. Contact printing, for example, will be either specular/diffuse or doubly diffuse depending upon the exposure source. Projection printing on the other hand will be some place between specular and diffuse depending on distance and format.

2.6.4.1 Collier's Q Factor

For any influx, diffuse density will be less than specular density and the relation between them is defined by Collier's Q factor as the ratio of specular density to diffuse density:

$$Q = D_s/D_D > 1. \tag{36}$$

Q is always greater than one and will be smaller for a fine-grain than for a coarse-grain emulsion.

2.6.4.2 Color

Color emulsions produce density primarily by absorbtion with dyes rather than by scattering and present a different kind of problem. In this case the measured density depends on the spectral composition of the influx. A red dye, for example, will yield a higher density measured with blue light, most of which will be absorbed, than it will measured with white light, most of which will be transmitted. A typical densitometer, such as the McBeth TD 504, is equipped with a turret for selecting standard color separation filters such as the Wratten 94 (blue), 93 (green), and 92 (red), thus permitting the determination of standardized color densities as well as black and white.

2.6.5 Reciprocity

The law of reciprocity for photographic emulsion states that image density (D) is a function only of the total exposure ($I \times t$) and is independent of the magnitude of either I or t. But because of the mechanism and kinetics of latent image formation, the reciprocity law does not hold true for exposures of high irradiance (short duration) and for exposures of low irradiance (long duration), and these two extremes are called high- and low-intensity reciprocity failure, respectively. In general, every emulsion has an optimum combination of $I \times t$ for producing a given density with all other combinations producing a lower density, but this condition is not as troublesome as it may seem, because for the modern emulsion likely to be encountered, reciprocity holds effectively over a broad range of $I \times t$ combinations. The reciprocity characteristics of

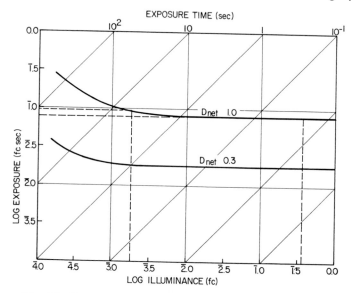

Fig. 12   Reciprocity failure curve for typical emulsion.

an emulsion are described graphically by plotting log E required to produce a given D_{net} against log I. A typical example is shown in Fig. 12 where constant exposure times are plotted as 45° lines intersecting the grid in such a way as to maintain the relation $E = I \times t$. Reciprocity failure curves for different wavelengths have essentially the same shape and are merely shifted along the time axis so that the total *energy* of exposure remains constant. Curves are found expressed in either photometric or radiometric units and are sometimes portrayed as a family of curves representing different wavelengths.

2.6.5.1   Intermittency Effect

Reciprocity failure also manifests in another manner. The image density produced by a total exposure E_t may be less if E_t is received as a series of intermittent exposures, E_i, rather than as a single continuous exposure. In other words, $D_i < D_t$ when $E_t = \Sigma\, E_i$. This intermittency effect is frequency dependent and occurs at both high and low irradiance as a direct consequence of the mechanism and kinetics of latent image formation.

2.6.5.2   Hypersensitization and Latensification

An emulsion which exhibits low- or high-intensity reciprocity failure can be used effectively in either of these regions by employing double exposure techniques referred to as hypersensitization and latensification, respectively.

111

2. Background

Low-intensity reciprocity failure can be largely overcome by first subjecting the emulsion to a uniform, blanket exposure with high irradiance of sufficiently short duration to produce no density of its own. This hypersensitized emulsion can then be subjected to a low irradiance, imagewise exposure and manifest little or no reciprocity failure. On the other hand, if very brief imagewise exposure to high irradiance is immediately followed by a uniform, blanket exposure to low irradiance (insufficient to produce density of its own) the otherwise attendant reciprocity failure can be largely overcome.

2.6.5.3 Using the Reciprocity Failure Curve

Returning now to Fig. 12 let us see how to use a photometrically expressed reciprocity curve to compensate monochromatic exposures. Suppose we have produced a D_{net} of 1.0 by an exposure of 6.31 erg cm^{-2} with a 488-nm source producing a continuous power output of 3.0 μW cm^{-2}. Our exposure time must therefore have been 0.21 sec ($t = E/I$). The reciprocity curve is expressed in illuminance values of foot-candles, so first we must convert. Using Table I we find that for 488-nm radiation, 1 fc equals 82.3 erg cm^{-2} sec^{-1}. Making the necessary conversions, we find

$$I \ (30 \ \text{erg cm}^{-2} \ \text{sec}^{-1}) = 0.36 \quad \text{fc},$$

$$\log I = \bar{1}.56 \quad \text{or} \quad -0.44,$$

$$E \ (6.31 \ \text{erg cm}^{-2}) = 7.70 \times 10^{-2} \quad \text{fc-sec},$$

$$\log E = \bar{2}.89 \quad \text{or} \quad -1.11.$$

Referring to the reciprocity curve, we find that log I versus log E falls well within the range of reciprocity.

Now suppose we substitute a 488-nm source producing a continuous power output of only 1.5×10^{-8} W cm^{-2} and want to know the exposure time required to produce the same 1.0 D_{net}. Were there no reciprocity failure we could simply use our original exposure of 6.31 erg cm^{-2} and calculate a new exposure time of 42 sec (6.31 erg cm^{-2}/0.15 erg cm^{-2} sec^{-1}). Our new irradiance is several orders of magnitude lower, however, and inspection of the reciprocity curve indicates we have entered the range where reciprocity failure must be taken into account. Thus, 1.5×10^{-8} W cm^{-2} converts to 1.83×10^{-3} fc, so log $I = -2.74$ or $\bar{3}.26$. Log E corresponding to this log I (for $D_{net} = 1.0$) is $\bar{2}.97$ and E is 9.33×10^{-2} fc-sec. Our exposure time is therefore 51 sec (9.33×10^{-2} fc-sec/1.83×10^{-3} fc), not 42 sec as would have been the case were there no reciprocity failure.

2.6.6 Image Characteristics

2.6.6.1   Halation

A portion of light incident on an emulsion passes through into the support where it can then be scattered and reflected back into the emulsion to produce unwanted, spurious exposures. Reflections can occur at both the emulsion/support interface and at the back side, support/air interface, and the result is degraded image quality, which is particularly undesirable where higher resolution is required. With coherent radiation these reflections produce unwanted wave-interference patterns. Point images produced by sufficiently high exposure can be seen surrounded by a halo, hence the origin of the term halation for describing this phenomenon.

Emulsions likely to be encountered are provided with antihalation layers which contain dyes or pigments to absorb the light which would otherwise be reflected. Black laquers are often applied to the back of the support and are removed during processing or by subsequent stripping. Antihalation layers may also be placed between emulsion and support and employ dyes which are bleached or removed during processing. The support material may also be colored, but in this case the color is a permanent contribution to the processed image.

2.6.6.2   Image Relief

The chemistries of both the development and bleaching processes produce biproducts which can harden the gelatin in the vicinity of the developing or dissolving silver image, respectively. This hardening results from chemical crosslinking of the gelatin which renders it less soluble or insoluble. This phenomenon, commonly referred to as tanning, varies in degree, depending upon the chemical nature of the developer or bleach. The hardened gelatin becomes less swollen and less permeable to water and processing chemicals, with the result that internal stresses are generated. These stresses are intensified during drying because the hardened gelatin contains less water and dries more rapidly than the more swollen, unhardened gelatin. The resulting strain manifests as a surface relief and as differences in internal refractive indices (Lamberto, 1972).

2.6.6.3   Grain

An undeveloped emulsion can be described as being fine, medium, or coarse grained according to the size of the silver halide crystals, but when developed and fixed, the emulsion contains only particles of metallic silver called developed grain. When speaking of graininess or granularity, we are describing a quality of the image. In Section 2.6.4 we discussed densitometry in terms of a uniformly exposed area containing a more or less uniform distribution of silver particles or grains. For standard densitometry this is the case because the aperture is very large compared to the grain size and distribution, so uniformity is guaranteed by large area integration. If this same area were scanned with the very small apertures afforded by a microdensitometer, the density of a single scan line might look like the one shown in Fig. 13 because the silver grains actually vary in size and shape, are distributed somewhat randomly, cluster together to form conglomerates and overlap one another to form agglomerates.

2.6.6.4 Graininess

Any silver picture which is sufficiently enlarged will appear grainy. This appearance is a subjective sensation produced by the irregularity or randomness in microscopic density as evidenced by the microdensitometer trace. This unquantified, subjective sensation is described as graininess.

2.6.6.5 Granularity

Variation in microscopic grain distribution depends on the particular emulsion and its development conditions, and the fluctuation in measured density depends on the size of the scanning aperture. The fluctuation being statistical in nature lends itself to analytical treatments, and the resulting quantified measures are described by "granularity" to distinguish them from the purely subjective description of graininess. Rms (root-mean-square) granularity σ_D is defined by

$$\sigma_D{}^2 = \sum_{i=1}^{N} [(D_i - \bar{D})^2/(N-1)], \tag{37}$$

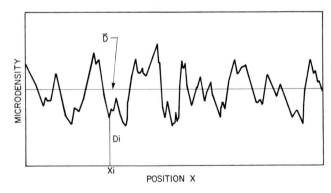

Fig. 13 Typical trace of microdensitometer.

where D_i is the microdensity at point i and $\bar{D}$ is the mean density over the scan. This value can vary, however, depending on the size of the scanning aperture, but Selwyn granularity circumvents this problem by defining granularity G as

$$G = (2A)^{1/2}\sigma_D, \tag{38}$$

where A is the area of the scanning aperture. This derives from $\sigma_D A^{1/2}$ being a constant if the number of silver particles in the scanning aperture is large, in which case Selwyn granularity is independent of the size of the scanning aperture. Kodak (1975) has more recently defined diffuse rms granularity by fixing the scanning aperture at 48 μm and applying a correction factor to equate microdensitometer values to standard diffuse densitometer values. A multiplication factor is then applied to produce a small whole number which defines a graininess class. For example, a diffuse rms granularity of 50 describes a very coarse grain, while a 6 describes an extremely fine grain.

2.6.6.6   Noise

The noise introduced by a photographic emulsion can be attributed to three sources:

(1)   defects and nonuniformities in the support,
(2)   random scattering of the input signal by the silver halide grains during exposure, and
(3)   random scattering of the output signal due to the granularity of the metallic silver image.

Present information theory has been unable to accurately describe the photographic process because of the complications arising from the output sign (density) being a nonlinear function of the input, and from the granularity (noise) being dependent on the input signal. In general, the signal-to-noise ratio and hence the information content of a silver halide emulsion can be improved by using a less sensitive, fine-grain emulsion with longer exposure times and by enhancing the output signal with a blanket post-exposure (see Section 2.6.5.2).

2.6.6.7   Resolution

The resolving power of a photographic emulsion is its ability to distinguish fine details of the subject or input signal and is expressed in line pairs per millimeter or cycles per millimeter, where one cycle equals one line pair. We would intuitively expect an emulsion to be unable to resolve detail finer than the size of the silver grains comprising the image, and this is so, but resolution is a far more complex subject since it is inherently dependent on factors such

2. Background

as granularity, contrast, sharpness or crispness of image, and clarity of the emulsion. The resolving power of an emulsion is determined experimentally, using a resolution target, typically a bar target like the one shown in Fig. 14. Resolution increases with increasing contrast between adjacent pictorial elements, so the contrast of the target must be specified if the specified resolution for a given emulsion is to be significant. The target is imaged by the emulsion, and the smallest set of bars which is just barely discernible defines the resolution or resolving power of that emulsion for specified exposure and development.

2.6.6.8 Modulation Transfer

The concept of modulation transfer was introduced to define resolving power in terms of spacial frequency only, independent of subject contrast. We saw in Section 2.6.3 that if an emulsion is exposed within the straight-line portion

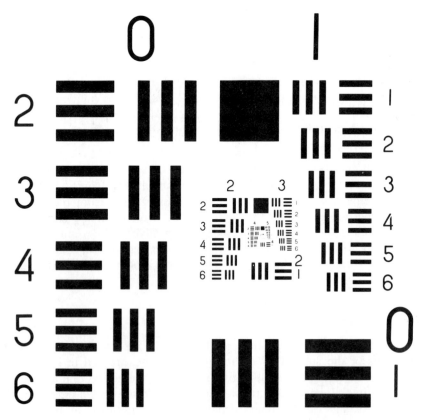

Fig. 14 Typical resolution target: *USAF TR1-BAR.*

of its characteristic curve, it will record differences in log I as proportional differences in density, with the proportionality defined by gamma (γ). However, this relationship defined by the characteristic curve only holds for low spacial frequencies, and as the frequency increases, the emulsion will record a lower contrast than defined by the curve. When the spacial frequency becomes sufficiently high, the recorded contrast will fall to zero because the emulsion can no longer differentiate the adjacent elements as a difference in density. Modulation transfer describes the ability of an emulsion to record the spacial frequency of a test subject and consequently defines the resolving power or resolution capability of that emulsion.

Modulation transfer is determined experimentally by exposing the emulsion with a spacially distributed sinusoidal flux of increasing frequency. The modulation of this flux M is defined by

$$M = (I_{min} - I_{max})/(I_{max} + I_{min}), \tag{39}$$

which, by reference to Fig. 15, is seen to define the ratio of flux amplitude to its mean value. Since contrast C is defined as I_{max}/I_{min} or $D_{max} - D_{min}$, we see that the modulation is indeed a function of contrast and is a number between zero and one:

$$M = (C - 1)/(C + 1) = (10^D - 1)/(10^D + 1). \tag{40}$$

The test object's modulation M_0 is a constant and is converted through the characteristic curve to a corresponding, expected modulation for the photographic image. Because of light scattering, the actually recorded modulation M_R will be lower than M_0 and this difference will increase with increasing spacial frequency. The ratio M_R/M_0 expresses the modulation transfer as a function of spacial frequency and is called the modulation transfer function or simply MTF. Figure 16 illustrates an MTF curve for a typical high-resolution emulsion.

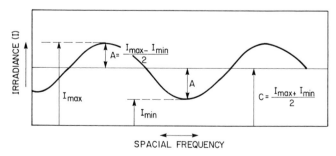

Fig. 15   Modulation of sinusoidal flux. $M = A/C = I_{max} - I_{min}/I_{max} + I_{min}$.

2. Background

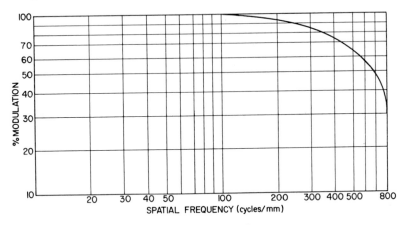

Fig. 16 Modulation transfer curve.

2.6.7 Environmental Changes and Lateral Dimensional Stability

Lateral dimensional stability is an important consideration for any photographic application where it is imperative to maintain accuracy of spacial relationships within the recorded signal, for example, the recorded fringe spacings of a hologram. Lateral dimensional changes are a function of both the emulsion type and the nature of its support (the base material) and are affected by thickness, temperature, relative humidity, pressure (vacuum), mechanical stress, and photographic processing. Absolute lateral dimensional stability is, of course, unknown and even glass plates will exhibit small dimensional changes with more drastic changes in the affecting parameters. But changes which occur under the environmental conditions likely to be encountered in recording applications such as holography are generally reversible; in other words, distortion of a recorded signal resulting from lateral dimensional changes of the recording medium can be negligible if readout is made with the same conditions, particularly temperature and relative humidity, with which the signal was recorded.

Materials which can or have been used as a support for the emulsion include glass, metals, paper, and a variety of plastic films including polyester, polystyrene, polycarbonate, and cellulose triacetate (often referred to as triacetate or simply acetate).

2.6.7.1 Glass

Compared to film supports, glass possesses rigidity and excellent dimensional stability, which is unaffected by humidity and by photographic pro-

118

cessing. Glass is practically unaffected by tensile forces, but of course it is brittle and breaks easily.

The manufacture of glass plate emulsions can cause the emulsion to exhibit anisotropic stresses, which if not relieved prior to exposure will cause permanent spacial distortions of the imagery after processing. These stresses can be relieved prior to exposure by soaking in water or by equilibrating the moisture content at high humidity, which provides the added advantage of enhancing the sensitivity of the emulsion (see Section 2.6.3.7).

2.6.7.2   Films

Polyester and cellulose triacetate are the two film supports encountered almost exclusively today and both exhibit greater dimensional changes than glass, although polyester is superior to triacetate in virtually every category.

(a) Mechanical Properties   Polyester possesses superior mechanical properties to triacetate; it is tougher, stronger, more resilient, and less adversely affected by temperature and relative humidity. Practically speaking, the environmental conditions and changes likely to be encountered in applications such as holography impose minimal concern for their effects on mechanical properties and consequent dimensional changes. Low temperatures or low relative humidity will cause brittleness, which can result in irreversible damage, and high temperatures can cause irreversible dimensional changes from stresses which would not otherwise have effect.

(b) Processing   Processing photographic films always causes some permanent change in lateral dimensions due to shrinkages. Very small dimensional changes always occur in the emulsion layer and typically amount to about 0.02% shrinkage. Under all processing conditions polyester films will show less shrinkage than triacetate simply because the triacetate also shrinks and polyester does not. Triacetate films can exhibit permanent shrinkages up to about 0.10%.

(c) Temperature   Temperatures and temperature changes likely to be encountered will primarily affect the support, emulsions being relatively unaffected whether virgin or processed. There are different coefficients of thermal expansion for the length and width of triacetate and they are two to three times greater than the coefficient of expansion for polyester. Extremes of either high or low temperature can adversely affect the emulsion, but such effects are not dimensional in the strict sense. Thus, low temperature can cause brittleness and irreversible damage on handling. Higher temperatures can cause sticking and result in mechanically induced, permanent distortions which would not otherwise be of concern. For example, incubation of gelatin-backed films (for anticurling) at temperatures above about 160°F can result in blocking (sticking of adjacent layers), the extent of which is dependent on the pressure on the contacting surfaces.

(d) Humidity Relative humidity affects moisture content and consequently the lateral dimensions of both support and emulsion. Both exhibit increasing size with increasing relative humidity, and the extent of change depends on the type and thickness of the emulsion and the chemical nature of the support. Triacetate expands more than polyester and to different extents for length and width. Gelatin swells with increasing moisture content and therefore increases in size. Unless the emulsion experiences severe, repeated cycling (see Section e below), the dimensions at any relative humidity can be maintained or restored by maintaining or reequilibrating to the same relative humidity. Low relative humidity can induce a particular type of fogging called streaking which results from static discharges and can be particularly troublesome in unwinding rolled film.

(e) Pressure Emulsions can be subjected to high vacuum (10^{-5} torr) for hours without blistering or cracking, but dimensional changes will naturally occur due to loss of moisture. A return to normal pressure and humidity will usually restore the original dimensions, but repeated cycling of pressure can induce permanent changes in the gelatin which will render it incapable of reabsorbing lost moisture, in which case irreversible shrinkage will occur.

2.6.8 Emulsion Storage Characteristics

The storage keeping qualities of an emulsion involve two considerations: firstly, the stability of the unexposed emulsion, usually referred to as its shelf-life or storage-life; and secondly, the stability of the finally processed imaged emulsion, usually referred to as its archival characteristics. Generally speaking, you should expect to encounter few if any problems with either shelf-life or archival characteristics by adhering to the manufacturer's specifications and recommendations for storing unopened and opened packages, and by religiously following fixing and washing instructions. Nevertheless, problems can be encountered, so a general understanding of the major factors which can affect keeping qualities can help to minimize these encounters as well as provide a greater appreciation for the specifications, recommendations, and instructions relating to the handling of photographic emulsions.

2.6.8.1 Shelf-Life Characteristics

Any specific emulsion varies slightly in its manufacture, especially among different production runs. Once manufactured, molecular diffusions and chemical reactions still occur and amount to continued ripening, but these changes are usually minimal and are generally accounted for in the specifications.

The environment from the time of packaging through the final processing is the critical factor affecting the results obtained with any specific emulsion. More specifically, the important factors are temperature, relative humidity, penetrating radiation, and chemical contamination. Although stabilizers are incorporated into emulsions to minimize change, these factors if not properly controlled will generally alter photographic sensitivity, increase fog and graininess, lower contrast, introduce spurious nonuniformities, and additionally in the case of color films, cause a change in color balance. The extent of such changes will increase with time and with the magnitude of the contributing factor(s) but not necessarily in linear fashion. Usually the higher speed, coarse-grain emulsions are more adversely affected than the slower, fine-grain emulsions.

Emulsions can be kept for at least several weeks at standard ambient conditions of 70°F and 45–55% relative humidity unless otherwise specified. Temperatures above 80°F are not recommended and should never exceed one week. Higher temperatures and longer times will severely increase fog, graininess, and sensitivity and reduce contrast and exposure latitude, while sufficiently high temperature will cause total fogging. Recommended temperatures for desirable and excellent keeping qualities are typically below 50 and 0°F, respectively. With rolled film, higher temperatures for short durations can cause significant variation in characteristics from outer to inner wrap and from center to edges because of poor thermal conductivity. Very low temperatures, below around −50°F, can cause the emulsion to temporarily lose its sensitivity because of a greatly reduced rate for latent image formation.

Packaged emulsions are sealed to exclude stray light and may be humectacally sealed to maintain proper moisture content and exclude gaseous contaminants. Once opened, however, precaution must be exercised to avoid stray light, chemical contamination, and extremes of relative humidity. Emulsions equilibrate rather quickly with the ambient humidity with the rate increasing with increasing humidity and temperature. Low relative humidity will retard changes in photographic sensitivity by slowing diffusion-controlled reactions but is not recommended because resulting brittleness can lead to mechanical damage (see Section 2.6.7.2.a). Rapid changes in relative humidity and temperature should be avoided to prevent moisture condensation, which will leave water marks and cause spurious sensitometric and densitometric changes through alteration of the chemistry of the emulsion. Always allow refrigerated emulsions to warm to ambient temperature before unpackaging.

High energy, penetrating radiation, and particles produce fog, increase graininess, lower gamma, and alter photographic sensitivity. Emulsions are typically more responsive to alpha and beta rays than to gamma and x rays.

2.6.8.2 Archival Characteristics

A finally processed, imaged emulsion is subject to loss of image density and other forms of deterioration by retained process chemicals and by external environmental factors. Deteriorations are accelerated by high humidity, especially at higher temperatures.

Residual thiosulfate fixer (hypo) will attack colloidal and fine-grain silver to produce yellow to brown silver sulfide, which weakens the image density. Subsequent aerial oxidation will convert the silver sulfide to colorless silver sulfate which will still further weaken the image. Residual silver thiosulfate complexes will slowly decompose to silver sulfide, causing an overall yellowing of the background. Proper adherance to manufacturer's recommendations for fixing and washing will avoid serious deterioration from these sources, but when archival storage must be assured, you should test for residual thiosulfate using one of the recommended procedures (Thomas, 1973). A thiosulfate (hypo) limit of 1 μg cm^{-2} is considered acceptable for archival storage of fine-grain emulsions, such as those used for microfilming. A hypo eliminator bath is recommended to assure complete removal of residual hypo (Thomas, 1973).

Oxidizing fumes such as nitric oxides, ozone, chlorine, and peroxides as well as sulfur dioxide, hydrogen sulfide, or ammonia in the presence of air and moisture will attack the silver image. Caution should be exercised to avoid common sources such as exhaust fumes and highly chlorinated water as well as fresh paints and treated woods, which can be a source of peroxides. Blemishes can result by contacting the emulsion with materials containing traces of reactive chemicals, such as cardboards, molded plastics, and rubber products including rubber bands. Emulsions can be made more resistant to external environmental attack by hardening the finally washed emulsion with formalin, but care must be taken to assure thorough removal of thiosulfates. The silver image can be protected by plating it with the more chemically resistant gold, and formulas for gold-protective solutions are available (Thomas, 1973).

High humidity storage can result in the growth of fungus which can attack the gelatin. Fungus can be removed with nonaqueous emulsion cleaners which are commercially available. It can be prevented by treating the emulsion with a fungicide when high humidity storage cannot be avoided.

2.6.9 Glossary of Additional Commonly Encountered Terms

Accelerator   An alkaline chemical added to the developer to accelerate the rate of development. Examples are sodium and potassium carbonates and hydroxides, and borax.

Actinic radiation   Any radiation which in relation to a given photographic emulsion is capable of producing a latent image.

Acutance   A measure of image sharpness produced by the spread of a knife-edge exposure. The acutance of an image is affected by the type of developer. See DIR and adjacency effects.

Adjacency effects The density of a uniformly exposed image area is higher at its edges compared to the interior (border effect), and the base plug fog immediately adjacent to the image edges is lower than the base plus fog measured at a greater distance (halo effect). This phenomenon results from a developer concentration gradient which is caused by the outward (from image area) lateral diffusion of spent developer encountering the inward (from outside the image area) lateral diffusion of fresh developer. Larger areas will show a greater effect which can be minimized by proper agitation.

Agitation The process of agitating the emulsion during processing steps. Insufficient or improper agitation can give pronounced adjacency effects, streaking, or staining and can result in unstable imagery or poor archival properties.

Antifoggant See restrainer.

Bleach–fix process See monobath.

Border effect Refers to the increased density at the edge of an image area resulting from adjacency effects.

Clearing bath Used before bleaching to remove residual developer which would otherwise be oxidized and cause staining.

Collagen The source of gelatin; the most abundant protein of higher animals constituting skin, bone, cartilage, tendon, and ligament.

Covering power The reciprocal of the photometric equivalent.

Densitometer A device for measuring the density of a photographic image.

Developer (developing agent) A chemical reducing agent capable of preferentially reducing exposed silver halide grains to metallic silver. Examples include hydroquinone paraphenylenediamine, Metol,* Phenidone,* and Amidol.*

DIR Development inhibitor releasing developer: a chemical developer which can produce sharp image edges by preventing development in unexposed regions. See border effect and adjacency effects.

Edge effects See adjacency effects.

Fog restrainer See restrainer.

Fringe effect Refers to the lower density base plus fog at the edge of an imaged area resulting from adjacency effects.

Gamma infinity Defines the highest gamma obtainable with a given emulsion when exposed to a specified light source and developed with a specified developer under specified conditions. An emulsion can have more than one gamma infinity when one or more of the above criterion are altered.

Halo effect See fringe effect.

Hypo eliminator A compound or formulated solution which removes traces of thiosulfate fixer, usually by oxiditive destruction.

---

* Registered trademark.

2. Background

Monobath A one-step process combining development and fixing or bleaching and fixing.

Photobleach positive Directly produced positive image resulting from exposing a prefogged emulsion followed by conventional processing. Phenomenon is closely related to solarization.

Photometric equivalent Defined as M/D where M is the mass of silver per unit area and D is density, usually measured as diffuse. See covering power.

Physical development The formation of silver image normally involves chemical reduction of the activated silver halide crystal. If sufficient silver ion is present in the developer bath, it can plate out on the developing silver grains to produce additional silver.

Point gamma The slope of the tangent (gradient) at any point along the characteristic curve. (See Section 2.6.3.3.d.)

Post-exposure See latensification (Section 2.6.5.2).

Preservative Commonly sodium sulfite, added to the developer to retard its deterioration by aerial oxidation and to hinder the formation of undesirable colored oxidation products which could stain the emulsion.

Print A photographic image viewed by reflected light.

Processing control latitude The time of development required to produce a given increase in gamma.

Psychophysical quantity An objective measurement of image quality that correlates with subjective judgements.

Reflection density Density as measured by reflection rather than by transmission. Reflection density is usually about twice the transmission density for a backed transparency.

Residual image Image formed by colored oxidation products of certain developers and which remains after bleaching of the silver image.

Restrainer Commonly potassium bromide added to the developer to retard formation of fog by decreasing the rate of fog formation to a greater extent than retarding image development. Other types may be incorporated in the emulsion during its manufacture.

Sensitometer A device for producing on a photographic emulsion a series of known exposures from which corresponding densities can be measured. Consequently, the characteristic curve can be plotted.

Solarization Describes the effect of decreasing density with overexposures, a phenomenon which occurs with some types of emulsions. See Fig. 2.

Stabilizer An agent added to the emulsion to retard changes in sensitometric characteristics. (See Section 2.6.8.1.)

Stabilizing bath Adjusts final pH of color emulsions for optimum dye stability.

Transparency A photographic image which can be viewed by transmission, such as those on clear film or glass supports.

REFERENCES

Cathey, W. T. (1974). "Optical Information Processing and Holography." Wiley, New York.

Collier, R. J., Burckhardt, C., and Lin, L. H. (1971). "Optical Holography." Academic Press, New York.

Graube, A. (1974). *Appl. Opt.* **13,** 2942.

James, T. H., and Higgins, G. C. (1960). "Fundamentals of Photographic Theory." Morgan & Morgan, New York.

James, T. H. (1977). "The Theory of the Photographic Process." Macmillan, New York.

Kodak (1965). *Tech. Bits* No. 1, 3.

Kodak (1975). *Tech. Bits* No. 1, 6.

Lamberto, R. L. (1972). *Appl. Opt.* **11,** 33.

Neblette, C. B. (1962). "Photography, Its Materials and Processes." Van Nostrand-Reinhold, New York.

Smith, H. M. (1977). "Topics in Applied Physics," Vol. 20, Holographic Recording Materials. Springer-Verlag, New York.

Thomas, W. (1973). "SPSE Handbook of Photographic Science and Engineering," pp. 531–535. Wiley (Interscience), New York.

Weast, R. C. (1976–1977). "CRC Handbook of Chemistry and Physics," 57th ed., p. E-247. CRC Press, Cleveland, Ohio.

3

Classification of Holograms

W. Thomas Cathey

3.1 INTRODUCTION

The purpose of this chapter is to provide a basic framework which will show the relations among different types of holograms. This is necessary because holograms differing by only one parameter can have vastly different properties. For example, two holograms of the same object, recorded on identical recording materials using light from the same source, developed identically, and illuminated in the same manner produce images having differeing fields of view and resolution if one has a reference wave coming from a point far away and the other has a reference wave produced by a point in the vicinity of the object.

Detailed treatments of some of the types of holograms appear in the following chapters. In this chapter, the types of holograms are listed, grouped according to type of reference wave and geometry used in recording, etc., and the main properties of each are given.

First, we discuss the type of recording medium and the recording techniques. The possibilities are not simply listed, but the implications and relation to other parameters of the hologram are laid out. Next, we point out the importance of the particular wave parameter being recorded; that is, we can record only the amplitude of the wave from the object, only the phase, or both. The implications of each choice and the associated type of hologram are presented. An important option closely related to the type of recording media chosen— the wave parameter modulated—is then discussed. This discussion deals with the wave parameter (amplitude or phase or both) of the hologram illuminating wave which is modified by the hologram and how the choice affects the images formed.

One of the more important parameters is the geometry used in recording the

HANDBOOK OF OPTICAL HOLOGRAPHY

hologram. Included are the categories of object location, type of reference wave used, and configuration of the recording material. These parameters affect the location of the image, the image detail, and the field of view.

The degree of coherence of the object illuminating or hologram illuminating (readout) wave has drastic effects on the image quality. If the coherence of the object and reference waves is low, less information is recorded. If the coherence of the readout wave is low, the image quality can be degraded.

Finally, suggestions are given for the use of a classification system to clarify discussions and papers. A table shows the relations between the parameters of some of the more common holograms.

3.2 RECORDING MEDIA AND USE

Rather than discuss materials suitable for holography (which are covered in Section 8.3), this section treats general characteristics applicable to almost any medium. First, the importance of the depth of the hologram medium is mentioned. Then the mode of illumination of the completed hologram is divided into two regions, reflection or transmission. Finally, it is noted that some holograms are not recorded but are computed.

3.2.1 Medium Thickness

If the recording uses only the surface of the medium in recording interference fringes, a thin, plane, or surface hologram is recorded. It is not the thickness of the recording medium but the effect that is the important aspect; even if a medium is thick, if the recording in depth is not used, the effect is that of a thin medium. We have a *thick* or *volume* hologram if the three-dimensional interference pattern is recorded and used in depth. It is the use of the volume which allows us to produce only one image rather than a primary and conjugate image. Fig. 1 shows three ways in which holograms can be recorded to produce surface and volume holograms.

3.2.2 Reflection or Transmission

This is a relatively simple distinction. In one case the light used to illuminate the hologram for wavefront reconstruction is reflected from the medium to form the image wavefront, and in the other, the light is transmitted. Less light is usually lost in the case of reflection. Other aspects are discussed in Section 3.3, where the effects of the medium on the hologram illuminating wave are treated.

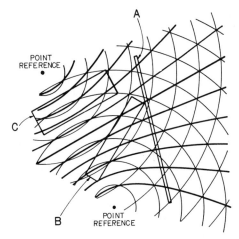

Fig. 1 Standing wave maxima produced by interference between waves from two point sources. Examples of the configuration and emulsion thickness for: A, plane, thin, or surface hologram; B, thick or volume hologram; C, volume hologram to be used in reflection mode.

3.2.3 Computer Generation of Holograms

In this case, the description of an object is given to a computer and the object wave is calculated. A reference wave could be added mathematically and the result obtained from a plotter would be analogous to the case of an optical recording. Generally, this is not done, but the computed hologram, when produced on a plotter, is a special arrangement of transparent apertures, coded to give the desired image wave. For more detail, see Huang (1971).

3.3 OBJECT-WAVE PARAMETERS RECORDED

The light wave from the object has variations in its amplitude and relative phase which can be described by

$$\mathbf{a}(x,y) = a(x,y) \cos[2\pi\nu t - \gamma z + \phi(x,y)], \qquad (1)$$

where $a(x,y)$ describes the amplitude variation across the hologram and $\phi(x,y)$ describes the relative phase. The terms ν and γ are the optical frequency and propagation constant. Normally, both the amplitude and phase of the object wave are preserved in the hologram as described in Chapter 1. If, however, the phase or amplitude information is removed, we have what is referred to as an *amplitude-information* or *phase-information* hologram. The term *phase-only* hologram can also be used when only the phase information $\phi(x,y)$ is

129

preserved, but this term sometimes leads to confusion with the phase-modulation hologram that is discussed in Section 3.4. The amplitude-information hologram is rarely used because of the poor quality of the image obtained (Powers *et al.*, 1971).

If the object is diffusly reflecting, most of the information is in the phase (Kermisch, 1970). In some cases, such as acoustical holography or computer-generated holograms, the amplitude information is discarded upon recording or computing the wavefront.

One phase-information device which discards amplitude information is the kinoform (Lesem *et al.*, 1969). In preparing a kinoform, the phase distribution of the object wave is calculated and the exposure of a photosensitive material is controlled so that the resulting transparency is a phase mask. When this mask is illuminated by a uniform plane wave, the phase distribution of the object wave is produced.

3.4 THE MODULATED PARAMETER

The hologram can modify either the amplitude or the phase of the hologram illuminting (readout) wave or both. The analogy to amplitude modulation (AM) or phase modulation (PM) of a temporal signal is helpful to those persons familiar with communication theory. The energy distribution at the hologram recording plane caused by the interference of the wave from the object and the reference wave is given by

$$
\begin{aligned}
I(x,y) = \; & a^2(x,y) + r^2(x,y) \\
& + 2a(x,y)r(x,y)\cos[2\pi(\xi_0 - \xi_r)x + \phi_0(x,y) - \phi_r(x,y)],
\end{aligned} \tag{2}
$$

where $a(x,y)$ and $r(x,y)$ are the amplitude variation of the object and reference wave, $\phi_0(x,y)$ and $\phi_r(x,y)$ are the phase variations of the object and reference waves. The parameters ξ_0 and ξ_r are defined as

$$
\xi_0 = (\sin \theta_0)/\lambda, \tag{3}
$$

$$
\xi_r = (\sin \theta_r)/\lambda, \tag{4}
$$

where λ is the wavelength of the light and θ_0 and θ_r are the angles at which the object and reference wave propagate with respect to the perpendicular to the plane of the hologram. Expression (2) is for a surface or thin hologram.

3.4.1 Amplitude Modulation

An *amplitude-modulation* hologram is formed when the amplitude of the hologram illuminating wave is modulated such that, after passing through the hologram, the amplitude of the wave is proportional to expression (2). This wave, after propagating a distance, gives rise to waves going in three directions. One of these waves is proportional to the original wave from the object. The amplitude modulation can be done by absorption of portions of the wave or by a hologram having a reflectivity which varies with x and y.

3.4.2   Phase Modulation

A *phase-modulation* hologram results when the hologram modulates the phase of the illuminating wave such that the resulting wave has relative phase shifts proportional to expression (2); that is, the wave can be described by $w(x,y)$, where

$$w(x,y) = \cos[2\pi\nu t - \gamma z + \phi_H(x,y)], \tag{5}$$

and

$$\phi_H(x,y) = pI(x,y). \tag{6}$$

The parameter p is an index of phase modulation. This wave will give rise to many waves, one of which is proportional to the wave from the object. If the value of p is small, the object wave is reconstructed with a minimum of noise. If p is not small, some of the other waves arising from the wave of (5) can contribute noise to the reconstructed object wave (Cathey, 1974). The phase modulation can be accomplished by causing the index of refraction or thickness of the hologram to change as a function of x and y or by changing the profile of the hologram and using it as a reflector.

3.4.3   Phase and Amplitude Modulation

Many holographic recording materials, such as photographic emulsions, produce amplitude and phase modulation of the illuminating wave; that is, the amplitude of the modulated wave is proportional to $I(x,y)$ and the phase to $\phi_H(x,y)$. Both the amplitude and phase of the wave have all the information recorded as described in Eq. (2). This effect occurs when a thin photographic emulsion is used. It has not been extensively studied and is not pursued here.

A very useful mode of phase and amplitude modulation is to arrange for the desired amplitude variation to be impressed onto the wave by amplitude modulation and the phase variation by phase modulation. This can be accomplished with a thick (or volume) hologram. Consider the hologram C of Fig. 1. The maxima of the energy distribution causes a deposition of silver following the contours shown. The hologram illuminating wave is reflected from these surfaces of deposited silver. The density of the deposited silver is related to the amplitude of the wave from the object. The reflectivity of the surface is determined by the density of silver deposited. Consequently, the wave reflected from the hologram has an amplitude variation proportional to that of the original wave from the object. The shape of the surface is dependent upon the relative phase of the interfering waves. Hence, when the illuminating wave is reflected, its phase is modulated in proportion to the phase of the original wave from the object. When the hologram is illuminated from the direction of the original reference wave and the amplitude and phase distributions are separately modulated onto the illuminating wave, only one wave is reconstructed, the original object wave.

3.5 CONFIGURATION

The term configuration is used to refer to the location of the object, the inclusion of lenses to form an image, or a Fourier transform using the object wave, the structure of the reference wave, and the shape and means of exposure of the holographic material. Differences in the path lengths for the object and reference waves are treated in Section 3.6. Details of the geometric optics of holography are found in Chapter 7.

3.5.1 Properties of the Object Wave

In general, if the object is close to the holographic recorder, we record what is known as a *Fresnel* hologram. For special reference-wave configurations, there are exceptions, but these are discussed in Section 3.5.2. The determination of "Where is close?" is relative. If the object is small, it can be within a few centimeters of the hologram and we would still record what is known as a *Fraunhofer* hologram. This is the case when the object is small enough or far enough from the hologram that the hologram is in what is known as the Fraunhofer region or far field of the object.

If the object is very close to the hologram or an image of the object is formed on or near the holographic recorder, we obtain an *image-plane* hologram. Because the resulting image is close to the hologram, rays of light of different wavelengths do not separate much before forming the image. This means that a source having a broad spectrum can be used to illuminate the hologram. This property makes an image-plane hologram particularly useful in displays (Cathey, 1974, Chapter 9).

If the lens is used to produce, in the hologram-recording plane, a two-dimensional spatial Fourier transform of the object-wave distribution, a *Fourier-transform* hologram is obtained. If a diffuse object and a point reference source are equidistant from the recording medium, we obtain a *quasi-Fourier-transform* hologram. Section 4.3 discusses both kinds of holograms in more detail.

3.5.2 Properties of the Reference Wave

The effects of the reference-wave configuration are much greater than at first would be imagined. It can affect the location and size of the image, the field of view of the image, the resolution of the image, and the resolution needed in the recording material.

If the source of the reference wave is a point, the same distance from the hologram as is the object, then the hologram has some of the same properties as the Fourier-transform hologram. Hence, the name *quasi-Fourier-transform* hologram has been applied. It has also been referred to as a *lensless Fourier-transform* hologram, but that term erroneously implies that a spatial Fourier transform has been obtained without the use of a lens (see Section 4.3).

The location of the point source of the reference wave has other effects. The limited resolution of the holographic recorder imposes limits on the field of view of the image, its resolution, or both. It is possible to compromise between field-of-view and resolution limits in the image by selection of the location of the reference-point source. If it is in the region of the object, image resolution is maximized at the expense of the field-of-view. If it is at infinity (plane reference wave), the field-of-view is maximized at the expense of the image resolution. If the reference point source is located between the object and infinitely far from the hologram, intermediate values of both field-of-view and image resolution are obtained. (Collier *et al.*, 1971; Cathey, 1974, Chapter 9).

If the reference wave comes from a point, the wave is a uniform spherical wave. The reference wave affects the amplitude and phase of the wave reconstructed by the hologram. See Eq. (2) which shows the energy distribution recorded by the hologram. If a reference wave having an arbitrary phase $\phi_r(x,y)$ is used, an identical hologram illuminating wave must be used to get a reconstructed-image wave with no distortions. Hence, the hologram can be a *coded* hologram, which requires that it be illuminated by a wave exactly like the reference wave if an image of the object is to be seen. The object can, of course, be a page of text or other material.

If the reference wave is not derived from a point source (spatially incoherent), the effect on the image is that its resolution is, in general, reduced by the size of the source. For a Fourier-transform hologram, the image distribution is given by the convolution of the object and source distributions.

Another form of extended source comes from what is called a *local-reference-beam* hologram. In this case, a portion of the object wave is taken and modified to form the reference wave. The incentive is to obtain a hologram even if the coherence of the waves is low. A discussion of the local-reference-beam hologram is given in Section 5.6.

3.5.3 Recording Material and Configuration

The recording material is normally flat photographic emulsion and exposed everywhere simultaneously. In this section, we see that this need not be the case.

3. Classification of Holograms

The recording material could be a thermplastic, giving a *thermoplastic hologram*. *Photochromic and dichromated-gelatin* holograms have been recorded. Almost any material capable of recording an image can be used to record a hologram. If the material is other than photographic emulsion, the name of the material is used to identify the type of hologram. See Section 8.3 for details on recording media.

If we form the hologram into a right circular cylinder, we have what is known as a *cylindrical* hologram (Jeong, 1967). Fig. 2 shows one means of exposing the hologram. For viewing the developed hologram, the arrangement is exactly the same except that the object is removed. When the hologram is illuminated by what was previously the reference wave, the object can be seen from every side. This makes the cylindrical hologram very useful in displays. See Section 10.3.

If the hologram is formed into a cone, we have a *conical hologram* with essentially the same properties as the cylindrical hologram.

3.5.4 Segmented and Multiple Recordings

There are many different ways in which the holographic recording can be made. We have assumed simultaneous exposure of the entire hologram. If we wish to record many images on the same hologram, we can do this in the manner described in Section 5.2 and obtain a *multiplexed* hologram.

In many applications, it is desirable to expose only a strip of the hologram to the interference of the reference and object waves. The same strip exposure is repeated across the entire hologram. The result is a hologram which presents three-dimensional information in only one direction—that along the width of the strip. The beneficial effect is that when the hologram is illuminated with a multiple-wavelength source, the spread of colors can be made to be up and down, which does not interfere with viewing the image. This is used in the case of what has commonly been called the *rainbow hologram* (Section 10.3).

One other case where the strip recording of holograms is useful is *synthetic*

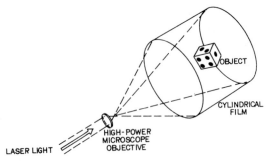

Fig. 2 Configuration for making a cylindrical hologram. [From Jeong (1967).]

holograms. In the first step of this two-step process many regular transparancies are made of different views of an object. In the second step, a coherent source is used to illuminate the transparencies and to provide a reference wave. Adjacent vertical slit exposures are made on the hologram, each using different transparencies showing views of the object from adjacent positions. When viewed, each eye sees a different aspect of the scene and the effect is similar to looking through a hologram of the object. A three-dimensional display can be made without making a hologram of the original object. This approach is also called *integral photography.*

3.6 SOURCE PROPERTIES

3.6.1 Coherence

We must distinguish between the properties of the object illuminating and reference waves on the one hand and the hologram illuminating wave on the other. The term *incoherent hologram* is usually reserved for holograms *recorded* using incoherent illumination. The interference fringes formed in recording an incoherent hologram are produced by interference of the light from one image point with itself. To do this, two images of the object are formed using an image-splitting device. The light from corresponding image points is coherent and can interfere. The light which does not interfere contributes to background exposure of the hologram (Kozma and Massey, 1969). Another means of obtaining interference fringes with a low-coherence source is by imaging a grating onto the hologram and inserting the object in one of the orders (Leith and Upanieks, 1967).

An extended reference source results in an incoherent reference wave, and that case has already been considered.

There are many different cases where the hologram is recorded using coherent light and illuminated with incoherent light. The name applied relates to characteristics of the hologram other than coherence. For example, a hologram recorded with coherent light but illuminated with *white light* is called a *white-light hologram.* This is possible because the thick hologram acts as a combination interference filter and hologram to reflect that portion of the illumination having the wavelength with which the hologram was recorded. A thin hologram can be used if a grating is employed to compensate for dispension. These have been referred to as *surface white-light holograms* (DeBetetto, 1966; Burckhardt, 1966). Image plane holograms and rainbow holograms can produce acceptable images with white-light illumination.

3.6.2 Polarization

The source is frequently polarized, especially if it is a laser. This means that the reference wave is polarized. In many cases, such as reflection of light off

3. Classification of Holograms

an object to produce an object wave, the object wave is randomly polarized. Because interference can occur only between waves having the same polarization, a portion of the wave is not recorded. Normally, no mention is made of this characteristic of hologram recording. If this property is exploited to examine some characteristic of the object by selection of the polarization of the reference wave, the process if referred to as *polarization holography* (Section 5.4).

TABLE I

	Recording				Configuration
				Object parameter recorded	object location
Common name	media	technique	Modulation		
Gabor			Amplitude	Amplitude and phase	
Leith–Upatnieks			Amplitude	Amplitude and phase	Near
Fresnel					NEAR
Fraunhofer					FAR
Image Plane					ON HOLOGRAM
Fourier Transform				Amplitude and phase	FOURIER TRANSFORM
Quasi-Fourier Transform				Amplitude and phase	Near
Local Reference beam				Amplitude and phase	
Coded Reference					
Polarization					
Reflection		REFLECTION	phase or both	Amplitude and phase	
Color	thick			Amplitude and phase	
White Light	THICK		phase or both	Amplitude and phase	
Surface White Light	thin		amplitude or phase	Amplitude and phase	
Phase			PHASE	Phase or both	
Phase Information (Phase only)			phase	PHASE	
Kinoform	thin	COMPUTER GENERATED	PHASE	PHASE	Fourier Transform
Computer Generated	thin	COMPUTER GENERATED			Fourier Transform
Multiplexed					
Rainbow	thin		phase		on hologram
Incoherent					
Synthetic	thin				
Cylindrical				Amplitude and phase	near
Thermoplastic		THERMOPLASTIC reflection	PHASE	Phase or both	

3.6.3   Wavelength

If multiple wavelengths are used in making the hologram, it is possible to record a *color hologram*. The hologram is not colored, of course, but when illuminated by multiple wavelengths, a color image is obtained (Section 5.3). Other labels concerning wavelength relate to the region of the spectrum or type of wave used; for example, *microwave hologram, acoustical hologram,* and *x-ray hologram*.

Relations between Parameters of Some Common Holograms

Configuration		Recording wave		Illumination wave	
reference wave	recording	coherence	wave-length	coherence	wave-length
point at infinity					
ON AXIS		partially coherent		coherent	
OFF AXIS		coherent		coherent	
PLANE WAVE					
PLANE WAVE					
off axis, plane wave		coherent		partially coherent	
off axis, plane wave		coherent		coherent	
off axis		coherent		coherent	
POINT IN OBJECT PLANE					
DERIVED FROM OBJECT					
WAVE, extended source					
SPECIAL PHASE					
selected					
polarization					
		coherent			
			TWO OR MORE		TWO OR MORE
				INCOHERENT	
		coherent	MANY	INCOHERENT	
				coherent	
MULTIPLEXED					
MULTIPLEXED		coherent		partially coherent	MANY
		INCOHERENT		coherent	
	MULTIPLEXED	INCOHERENT		coherent	
	CYLINDRICAL	coherent			

3.7 DESCRIPTION OF A HOLOGRAM

The adjectives discussed are used when the hologram is an exception to the normal form. If one says he is going to record a hologram, it would probably be assumed that he plans to use a laser, place the hologram in the Fresnel region of the object, have an off-axis point reference source at least as far as the object distance from the hologram, use flat photographic emulsion, and record a surface hologram. We might think of this as being "standard" hologram. If there are exceptions, they would be specified, such as *image-plane hologram*. We would continue to make the same assumptions concerning the other parameters. If we are told that a thermoplastic hologram is being made, we know that photographic film is not being used, but we would continue to assume the other standard parameters.

Table I shows some of the names applied to holograms and the relevant parameters. If the stipulation in a parameters column is in all caps, it is a necessary or important specification for that type of hologram. If lower case is used, that is the usual or historical specification for the parameter. If the column is left blank, the name implies nothing concerning that parameter. The table shows how two adjectives may be used to more fully describe a hologram. For example, a white-light image-plane hologram has the properties of both of the associated rows. If the same column of both rows has conflicting all caps specifications, the adjectives are incompatible. If only one specification is all caps, the other can adapt. This is only one example where all adjectives are not necessary. All image-plane holograms can be illuminated by white light.

REFERENCES

Burckhardt, C. B. (1966). *Bell Sys. Tech. J.* **45**, 1841–1844.
Cathey, W. T. (1974). "Optical Information Processing and Holography," Chapter 3. Wiley (Interscience), New York.
Collier, R. J., Burckhardt, C. B., and Lin, L. H. (1971). "Optical Holography," Chapter 8. Academic Press, New York.
DeBetetto, D. J. (1966). *Appl. Phys. Lett.* **9**, 417–418.
Huang, T. S. (1971). *Proc. IEEE* **59**, 1335–1346.
Jeong, T. H., (1967). *J. Opt. Soc. Am.* **57**, 1396–1398.
Kermisch, D. (1970). *J. Opt. Soc. Am.* **60**, 15–17.
Kozma, A., and Massey, N. (1969). *Appl. Optics* **8**, 393–397.
Leith, E. N., and Upatnieks, J. (1967). *J. Opt. Soc. Am.* **57**, 975–980.
Lesem, L. B., Hirsch, P. M., and Jordan, J. A., Jr. (1969). *IBM J. Res. Dev.* **13**, 150–155.
Powers, J., Landry, J., and Wade, G. (1971). In "Acoustical Holography." (A. F. Metherell, ed.), Vol. 2. Plenum Press, New York.

4

Major Hologram Types

4.1 FRESNEL HOLOGRAPHY

John B. DeVelis
George O. Reynolds

4.1.1 Introduction

Fresnel holography, first introduced by Gabor (1948, 1949, 1951) and later revitalized by Leith and Upatnieks (1962, 1963, 1964), is a two-step imaging process.

The first step, formation of the hologram, consists of photographically recording the interference pattern between the Fresnel diffraction wave of an object and a reference wave. The second step, reconstruction of the hologram, consists of illuminating the hologram with a replica of the reference wave to reproduce an image of the object. This reconstructed image possesses the three-dimensional characteristics of the original object, and its image quality depends upon the angle between the reference wave and the diffraction wave. Gabor worked with in-line holograms for which this angle was zero degrees (i.e., the reference wave and the diffraction wave were coaxial). Upon reconstruction, these holograms produced two conjugate images and coherent background noise centered on-axis. This resulted in image deterioration due to interference between the desired focused image with the background noise and the out-of-focus image. Leith and Upatnieks experimentally introduced an off-axis reference wave which behaved as a carrier wave modulated by the object information. Upon reconstruction, these holograms also produced two conjugate images and background noise; however, the two images which may be separately focused on were angularly separated in space from each other and the on-axis background noise. This resulted in good reconstructed-image quality without interference from the other distributions created by the hologram process.

HANDBOOK OF OPTICAL HOLOGRAPHY

4. Major Hologram Types

We will discuss Fresnel holography with a series of special cases which minimize mathematical detail, while simultaneously emphasizing the important physical characteristics of the process. We start with the general equations for off-axis Fresnel holography. In this approach, on-axis holography occurs when the reference angle is zero degrees in the general equations. For both on-axis and off-axis Fresnel holography we define four cases of increasing complexity in order to obtain the physical characteristics necessary to deduce the space bandwidth product and information capacity for such holograms. Magnification effects resulting from spherical-wave illumination are discussed through their impact on the space bandwidth product. These physical characteristics are then tabulated for use in parametric studies in system design. In Section 4.1.2.1, we consider a point object with no limitations on film size or resolution in order to emphasize the imaging properties of the two Fresnel holographic processes. In Section 4.1.2.2, introduction of a finite object demonstrates that the imaging properties remain the same even though the noise characteristics of the reconstructed images change. The case in Section 4.1.2.3 utilizes a point object and a recording medium of finite size and resolution to demonstrate the resolution limitations of these holographic processes and enable us to introduce the concept of the space bandwidth product. In Section 4.1.2.4, we consider objects of finite size and films of finite size and resolution to combine the results of the previous cases. In this case, realistic film transfer functions are considered. Noise effects are treated by introducing the signal-to-noise ratio and then combining it with the space bandwidth product to define the information capacity of Fresnel holographic systems. Finally, as an example, a parametric experimental design comparing on-axis and off-axis Fresnel holography is discussed in Section 4.1.3.

4.1.2 Mathematical Analysis

In this section we present a description of Fresnel holography (assuming thin-film recording) in order to determine the major properties of the process. We will start with a treatment of off-axis holography and consider in-line holography as the limiting case in which the reference angle is zero. The hologram is formed by interfering the Fresnel diffraction wave from the object with an off-axis reference plane wave as shown geometrically in Fig. 1a. The intensity distribution in the hologram plane is given by (DeVelis and Reynolds, 1967)

$$I_H(x_2, y_2) = |K \exp(i\mathbf{k}_1 \cdot \mathbf{r}_2) + D(x_2, y_2)|^2, \tag{1}$$

where

$$\mathbf{k}_1 \cdot \mathbf{r}_2 = |\mathbf{k}_1|(x_2 \sin \theta + y_2 \cos \theta \cos \phi + z_1 \cos \theta \sin \phi),$$

$$\mathbf{r}_2 = \hat{i}x_2 + \hat{j}y_2 + \hat{k}z_1,$$

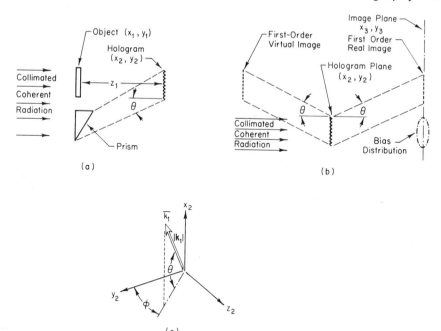

Fig. 1 Schematic of system for formation and reconstruction of off-axis Fresnel holograms: (a) formation, (b) reconstruction, (c) coordinate geometry.

$$\mathbf{k}_1 = \hat{i}|\mathbf{k}_1| \sin \theta + \hat{j}|\mathbf{k}_1| \cos \theta \cos \phi + \hat{k}|\mathbf{k}_1| \cos \theta \sin \phi,$$

$$|\mathbf{k}_1| = 2\pi/\lambda_1 = k_1,$$

K is the amplitude of the reference wave, and the boldface symbols represent vector quantities. The Fresnel diffraction wave from the object $D(x_2, y_2)$ is given by

$$D(x_2, y_2) = C \exp\left[\frac{ik_1(x_2^2 + y_2^2)}{2z_1}\right] \int_{z_1=0} s(x_1, y_1)$$

$$\times \exp\left[ik_1 \frac{(x_1^2 + y_1^2)}{2z_1} - \frac{x_1 x_2 + y_1 y_2}{z_1}\right] dx_1 \, dy_1, \qquad (2)$$

where C is a complex constant and $s(x_1, y_1)$ denotes the amplitude transmittance of the object. Equations (1) and (2) describe the intensity distribution in the hologram. If this intensity distribution is recorded on film and processed so that the amplitude transmittance of the film is linearly proportional to the exposure distribution (Goodman, 1968), then the reconstructed image can be observed in the focused-image plane by illuminating the processed film with

141

4. Major Hologram Types

a replica of the reference wave. A detailed analysis of this reconstruction process is given by DeVelis and Reynolds (1967). In order to reconstruct the Fresnel hologram we will choose a series of mathematically simple cases which demonstrate the essential features of the process.

4.1.2.1 Point Object, On-Axis Reference Wave, Large Recording Film Plane

This approximation corresponds to the in-line process with a point object. For this case, $s(x_1, y_1)$ in Eq. (2) becomes $\delta(x_1, y_1)$ with $\theta = 0°$ and $\phi = \pi/2$ in Eq. (1). The resulting intensity distribution in the hologram from Eq. (1) becomes

$$I_H(x_2, y_2) = |K + C \exp[ik_1(x_2{}^2 + y_2{}^2)/2z_1]|^2. \tag{3}$$

Equation (3) is the paraxial approximation for the interference between a plane wave with a coaxial spherical wave. Reconstruction of this hologram with a plane wave of wavelength λ_2 will result in two conjugate images similar to the primary foci of a Fresnel zone plate. This is easily seen mathematically by reconstructing the hologram given in Eq. (3). Illumination of the hologram with a plane wave, as indicated in Fig. 1b, produces an amplitude transmittance proportional to Eq. (3). This wave field is made up of four terms, two constants, and two spherical wavefronts, in the direction of propagation. One of the spherical wavefronts is diverging from a virtual point located behind the hologram, while the other is converging toward a focal point in the direction of propagation. The wavefield in the observation plane at a distance z_2 from the hologram is given by

$$A(x_3, y_3) = C' \int_{-\infty}^{\infty} \left| K + C \exp\left[ik_1\left(\frac{x_2{}^2 + y_2{}^2}{2z_1}\right)\right]\right|^2$$

$$\times \exp\left\{ \frac{ik_2}{2z_2}[(x_3 - x_2)^2 + (y_3 - y_2)^2]\right\} dx_2\, dy_2, \tag{4}$$

where C' is a complex constant. Examination of the term representing the converging spherical wave (quadratic in x_2 and y_2) yields a delta function image when

$$k_1/z_1 = k_2/z_2. \tag{5}$$

Equation (5), which is known as the focusing condition, is a necessary constraint to obtain the image. Since the magnification of the process is given by

$$m = k_1 z_2 / k_2 z_1, \tag{6}$$

the focused image always has unit magnification for plane-wave holograms. The other three terms from Eq. (4) yield coherent background fields.

142

If we had examined the term representing the diverging spherical wave in Eq. (4), a focused point image at a distance z_2 units behind the hologram would result when

$$k_1/z_1 = -k_2/z_2. \tag{7}$$

Again the other three terms would yield coherent background fields. Thus, we see that two conjugate images exist within the hologram, each obeying a different focusing condition as demonstrated by Eqs. (5) and (7). The minus sign in Eq. (7) also results in a conjugate image that is inverted relative to the real image. The focusing constraints are fundamental to all holographic systems and tend to be a limiting parameter on the information capacity of the process through their effect on the system magnification. This will become apparent in the discussion of the space bandwidth product of holographic systems.

4.1.2.2 Finite Object, On-Axis Reference Wave, Large Recording Film Plane

If the object under discussion has a finite size, then a reconstruction of the object can still be obtained with the same focusing conditions, except that the coherent interference between the focused image and the background fields gives rise to deterioration of the image. The focused-image term from the on-axis limit of Eqs. (1) and (2) gives rise to a distribution of the form

$$KC' \int_{-\infty}^{\infty} R^*(x_2, y_2) \exp\{ik_2[(x_2 - x_3)^2 + (y_2 - y_3)^2]/2z_2\} \, dx_2 \, dy_2, \tag{8}$$

where $R(x_2, y_2)$ is given by Eq. (2), C' is a complex constant, and $*$ denotes a complex conjugate. Use of the focusing condition given in Eq. (5), which removes the quadratic phase factor in (x_2, y_2), yields

$$\exp[ik_2(x_3^2 + y_3^2)/2z_2]KC^*C' \int_{-\infty}^{\infty} s^*(x_1, y_1) \exp[-ik_1(x_1^2 + y_1^2)/2z_1]$$

$$\times \left[\int_{-\infty}^{\infty} \exp\left[ix_2\left(\frac{k_1 x_1}{z_1} - \frac{k_2 x_3}{z_2}\right) \right] \right. \tag{9}$$

$$\times \exp\left[iy_2\left(\frac{k_1 y_1}{z_1} - \frac{k_2 y_3}{z_2}\right) \right] dx_2 \, dy_2 \Bigg] \, dx_1 \, dy_1.$$

Recognizing the x_2 and y_2 integrals as delta functions and that the system magnification is unity subject to the focusing condition, the reconstructed wavefield of the focused-image term is given by

$$A'(x_3, y_3) = KC'C^*s^*(x_3, y_3). \tag{10}$$

Equation (10) represents a focused, erect, real image of the object, subject to the constraint that the film is large enough to assume infinite limits of integration. This image is deteriorated due to the presence of the other three terms in the on-axis approximation of Eq. (1). One of these terms represents the out-of-focus conjugate image, while the other two are bias terms resulting from the square law recording process. The coherent interference of these background fields with the focused image was encountered by Gabor. Gabor's experiments (1948, 1949, 1951) were of limited practical value because of this effect. Two techniques for circumventing this problem are off-axis reference waves (Section 4.1.2.5–4.1.2.8) and Fraunhofer diffraction (which will be discussed in Section 4.2).

4.1.2.3 Point Object, On-Axis Reference Wave, Finite Recording Film Plane

The reconstruction of a point image from Eq. (4), using the focusing condition of Eq. (5), assumed an infinitely large hologram as demonstrated by the infinite limits of integration in Eq. (4). In actuality, the finite resolving power of photographic film limits the maximum Fresnel fringe frequency which can be recorded and thereby defines the limits of integration in Eq. (4). If we assume that the resolution limit (RL) of the film having half-width "$L/2$" is l_1 line pairs/mm and that its modulation transfer function (MTF) is uniform up to cutoff, then the point image amplitude, reconstructed from Eq. (3), is

$$A'(x_3, y_3) = KC^*C' \exp[ik_2(x_3{}^2 + y_3{}^2)/2z_2]$$
$$\times L^2 \operatorname{sinc}(\pi L x_2/\lambda_2 z_2) \operatorname{sinc}(\pi L x_2/\lambda_2 z_2), \tag{11}$$

where $\operatorname{sinc} x = (\sin x)/x$.

The limiting film size in one dimension is set by the ability of the film to record the Fresnel fringes in the hologram. The point-to-point spatial frequency of the Fresnel fringes is found by differentiating the phase of the holographic fringes described by Eq. (3) and evaluating the derivation at the film cutoff frequency l_1. This gives

$$l_1 = L/2\lambda_1 z_1. \tag{12}$$

If we use the inverse of the diffraction spot radius in the focused-image plane as a resolution criterion (Rayleigh criterion), then from Eq. (11) we get

$$(RL)_{im} = 1/X_0 = L/\lambda_2 z_2, \tag{13}$$

where X_0 is the radius of the diffraction spot. Combining Eqs. (12) and (13) and using Eq. (6), we get

$$(RL)_{im} = 2\lambda_1 z_1 l_1/\lambda_2 z_2 = 2l_1/m, \tag{14}$$

where m is the system magnification. Multiplying Eq. (14) by the system magnification gives the object resolution of the system as

$$(\text{RL})_{\text{ob}} = 2l_1. \tag{15}$$

Equation (15) shows that the ability of a Fresnel hologram to resolve information in the object is determined primarily by the resolution of the film used to make the hologram.

4.1.2.4   Finite Object, On-Axis Reference Wave, Finite Recording Film Plane

It has been shown that a linear relationship exists between the object amplitude and the reconstructed focused image amplitude (Goodman, 1968, pp. 225–230). This linear process directly implies that the system possesses a coherent transfer function of the form

$$Y(\xi/\lambda_1 z_1, \eta/\lambda_1 z_1) = P(\xi, \eta)H(\xi/\lambda_1 z_1, \eta/\lambda_1 z_1), \tag{16}$$

where Y is the coherent transfer function of the linear process, P is the pupil function of the hologram, H represents the film MTF, and $\nu_x = \xi/\lambda_1 z_1$, $\nu_y = \eta/\lambda_1 z_1$. Equation (16) shows that the film MTF and the hologram pupil function multiply to yield the coherent transfer function of the process. This further implies that the amplitude in the focused reconstructed image is of the form

$$A'(x_3, y_3) = s(x_3, y_3) * p(x_3, y_3) * h(x_3, y_3), \tag{17}$$

where p is the amplitude spread function of the hologram, h is the spread function of the film, and $*$ denotes the convolution process. Equation (17) indicates that the object amplitude is convolved with the amplitude spread function of the hologram and the spread function of the film. The previous cases are obtained as limiting forms of Eq. (17).

(a)   **On-Axis Space–Bandwidth Product**   The one-dimensional space–bandwidth product of this holographic process, which will define the number of resolution elements of the holographic system, can only be determined by giving the object a finite size. If we equate the object size with the hologram size L, the one-dimensional space–bandwidth product (SBP) for the plane-wave process is given by

$$\text{SBP} = 2l_1 L. \tag{18}$$

If the Fresnel hologram is formed with a spherical reference wave of radius R_1 and reconstructed with a spherical wave of radius R_2, nonunit magnification given by

$$m = \left(\frac{R_1}{R_1 + z_1} - \frac{k_2 z_1}{k_1 R_2} \right)^{-1} \tag{19}$$

will result.

4. Major Hologram Types

If we combine the magnification of Eq. (19) and the appropriate focusing constraint with the resolution limit of the magnified object, the same one-dimensional holographic space–bandwidth product, described by Eq. (18), is obtained. Thus, consistent with the stated assumptions, the one-dimensional space–bandwidth product of the hologram is fixed by the film parameters independent of the magnification. The extension of this analysis to two dimensions yields an equation of the form

$$(\text{SBP})_{2D} = (\text{const})l_1{}^2 A, \tag{20}$$

where the constant is determined by the resolution criterion and A is the area of the hologram.

(b) Information Content The preceding analysis assumed a film transfer function which was flat up to cutoff (a nonrealizable assumption for photographic film), rectangular geometry, and the Rayleigh resolution criterion. These assumptions determine the constant factor of 2 in Eq. (18).

The inclusion of a realistic film MTF limits the cutoff spatial frequency of the process as demonstrated by Eq. (15) and further limits the process in that the Fourier spectrum of the reconstructed focused image is a product of the Fourier spectrum of the object with the film MTF (DeVelis and Reynolds, 1967, Chapter 5). This product, involving the film MTF as one term, reduces the object modulation as a function of spatial frequency. In essence, this is a noise effect which limits the measurable number of gray levels per resolution element passed by the system. In order to consider this noise effect on the holographic process, the concepts of information theory can be used (Yu, 1973; Smith, 1969). To a first-order approximation, the number of resolvable gray levels within a given resolution element can be used to define the signal-to-noise ratio of the holographic process (Yu, 1973; Jones, 1961).

Thus, the space–bandwidth product defines the number of resolution elements in the system, and the signal-to-noise ratio (Yu, 1973; Smith, 1969; Jones, 1961; Fellgett and Linfoot, 1955) quantizes the number of gray levels for each resolution element. If the amplitude transmittance of the hologram is assumed to be gaussianly distributed about its mean value, the maximum channel capacity C_m of the system (Jones, 1961) is

$$C_m = 2(\text{SBP}) \log_2[1 + (S/N)]^{1/2}, \tag{21}$$

where S is the mean square signal amplitude, and N is the mean square noise amplitude. An alternate quantization method is realized by considering the photographic grain noise as a continuous parameter Markov chain (Yu, 1973).

Both of these quantization approaches assume a constant signal-to-noise ratio for the process, ignoring the variation of the signal-to-noise ratio with spatial frequency. Fellgett and Linfoot (1955) defined signal-to-noise ratio in frequency space and thereby obtained a channel capacity per unit area for film

146

of the form

$$C_0 = \frac{1}{2} \int\limits_{-\infty}^{\infty}\!\!\int \log_2\left(1 + \frac{S(\nu_x, \nu_y)}{N(\nu_x, \nu_y)}\right) d\nu_x \, d\nu_y, \qquad (22)$$

where $S(\nu_x, \nu_y)$ is the signal power spectrum, and $N(\nu_x, \nu_y)$ is the noise power spectrum. Equation (22) includes the total effect of signal-to-noise ratio as a function of spatial frequency compensating for the limitations of Eq. (21). Since the signal-to-noise power spectral ratio of Eq. (22) is difficult to measure, in practice the approximate channel capacity of Eq. (21) is generally used and found to be adequate for determining system performance.

4.1.2.5   Point Object, Off-Axis Reference Wave, Large Recording Film Plane

The implementation of an off-axis reference wave by Leith and Upatnieks (1962, 1963, 1964) resolved the problem of interference between the focused image and the coherent noise background information inherently present in the Gabor in-line process (Gabor, 1948, 1949, 1951). The off-axis reference wave introduces an optical carrier into the hologram process. The corresponding spatial frequency of the carrier is proportional to the angle between the object wavefront and the reference wavefront. Upon reconstruction, this carrier produces an angular separation between the conjugate images in their respective planes and the on-axis noise distribution. Focusing on either of the off-axis conjugate images yields a good reconstructed image without interference from the other distributions present in the reconstructed-image plane.

These effects are best demonstrated by using a point object in Eq. (2). If we further restrict ourselves to a one-dimensional analysis [i.e., $\phi = \pi/2$ in Eqs. (1) and (2)], the hologram is represented by

$$I_H(x_2, 0) = \big| K \exp[i|k_1|(x_2 \sin\theta + z_1 \cos\theta)]$$

$$+ C[\exp(i|k_1|x_2{}^2/2z_1)]\big|^2. \qquad (23)$$

Reconstruction of the hologram in Eq. (23), by using the focusing constraint of Eq. (5), results in a reconstructed point image intensity given by

$$I(x_3) \cong \delta(x_3 - z_1 \sin\theta), \qquad (24)$$

where the system magnification is unity because plane waves were used. Equations (5) and (24) show that the point image is in focus in a plane located at a distance $z_2 = k_2 z_1/k_1$ units from the hologram plane. The point image is located at a distance $z_1 \sin\theta$ units from the optical axis as shown in Fig. 2.

In Fig. 2 the conjugate image is in focus at a distance z_2 units behind the hologram as given by Eq. (7). Since this image is propagating in the $(-\theta)$ direction, it will not interfere with the focused real image.

4. Major Hologram Types

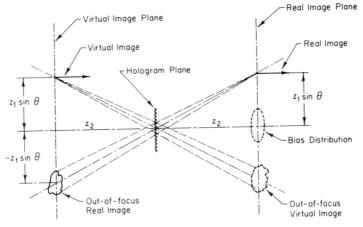

Fig. 2 Graphic illustration, for point objects, of spatial separation of the focused real image from the out-of-focus virtual image and the on-axis bias distribution.

4.1.2.6 Finite Object, Off-Axis Reference Wave, Large Recording Film Plane

For a finite object, the focusing condition, Eq. (5), still applies and a reconstructed image of the object is obtained in the focused-image plane (De Velis and Reynolds, 1967, pp. 50–52). There will be no deterioration of the focused image due to bias and conjugate image terms which are physically separated in space as shown in Fig. 3.

The zero-order bias terms which appear on axis do not interfere with the focused real image, provided that the distance $x_3 - z_1 \sin \theta$ is greater than or equal to 1.5 times the object dimension (Goodman, 1968, Chapter 8). An equivalent interpretation of this effect in the frequency domain is shown in Fig. 4. The one-dimensional object has a Fourier spectrum $S(\xi/\lambda_1 z_1)$ of bandwidth $2B$ as shown in Fig. 4(a). The Fourier spectrum of the hologram, $S_H(\xi/\lambda_1 z_1)$, is shown in Figure 4(b). This bias term, which involves an autocorrelation of the object, has a maximum bandwidth of $4B$ in its Fourier spectrum, $S_B(\xi/\lambda_1 z_1)$. Therefore, the spatial frequency of the holographic carrier must satisfy the condition that

$$\xi_0/\lambda_1 z_1 \geq 3B, \tag{25}$$

or equivalently, that

$$\theta_{\min} = \sin^{-1} 3\lambda_1 B. \tag{26}$$

Equations (25) and (26) specify the condition necessary to guarantee that the images will be separated in space and therefore will not interfere with each other.

148

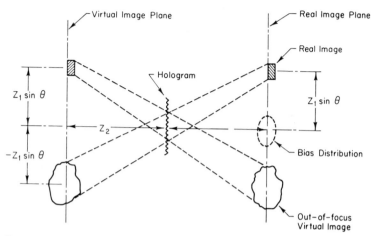

Fig. 3 Graphical illustration, for extended objects, of spatial separation of the focused real image from the out-of-focus virtual image and the on-axis bias distribution.

4.1.2.7 Point Object, Off-Axis Reference Wave, Finite Recording Film Plane

The reconstruction of a point image with a finite recording film plane is obtained from Eqs. (1) and (2), where the limits of integration are $-L/2$ to $L/2$, and the focusing condition given by Eq. (5) is used. The intensity distribution in the hologram is a coherent superposition of a spherical wave from a point scatterer and an off-axis plane wave making an angle θ with the optical axis (DeVelis and Reynolds, 1967, pp. 95–97). Reconstruction of this hologram yields a diffraction spot, characteristic of the hologram diameter, as the re-

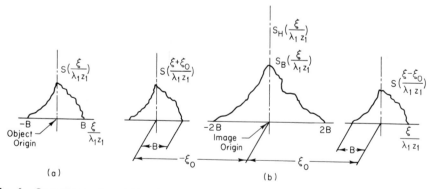

Fig. 4 One-dimensional Fourier spectra of (a) the object and (b) the hologram.

4. Major Hologram Types

constructed image. The resolution limit of the system in object space, determined by using the Rayleigh resolution criterion and the approach described in Section 4.1.2.3 to derive Eq. (15), is

$$RL_{ob} = 2\left(l_1 - \frac{\sin\theta}{\lambda_1}\right). \tag{27}$$

If the hologram is formed with a spherical wave of radius R_0 and reconstructed with a spherical wave of radius R_2, reconstructed images having magnification given by (DeVelis and Reynolds, 1967)

$$m = \left(1 - \frac{z_1}{R_0} - \frac{\lambda_1 z_1}{\lambda_2 R_2}\right)^{-1} \tag{28}$$

are obtained.

The resulting resolution limit in object space is

$$RL_{ob} = 2\left(l_1 - \frac{\sin\theta}{\lambda_1}\right)\bigg/\left(1 - \frac{z_1}{R_0}\right). \tag{29}$$

Equations (28) and (29) show that meaningful magnification can be obtained from the holographic process up to the limit set by the focusing constraint.

4.1.2.8 Finite Object, Off-Axis Reference Wave, Finite Recording Film Plane

The existence of the linear relationship between the object amplitude and the amplitude in the reconstructed focused image (Goodman, 1968, pp. 225–230) implies that the coherent transfer function of the process is

$$Y\left(\frac{\xi}{\lambda_1 z_1}, \frac{\eta}{\lambda_1 z_1}\right) = P(\xi, \eta)H\left(\frac{\xi}{\lambda_1 z_1} - \frac{\sin\theta}{\lambda_1}, \frac{\eta}{\lambda_1 z_1}\right), \tag{30}$$

where Y is the coherent transfer function of the linear process, P is the pupil function of the hologram, and H represents the film MTF.

Equation (30) implies that the focused-reconstructed-image amplitude resulting from holographically recording an object of finite size with an off-axis reference wave onto a film having a finite resolution limit and a finite size is given by

$$A'(x_3, y_3) = s\left(x_3 - \frac{\sin\theta}{\lambda_1}, y_3\right) * p(x_3, y_3) * h(x_3, y_3), \tag{31}$$

where p is the amplitude point spread function of the hologram, h is the point spread function of the film, and $*$ denotes the convolution process.

Equation (31) shows that the reconstructed focused image located off-axis consists of the object convolved with the amplitude spread function of the hologram and the spread function of the film. The previous off-axis cases are limiting forms of Eq. (31).

(a) Off-Axis Space–Bandwidth Product The one-dimensional space bandwidth product for off-axis holograms will define the number of resolution elements contained in the reconstructed image. In order to determine this number of resolution elements we must specify the film size L and the film resolution limit l_1. The resulting space–bandwidth product is obtained by multiplying Eq. (27) by the appropriate film size. This yields

$$SBP = 2\left(l_1 - \frac{\sin\theta}{\lambda_1}\right)L. \tag{32}$$

If spherical waves are used in the process, Eqs. (28) and (29) can be combined to yield the same space–bandwidth product as given by Eq. (32), showing that the field of view and object resolution can be varied in any given situation.

The extension of this analysis to two dimensions with a reference wave incident at the angle $(\theta, \phi = \pi/2)$ will give a space–bandwidth product of the form

$$(SBP)_{2D} = (const)\left(l_1 - \frac{\sin\theta}{\lambda_1}\right)l_1A, \tag{33}$$

where the constant is determined by the resolution criterion and A is the hologram area.

(b) Information Content The inclusion of a realistic film MTF into the process yields a reconstructed focused image whose amplitude spectrum is shaded in a one-sided fashion and whose phases are shifted linearly by the film transfer function as depicted in Fig. 5 (Goodman, 1968).

The film MTF also determines the number of resolvable gray levels within a given resolution element. Quantization of the gray scale by the signal-to-noise ratio for an amplitude transmittance which is gaussianly distributed about its mean value gives the maximum channel capacity of Eq. (21), where the space–bandwidth product of Eq. (32) or (33) must be used.

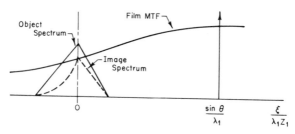

Fig. 5 Effects of film MTF in off-axis holography with plane waves showing the one sided spatial frequency attenuation effect.

4. Major Hologram Types

Measurements of the S/N ratio for various films used in off-axis holography as a function of reference angle, beam intensity ratio, diffraction efficiency, and film resolution have been performed (Zech, *et al.*, 1974). For well-resolved carrier frequencies, signal-to-noise ratios in the range 10–25 dB were measured, depending upon the combination of system parameters used in the experiment.

4.1.3 Results and Examples

The limitations of on-axis and off-axis Fresnel holography are summarized in Table I. This table shows how the maximum channel capacity depends on

(1) the product of the SBP of the holographic process under consideration times the quantization level of the gray scale, $\log_2[1 + (S/N)]$, in bits
(2) the resolution limit of the film l_1,
(3) the size of the film L,
(4) the angle θ of the reference wave,
(5) the magnification m of the process.

TABLE I

Limitations of Fresnel Holographic Systems

Type of hologram	On-axis Fresnel	Off-axis Fresnel
Plane-wave magnification	$m = 1$	$m = 1$
Spherical-wave magnification	$m = \left(\dfrac{R_1}{R_1 + z_1} - \dfrac{k_2 z_1}{k_1 R_2} \right)^{-1}$	$m = \left(1 - \dfrac{z_1}{R_0} - \dfrac{\lambda_1 z_1}{\lambda_2 z_2} \right)^{-1}$
Plane-wave resolution limit	$\mathrm{RL_{ob}} = 2 l_1$	$\mathrm{RL_{ob}} = 2 \left(l_1 - \dfrac{\sin\theta}{\lambda_1} \right)$
Spherical-wave resolution limit	$\mathrm{RL_{ob}} = \dfrac{2 l_1}{1 - [z_1/(R_1 + z_1)]}$	$\mathrm{RL_{ob}} = \dfrac{2 \left(l_1 - \dfrac{\sin\theta}{\lambda_1} \right)}{1 - z_1/R_0}$
Space–bandwidth product		
One-dimensional	$\mathrm{SBP} = 2 l_1 L$	$\mathrm{SBP} = 2 \left(l_1 - \dfrac{\sin\theta}{\lambda_1} \right) L$
Two-dimensional	$(\mathrm{SBP})_{2D} = (\mathrm{const}) l_1^2 A$	$(\mathrm{SBP})_{2D} = (\mathrm{const}) \left(l_1 - \dfrac{\sin\theta}{\lambda_1} \right) l_1(A)$
Channel capacity		
(Gaussian assumption)	$C_m = 2(\mathrm{SBP}) \log_2 \left(1 + \dfrac{S}{N} \right)^{1/2}$	$C_m = 2(\mathrm{SBP}) \log_2 \left(1 + \dfrac{S}{N} \right)^{1/2}$

These relationships are useful in designing any Fresnel holographic system of interest.

As an example demonstrating the use of on-axis Fresnel holography we will consider the design of a particle sizing experiment. Even though on-axis Fresnel holography is not the optimum technique for particle sizing because of the presence of conjugate image noise, the example illustrates a typical parametric experimental design. For stationary particle diameters of 1 mm illuminated with a plane wave from a HeNe laser of wavelength 6328 Å, we first determine the hologram formation distance, z_1. We will pick $z_1 = 300$ mm, which is well within the Fresnel zone for this 1-mm-diameter object. The size of the localized hologram of the particle is determined by equating the spatial frequency of the Fresnel diffraction pattern with the frequency of the film such that the signal-to-noise ratio is 10 or greater. Experimental results indicate that signal-to-noise ratios of 10 or greater exist at those spatial frequencies when the film MTF has a value of approximately one-half (Zech *et al.*, 1974). Therefore, the criterion to be used for choosing the film in the experiment is (DeVelis and Reynolds, 1967)

$$\text{Fresnel fringe frequency} = x_2/\lambda_1 z_1 = l_1/2. \tag{34}$$

In Eq. (34), if x_2 is chosen larger than three Airy disk diameters ($x_2 \cong$ 1.39 mm), then the film resolution requirement is given by

$$l_1 = 14.60/d = 14.60 \quad \text{line pairs/mm,} \tag{35}$$

where d is the particle diameter. Since the localized hologram has a diameter of approximately 2.78 mm, a 70-mm film will record holograms of many such particles ($\sim$4900) in the sample volume. This system has a one-dimensional space–bandwidth product of 2044. Upon reconstruction with plane waves from a HeNe laser, the reconstructed images will be found at a distance $z_2 = z_1$ from the hologram.

For comparison purposes, the same experiment will be parametrically designed using off-axis holography. Plane wave radiation having a wavelength of 6328 Å is assumed in both steps of the process. The resolution necessary to resolve a 1-mm-diameter particle from the Rayleigh resolution criteria is 1 line pair/mm. From Eq. (26) the angle between the diffraction and reference waves necessary to just separate the reconstructed focused image spectrum from the background bias spectrum is $\theta_{\min} = 0.11°$. From Eq. (24) the center of the reconstructed image must satisfy the condition that

$$z_1 \sin \theta = 105 \quad \text{mm,} \tag{36}$$

or that $\theta \geq 20.49°$. The carrier frequency necessary for separating both the images and their spectra in this experiment, obtained by choosing the larger of the two angles, is given by

$$\nu_c = (\sin \theta)/\lambda_1 = 550 \quad \text{line pairs/mm.} \tag{37}$$

4. Major Hologram Types

Assuming a signal-to-noise ratio of 10 so that the sum of the Fresnel fringe frequency plus the carrier frequency is equated to one-half the film cutoff frequency, we obtain

$$(x_2/\lambda_1 z_1) + \nu_c = l_2/2. \tag{38}$$

Substitution into Eq. (38) yields a film resolution requirement of $l_2 = 1114.64$ line pairs/mm. This system will also record holograms of many such particles in the sample volume (~ 4900) and has a one-dimensional space–bandwidth product of 79,050.

A comparison of the on-axis and off-axis results shows that a much higher resolution film is needed to store the information from the volume of particles when an off-axis hologram is used. If the same film, having a resolution $l_2 = 1114.64$ line pairs/mm, were used for recording the Fresnel on-axis hologram, a space–bandwidth product of 156,050 would result. This means that more Fresnel fringes would be recorded using such film; hence, the reconstructed image would have higher resolution. However, the presence of the hologram noise causing image deterioration in this case far exceeds the resolution gains, so that in practice the use of film bandwidth for storing the carrier frequency is well worthwhile.

4.1.4 Conclusions

We have examined on-axis and off-axis Fresnel holography with a series of different cases to minimize mathematical complexity without loss of physical results. The properties of these two systems have been compared, tabulated, and illustrated with an example. The superiority of the focused reconstructed image in the case of off-axis Fresnel holography results from physically separating it from all other energy distributions arising from the holographic process.

REFERENCES

DeVelis, J. B., and Reynolds, G. O. (1967). "Theory and Applications of Holography." Addison-Wesley, Reading, Massachusetts.
Fellgett, P. B., and Linfoot, E. H. (1955). *Phil. Trans. Roy. Soc. (London)* **A247**, 369–407.
Gabor, D. (1948). *Nature* **161**, 777.
Gabor, D. (1949). *Proc. Roy. Soc. (London)* **A197**, 454.
Gabor, D. (1951). *Proc. Phys. Soc.* **B64**, 449.
Goodman, J. W. (1968). "Introduction to Fourier Optics." McGraw-Hill, New York.
Jones, R. C. (1961). *J. Opt. Soc. Amer.* **51**, 1159.
Leith, E. N., and Upatnieks, J. (1962). *J. Opt. Soc. Amer.* **52**, 1129.
Leith, E. N., and Upatnieks, J. (1963). *J. Opt. Soc. Amer.* **53**, 1377.
Leith, E. N., and Upatnieks, J. (1964). *J. Opt. Soc. Amer.* **54**, 1295.
Smith, H. M. (1969). "Principles of Holography." Wiley (Interscience), New York.

Yu, F. T. S. (1973). "Introduction to Diffraction, Information Processing and Holography." MIT Press, Cambridge, Massachusetts.

Zech, R. G., Ralston, L. M., and Shareck, M. W. (1974). Realtime Holographic Recording Materials, Contract No. F30602-74-C-0030, Rome Air Development Center, Griffis AFB, New York 13441, November 1974, AD/A-002 849.

4.2 FRAUNHOFER HOLOGRAMS

Brian J. Thompson

4.2.1 Formation of the Hologram

It has been established that the nature of the hologram, and to some extent the properties of the resultant image, are dependent upon the actual diffracted field associated with the object or signal of interest. An important category of holograms are those formed with in-line systems in which the object is transparent enough to allow for sufficient undiffracted light to be present to provide the background. Furthermore, the plane in which the hologram is recorded is in the far field of the object of interest. Thus the hologram is actually a record of the interference pattern of the Fraunhofer diffraction pattern of the object formed in the far field and the collinear coherent background.

This type of hologram was developed for the particular application of particle size analysis and was introduced by Thompson (1963) and Parrent and Thompson (1964). Theoretical analysis of the process continued (see, e.g., DeVelis and Reynolds, 1967) and recently a detail reassessment of the process was given by Tyler and Thompson (1976). At the same time, considerable insight into Fraunhofer holography was developed by its application to significant problems (see, e.g., Thompson, 1974; Trolinger, 1975).

The object, then, is placed in the $x_1 y_1$ plane and is illuminated with a collimated beam of coherent light (a collimated beam is used here for the discussion, but a noncollimated beam can be used provided the normal far field conditions are met). The hologram is recorded in the $x_2 y_2$ plane a distance z away (see Fig. 1). For this discussion, we will consider the object distribution to be described by an amplitude transmittance $S(x_1, y_1)$ and the illumination to be of unit amplitude and wavelength λ. [The analysis here will follow that given by Tyler and Thompson (1976).] The field distribution $R(x_2, y_2)$ is then given by the Huygens–Fresnel principle

$$R(x_2, y_2) = \frac{-i}{\lambda z} \exp[ikz] \int_{-\infty}^{\infty} \int_{-\infty}^{\infty} [1 - S(x_1, y_1)]$$

$$\times \exp\left\{ \frac{ik}{2z} [(x_2 - x_1)^2 + (y_2 - y_1)^2] \right\} dx_1 \, dy_1. \tag{1}$$

HANDBOOK OF OPTICAL HOLOGRAPHY
Copyright © 1979 by Academic Press, Inc.
All rights of reproduction in any form reserved.
ISBN-0-12-165350-1

4. Major Hologram Types

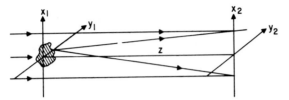

Fig. 1 Schematic diagram for the formation of a Fraunhofer hologram.

The resultant intensity $I_H(x_2, y_2)$ under the far-field condition is then

$$I_H(x_2, y_2) = 1 - \frac{2}{\lambda z}\left[\sin\left(\frac{\pi r^2}{\lambda z}\right)\text{Re}\,\tilde{S}\left(\frac{x_2}{\lambda z}, \frac{y_2}{\lambda z}\right)\right.$$

$$+ \cos\left(\frac{\pi r^2}{\lambda z}\right)\text{Im}\,\tilde{S}\left(\frac{x_2}{\lambda z}, \frac{y_2}{\lambda z}\right)\Bigg]$$

$$+ \frac{1}{\lambda^2 z^2}\tilde{S}\left(\frac{x_2}{\lambda z}, \frac{y_2}{\lambda z}\right)\tilde{S}^*\left(\frac{x_2}{\lambda z}, \frac{y_2}{\lambda z}\right), \qquad (2)$$

where $r^2 = x_2{}^2 + y_2{}^2$, Re and Im denote real and imaginary parts, respectively, and $\tilde{S}(x_2/\lambda z, y_2/\lambda z)$ is the Fourier transform of the object distribution defined by

$$\tilde{S}\left(\frac{x_2}{\lambda z}, \frac{y_2}{\lambda z}\right) = \int_{-\infty}^{\infty}\int_{-\infty}^{\infty} S(x_1, y_1)$$

$$\times \exp\left\{-2\pi i\left[x_1\left(\frac{x_2}{\lambda z}\right) + y_2\left(\frac{y_1}{\lambda z}\right)\right]\right\}dx_1\,dy_1. \qquad (3)$$

As an illustrative example, we will consider that the object is a wire of width $2a$ so that the problem reduces to a one-dimensional analysis (the far-field condition only applies in one direction). The object distribution is now $S(x_1)$ and Eq. (2) becomes

$$I_H(x_2, y_2) = 1 - \frac{2}{\sqrt{\lambda z}}\left[\cos\left(\frac{\pi x_2{}^2}{\lambda z} - \frac{\pi}{4}\right)\text{Re}\,\tilde{S}\left(\frac{x_2}{\lambda z}\right),\right.$$

$$\left. - \sin\left(\frac{\pi x_2{}^2}{\lambda z} - \frac{\pi}{4}\right)\text{Im}\,\tilde{S}\left(\frac{x_2}{\lambda_2}\right)\right] + \frac{1}{\lambda z}\tilde{S}\left(\frac{x_2}{\lambda z}\right)\tilde{S}^*\left(\frac{x^2}{\lambda z}\right), \qquad (4)$$

where

$$\tilde{S}(x_2/\lambda z) = 2a\,\text{sinc}(kax_2/z). \qquad (5)$$

Finally, we may note that the amplitude transmittance of this hologram is

158

given by

$$I(x_2) = 1 - \frac{4a}{\sqrt{\lambda z}} \cos \frac{\pi x_2{}^2}{\lambda z} \operatorname{sinc} \frac{k a x_2}{z} + \frac{4a^2}{\lambda z} \left[\operatorname{sinc} \frac{k a x_2}{z} \right]^2. \tag{6}$$

Fig. 2 illustrates this result for a wire of diameter $2a = 100$ μm that was illuminated with light from an argon-ion laser at $\lambda = 0.5146$ μm; $z = 60$ cm. The hologram was recorded on a Kodak 649F plate and is illustrated here in Fig. 2a as a positive print. Figure 2b shows a microdensitometer trace across

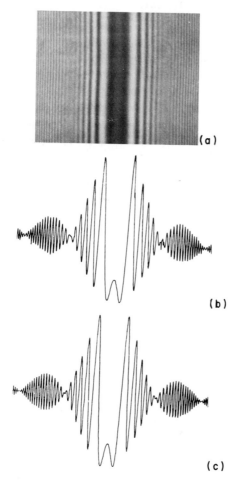

Fig. 2 Fraunhofer hologram of a wire of diameter $2a = 160$ μm formed at a distance of $z = 60$ cm with light of wavelength 0.5145 μm is (a) a positive print of a portion of the hologram, (b) a microdensitometer trace across the hologram, and (c) a computer plot of Eq. (6). (After Tyler and Thompson, 1976.)

the hologram and Fig. 2c a computer plot from Eq. (6). It is interesting to note that a similar plot to that shown in Fig. 2c could be obtained by treating the total aperture function as a single function and then performing the calculation of the appropriate Fresnel diffraction pattern of the whole field. For example, a similar plot can be found in many texts that discuss Fresnel diffraction (see, e.g., Jenkins and White, 1957). Of course, the point is then missed that this is a hologram and the physical understanding is lost.

4.2.2 Formation of the Image

The hologram is illuminated again with a plane wave of unit amplitude and produces a real image in the $x_3 y_3$ plane a distance z from the hologram. (The effect of spherical-wave recording and reconstruction may be included by allowing the zs to be different.) The image field amplitude $A(x_3, y_3)$ is then given by

$$
\begin{aligned}
A(x_3, y_3) = \exp[ikz] \Bigg\{ & 1 - S^*(x_3, y_3) - \frac{1}{2\lambda z} \exp\left[\frac{i\pi p^2}{2\lambda z} - \frac{\pi}{4}\right] \tilde{S}\left(\frac{x_3}{2\lambda z}, \frac{y_3}{2\lambda z}\right) \\
& + \frac{1}{\lambda^2 z^2} \tilde{S}\left(\frac{x_3}{\lambda z}, \frac{y_3}{\lambda z}\right) \tilde{S}^*\left(\frac{x_3}{\lambda z}, \frac{y_3}{\lambda z}\right) \Bigg\},
\end{aligned}
\tag{7}
$$

where $p^2 = x_3^2 + y_3^2$. The resultant intensity can then be formed from Eq. (7).

For the purpose of illustration and discussion, we will consider the case of the image formed from a hologram of the wire. The intensity in the image is given by

$$
\begin{aligned}
I(x_3) = 1 - \left(\frac{2}{\lambda z}\right)^{1/2} & \left[\cos\left(\frac{\pi x_3^2}{2\lambda z} - \frac{\pi}{4}\right) \operatorname{Re} \tilde{S}\left(\frac{x_3}{2\lambda z}\right) \right. \\
& \left. - \sin\left(\frac{\pi x_3^2}{2\lambda z} - \frac{\pi}{4}\right) \operatorname{Im} \tilde{S}\left(\frac{x_3}{2\lambda z}\right)\right] \\
& + \frac{1}{2\lambda z} \tilde{S}\left(\frac{x_3}{2\lambda z}\right) \tilde{S}^*\left(\frac{x_3}{2\lambda z}\right) + S(x_3)S^*(x_3)
\end{aligned}
$$

$$
+ \text{ other terms.}
\tag{8}
$$

Some considerable insight can be gained by looking at Eq. (8). The second term is the reconstructed image and the third term is the field propagating from the virtual image, and thus when the detected intensity is formed, the first and third terms are squared to produce an expression like Eq. (2) but with z replaced by $2z$. That is, it is a hologram made at a distance $2z$. The result of this is that the real image is superimposed on a hologram of itself, formed at twice the original distance. The image then falls in a region that is essentially constant and hence little interference results.

This process is illustrated in Fig. 3 for the wire example. Fig. 3a shows a photograph of the image formed from the hologram of Fig. 2a. The image is quite good with the hologram formed from the virtual image quite clearly visible in the background. For comparison, Figs. 3b and 3c show a microdensitometer plot of "a" and a computer plot from the appropriate theoretical analysis.

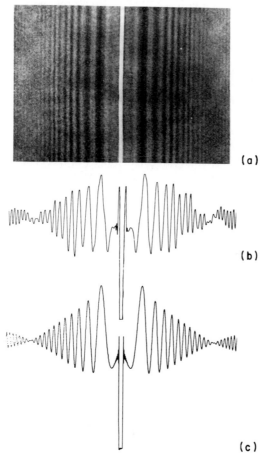

(a)

(b)

(c)

Fig. 3 Real image formed from the Fraunhofer hologram of the wire shown in Fig. 2. (a)A positive print of the record of the image plane, (b) a microdensitometer trace of the negative, and (c) the theoretical plot. (After Tyler and Thompson, 1976.)

4.2.3 Properties of Fraunhofer Holograms

Some of the examples described in Section 1.2 in the discussion of Fresnel holograms with on-axis reference beams are, in fact, Fraunhofer holograms because of the object chosen. In particular, Case 1 is for a point object. Naturally it is impossible not to be in the far field of a point object. The image of such a point object formed by the hologram is a measure of the impulse response of the overall system. This response function will, since a large recording film format is used, be determined by the resolution limit of the film and/or the lack of perfect coherence in the illuminating beam. Case 3 of Section 1.2 examines the effect of the finite size of the recording film and since the object considered is, again, a point, the analysis applies directly to the Fraunhofer hologram.

The results given in Table I of Section 4.1 for the on-axis Fresnel holographic system apply to the in-line (on-axis) Fraunhofer holograms discussed in this section. The important difference, however, is that the two image fields do not interfere significantly with each other as they do in the Fresnel case and hence significant use has been made of the Fraunhofer process, whereas essentially no practical use has been made of the on-axis Fresnel process.

The one parameter that has not yet been discussed in the holographic process is the effect of the finite spatial and temporal coherence that the illuminating beam might possess. This is not often a particularly serious limitation but sometimes needs to be considered. We will assume for this brief discussion that the coherence is going to be the limiting process. Let us assume that for the resolution required in the process, the hologram must be recorded out to the point P (see Fig. 4). The diffracted component of the light from the object centered at O has a path approximately equal to OP. The undiffracted light that interferes with the diffracted light at P came from point A in the object plane. Hence the light at O and A must have a significant spatial coherence for good interference at P. The temporal coherence is determined by the path difference experienced by the two beams in traveling to P (i.e., OP–AP). This path difference must be well within the coherence length of the radiation. In practice, the system design should be such that the spatial and temporal coherence are not the limiting factors.

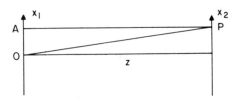

Fig. 4 Illustrating the spatial and temporal coherence requirements of the illumination.

REFERENCES

DeVelis, J. B., and Reynolds, G. O. (1967). "Theory and Application of Holography." Addison-Wesley, Reading, Massachusetts.

Jenkins, F. A., and White, H. E. (1957). "Fundamentals of Optics," 3rd ed., p. 376. McGraw-Hill, New York.

Parrent, G. B., and Thompson, B. J. (1964). *Optica Acta* **11**, 183.

Thompson, B. J. (1963). *J. Soc. Photo-Optical Instr. Engrs.* **2**, 437.

Thompson, B. J. (1974). *J. Phys. E.* **7**, 781.

Trolinger, J. (1975). *Opt. Eng.* **14**, 383.

Tyler, G. A., and Thompson, B. J. (1976). *Optica Acta* **23**, 685.

4.3 FOURIER HOLOGRAPHY

Henri H. Arsenault
Gilbert April

4.3.1 Introduction

Fourier holograms may be defined as holograms recorded with the planar object and the reference source in the same plane, parallel to the hologram plane. Strictly speaking, this analysis is thus intended to apply only to two-dimensional objects and is less applicable as the object extends out of the input plane. There are a variety of Fourier hologram types, depending on whether the holograms are recorded with or without lenses and on how the object is illuminated, but they all have some similarities and very useful properties.

Fourier holograms draw their name not from the fact that the Fourier transform of the object is recorded on the hologram, but from the property that an image of the object may be obtained by Fourier transforming the hologram.

One way to discuss Fourier holography is by using the Fourier-transformation properties of lenses; these properties are essential to an understanding of the spatial filtering properties of optical processors used with nonholographic filters, however, they are not essential to understanding the properties of Fourier holograms. So we shall use an alternate approach to Fourier holography, where lenses (when they are used) serve only their usual function of mapping an object space into an image space. All Fourier holograms may be shown to be a special case of a lensless Fourier-transform hologram recorded with the object illuminated with noncollimated light.

4.3.2 Mathematical Preliminaries

The Fourier transform of a two-dimensional function of space $f(x, y)$ is equal to

$$F(u, v) = \int\limits_{-\infty}^{\infty}\!\!\int f(x, y) \exp\{-2\pi i(ux + vy)\}\, dx\, dy, \qquad (1)$$

where u, v are spatial frequencies. In order to lighten the notation, let us use

HANDBOOK OF OPTICAL HOLOGRAPHY
Copyright © 1979 by Academic Press, Inc.
All rights of reproduction in any form reserved.
ISBN-0-12-165350-1

4. Major Hologram Types

the vector notation $\mathbf{x} = (x, y)$ and $\mathbf{u} = (u, v)$ so that the preceeding definition can now be written

$$F(\mathbf{u}) = \int_{-\infty}^{\infty} f(\mathbf{x}) \exp\{-2\pi i \mathbf{u} \cdot \mathbf{x}\} \, d^2\mathbf{x}. \tag{1}$$

We shall use the optical diffraction propagator expression in which the light diffracted into a plane P_2 from a plane P_1 separated by a distance d is written as a convolution between the complex light amplitude $a_1(\mathbf{x})$ in plane P_1 and the propagator $\psi(\mathbf{x}; d)$, which is defined (in the Fresnel approximation) as

$$\psi(\mathbf{x}; d) = \exp\{(i\pi/\lambda d)\mathbf{x}^2\}. \tag{2}$$

The complex amplitude in plane P_2 is expressed as

$$a_2(\mathbf{x}) = a_1(\mathbf{x}) * (1/i\lambda d)\psi(\mathbf{x}; d), \tag{3}$$

where $*$ means convolution and λ is the wavelength of the light.

Some of the properties of the optical propagator $\psi(\mathbf{x}; d)$ may be found in Collier *et al.* (1971) and in Vander Lugt (1966). These properties are summarized in Section 4.3.6.

4.3.3 Recording and Reconstruction Geometries

The earliest use of Fourier holography was by Vander Lugt (1964) who used a Mach–Zehnder interferometer with lenses in what was equivalent to the Fourier–Fraunhofer configuration (Leith, 1964); this configuration, which remains to this day the most popular, will be discussed in Section 4.3.3.4. All Fourier holograms may be considered as particular cases of the type of hologram discussed next: the lensless Fourier-transform hologram.

4.3.3.1 The Lensless Fourier-Transform Hologram

Stroke (1965) showed that holograms having properties similar to the Fourier holograms previously recorded with lenses could be obtained without the use of lenses. To record a lensless Fourier-transform hologram, the reference source is placed in the same plane as the object. Let us assume for the moment that the object is a point. The interference pattern recorded on the hologram will be a family of equally spaced fringes, in contrast to the case where the reference source is not at the same distance from the hologram as the object; in the latter case, the fringes crowd together as they get farther away from the axis of symmetry.

Consider now the general case described in Fig. 1. The object, a complex transmittance $t(\xi)$, is illuminated by a point source placed at a distance d_1 from the object. In the figure, the object is illuminated by a divergent beam,

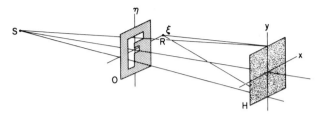

Fig. 1 Recording a lensless Fourier-transform hologram.

but it could as well be illuminated by a convergent beam, without loss of generality (in fact, Fourier holograms are sometimes recorded this way). The object is assumed to be at a distance d_2 from the hologram. The reference source R is in the same plane as the object $t(\xi)$. The complex amplitude of the reference point source may be written as

$$r_0(\xi) = \delta(\xi - \xi_0). \tag{4}$$

From Eqs. (3) and (4), the complex amplitude of the reference wave at the hologram plane is

$$r(\mathbf{x}) = (1/i\lambda d_2)\psi(\mathbf{x} - \xi_0; d_2) \tag{5}$$

or

$$r(\mathbf{x}) = (1/i\lambda d_2)\psi(\mathbf{x}; d_2) \exp\{-(2\pi i/\lambda d_2)\mathbf{x}\cdot\xi_0\} \tag{6}$$

apart from a constant complex phase factor $\psi(\xi_0; d_2)$ whose modulus is unity.

The light arriving at the hologram from the object is diffracted from the point source to the object where it is multiplied by the complex transmittance $t(\xi)$ of the object, then diffracted from the object to the hologram. The complex amplitude $a(\mathbf{x})$ of the light from the object at the hologram plane is therefore

$$a(\mathbf{x}) = [\psi(\mathbf{x}; d_1)t(\mathbf{x})] * (1/i\lambda d_2)\psi(\mathbf{x}; d_2). \tag{7}$$

Using the properties of the ψ function, Eq. (7) becomes

$$a(\mathbf{x}) = [d_1/(d_1 + d_2)]\psi(\mathbf{x}; d_2)[T(\mathbf{u}) * \psi^*(\lambda d'\mathbf{u}; d')], \tag{8}$$

where $d' = d_1 d_2/(d_1 + d_2)$ and $T(\mathbf{u})$ is the Fourier transform of $t(\xi)$ with spatial frequency

$$\mathbf{u} = \mathbf{x}/\lambda d_2.$$

The intensity of the interference pattern formed by the object and reference waves as recorded in the hologram plane is

$$I(\mathbf{x}) = |a(\mathbf{x}) + r(\mathbf{x})|^2, \tag{9}$$

$$I(\mathbf{x}) = a(\mathbf{x})r^*(\mathbf{x}) + a^*(\mathbf{x})r(\mathbf{x}) + |a(\mathbf{x})|^2 + |r(\mathbf{x})|^2. \tag{10}$$

4. Major Hologram Types

The terms of interest are the first two terms of Eq. (10). We shall consider only the first term. Similar considerations may be applied to the second term which will be discussed later.

In the usual fashion let us assume that the photographic plate is developed to yield a transparency with amplitude transmittance proportional to exposure. The direct image-forming term of the hologram transmittance is thus represented by

$$g_d(\mathbf{x}) = a(\mathbf{x})r^*(\mathbf{x}), \tag{11}$$

$$g_d(\mathbf{x}) = \frac{i\lambda d'}{(\lambda d_2)^2} [T(\mathbf{u}) * \psi^*(\lambda d'\mathbf{u}; d')] \exp\{2\pi i\mathbf{u}\cdot\boldsymbol{\xi}_0\}. \tag{12}$$

This is the complex amplitude of the diffracted wave just behind the hologram when it is illuminated by a plane wave of unit amplitude, if the appropriate holographic recording parameters are chosen.

Now consider the Fourier transform of Eq. (12). The terms within the square brackets will yield the complex transmittance of the object, multiplied by a complex phase factor. The exponential term of the right will cause the reconstructed image to be shifted by an amount proportional to $\boldsymbol{\xi}_0$, the offset distance of the reference beam. If the image is observed, the complex phase factor will disappear and only the irradiance of the original object will remain. So, except for what becomes a multiplicative phase factor upon reconstruction, Eq. (12) represents the Fourier transform of the object. That is why a hologram recorded as in Fig. 1 is called a lensless Fourier-transform hologram. The properties of such holograms will be discussed in Section 4.3.4.

4.3.3.2 The Equivalent Lensless Fourier-Transform Hologram

In this section, we shall show that any Fourier hologram is equivalent to a lensless Fourier-transform hologram. Consider the general case of a Fourier hologram where a lens is assumed to operate on the light from both the object and the reference source; the recording geometry to be considered here is illustrated in Fig. 2, the object is shown illuminated with divergent light from a point-source S located some distance d_s before it. The reference source is again a point source R coplanar with the object with coordinates $\boldsymbol{\xi}_0$. Let I be the geometrical image of the object O in the lens and let H define the hologram plane. We assume that the lens is large enough so that we can neglect the effects of vignetting.

If we denote by d_o and d_h the respective positions of the object and the hologram relative to the lens, the complex amplitude of the object light in the hologram plane can be written as

$$a(\mathbf{x}) = -(1/\lambda^2 d_o d_h)[\{[\psi(\mathbf{x}; d_s)t(\mathbf{x})] * \psi(\mathbf{x}; d_o)\}\psi^*(\mathbf{x}; f)] * \psi(\mathbf{x}; d_h). \tag{13}$$

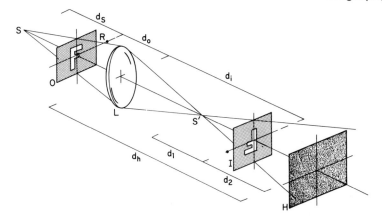

Fig. 2 The equivalent lensless Fourier-transform hologram.

Similarly, the complex amplitude of the reference wave originating from a point source $\delta(\boldsymbol{\xi} - \boldsymbol{\xi}_0)$ in the object plane is given in the hologram plane (apart from a constant phase factor) by

$$r(\mathbf{x}) = -(1/\lambda^2 d_o d_h)[\psi(\mathbf{x} - \boldsymbol{\xi}_0; d_o)\psi^*(\mathbf{x}; f)] * \psi(\mathbf{x}; d_h). \qquad (14)$$

In terms of the distance d_1 and d_2 relating the relative positions of the hologram plane, the image S′ of the source and the geometrical image I of the object, Eqs. (13) and (14) can be shown to be equal to

$$a(\mathbf{x}) = \frac{-d_1}{M(d_1 + d_2)} \psi(\mathbf{x}; d_2)[T(\mathbf{u}) * \psi^*(\lambda d'\mathbf{u}; d')], \quad \mathbf{u} = -\frac{M\mathbf{x}}{\lambda d_2}, \qquad (15)$$

$$r(\mathbf{x}) = -\frac{M}{i\lambda d_2} \psi(\mathbf{x}; d_2) \exp\{-2\pi i \mathbf{u}\cdot\boldsymbol{\xi}_0\}, \qquad (16)$$

where $d' = M^{-2}[d_1 d_2/(d_1 + d_2)]$ and $M = d_i/d_o$ is the magnification of the imaging system. The direct-image-forming term of the hologram transmittance is therefore given as in Eq. (11) by

$$g_d(\mathbf{x}) = (M/\lambda d_2)^2 i\lambda d'[T(\mathbf{u}) * \psi^*(\lambda d'\mathbf{u}; d')] \exp\{2\pi i \mathbf{u}\cdot\boldsymbol{\xi}_0\}. \qquad (17)$$

Comparing this expression with Eq. (12), we see that the two expressions are formally identical, except for a constant factor. The equivalence of Eqs. (12) and (17) leads to the following conclusion: *Any Fourier hologram is equivalent to a lensless Fourier-transform hologram. The recording configuration for the equivalent lensless Fourier-transform hologram is found in the following way:*

(1) Find the image I of the object in all the lenses between the object and the hologram plane.

4. Major Hologram Types

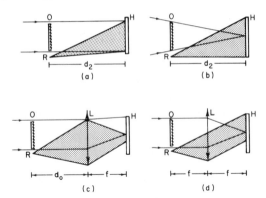

Fig. 3 Equivalent lensless Fourier-transform hologram parameters for various recording geometries: (a) lensless Fourier-transform hologram (parallel illumination), $M = -1$, $d_1 = \infty$; (b) convergent illumination, $M = -1$, $d_1 = -d_2$; (c) object in front of lens, $d_1 = -d_2 = f[(d_o/f) - 1]^{-1}$, $M/d_2 = -1/f$; (d) Fourier–Fraunhofer hologram, $d_1 = -d_2 = \infty$, $M/d_2 = -1/f$

(2)   Find the image S' of the illuminating source in all the lenses between the source and the hologram.

The hologram is equivalent to lensless Fourier transform hologram recorded by illuminating an object identical to the geometrical image I with an illuminating source identical to S'.

For the case corresponding to Fig. 2, the equivalent lensless Fourier-transform hologram is that recorded by illuminating I with the source S'. This concept of equivalence allows important simplifications for several special cases of interest (Fig. 3).

The direct-image forming terms for the cases illustrated in Fig. 3 are the following:

(a)   lensless Fourier-transform hologram (parallel illumination):

$$d_1 = \infty, \qquad M = -1, \qquad d' = d_2,$$

$$g_d(\mathbf{x}) = (1/\lambda d_2)[T(\mathbf{u}) * \psi^*(\lambda d_2 \mathbf{u}; d_2)]\exp\{2\pi i \mathbf{u}\cdot\boldsymbol{\xi}_0\}, \qquad \mathbf{u} = \mathbf{x}/\lambda d_2;$$

(b)   convergent illumination:

$$d_1 = -d_2, \qquad M = -1, \qquad d' = \infty,$$

$$g_d(\mathbf{x}) = (1/\lambda^2 d_2{}^2)T(\mathbf{x}/\lambda d_2)\exp\{(2\pi i/\lambda d_2)\mathbf{x}\cdot\boldsymbol{\xi}_0\};$$

(c)   object in front of lens:

$$M/d_2 = -1/f, \qquad d_1 = -d_2 = f[(d_0/f) - 1]^{-1},$$

$$g_d(\mathbf{x}) = (1/\lambda^2 f^2)T(\mathbf{x}/\lambda f)\exp\{(2\pi i/\lambda f)\mathbf{x}\cdot\boldsymbol{\xi}_0\};$$

170

(d)   Fourier–Fraunhofer hologram:

$$M/d_2 = -1/f, \qquad d_1 = -d_2 = \infty,$$

$$g_d(\mathbf{x}) = (1/\lambda^2 f^2)\,T(\mathbf{x}/\lambda f)\,\exp\{(2\pi i/\lambda f)\mathbf{x}\cdot\boldsymbol{\xi}_0\}.$$

4.3.3.3   Reconstruction of Fourier Holograms

We now discuss the reconstruction of lensless Fourier-transform holograms. We have shown in the previous section that all Fourier holograms have an equivalent lensless Fourier-transform hologram, so this section applies to all Fourier holograms.

If the developed hologram has an amplitude transmittance proportional to exposure in the recording step, then according to Eq. (10), the complex amplitude of the diffracted wave just behind the hologram when illuminated with a normally incident plane wave in the reconstruction process is

$$g(\mathbf{x}) = a(\mathbf{x})r^*(\mathbf{x}) + a^*(\mathbf{x})r(\mathbf{x}) + |a(\mathbf{x})|^2 + |r(\mathbf{x})|^2. \qquad (18)$$

The two image-forming terms are $g_d(\mathbf{x}) = a(\mathbf{x})r^*(\mathbf{x})$ and its complex conjugate.

In order to reconstruct the lensless Fourier-transform hologram, we must produce the Fourier transform of the hologram. This may be done by observing the Fraunhofer diffraction pattern of the hologram. The Fourier transformation may also be observed at the focal plane of lens illuminated with collimated light: the hologram is then placed in the beam of light, either before or after the lens. For instance, if the hologram is placed immediately after a lens of focal length f, as shown in Fig. 4, the zero-order terms in Eq. (18) will be focused about the origin of the focal plane. Using the usual Fourier transformation properties of lenses, the direct and conjugate image-forming terms will

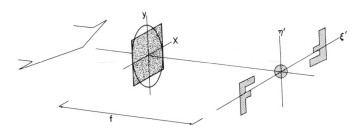

Fig. 4   Reconstructing the real images of a Fourier hologram.

4. Major Hologram Types

give rise to complex amplitude distributions $w_d(\xi')$ and $w_c(\xi')$ given by

$$w_d(\xi') = Kt\left[-\left(\frac{d_2}{f}\xi' - \xi_0\right)\right]$$

$$\times \psi\left[-\left(\frac{d_2}{f}\xi' - \xi_0\right); \frac{d_1 d_2}{d_1 + d_2}\right]\psi(\xi'; f), \qquad (19)$$

$$w_c(\xi') = Kt^*\left(\frac{d_2}{f}\xi' + \xi_0\right)$$

$$\times \psi^*\left(\frac{d_2}{f}\xi' + \xi_0; \frac{d_1 d_2}{d_1 + d_2}\right)\psi^*(\xi'; f), \qquad (20)$$

where K is a complex constant.

If the intensity of the light distribution in the focal plane of the lens is observed, all phase factors drop out and the intensity of the images is proportional to

$$I(\xi') = \left|t\left[-\left(\frac{d_2}{f}\xi' - \xi_0\right)\right]\right|^2 + \left|t\left(\frac{d_2}{f}\xi' + \xi_0\right)\right|^2. \qquad (21)$$

There are two reconstructed images, on either side of the optical axis. The two images are symmetrical with respect to the optical axis: the direct image, represented by the first term of Eq. (21) appears inverted and centered at $\xi_d' = (f/d_2)\xi_0$ and the conjugate image is upright and centered at $\xi_c' = -(f/d_2)\xi_0$. With this reconstruction scheme, both images are real.

Virtual images may be reconstructed by simply illuminating the hologram with a point source, as shown in Fig. 5, and looking at the source through the hologram. The reconstructed virtual images appear on either side of the source. The location of the images may also be understood from the geometrical considerations of Section 4.3.4.1.

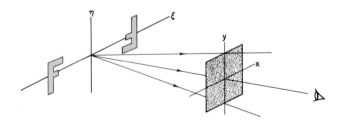

Fig. 5 Reconstructing the virtual images of a Fourier hologram.

4.3.3.4   The Fourier–Fraunhofer Hologram

There is an important special case of Fourier hologram which deserves mention here. A hologram which records the interference of two waves whose complex amplitudes at the photographic plate are the Fourier transforms of both the object and the reference point source is called a *Fourier–Fraunhofer hologram*. One frequently used recording geometry, also known as the *f–f* configuration, producing such a hologram is shown in Fig. 6. The object and the reference point source are in the front focal plane of a lens and the photographic plate is located in the back focal plane. Each point in the object gives rise to a parallel beam of light incident on the photographic plate. The off-axis reference point source is also transformed by the lens, yielding a collimated reference beam at some angle with the optical axis. As seen by the hologram, both the object and the reference source are effectively at infinity. When discussing hologram aberrations, this latter property will be proven significant.

Noting that the object is illuminated with a plane wave, the equivalent lensless Fourier-transform hologram has $d_s = \infty$, $d_2 = -d_1$. If the lens has a focal length f, the direct image-forming term as expressed by Eq. (17) becomes

$$g_d(\mathbf{x}) = (1/\lambda^2 f^2) T(\mathbf{x}/\lambda f) \exp\{(2\pi i/\lambda f)\mathbf{x}\cdot\boldsymbol{\xi}_0\}, \qquad (22)$$

where we used the fact that under these conditions

$$M/\lambda d_2 = -1/\lambda f \qquad (23)$$

and

$$\delta(\mathbf{u}) = \lim_{d\to\infty} i\lambda d\psi^*(\lambda d\mathbf{u}; d). \qquad (24)$$

Note also that the complex amplitude in the hologram plane is given by the product of the exact Fourier transforms of the object transmittance function

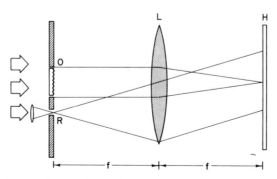

Fig. 6   Recording a Fourier–Fraunhofer hologram.

4. Major Hologram Types

and the reference source, that is,

$$a(\mathbf{x}) = (1/\lambda f) T(\mathbf{x}/\lambda f) \tag{25}$$

and

$$r(\mathbf{x}) = (1/\lambda f) \exp\{-(2\pi i/\lambda f)\mathbf{x}\cdot\boldsymbol{\xi}_0\}. \tag{26}$$

When this hologram is reconstructed with the same configuration, the complex amplitude of the object will be reconstructed without any multiplicative phase factor. If in the reconstruction process the focal length f_2 of the lens is different from the one used in the recording step (say f_1), both the direct and the conjugate images formed in the focal plane of the lens will be magnified by a factor f_2/f_1.

The Fourier–Fraunhofer hologram, introduced by Leith and Upatnieks (1964), has several advantages that will be discussed in Section 4.3.4.

4.3.3.5 The Quasi-Fourier–Fraunhofer Hologram

Because holograms are recorded on materials that have a limited dynamic range, it is usually desirable to reduce the range of exposures on the hologram. The Fourier transform of a continuous tone object may have an extremely intense central order, whereas the high-frequency information is of low intensity. The holographer is thus forced to choose between overexposing the plate for low-frequency regions or underexposing the high frequencies. In both cases, the diffraction efficiency of the developed hologram is affected.

The dynamic range may be improved by defocusing the hologram slightly in the recording step, as shown in Fig. 7. In this case, the transmittance of the direct image-forming term of the hologram is

$$g_{\rm d}(\mathbf{x}) = (1/\lambda\epsilon) \, [T(\mathbf{u}) * \psi^*(\lambda(f^2/\epsilon)\mathbf{u}; \ f^2/\epsilon)] \exp\{2\pi i\mathbf{u}\cdot\boldsymbol{\xi}_0\}, \tag{27}$$

where ϵ is the amount of defocusing and $\mathbf{u}$ is the spatial frequency vector $\mathbf{u}$

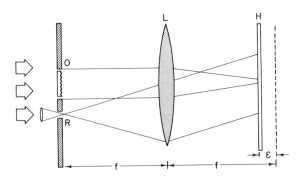

Fig. 7 Recording a quasi-Fourier–Fraunhofer hologram.

$= x/\lambda f$. The "smoothing" operator $\psi^*(\lambda(f^2/\epsilon)\mathbf{u}; f^2/\epsilon)$ will also act on the undesired on-axis light, reducing the strength of the high-intensity peak at the origin of the hologram plane. It is possible to make ϵ large enough to reduce significantly the dynamic range of the intensity incident on the hologram while maintaining it small enough to retain the advantages of Fourier–Fraunhofer holograms. It is usually sufficient to set ϵ approximately equal to 5 or 10% of the focal length of the transforming lens. This is a frequently used technique for applications such as holographic character recognition and mass information storage.

4.3.4   Properties of Fourier Holograms

4.3.4.1   Geometrical Properties

We have discussed some of the geometrical properties of Fourier holograms in the previous section. The main property of Fourier holograms is that the direct and the conjugate images are both in a plane containing the reconstruction source or its image. This property may be deduced from the mathematical analysis, or more simply, by means of the holographic conjugate relations given in Chapter 7. From Table III of Chapter 7, setting $y_r = y_0$ yields for the coordinates of the direct image (setting $x_c = 0$)

$$(x_{ds}, y_{ds}) = [(x_0/y_0)y_c, -y_c], \tag{28}$$

and for the conjugate image

$$(x_{cs}, y_{cs}) = [-(x_0/y_0)y_c, -y_c], \tag{29}$$

where (x_0, y_0), (x_r, y_r), and (x_c, y_c) are the cartesian coordinates of a point in the object, the reference source, and the reconstruction source, respectively. The direct and conjugate images are on either side of the reconstruction source R_c, symmetrical with respect to it, at distance

$$x = \pm[(x_0 y_c/y_0) - x_c]. \tag{30}$$

The magnification of the image is therefore

$$M = \partial x/\partial x_0 = y_c/y_0. \tag{31}$$

The most important property of Fourier holograms is a consequence of the object and the reference sources being in a plane parallel to the hologram, in the recording step. Under this condition, the phase factor $\psi(x; d_2)$ that multiplies the object transmittance in the recording step represented in Eq. (7) cancels out in the image-forming term of the hologram transmittance described by Eq. (12). This results in an important geometrical property: the location of the reconstructed image is invariant under a lateral translation of the hologram, an interesting property for some applications such as holographic movies. This property may also be seen from Eq. (30).

However, in some applications such as correlation filtering, where the position in the input plane of the object to be detected is unknown, the remaining phase factor $\psi^*(\lambda d'\mathbf{u}; d')$ in Eq. (12), which yields a multiplicative phase factor in the reconstructed image, must also be eliminated. Otherwise the correlation obtained will be between the input object, and the recorded object multiplied by the spherical phase factor.

The recording configurations that will assure the removal of this phase factor are seen from Fig. 3 to be those Fourier holograms recorded in such a way that the hologram is in a plane that contains the Fourier transform of the object.

4.3.4.2 Effects of Space Variance

In the previous discussions, no account was taken of the effects caused by the finite extent of the lenses. If there are no lenses between the object and the hologram, other lenses present in the recording step will have no effect, except perhaps by introducing aberrations if they are not of sufficient quality. If there is a lens (or lenses) between the object and the hologram, however, as in the recording of a Fourier–Fraunhofer hologram, the lens may cut out some of the higher frequencies in the edge of the object field, especially if the lens is not large relative to the size of the object. This effect has been discussed extensively by Goodman (1968). Arsenault and Brousseau (1973) have shown that if the lens diameter is at least twice the diameter of the object field, a space-invariant Fourier transform of the object may be obtained, provided that no spatial frequencies greater than $R/2\lambda d$ are present in the object (R is the radius of the lens and d is the distance from the object to the lens). Under these conditions, the maximum two-dimensional space–bandwidth product of the system is equal to

$$\text{SW}_{\text{max}} = (R^2/4\lambda d)^2. \tag{32}$$

Similar considerations apply to the reconstruction of the images if a lens is placed after the hologram.

If the lens placed after the object is a simple lens, this effect can drastically limit the information content of the hologram. But complex lenses designed specifically for Fourier transform work have their focal planes relatively close to the entrance pupil, so space-variance effects are not a predominant limiting effect with such lenses.

4.3.4.3 Effects of the Recording Medium

In the foregoing analysis, we have assumed that the recording medium was able to resolve the whole spectrum of spatial frequencies of interest, except possibly for a cutoff caused by the finite extent of the hologram or the lens used in the recording process. Of course, in any practical situation, the recording medium (a photographic plate for instance) has a finite resolution and the modulation transfer function (MTF) of the recording medium is a useful measure of the range of frequencies over which significant response is obtained. For a Fourier-transform hologram, the effect of limited MTF of the recording medium is not degradation of image resolution, but rather restriction of field of view about the reference point. A general study of such effects has been made by Van Ligten (1966).

To explain the effect of the recording medium on the reconstructed image, we consider for simplicity the Fourier–Fraunhofer configuration and a single point source object $\delta(\xi - \xi_1)$ with coordinates ξ_1. The intensity incident on the recording medium during exposure is given from Eq. (10) as

$$I(\mathbf{x}) = a(\mathbf{x})r^*(\mathbf{x}) + a^*(\mathbf{x})r(\mathbf{x}) + |a(\mathbf{x})|^2 + |r(\mathbf{x})|^2, \tag{33}$$

where

$$a(\mathbf{x}) = (1/i\lambda f)\exp\{-(2\pi i/\lambda f)\mathbf{x}\cdot\xi_1\}, \tag{34}$$

$$r(\mathbf{x}) = (1/i\lambda f)\exp\{-(2\pi i/\lambda f)\mathbf{x}\cdot\xi_0\}. \tag{35}$$

Thus Eq. (33) can be written

$$I(\mathbf{x}) = \frac{2}{\lambda^2 f^2} + \frac{2}{\lambda^2 f^2}\cos\left[\frac{2\pi}{\lambda f}(\xi_0 - \xi_1)\cdot\mathbf{x}\right]. \tag{36}$$

This shows that in Fourier holography, each point of the object generates on the hologram a set of equidistant fringes whose spatial frequency is proportional to the distance between the point and the reference source.

By definition of the film MTF, the effective intensity recorded will be

$$I'(\mathbf{x}) = \frac{2}{\lambda^2 f^2} + \frac{2}{\lambda^2 f^2} M\left(\frac{\xi_0 - \xi_1}{\lambda f}\right)\cos\left[\frac{2\pi}{\lambda f}(\xi_0 - \xi_1)\cdot\mathbf{x}\right] \tag{37}$$

where $M(\mathbf{u})$ is the MTF at spatial frequency $\mathbf{u}$ for the light contributing to the image of the point ξ_0. Since object points farthest from the reference point source generate fringes with the highest spatial frequencies on the hologram, the images of such points suffer the most attenuation. Goodman (1968) has shown that if the frequency response at the film is negligible beyond a maximum frequency $\mathbf{u}_{max}$, only those points ξ_1 of the object with coordinates satisfying the relation

$$|\xi_1 - \xi_0| \le |\mathbf{u}_{max}|\lambda f \tag{38}$$

will appear in the image. The reader may find a more detailed discussion of MTF effects in the book by Smith (1969).

4.3.4.4 Aberrations of Fourier Holograms

Aberrations of holograms are discussed in Section 2.5. From the present section, the reader can see that Fourier holograms in general have fewer aberrations than Fresnel holograms. For instance, spherical aberration may always be eliminated. The Fourier–Fraunhofer hologram, in which the object and the reference source are at infinity in the recording step, is the only practical case where an aberration-free image may be reconstructed from a plane hologram when the reconstruction source is not at the same place as the reference source relative to the hologram. The reason for this is that if a hologram must change the curvature of an incident wave, aberrations are introduced. When the object and the reference source are projected to infinity, all the waves incident on the hologram in the recording step are plane waves. If the hologram is reconstructed with any plane wave, an aberration-free image will result, because a hologram can transform an incident plane wave into another plane wave without introducing aberrations.

We have not considered effects of reconstructing holograms with a wavelength different from the recording wavelength, or the effects of emulsion shrinkage during the development process, which is equivalent to a change of wavelength. For the Fourier–Fraunhofer hologram, no additional aberrations are caused by those effects if a plane wave is used for reconstruction.

Because it is usually holographic aberrations that limit image quality in holography, the Fourier–Fraunhofer hologram holds a privileged status among all hologram types.

4.3.4.5 Information Content of Fourier Holograms

When the maximum storage density of the recording medium is used, Fourier holograms have a much higher information content than Fresnel holograms. Let us assume an object field of length L_0. From the sampling theorem, the Fourier transform of the object is completely determined by its equidistant samples separated by a distance $\lambda f/L_0$ if the object is Fourier transformed with a lens of focal length f. If the Fourier transform has a spatial extent L_f, the number of samples in length L_f is $L_0 L_f/\lambda f$, which is called the space-bandwidth product. For a two-dimensional object, the number of independant samples in the hologram is

$$N = (L_0 L_f/\lambda f)^2. \tag{39}$$

For Fresnel holograms, however, the recorded intensity has an additional multiplicative term of the form $\cos(\pi x^2/\lambda d)$ which is the lens-like term that allows the reconstruction of holographic images without lenses. The number of samples required to record this lens function increases rapidly with the size of the hologram. Typical factors between the storage capacity of Fourier and

Fresnel holograms vary from 4 to 100. In practical cases, other considerations such as the signal-to-noise ratio must be taken into account and usually preclude using the maximum storage density. A detailed discussion of these questions, including a numerical example, may be found in the appendix of a report by Kozma *et al.* (1971).

For practical systems, a storage capacity of 1.5×10^4 bits/mm² has been reported by Bestenreiner *et al.* (1972) for a hologram matrix of full-page texts recorded with slightly defocused Fourier–Fraunhofer holograms having a diameter of 1.6 mm. Hologram matrices are discussed further in Section 10.1.

4.3.5   Conclusion

We have seen from Eq. (17) that any Fourier hologram is formally equivalent to a lensless Fourier-transform hologram. This useful property should not obscure the fact that there are important differences among various types of Fourier holograms.

When the Fourier hologram is to be used as a filter (in applications such as correlation and Wiener filtering), it is usually necessary to use one of the recording configurations where the Fourier transform of the object coincides with the plane of the hologram. Although the Fourier–Fraunhofer hologram is theoretically the best choice because it allows holographic aberrations to be minimized, the degree of aberration correction of the lens used is severe, and if high resolution is required, the cost of the lens may be prohibitive for some applications.

When a diffuser is used behind the object to spread out the light in order to use the area of the hologram more efficiently, it is still necessary to locate the hologram properly, according to the preceding considerations.

Appendix: Properties of the Optical Propagator

The optical diffraction propagator is defined as

$$\psi(\mathbf{x}; d) = \exp\left\{ \frac{i\pi}{\lambda d} \mathbf{x}^2 \right\} . \tag{A1}$$

The usefulness of this operator comes from the fact that propagation of a coherent light wave through a distance d in space can be described by a convolution of the complex amplitude $a(\mathbf{x})$ with $(1/i\lambda d)\psi(\mathbf{x}; d)$. Moreover, passage of a wave through a lens of focal length f can also be described as a multiplication of the complex amplitude incident on the lens by $\psi^*(\mathbf{x}; f)$.

Several important properties of the operator $\psi(\mathbf{x}; d)$ are listed below; their

4. Major Hologram Types

verification can be easily performed from definition (A1):

$$\psi^*(\mathbf{x}; d) = \psi(\mathbf{x}; -d), \tag{A2}$$

$$\psi(-\mathbf{x}; d) = \psi(\mathbf{x}; d), \tag{A3}$$

$$\psi(c\mathbf{x}; d) = \psi(\mathbf{x}; d/c^2), \tag{A4}$$

$$\psi(\mathbf{x}; d_1)\psi(\mathbf{x}; d_2) = \psi[\mathbf{x}; d_1 d_2/(d_1 + d_2)], \tag{A5}$$

$$\psi(\mathbf{x}_1 - \mathbf{x}_2; d) = \psi(\mathbf{x}_1; d)\psi(\mathbf{x}_2; d) \exp\{-(2\pi i/\lambda d)\mathbf{x}_1 \cdot \mathbf{x}_2\}, \tag{A6}$$

$$\psi(\mathbf{x}; \infty) = 1, \tag{A7}$$

$$\mathrm{FT}\{\psi(\mathbf{x}; d)\} = i\lambda d\psi^*(\lambda d\mathbf{u}; d), \tag{A8}$$

$$\psi(\mathbf{x}; d_1) * \psi(\mathbf{x}; d_2) = i\lambda[d_1 d_2/(d_1 + d_2)]\psi(\mathbf{x}; d_1 + d_2), \tag{A9}$$

$$\lim_{d \to 0} (1/i\lambda d)\psi(\mathbf{x}; d) = \delta(\mathbf{x}). \tag{A10}$$

REFERENCES

Arsenault, H. H., and Brousseau, N. (1973). *J. Opt. Soc. Amer.* **63**, 555–558.

Bestenreiner, F., Greis, U., and Weirshausen, W. (1972). *Phot. Sci. Eng.* **16**, 420–431.

Collier, R. J., Burckhardt, C. B., and Lin, L. H. (1971). "Optical Holography." Academic Press, New York.

Goodman, J. W. (1968). "Introduction to Fourier Optics." McGraw-Hill, New York.

Kozma, A., Peters, P., Vander Lugt, A., Lee, W. H., and Rotz, F. (1971). Holographic Storage and Readout techniques. Rome Air Development Center Final Tech. Rep. RADC-75-71-54.

Leith, E., and Upatnieks, J. (1964). *J. Opt. Soc. Amer.* **54**, 1295–1301.

Smith, H. M. (1969). "Principles of Holography." Wiley (Interscience), New York.

Stroke, G. W., Brumm, D., and Funkhouser, A. (1965). *J. Opt. Soc. Amer.* **55**, 1327.

Vander Lugt, A. (1964). *IEEE Trans. Inf. Theor.* **IT-10**, 139–145.

Vander Lugt, A. (1966). *Proc. IEEE* **54**, 1055–1010.

Van Ligten, R. F. (1966). *J. Opt. Soc. Amer.* **65**, 1009–1014.

5

VARIATIONS

5.1 REFLECTION HOLOGRAMS

H. J. Caulfield

5.1.1 Motivations

Why would we want to record a reflection hologram? There are many reasons which may apply. Not all of them are mutually compatible, but all of them give insight into some of the uses of reflection holograms.

First, reflection holograms are easier to use in many applications than transmission holograms. If we had a white-light viewable reflection hologram, we could treat it somewhat like we treat an ordinary photograph. We could nail it to a wall, place it on a book cover, etc. The only restrictions are on illumination angle and viewing angle.

Second, if the hologram is encoded by surface relief, it can be coated with a thin reflective material (aluminum, silver, etc.) and thus be usable at any wavelength. This allows us to record holograms in the visible and use them in the infrared. This is particularly useful for holographic diffraction gratings.

5.1.2 Basic Geometry

We return to the basic analysis of holography (see Chapter 1) in which coherent reference and object beams, R and O, are recorded to form a hologram of transmission $T = |R^2| + |O|^2 + RO^* + R^*O$. For an incident reconstructing beam R, the hologram produces wavefronts $(|R|^2 + |O|^2)R$, R^2O^*, and $|R|^2O$. For the primary image to be reflective, O must be so aligned with respect to the hologram that it could represent (in holographic reconstruction) a reflection from R. Figure 1 shows some cases of proper and improper geometries. It is clear that a geometric relationship is required among R, O, and the photographic plate P. What we have reasoned out from geometric considerations could just as easily have been seen from interference fringe considerations. Figure 2a shows that for two plane waves the interference fringes lie parallel to the bisector of the two beams. We can then form a

HANDBOOK OF OPTICAL HOLOGRAPHY

5. Variation

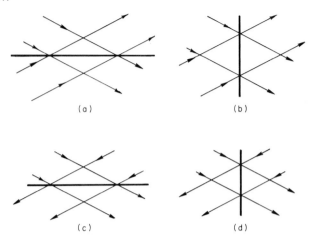

Fig. 1 The same pattern of incident beams can lead to (a) and (d) a reflection hologram or (b) and (c) a transmission hologram. What counts is the mutual alignment among the two beams and the hologram.

hologram with either of the two configurations shown in Figs. 2b and 2c. It is easy to see that the case in which the fringes run more or less parallel to the photographic plate (Fig. 2b) will lead to a reflective hologram. The recorded fringes act like mirrors. The other hologram (Fig. 2c) clearly must work by diffraction. In general, reflective holograms have the property of having fringes more or less parallel to the hologram. The general relationship among R, O, and P is that R and O must come from different sides of P to lead to a reflective hologram.

The fringes lie in exactly the angle a mirror would be oriented to convert R into O. Successive "mirror planes" are separated by

$$d_z = \lambda/2 \cos \psi, \tag{1}$$

where λ is the recording wavelength and ψ the angle of incidence (and, of course, angle of reflection) at the mirror.

5.1.3 Types of Reflection Holograms

We can classify reflection holograms according to whether they are coated or not coated and according to whether they are thick or thin.

Uncoated thin reflection holograms must be very thin indeed. The total thickness must be comparable to or less than d_z. Such a hologram would have a very low diffraction efficiency but would behave in the same way for all reconstructing wavelengths, just like an ordinary mirror.

Uncoated thick reflection holograms have many depth layers, i.e., thickness $T \gg d_z$. These can be further subdivided according to whether the hologram

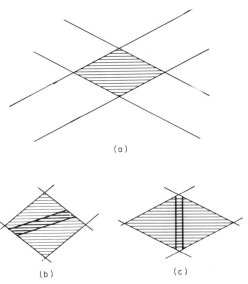

(b)   (c)

Fig. 2   (a) The fringe pattern is fixed by the beam geometry. We obtain a reflection hologram if the fringes are more (b) parallel than (c) perpendicular to the hologram.

is encoded by amplitude or phase modulation. The analysis of the behavior of these holograms is very complicated and is given by Collier *et al*. (1971, Chapter 9). The results are shown below for the case of a two-plane-wave hologram exactly aligned with the fringes. The modulation of the dielectric constant is of the form

$$\epsilon = \epsilon_0 + \epsilon_1 \cos(\mathbf{K} \cdot \mathbf{r}), \tag{2}$$

and the modulation of the conductivity (which controls absorption) is of the form

$$\sigma = \sigma_0 + \sigma_1 \cos(\mathbf{K} \cdot \mathbf{r}), \tag{3}$$

where $\mathbf{K}$ is the "grating vector" for the interference pattern in the hologram and $\mathbf{r}$ is a unit position vector. We shall find it useful to use the quantities

$$\alpha = c\mu_0\sigma_0/2(\epsilon_0)^{1/2}, \tag{4}$$

and

$$\alpha_1 = c\mu_0\sigma_1/2(\epsilon_0)^{1/2}. \tag{5}$$

Here α is the absorption coefficient and α_1 the modulation depth of the absorption coefficient.

For a thick absorption hologram, the amplitude of the reflected object wave

5. Variation

is

$$S(O) = - \left\{ \frac{\xi_{ra}}{\nu_{ra}} + \left[\left(\frac{\xi_{ra}}{\nu_{ra}} \right)^2 - 1 \right]^{1/2} \coth(\xi_{ra}^2 - \nu_{ra})^{1/2} \right\}^{-1}, \tag{6}$$

where

$$\nu_{ra} = \frac{\alpha_1 T}{2 \cos \psi_0}, \tag{7}$$

$$\xi_{ra} = \frac{\alpha T}{\cos \psi_0} + \frac{i \Gamma T}{2 \cos \psi_0}, \tag{8}$$

ψ_0 is the angle between the incident electric field vector $\mathbf{P}$ and the z axis (normal to the hologram) for Bragg incidence, θ_0 the Bragg angle,

$$\theta = \theta_0 + \delta = \text{angle of incidence}, \tag{9}$$

$$\Gamma = \frac{2\pi(\epsilon_0)^{1/2} \delta \sin 2\theta_0}{\lambda_a}, \tag{10}$$

and λ_a is the wavelength in air. The diffraction efficiency is

$$\eta = |S(O)|^2. \tag{11}$$

At Bragg incidence $\delta = 0$, so $\Gamma = 0$ and

$$\xi_{ra}/\nu_{ra} = 2\alpha/\alpha_1 = 2\sigma_0/\sigma_1. \tag{12}$$

The depth of modulation in σ is

$$D(\sigma) = \frac{(\sigma_0 + \sigma_1) - (\sigma_0 - \sigma_1)}{(\sigma_0 + \sigma_1) + (\sigma_0 - \sigma_1)} = \frac{\sigma_1}{\sigma_0}. \tag{13}$$

In Fig. 3 we show a plot of η versus $\alpha_1 T/\cos \psi_0$ for various values of $D(\sigma)$. The peak (asymptotic) occurs at $\eta = 0.072$.

For a thick dielectric ($\alpha = \alpha_1 = 0$), a two-plane-wave hologram produces a primary image of amplitude

$$S(O) = \frac{-i}{(i\xi_r/\nu_r) + [1 - (\xi_r/\nu_r)^2]^{1/2} \coth(\nu_r^2 - \xi_r^2)^{1/2}}, \tag{14}$$

where

$$\xi_r = \Gamma T/(2 \cos \psi_0), \tag{15}$$

and

$$\nu_r = \pi \epsilon_1/(2\lambda_a(\epsilon_0)^{1/2} \cos \psi_0). \tag{16}$$

For Bragg incidence, $\Gamma = 0$, so $\xi_r = 0$ and

$$S(O) = -i/\coth \nu_r. \tag{17}$$

184

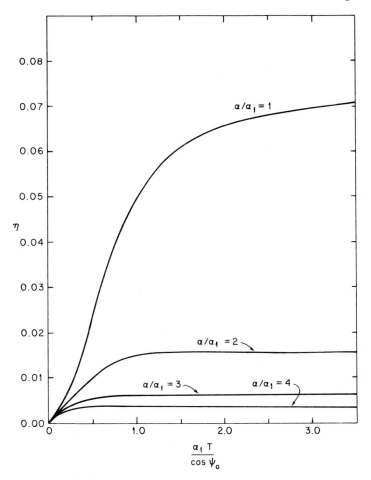

Fig. 3 Diffraction efficiency for thick reflection amplitude holograms.

Of course coth ν_r approaches unity as ν_r approaches infinity. Thus $\eta = |S(O)|^2$ can approach unity. For nonzero ξ_r we obtain less than maximum η. Indeed, η goes to zero for some ξ_r for each ν_r. Calling $\eta(\xi_r = 0) \triangleq \eta_0$, we can plot η versus ξ_r for various values ν_r. Figure 4 shows the result.

Changing from λ_a to $\lambda_a + \Delta\lambda$ ($\Delta\lambda/\lambda_a \ll 1$), the Bragg diffraction equation leads to

$$\delta = (\Delta\lambda \tan \theta_0)/\lambda_a \tag{18}$$

which, in turn, leads to

$$\xi_r = \frac{\Delta\lambda(2\pi)(\epsilon_0)^{1/2}T \sin \theta_0}{\lambda_a^2}. \tag{19}$$

185

5. Variation

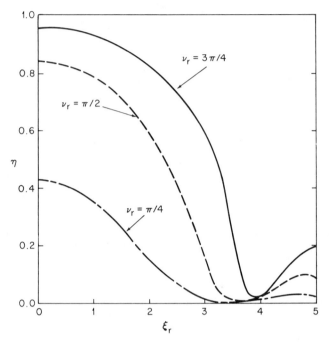

Fig. 4 Diffraction efficiency for thick reflection phase holograms.

Using this and Eq. (14), we can calculate the $\Delta\lambda$ dependence of η. In general the angular sensitivity is very great. Collier *et al*. (1971) calculate that for the reasonable case $\lambda_a = 0.488$ μm, $T = 15$ μm (typical photographic film), $\eta_0 = 1.52$, and $\theta_0 = 80°$; a $\Delta\lambda$ of 0.0059 μm leads to $\eta = 0$. This high wavelength sensitivity leads to the color reflection holograms of Section 5.3.

A thin-coated reflection hologram would have to have the shape of a single interference maximum. It would be a "bumpy" mirror. We know of close approximations in the holograms recorded on thermoplastic materials, on ruticons, etc. (see Section 8.3).

A thick-coated reflection hologram was introduced by Sheridon (1968). The hologram is recorded on thick ($T \gg dz$) photoresist, which (upon development) produces a deep "blazed" hologram. Sheridon (1968) obtained $\eta = 0.73$ for a two-plane-wave aluminum-coated blazed hologram. Kermish (1970) has shown that amplitude variations across the object wavefront lead to phase errors in these blazed holograms.

5.1.4 Deriving the Reference and Object Beams

We want to derive the reference and object beams from the same laser beam, balance their paths, control their power ratio K, and control their

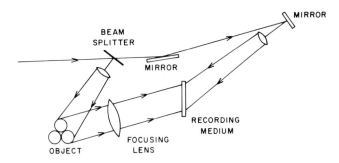

Fig. 5 With an object a long distance from the hologram plane, independent object and reference beams are easy to construct.

directions of incidence onto the recording medium. It turns out that this is easy when the object is far enough away from the recording medium and very difficult otherwise.

5.1.4.1 Distant Scattering Object

We show a suitable illumination geometry in Fig. 5. The lens, making this a focused-image-hologram setup, is not the important point to notice. It is important to notice that the object is far enough away that it can be illuminated from one angle and viewed from another. For head-on illumination, we could use a beamsplitter as shown in Fig. 6. In either case the key requirement is speparation between the object and the recording medium. The rule of thumb is that the object distance from the recording medium be at least equal to the object width.

5.1.4.2 Closer Scattering Objects

The problem with closer objects is now apparent. How do we illuminate them? We present here some imperfect solutions.

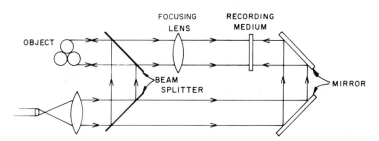

Fig. 6 Even at moderate distances we can achieve separate illumination using a beamsplitter.

5. Variation

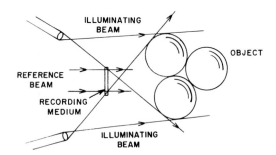

Fig. 7 The object may be illuminated through the recording medium to obtain a very simple reflection hologram. Auxiliary lighting can be added (as shown) for large objects.

The first solution, by Denisyuk (1962), is the simplest and most popular. We illuminate through the recording medium as shown in Fig. 7. The illuminating beam provides the reference. The primary drawback of this method is that it does not provide for independent control of the beam ratio K. In general we have $K \geqslant 1$, but the ratio depends on the reflectivity and distance of the object.

We can supplement the object illumination by (1) using a small part or many small parts of the recording medium very inefficiently (very high K) and (2) illuminating those areas with very bright, diverging light. Figure 8 shows one way of doing this. The hologram will have a central "dead" spot. Of course the spot need not be central. This type of solution is limited by user ingenuity and tolerable engineering complexity.

If we are willing to sacrifice vertical parallax, a single horizontal strip hologram can be used (see Sections 5.5 and 10.9). We can then illuminate the object very conveniently. Figure 9 shows one suitable arrangement. Of course we must then synthesize a vertical array of these strip holograms to obtain an easy-to-view, no-vertical-parallax hologram. The required copying techniques are discussed in Section 9.3.

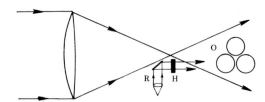

Fig. 8 By sacrificing vertical parallax we can let the illuminating beam bypass the hologram (medium H). R and O indicate the reference- and object-illuminating beams.

188

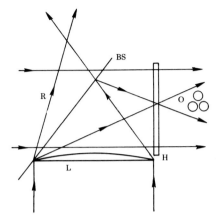

Fig. 9 The reference beam *R* is incident from the rear as in Fig. 5. It provides some object illumination. Supplementary object illumination can pass through a small region of the hologram medium *H*.

5.1.4.3 Nonscattering Objects

Nonscattering objects can be back-illuminated, so the close-object illumination problems do not arise. This case is nontrivial in that it is useful in such important applications as holographic optical elements (Section 10.8) and holographic diffraction gratings.

REFERENCES

Collier, R. J., Burckhardt, C. B., and Lin, L. H. (1971). "Optical Holography." Academic Press, New York.
Denisyuk, Y. U. (1962). *Sov. Phys. Dok.* **7**, 543.
Kermish, D. (1970). *J. Opt. Soc. Amer.* **60**, 782.
Sheridon, N. K. (1968). *Appl. Phys. Lett.* **12**, 316.

5.2 MULTIPLEXED HOLOGRAMS

W. Thomas Cathey

A multiplexed hologram is one where many images are stored simultaneously, parts of one image are stored separately, or a single image is stored many times. We shall treat these in four sections. The first involves partitioning of the hologram and can be called spatial multiplexing. The second is a composite-image hologram made up of a number of points. Next, scanned holograms are treated. They occur when the object illuminating beam is smaller than the object and is scanned across it during the recording of the hologram. Finally, multiple image storage is discussed. In this case, many complete images are recorded on the same hologram.

5.2.1 Spatial Multiplexing

In the storage of data, a single photographic plate or other material may be used to record many holograms, each of which can form images of data. The holograms may be arranged in a checkerboard array with a laser beam scanned across the array to read out each hologram. This is treated in Section 10.1.

Another spatial division of a hologram occurs when the same object wave or different views of the same object are recorded in strips. In the first case, the strip hologram is merely replicated so that the entire hologram can produce the image. The second case occurs in the construction of synthetic holograms for display purposes. For a detailed discussion, see the parts on display (Section 10.3) and synthetic holograms (Section 5.5).

5.2.2 Compound Images

By compound-image holograms, we mean holograms producing images made up of parts which were recorded separately. A common example is the separate recording of the waves from a number of point objects to make up an image comprising point images. This example, being easy to analyze, has provided answers to questions concerning multiple exposures. Figure 1 shows two views of the image of an object made up of 4000 points.

If m object waves are recorded sequentially, the distribution recorded by

HANDBOOK OF OPTICAL HOLOGRAPHY
Copyright © 1979 by Academic Press, Inc.
All rights of reproduction in any form reserved.
ISBN-0-12-165350-1

5. Variation

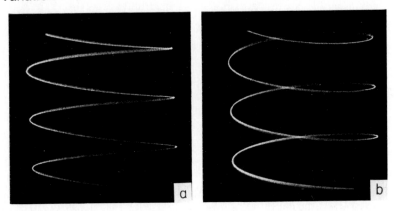

Fig. 1 Two views of the image of a helix made up of 4000 point objects. (From Caulfield *et al.*, 1968)

the hologram is proportional to

$$\sum_{j=1}^{m} \{|a_j|^2 + |a_r|^2 + a_j a_r^* + a_j^* a_r\}, \tag{1}$$

where a_j is the complex amplitude of the jth object wave and a_r is the complex amplitude of the reference wave. It is assumed that the same reference wave is used for all recordings. If all points of the object were recorded simultaneously, the hologram distribution would be proportional to

$$\left|\sum_{j=1}^{m} a_j\right|^2 + |a_r|^2 + a_r \sum_{j=1}^{m} a_j^* + a_r^* \sum_{j=1}^{m} a_j. \tag{2}$$

There are two essential differences between the recordings described by (1) and (2). In the first, the total exposure of the hologram due to the reference wave is related to the number of object points recorded and hence the number of exposures. The bias of the hologram is increased; the visibility of the recorded fringes and efficiency of the hologram are decreased. The diffraction efficiency is higher if phase holograms are constructed. The average background no longer absorbs a portion of the wave; it simply adds a uniform phase shift. The limit on the bias is then the allowable range of total phase shift, and the total shift available must be allocated to each exposure. As the number of exposures increases, the allowed exposure for each image decreases.

An advantage to the recording technique related to (1) is that the on-axis distribution is quite narrow. The on-axis distribution related to the first term of (2) is spread because of the mutual interference of the object waves. Very coarse fringes are produced on the hologram which diffract energy into a region about the axis of illumination.

The most important characteristic of a sequentially exposed hologram is the build up of the exposure or bias of the hologram. A prebias or preexposure of photographic film can be used to increase the sensitivity of the film to a later exposure. Reference to a transmittance versus exposure curve shows that the preexposure causes the hologram to be recorded on a higher part of the curve where there is a greater change in transmittance with exposure. The bias provided by the multiple exposures of the references has a similar effect. Images which cannot be recorded when only one object point is recorded can be formed if more image points are recorded on the same hologram (Caulfield *et al.*, 1968).

The number of exposures has a direct effect upon the optimum procedure for exposing the hologram. If N object points are used to simultaneously expose a hologram, the efficiency of the hologram in directing energy to any one image point is reduced by N when compared to the efficiency if only one image point is recorded. If the object points sequentially expose the hologram, the efficiency is reduced by N^2 (Smith, 1969; LaMacchia and Vincelette, 1968).

The order of exposure also has a dramatic effect. If all sequential recordings expose the hologram equally, the image of the first one recorded will be much brighter. Nishida and Sakaguchi (1971) have published results which show that the relative exposure of later recordings must increase if the energies in the image points are to be equal.

5.2.3 Scanned Holograms

A scanned hologram is one that has a scanning beam illuminating the object or a scanning reference beam illuminating the hologram. We shall discuss separately the cases of scanning object and scanning reference beams. The techniques for doing each are given and the advantages and limitations are presented.

5.2.3.1 Scanning Object Beam

The object illumination beam is sometimes reduced in diameter so that it can no longer illuminate the whole object at once, but must be scanned across the object. The result is a multiple-exposure hologram in which an image of each illuminated spot of the object is stored separately. The conclusions concerning number of exposures, background bias on the hologram, and the sequence of exposures have already been discussed. What remains to be pointed out is why the scanning approach may be desirable when the object is a physical object, not one to be synthesized using point sources.

If the object is large, we may wish to narrow and scan the object illuminating

5. Variation

beam so that a brighter object wave will fall on the hologram. This will reduce the exposure time needed to record a hologram *of that portion of the object*. The total exposure may not be reduced. The reduction for each portion, however, will allow a hologram to be recorded of an object which is moving slightly or when there is motion somewhere else in the system. For example, assume that a 1 sec or shorter exposure is required to record a hologram because of motions in the system somewhere. If the power meter indicates that a 20-sec exposure would be needed to make a regular hologram, 20 or more sections of the object could be illuminated sequentially. The movement during each exposure would be tolerable. The only restriction on movement between exposures is that similar to photography.

A second advantage is that the width of the on-axis distribution due to the object is narrowed. This is because of the same reasons given in the discussion following Eqs. (1) and (2). The result is a reduction in what is called flare light, coming from the on-axis distribution. If the recording process is nonlinear (which is frequently the case), light related to the on-axis distribution appears in the region of the image. (Cathey, 1974). Reduction of the width of the on-axis distribution, by scanning the object illumination, reduces the flare light appearing about the image (DeBitetto and Dalisa, 1971).

The disadvantage of using an object scanning system is that in addition to requiring a more complex system the diffraction efficiency of the hologram is reduced. The reduction is caused by the increased background exposure which occurs when multiple exposures are used.

5.2.3.2 Scanning Reference Beam

In the scanning reference beam approach, the entire object is illuminated, but the reference beam is scanned across the hologram. The total amount of light falling on that portion of the hologram can therefore be increased and the exposure time of a portion of the hologram reduced. The result is that an object having some motion can be used (Palais, 1970). The diffraction efficiency is reduced because of the increase in the reference to object beam energy ratio. The width of the on-axis beam is the same as with a normal hologram.

5.2.4 Separate Image Storage

We are now concerned with the storage of many images on a single hologram. The difference is that these images are not all retrieved at the same time. For example, we may wish to store many pages of data on a single hologram or many views of the same object. The angle of the reference beam is changed between each exposure so that a different fringe pattern is recorded for each image. When the hologram illuminating wave comes from the same direction as one of the reference waves, the corresponding object wave is reconstructed, producing the desired image. The limits on the number and field of view of the images depend upon the thickness of the recording medium and the type modulation used. More images can be stored if phase modulation holograms are used.

5.2.4.1   Thin Recording Media

If the recording medium is thin, the restrictions discussed in the previous sections hold. The thin hologram has no way of distinguishing between two images that were intended to be combined into one and two images which are to be considered separately. All images are formed simultaneously. This is not true if the recording medium is thick.

5.2.4.2   Thick Media

When depth is added to the recording medium, the interference fringes are recorded as shown in Fig. 2. When the hologram illuminating wave is introduced at the Bragg angle (where the waves reflected from each reflecting layer

Fig. 2   Cross section of a hologram recorded on Kodak 649F showing the recording of the fringes in depth. (From Akagi *et al.*, 1972)

add in phase), the brightness of the image is maximized. As the thickness of the recorder increases, the allowable deviation from the Bragg angle in illuminating the hologram decreases. For example, if Kodak 649F plates with an emulsion thickness of 15 μm are used, the brightness of the image is reduced by 10 dB if the angle at which the hologram-illuminating wave is introduced deviates from the Bragg angle by $\pm 5°$. If the recording medium is 1500 μm thick, the image brightness is reduced by more than 10 dB if the deviation is ± 2 min of arc.

The limited range of illumination angles for which an image can be obtained means that many images can be recorded using different angles between the object and reference and that they can be read out one at a time. If the Bragg angle requirement is satisfied for one image, it is not for the others.

The maximum diffraction efficiency for a hologram which modulates the readout wave by absorption is low if only one hologram is recorded. When many are recorded on the same recorder, the efficiency rapidly becomes so low that absorptive materials are not practical for making multiplex holograms.

If the wave from the object is a plane wave, the diffraction efficiency of a thick phase modulation hologram can, in theory, be 100% (Kogelnik, 1969; Collier *et al.*, 1971; Cathey, 1974). In practice, the wave from the object is not a plane wave and it varies in amplitude across the hologram, meaning that the optimum reference to object wave energy ratio can not be obtained everywhere. In addition, if multiple images are recorded, the efficiency is reduced still further. It is not reduced by accumulating background, as in the case of absorption holograms, but by other effects. If the multiple exposures are made so that the images always appear along the same axis (the objects are always placed in the same location; only the angle of the reference wave changes), the reconstructed-image wave interacts with all the recorded gratings, which direct light in the directions from which the reference waves came. In this manner, energy is taken from the image wave. This effect can be eliminated by introducing each pair of object and reference waves at completely different angles. A given image wave can no longer interact with the recorded gratings, thereby losing energy. The disadvantage is that the images either all appear in different locations or the hologram must be rotated as well as the readout wave redirected (Collier *et al.*, 1971).

5.2.4.3 Coded Reference Waves

The reference waves for each object wave can all be coded differently. That is, the phase of each one can be made different by, for example, directing each through different parts of a piece of ground glass. The coupling between holograms for each image is consequently very small. This procedure has been used by LaMacchia and White (1968) to obtain a signal-to-noise ratio of more than 20 dB for 1000 superimposed exposures. The objects for each exposure were point sources, however. Fewer superimposed exposures could be made using other object distributions. Krile *et al.* (1979) have shown that a chirp-modulated binary phase coding can produce improved correlation properties.

5.2.4.4 Uses of Multiple-Image Holograms

These are simply listed here because other chapters treat the applications. One obvious use is data storage. Many pages of data could be read out from one hologram. If the multiple images are sequential views of a moving object, rapid switching from one image to another results in a holographic movie. Finally, if two images are formed in the same space, one before a slight strain and the other after, holographic interferometry can be used to measure the strain.

REFERENCES

Akagi, M., Kaneko, T., and Ishiba, T. (1972). *Appl. Phys. Lett.* **21**, 93–95.

Cathey, W. T. (1974). "Optical Information Processing and Holography," Section 6-3. Wiley (Interscience), New York.

Caulfield, H. J., Lu, S., and Harris, J. L. (1968). *J. Opt. Soc. Amer.* **58**, 1003–1004.

Collier, R. J., Burckhardt, C. B., and Lin, L. H. (1971). "Optical Holography," Chapter 9. Academic Press, New York.

DeBitto, D. J., and Dalisa, A. L. (1971). *Appl. Opt.* **10**, 2292–2296.

Kogelnik, H. (1969). *Bell Syst. Tech. J.* **48**, 2909–2947.

Krile, T. F., Hagler, M. O., Redus, W. D., and Walkup, J. F. (1979). *Appl. Opt.* **18**, 52–56.

LaMacchia, J. T., and Vincelette, C. J. (1968). *Appl. Opt.* **7**, 1857–1858.

Nishida, N., and Sakaguchi, M. (1971). *Appl. Opt.* **10**, 439.

Palais, J. C. (1970). *Appl. Opt.* **9**, 709–711.

Smith, J. M. (1969). "Principles of Holography," pp. 217–220. Wiley (Interscience), New York.

5.3 COLOR HOLOGRAMS

W. Thomas Cathey

The term color hologram refers to a hologram that produces color images. Color holograms are simply multiplex holograms that have overlapping images, each of a different color. Consequently, the discussion in this section leans heavily on Section 5.2. As with other forms of multiplexed holograms, different problems arise when thin or surface holograms are used and when the recording material has appreciable depth. Those recorded with a thin medium produce multiple images, which corresponds to multiple diffraction orders. There are many approaches to eliminating the unwanted ones. Holograms recorded with a thick medium may not respond to illumination with the original wavelengths because of a shrinkage or swelling of the emulsion. If black and white images, as opposed to red and black, for example, are included in the discussion, the problem of dispersion must be included. An image plane hologram has fewer of these problems because of the shorter distance between the hologram and the image. Much of the earlier work with color holograms is nicely summarized by Collier *et al*. (1971).

5.3.1 White-Light Hologram

A hologram is a coded diffraction grating. Consequently, when a hologram is illuminated with white light, the waves are diffracted with the longer wavelengths being diffracted further than the shorter wavelengths from the axis of the hologram-illuminating wave. The result is a smeared image. This can be compensated somewhat by using a diffraction grating having a line spacing equal to the average spacing of the fringes on the hologram. The hologram produces dispersion, but the grating intercepts the first order and causes dispersion in the other direction (see Fig. 1). The on-axis light from the hologram is avoided by either sufficient separation of the hologram and grating (DeBitetto, 1966) or by insertion of a blocking structure similar to a Venetian blind (Burkhardt, 1966).

The preceding discussion applies to thin holograms. Volume holograms have wavelength selectivity and will reflect or transmit only a narrow band of wavelengths due to the Bragg effect.

HANDBOOK OF OPTICAL HOLOGRAPHY   199

5. Variation

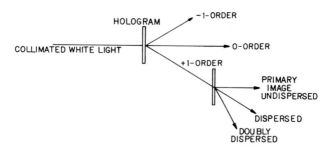

Fig. 1 The dispersion produced by the hologram is compensated for by the grating. (DeBitetto, 1966.)

5.3.2 Thin Holograms

In making a color hologram with a thin recording material, multiple exposures are made for each color. So few exposures are made that the reduction in efficiency is not a problem, but the effects described in Section 5.2 still apply. The greatest difficulty comes from crosstalk.

Crosstalk occurs when, for example, the hologram is recorded using two wavelengths. Because the wavelengths are different, the fringe spacing of the two holograms is different, even though the angles between object and reference waves are the same for both wavelengths; that is, the fringe spacing is $(1/\lambda) \sin \theta$, where θ is the angle between the waves. When the hologram is illuminated with two waves having different wavelengths, each of them see two holograms. The first wave interacts with the two holograms, producing one image in the proper location and another displaced because of the different fringe spacings. Similarly, the second wave produces one image in the desired location, which overlaps the image of the first wave, and a second one displaced. Figure 2 illustrates the problem where, for example, the waves are labeled red and blue. The images labeled R, R and B, B are the desired red and blue images which overlap giving a two color image. There is a similar set of conjugate images on the other side of the on-axis distribution. It is obvious that a three-color hologram would have nine images in the region where Fig. 2 shows four.

5.3.2.1 Spatial Filtering

The waves that emerge from the hologram of Fig. 2 can be collected by a lens and, in the back focal plane, an aperture used to remove the unwanted images leaves only B, B and R, R to produce a two-color image. This technique works best with a two-dimensional object or a three-dimensional object of limited depth. The resolution or spatial frequency spectrum of the image is limited by the aperture. This is a drawback for data storage, but the resolution

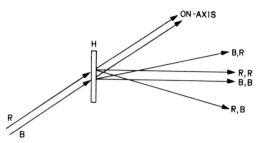

Fig. 2 Effect of illuminating a thin, two-color hologram with two wavelengths. *H*—hologram; *R*—red illuminating wave; *B*—blue illuminating wave; *B, R*—blue image produced when the blue light illuminates the hologram originally recorded with red light; *R, R*—red image produced when red light illuminates the proper hologram; *B, B*—blue image produced when blue light illuminates the proper hologram; and *R, B*—red image produced when red light illuminates the hologram originally recorded with blue light.

of about 400 lines/mm is more than adequate for image formation (Ih, 1975; Collier *et al.*, 1971).

5.3.2.2 Spatial Multiplexing

The separate holograms can be recorded on different portions of the recording material. This can be accomplished, for example, by placing a mosaic of color filters over the hologram. Only those regions behind the blue filter would record the hologram of the object as seen in blue light. Only those behind the red filter would record the hologram of the object as seen in red light, etc. The result is as if a hologram for blue light were recorded with certain regions covered and then a hologram for red light recorded in the previously unexposed regions. The hologram must be illuminated through a mosaic of color filters corresponding to the appropriate regions of the composite hologram. The result is a high-quality color image without the extra images discussed previously. The array of holograms presents no difficulty if the real image is to be projected on a screen. For comfortable viewing of the virtual image, however, the separate holograms must be small. Otherwise, the viewer is distracted when viewing an image through the composite hologram.

5.3.2.3 Coded Reference Waves

Coded reference waves can be used to eliminate the unwanted images of thin color holograms. One method of coding the reference waves is to send the light, which contains the wavelengths being used, through a diffuser. Even though all the reference waves go through the same diffuser, the amplitude

5. Variation

and phase structure of each differs at the hologram because of their different wavelengths. The distribution of amplitude and phase of each reference is approximately random and differs from the others. When the diffuser is kept in place and the developed hologram replaced in its original position, each hologram is illuminated by the appropriate wave for the color image to be formed. The result is that each of the color images are superimposed. In addition, however, each illumination wave illuminates the holograms formed by light of other wavelengths. As in the case of an extended reference hologram, the resulting wave is as if the image were viewed through a diffuser having phase shifting properties identical to the difference in phase between the reference and hologram-illuminating wave. The extra, unwanted, images are therefore smeared, giving background noise. This is sometimes objectionable. A more serious problem is that the relative positions of diffuser, hologram, and light source during the hologram recording must be maintained to a high degree of accuracy during hologram illumination.

5.3.3   Volume Holograms

A thick or volume hologram can be made to serve as a filter as well as a hologram. In Section 5.2, it is shown that a hologram recorded in a thick medium produces surfaces within the recorder rather than simple fringes. The optimum angle for illuminating volume holograms is the same angle at which the reference wave was introduced. In a volume hologram this is the same as the Bragg angle if the hologram material does not change shape or shrink between the times of recording and using the hologram and the same wavelength is used. The diffraction efficiency of a hologram not only falls off when the angle of the illuminating wave deviates from the angle used in recording, but also when the wavelength changes. That is, the Bragg angle is determined by wavelength as well as geometry. Change the wavelength, and the angle at which all reflected waves add in phase changes. This effect eliminates the extra images found in surface color holograms. A hologram will produce an image with high efficiency only when it is illuminated at the proper angle and with the wavelength used in recording. See Section 5.1 for details on thick reflection holograms.

Volume color holograms are used in two different modes, transmission or reflection, depending upon whether the hologram is to be illuminated with the laser light used to make it or with white light. The reason for the difference is that allowable deviation of the wavelength becomes very small for large Bragg angles. See Fig. 3, which shows the illuminating configurations for transmission and reflection holograms. The Bragg angle θ_B is related to the normal spacing of the reflecting surfaces d, the mean index of refraction n_0, and the illuminating wavelength λ. The reflecting surfaces lie along the bisector of the angle between the object and reference waves during recording. Consequently,

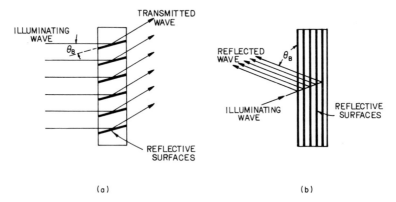

Fig. 3   (a) Small Bragg angle θ_B for a transmission hologram. (b) Large Bragg angle θ_B for a reflection hologram.

to get a reflection hologram, the reference wave must be introduced from the back of the recorder. From the geometry of Fig. 3, we can see that for all reflections to be in phase the tolerances on the spacing of the reflecting surfaces, as measured in wavelengths, must be more stringent when the light is reflected almost directly back from the surfaces. This is either a blessing or a curse, depending upon your intentions.

If the plan is to illuminate the hologram with the same waves as used for a reference, the transmission hologram is easier to use. One reason is that alignment is not as critical. Recall that if the angle of illumination is not correct, low efficiency results. Another reason is that frequently the recording material shrinks during processing. The value of d then changes dramatically in the reflection hologram but is not affected severely in the transmission hologram because it is difficult for the material to shrink along the direction of the substrate, which is usually glass. Shrinkage makes it difficult to change the angle of illumination so that the Bragg condition is satisfied simultaneously for all wavelengths.

If the hologram is to be illuminated with white light, a high wavelength discrimination is desirable. The reflection hologram strongly reflects only those wavelengths with which it was made and waves of other colors are absorbed or transmitted. At least that is the case if the illumination angle is the same as the angle of the reference wave used in recording and if the material did not shrink. If the material shrank, a different wavelength of light will add in phase in the direction of the image and the color will shift toward the blue. If the angle of illumination and viewing differs from the Bragg angle for the recording wavelength, the new angle will be a Bragg angle for a different wavelength, causing a shift in the color of the image. To see the proper colors, shrinkage

5. Variation

of the recorder must be prevented or the emulsion must be reswelled (Nishida, 1970; Section 9.1 of this volume) and the viewing angle must be controlled.

Of course, if only one wavelength were used in the recording, the color of the image would not matter.

REFERENCES

Burckhardt, C. B. (1966). *Bell Syst. Tech. J.* **45,** 1841–1844.
Collier, R. J., Burckhardt, C. B., and Lin, L. H. (1971). "Optical Holography," Chapter 17. Academic Press, New York.
DeBitetto, D. J. (1966). *Appl. Phys. Lett.* **9,** 417–418.
Ih, C. S. (1975). *Appl. Opt.* **14,** 438–444.
Nishida, N. (1970). *Appl. Opt.* **14,** 238–240.

5.4 POLARIZATION HOLOGRAMS

W. Thomas Cathey

The name "polarization hologram" is used when particular attention is given to the polarization of the object or image wave. In a sense, a *polarization* hologram always is recorded because only components of the object waves having the same polarization as the reference wave are recorded on the hologram. We first review why the object wave has particular polarization characteristics and then note the effect on the hologram recording. A first step toward a more realistic image is to record a view of the object as would be seen by a person. For example, specular reflections could be lost if the polarization of the reflection were orthogonal to that of the reference wave. Finally, we consider ways to preserve, in the image, the polarization of the object wave. If this is done, the image can be viewed through a polarizer and polarization studies, such as are done in photoelasticity, can be made using the holographic image. Section 2.3 provides a vital background for this discussion.

5.4.1 Polarization of the Object Wave

Frequently, the object being used in holography is a diffuse reflector. This means that a light ray, before being reflected to the viewer, is reflected several times off the microscopic structure of the object. One result is the depolarization of the object wave even if the illumination was polarized. Alternatively, some surfaces, because the coefficients of reflection vary with polarization, polarize a wave reflected off them. One such surface is water. Consequently, objects can either polarize or depolarize an illuminating wave and the effect may be important. Other objects change the polarization of waves transmitted through them. If these are to be studied holographically, the polarization must be preserved.

5.4.2 Effect of Reference Wave Polarization on the Holographic Recording

The first thing to remember is that for interference and, consequently, holographic recording to take place, the electric vectors of the two waves must be aligned. Cross polarized waves do not interfere. If, for example, one wave

5. Variation

is randomly polarized, as will occur after reflection from a diffuse object, only that component of the electric vector which is parallel to the electric vector of the reference wave will be recorded with a hologram. The other portion of the object wave will simply increase the background exposure of the hologram. This should be taken into account when measuring the ratio of reference to object wave energy for optimum performance. If the two waves are polarized but do not have the same polarization, only the components in the same direction will interfere.

The experimenter must not devote all his attention to the direction of the electric vector in a plane perpendicular to the axis of propagation. Refer to Fig. 1 which shows the directions of propagation of object and reference waves. If the electric vectors of the two waves are perpendicular to the paper, they are aligned and maximum interference can take place no matter what the value of θ. If, however, the electric vectors are as shown, the allowed interference is reduced by $\cos \theta$. If θ is 90°, the two electric vectors are orthogonal and no interference takes place. Hence, even if the two waves are linearly polarized, the interference terms in the hologram equation can be reduced by the cosine of the angle between the direction of propagation of the waves.

Another practical problem concerning polarization is that beamsplitters may not have the same splitting ratio for different polarizations. This should be taken into account when planning experiments.

5.4.3 Recording the Total Object Wave

We have seen that if we wish to record a selected linear polarization, for example, we need only provide a reference wave with the desired polarization. Now let us consider the problem of recording the total wave. If the only requirement is that we form an image which would be viewed by the unaided eye as being the same as the object, we need only provide a reference wave having two orthogonal polarizations. This can be done sequentially by making

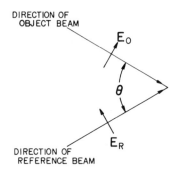

Fig. 1 Effect of polarization and the angle between object and reference waves.

206

two exposures, one with the reference of one polarization and one with the polarization of the reference rotated. A hologram having all polarization information can also be formed by using a laser operating in two orthogonally polarized modes. The fraction of power in each mode may vary with time, however, depending upon the design of the laser. This and the beamsplitter ratio for each polarization must be taken into account.

The resulting hologram will produce an image which has the same energy distribution as if the object were viewed directly. Two holograms are recorded, one for each orthogonal polarization, so that features such as specular reflections or glints off cut surfaces will show up. The image, however, will have only the polarization of the hologram-illuminating wave. The energy distribution is the same as if the polarization information were there, but we can tell nothing about the polarization of the object wave.

5.4.4   Preservation of Polarization throughout the Holographic Process

It is possible to use multiplexed holograms (see Section 5.2) and preserve the polarization of the object wave. Two holograms are recorded, one with each polarization reference wave. For example, the reference waves may be introduced at different angles, so that two hologram illumination waves can be used, one for each polarization. The process is similar to that of making and using a two-color hologram (see Section 5.3). Each illuminating wave interacts with two holograms so that a total of four images are produced (Lohmann, 1965). Two of the images overlap to produce an image having the correct polarization. The advantage of such an image is that the full properties of the object wave are preserved. For example, images of objects which exhibit polarization effects can be studied through a polarizer after the object has been removed (Bryngdahl, 1967; Fourney et al., 1968; Kubo et al., 1975, 1976).

In one important respect, full polarization holography is much more difficult than color holography. The relative phase of the reference waves must be the same as the relative phase of the two hologram-illuminating waves. If the phase is not maintained, erroneous results are obtained. For example, a phase shift of only one-quarter wavelength would change circular polarization into linear polarization. Even less changes it from circular to elliptical. Maintaining such stringent conditions are difficult.

Two particularly useful means of maintaining the relative phase have been used. One, suggested by Kurtz (1969), is to use coded reference waves. If a diffuser is placed in the reference beam, the two orthogonal polarizations are affected differently. Two unique reference waves are produced. Two holograms, one from each polarization, are subsequently recorded. When these holograms are illuminated through the same diffuser, they are illuminated by

5. Variation

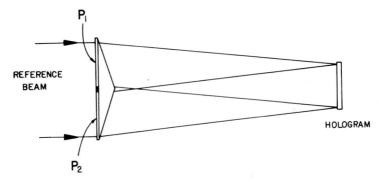

Fig. 2 A reference beam, polarized at 45°, is used to make two reference beams, polarized at 0 and 90°. P_1, 0° polarizer; P_2, 90° polarizer.

both polarizations. The interaction with only one hologram for each polarization produces one image; the "wrong" hologram produces only background scatter. If removed for development, the hologram must be repositioned accurately for the phase of the illuminating waves to be the same as the phase of the reference waves.

An alternative approach is to use a single reference wave, polarized at 45°, which is then split into two by use of a biprism as shown in Fig. 2 (Gåsvik, 1975). Polarizers at 0 and 90° are on the back of the biprism. Two reference waves result which transverse almost the same path so that the relative phase can be maintained when the same waves are used to illuminate the hologram. One disadvantage of this approach is that the extra images lie close to the desired one.

REFERENCES

Bryngdahl, O. (1967). *J. Opt. Soc. Amer.* **57**, 545–546.
Fourney, M. E., Waggoner, A. P., and Mate, K. V. (1968). *J. Opt. Soc. Amer.* **58**, 701–702.
Gåsvik, K. (1975). *Optica Acta* **22**, 189–206.
Kubo, H., and Nagata, R. (1976). *Jap. J. Appl. Phys.* **15**, 1095–1099.
Kubo, H., Iwata, K., and Nagata, R. (1975). *Optica Acta* **22**, 59–70.
Kurtz, C. N. (1969). *Appl. Phys. Lett.* **14**, 59–61.
Lohmann, A. W. (1965). *Appl. Opt.* **4**, 1667–1668.

5.5 SYNTHETIC HOLOGRAMS

H. J. Caulfield

5.5.1 Introduction

It is often desirable to display holographically an image of an object which does not exist or which cannot be recorded by normal holographic methods. Thus we might want to display a three-dimensional model of a molecule without ever having to build such a model. We might want to display "slices" through an object (such as might be produced by ultrasonic B scans) in their proper three-dimensional relationships. We might wish to record a small hologram of a large object without having the image far away from the hologram. For these and many other purposes a variety of techniques for creating "synthetic" images have been devised. We describe some of these techniques here. Conspicuously absent will be a discussion on computer-generated holograms. Discussion of that well-developed field would carry us out of the domain of fully optical holography. Computer-generated holograms are discussed by Collier *et al.* (1971, Chapter 19). Many of the basic ideas of synthetic images are discussed in Chapter 18 of the same book. Wherever references are omitted, interested readers should turn to that chapter to find further discussion, illustration, and references.

5.5.2 Point-by-Point Image Synthesis

If the scene to be displayed comprises points at well-defined xyz coordinates, e.g., the function $z = f(x,y)$, then it is practical to consider storing that image point-by-point by multiple exposure holography.

A typical recording geometry is shown in Fig. 1. The collimated light path matched to the reference beam for good coherence is incident on an xyz translatable mask containing a lens. The lens forms a point image which serves as the object source for the hologram. The position of the xyz object beam source tracks the position of the mask perfectly so long as the lens does not go out of the illuminating beam. It may be desirable to attach a holder to the mask which maintains a weak diffuser at the focal point. This is simply a means to increase the divergence of the light from the object point. For most purposes, using a lens with the proper divergence to start with is a better

5. Variation

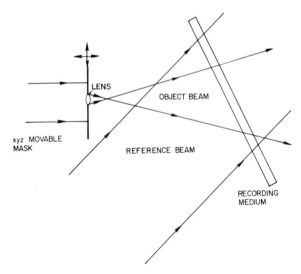

Fig. 1 Recording a hologram of the orbit of a real point image.

approach. Low-divergence point sources are hard to localize visually. Of course many alternative configurations are conceivable. The xy translation can be eliminated and replaced by a two-axis beam stearer, followed by the lens. The angles the incident beam makes with the axis of the lens determine the xy location of the object source. The z translation must still be a real translation. However, it is easy to "paint" the image one z depth at a time unless the data are being recorded holographically as it is taken, and thus the three values are unpredictable.

The primary problem with point-by-point synthetic imaging is that the signal-to-noise ratio associated with any point in the image is about N^{-2}, that which would be associated with an $N = 1$ hologram (see Section 5.2). To our knowledge the largest number of points ever used in point-by-point synthetic image holography was 4000, by Caulfield $et\ al.$ (1968).

To maximize the signal-to-noise ratio we seek equal intensity reference and object beams ($K = 1$) and try to make certain that the reference beam is never on when no object beam is on. To accomplish this, Caulfield $et\ al.$ (1968) chopped the laser beam prior to the beam splitting and moved the xyz position continuously. The possibility of recording holograms of continuously moving point sources had been shown previously by Lu $et\ al.$ (1967), but this low-duty-cycle continuous movement was effective in increasing the length of the source orbit which could be recorded with a good signal-to-noise ratio.

By placing the synthetic object near the focal plane, we can reduce the number of holograms which overlap and thus record even longer orbits.

210

5.5.3   Fixed-Alphabet Synthetic Images

A simple variation on point-by-point image synthesis is fixed-alphabet image synthesis. Here the final image is composed of predetermined shapes at variable locations, extents, and magnifications rather than of points. Molecular models are a good example. Note that such models do not look like physical models because images of more distant objects are not blocked by closer ones. In this sense they convey more information than typical ball-and-stick models.

5.5.4   Multiple Photographs

5.5.4.1   Varying the Viewing Direction

The next step in the evolution of synthetic images is to use photographs of objects rather than the real objects in image synthesis. There are two primary motivations for this. First, photographs are easier to take than holograms. Photography, in most cases, is passive. The equipment is superb and, relative to holographic equipment, inexpensive. Second, images can be magnified or demagnified. Thus a 10×10 cm hologram could contain a near-hologram image (about 10×10 cm) of a fly (requiring magnification) or a man (requiring demagnification).

The problem is, How do we make the image look three-dimensional using ordinary two-dimensional photographs? The solution is simple in concept: We cause each eye to see appropriately different views of the object. This concept predates holography and goes back to stereophotography and its ourgrowth—integral photography. Indeed, it is possible to use holography to record and replicate stereophotographs (see Section 10.14) and integral photographs, as we shall see.

We must provide some background on stereo- and integral photography to set the stage for what follows. Suppose we view a three-dimensional object from some location. We obtain three primary kinds of depth cues. First, the image formed on the retina by each eye is different. These perspective changes allow us to make very accurate depth judgments. This is why eyes capable of such judgment have been favored in the evolution of both hunting and hunted animals. Second, the focusing of the lens required to obtain a sharp image of the object provides good depth information. One-eyed people still see the world in depth. Third, learned relationships of image size, object size, shadowing, foreshortening, etc., give useful depth information. Artists know and exploit these cues to give their paintings realism or its opposite (e.g., Escher). If we could cause our eyes to have all of these cues, then our brain would surely cause us to see a three-dimensional object whether or not that object exists. Stereophotography is the most straightforward implementation of this concept. Two cameras are used to produce two images. The images are presented to the viewer in such a way that each eye sees one and only one image

5. Variation

when he looks toward the "object." The brains of a great majority of two-eyed people dutifully "fuse" these disparate images into a single three-dimensional image. Two problems arise. First, finding a "trick" to cause the left and right eyes to see different views is sometimes difficult or clumsy (red and green spectacles, etc.). Second, the perceived image resembles the actual object only to the extent that the eyes-image geometry duplicates angularly the cameras-object geometry. For instance, consider a stereoimage of a tall tree recorded from directly above it. As we move in such a way that we view the stereoimage from the side, the treetop will follow us; that is, it will always appear to point toward our eyes (cameras). Thus stereophotography can give very distorted images. To avoid this it is possible to use many more than two images to give a proper image from any viewing angle. The classical way to do this is called "integral photography" because the camera, film, and projector are all integrated into the same structure. Multiple lenses are built into a unit with a photographic film. The lenses are a focal length away from the film. A photograph taken by exposing such an integral camera to light (Fig. 2a) is really many photographs—one from each direction. Illuminating the positively developed integral photograph diffusely (e.g., as in Fig. 2b) produces the proper view for each eye regardless of viewer position. The viewer then sees a three-dimensional image which can be viewed from many perspectives. The image so formed is pseudoscopic.

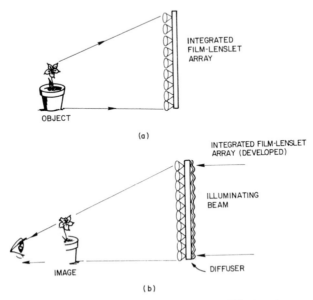

Fig. 2 (a) Recording an integral photograph and (b) using it to produce a real image.

212

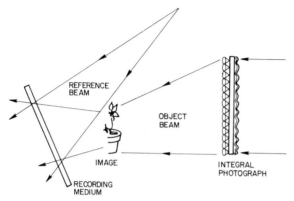

Fig. 3 Recording a hologram using the laser-produced image beam from an integral photograph as an object beam.

Holography can be used to record the image produced by such an integral photograph as shown in Fig. 3. Of course by proper illumination of the hologram we can cause the viewer to see an orthoscopic image. The advantage holography offers here is that of simple viewing and simple copying.

With only one more step we can proceed to a very powerful form of holographic image synthesis. We take photographs from many directions around the object. A popular way of doing this is to revolve the object on a turntable and record it with a motion picture camera. If the period of revolution is T and the time between frames is t, each image represents an average $360(t/T)$-degree view of the object. Since automobile showroom turntables are very large and powerful, living people and large objects can be used. The image can be synthesized as shown in Fig. 4. Each vertical "stripe" hologram contains information about one and only one frame of the motion picture. The developed hologram is then assembled into a cylinder with the reference source on its axis. That hologram is illuminated from above by an on-axis point source of light. A viewer walking around the hologram sees the three-dimensional image from all sides as does a viewer who watches the turning hologram. This type of hologram evolved from work by King (1970) and Jeong *et al.* (1966). It reached fruition in unpublished work by Lloyd Cross. See Section 10.3.

Many variations of this technique have been demonstrated. The object can undergo movements not too fast with respect to t as it is being recorded. The perceived image then emulates that motion as the viewer or the hologram moves. The illumination can be white light and the source can be extended (especially along the axis) without much image degradation. As yet no adequate camera positioning method has been devised to allow recording of nonrotating outdoor objects routinely. Without smooth variation from frame to frame, the image will appear to rotate in a jerky way.

5. Variation

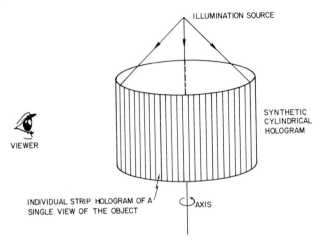

Fig. 4 A cylindrical synthetic-image hologram which allows a human viewer to synthesize a three-dimensional image. No three-dimensional recording of the object is necessary.

The required size of the individual stripe hologram has been worked out by Collier *et al.* (1971) by balancing the desire for large aperture (for resolution) and the desire for small aperture (for large T/t). Their geometry is shown in Fig. 5. Their conclusion is that the optimum choice is

$$D_h = D_e d_h / d_e,$$

where D_h is the width of the elemental hologram, d_h the diameter of the cylinder, D_e the diameter of the eye, and d_e the viewing distance of the eye. Cross normally uses $T/t = 180$. These are $360/180 = 1/3°$ wide. We can use $D_e = 0.3$ cm as an average. Using a comfortable value of $d_h = 20$ cm and $D_h = \pi d_h/180 = 0.058$ cm, we have a reasonable optimum viewing distance of $d_e = 1$ m. The stereo effect vanishes when $D_h d_e / d_h$ equals or exceeds S_e, the eye separation. For $D_h = \pi d_\pi/180$, and $S_e = 6$ cm, we have to restrict the viewing

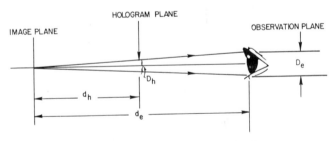

Fig. 5 Geometry of hologram viewing.

214

distance to $d_e < 20.6$ m. At a greater distance we see the same view with both eyes.

5.5.4.2   Varying the Scene Depth, Inverse Tomography

It often happens that only one depth plane of an object is subject to observation at any instant. We give some examples. Ultrasonic B scans give us views of "slices" or "tomographic views" in depth through the object by echo soundings along a line across the object. Transaxial tomography gives us a transaxial cross section across the object by x-ray analysis. Coded aperture imaging gives us the ability to see any depth plane in the object as viewed by γ rays. Ultrasonic holography does the same. In all of these cases we have N images to be recorded at N depths. Again, holographic multiplexing provides an easy way to view all of those images simultaneously at their appropriate depths. This matter has been reviewed by Caulfield (1978). The recording geometry is shown in Fig. 6. Instead of moving the recording material between exposures and using only a thin slice of it at a time, we move the diffuse screen between exposures and use the whole hologram for each exposure. When many depths are to be recorded, we may wish to resort to multiplexing methods more exotic than simple multiple exposure to avoid the N^{-2} signal-to-noise effect (see Section 5.2). While these synthetic images are useful, they never give the viewer the sensation that the solid object is there. Presumably, if N were large enough, the depth spacings between successive photographs small enough, and the alignment good enough, we could be caused to see a full semitransparent image of the object. Thus, as "tomography" is the art of

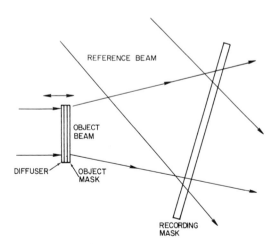

Fig. 6   To record noninterfering images at different depth planes, we multiplex the individual holograms onto one photographic plate as shown.

5. Variation

recording slices of three-dimensional objects, this can be called "inverse tomography"—the art of synthesizing three-dimensional images from tomographic slices.

5.5.5 Photographic Quality Required for Holographic Image Synthesis

The most important property of the photographic images used for holographic image synthesis is image contrast. The black should be quite black (densities in excess of 2 are best). Frame-to-frame uniformity is critical for cosmetic quality. For photographic images not automatically aligned by sprockets on the film, it is necessary to place alignment marks on each image. These marks should be at opposing corners for good alignment and to make them as inconspicuous as possible. Corresponding alignment points on the diffusing screen can be used for quick, accurate registration.

5.5.6 Geometric Exaggeration by Synthetic Image Holography

One can exaggerate the rotation rate by systematically rotating the object more between frames of the photographs by more than would be necessary for an accurate rotation record. This concept was used by King (1968) to record 360° of holographic images on a flat hologram. The photographic plate subtended about 90° from the center of the turntable on which the object was located. Then N strip holograms were made sequentially across the photographic plate as the object was rotated by $360°/N$ between exposures.

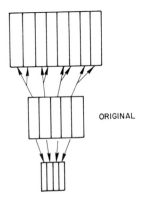

ORIGINAL

Fig. 7 An original Fourier transform may be physically expanded or collapsed in either direction by "cut and paste" methods. By expanding or decreasing the Fourier coordinates, we expand or decrease the image parallax in that direction.

Holography can be used to heighten or minimize the apparent depth variations across the object. This is done by changing the binocular parallax (see Lin, 1968). We record a Fourier-transform hologram of the object and change the spatial frequency content in either the horizontal or vertical direction by physically collapsing or expanding the hologram in that direction. Figure 7 shows some ways this can be done. Because the images overlap regardless of the positions of these subholograms in the Fourier-transform image reconstruction, these manipulations in the hologram do not affect the registration in the image of contributions from the various subholograms.

5.5.7 Special Effects

Synthetic imaging should be interpreted in the broadest sense as the creation of images that could not be recorded by ordinary direct holography. The superposition of interpenetrating, normally incompatible scenes is certainly possible. Schinella (1976) used the absence of a scene as a component scene. He used a back-illuminated diffuser with a black cube in front of it as an object. In a second component scene he occupied that space by a front-lighted object. Clearly, we are imagination-limited, not technology-limited, in synthetic image holography.

REFERENCES

Caulfield, H. J. (1978). *In* "Topics in Applied Physics," Vol. 23, Optical Data Processing, Applications (D. Casasent, ed.). Springer-Verlag, New York.
Caulfield, H. J., Lu, S., and Harris, J. L. (1968). *J. Opt. Soc. Amer.* **58**, 1003.
Collier, R. J., Burckhardt, C. B., and Lin, L. H. (1971). "Optical Holography." Academic Press, New York.
Jeong, T. H., Rudolf, P., and Luckett, A. (1966). *J. Opt. Soc. Amer.* **56**, 1263.
King, M. C. (1968). *Appl. Opt.* **7**, 1641.
Lin, L. H. (1968). *J. Opt. Soc. Amer.* **58**, 1539.
Lu, S., Hemstreet, H. W., and Caulfield, H. J. (1967). *Phys. Lett.* **25A**, 294.
Schinella, J. (1976). Communication to the author.

5.6 LOCAL REFERENCE BEAM HOLOGRAMS

W. Thomas Cathey

A local reference beam hologram is one in which the reference wave is derived in the region of the hologram. The reference wave is obtained from a portion of the wave from the object. It is the purpose of this discussion to describe the local-reference-beam (LRB) hologram, discuss its uses, and to discuss various implementations and the characteristics of each.

5.6.1 Description

The reference wave can be obtained from the object wave as shown in Fig. 1. The wave from the object is focused to a small image in the plane of the iris. The iris can be adjusted to pass the entire wave or can be reduced in size to remove the higher spatial frequency components of the wave. If a small hole is used, a high-quality, but low-energy wave is produced. The effects of the options are discussed in Section 5.6.3. After passing the iris, the wave passes through a beamsplitter and interferes with the remainder of the object wave on the hologram plate. Other configurations are possible. The wave from the object could be split with a beamsplitter, or the hologram plate could be positioned so that a reflection hologram is recorded (Caulfield *et al.*, 1967; Cathey, 1968).

5.6.2 Special Aspects of the LRB Hologram

5.6.2.1 Advantages

The advantages of the LRB hologram stem from the fact that the reference wave originates from the object wave. This means that the reference and object waves both travel the same distances from the source, and the requirements on temporal coherence of the source can be relaxed. A hologram can be made of a distant object without having to provide an equally long path for the reference wave or requiring a very long coherence time for the illuminating source.

The reference and object waves also have the same frequency and phase, unaffected by motions of the object. If the object moves, imparting a Doppler

5. Variation

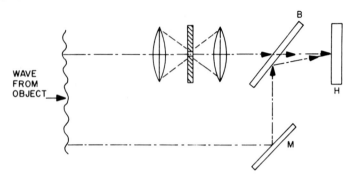

Fig. 1 General method of recording a local-reference-beam hologram. M, mirror; B, beamsplitter; H, hologram recording material.

shift to the object wave, the frequency of the reference wave is shifted an equal amount. In normal holographic recordings, if the object moves as much as a fraction of a wavelength during the exposure of the hologram, the interference fringes move and are not recorded. With a LRB hologram, the fringes are stationary even if the object moves slightly. The limitations on object motion become the same as for normal photography.

The object and reference waves are affected in the same way by the medium between the object and the hologram. Any phase distortions, for example, will appear on both waves and will cancel when the waves interfere on the hologram. The cancellation of frequency shifts can be shown by the equation giving the exposure of the hologram in terms of the sum of the object and reference waves:

$$E(x, y) = \int_T [a_o(x, y) \cos \omega_o t + a_r(x, y) \cos \omega_r t]^2 \, dt, \tag{1}$$

where $a_o(x, y)$ and $a_r(x, y)$ are the complex wave amplitudes for the object and reference waves, respectively, and ω_o and ω_r are the frequencies of the object and reference waves. The integration is over the recording time of the hologram. Equation (1) becomes

$$
\begin{aligned}
E(x, y) = \int_T \bigg\{ & \frac{|a_o|^2 + |a_r|^2}{2} + \frac{|a_o|^2}{2} \cos 2\omega_o t + \frac{|a_r|^2}{2} \cos 2\omega_r t \\
& + \frac{1}{2}|a_o||a_r| \cos[(\omega_o + \omega_r)t + \phi_o + \phi_r] \\
& + \frac{1}{2}|a_o||a_r| \cos[(\omega_o - \omega_r)t + \phi_o - \phi_r] \bigg\} \, dt,
\end{aligned} \tag{2}
$$

where ϕ_o and ϕ_r are the phases of the object and reference waves, including linear phase shifts caused by tilts in the wavefront. The integration time is long

220

with respect to the optical period so that all terms which vary with time are removed. If ω_0 and ω_r are the same, as in LRB holograms, the exposure is

$$E(x, y) = (T/2)[|a_0|^2 + |a_r|^2 + |a_0||a_r|\cos(\phi_0 - \phi_r)], \tag{3}$$

which is the same form as obtained when the hologram is recorded of a stationary object using normal procedures. Limitations on object depth imposed by the source coherence length also are removed because the wave from each point of the object is coherent with respect to the corresponding portion of the derived reference wave. If the source coherence length is very short, however, the effect on the hologram is similar to that of superimposing holograms on the same recorder—the number being equal to the object depth divided by the source coherence length.

5.6.2.2 Problems

If the iris has a small hole to provide a reference wave with a uniform phase, the energy in the reference wave is reduced. Unless more of the object wave is used to make the reference wave or unless amplification is provided for the reference wave, this reduction causes the fringes to have low contrast. Low contrast fringes mean that the hologram will form dim images. In addition, the image is likely to be distorted when the reference wave is not brighter than the wave from the object (Goodman, 1968; Cathey, 1974).

If the iris of Fig. 1 has a large hole, the reference source is a small image of the object. The reference wave is then bright, but this means that the phase of the wave illuminating the hologram, in general, will not match the phase of the reference wave. The result is a reduction in the image quality. Further discussion of this problem and some solutions are presented in Section 5.6.3.

5.6.2.3 Resolution Required of Recording Material

Local reference wave holograms can be made using almost any configuration (Fourier transform, quasi-Fourier transform, Fresnel, etc.), and the resolution required is determined by the configuration chosen. By selection of the location of the reference source, minimum image resolution or maximum image field of view can be selected for a given recorder resolution. Because the quality of the reference source may be low in LRB holograms, thereby causing lower quality images anyway, it may be desirable to opt for a large field of view. The image resolution will be acceptable for display. The term "lower quality" is in comparison to micrometer resolutions obtainable in high-quality images. For display, anything better than the eye's resolution is probably wasted.

5. Variation

5.6.2.4 Situations in Which the LRB Hologram Is Useful

Let us summarize properties of a LRB hologram by giving a few cases where it is of use. One is when it is impossible or inconvenient to provide a separate reference wave in the normal manner. For example, the object may be far away, we do not have a laser with very long coherence time, and we do not wish to place a mirror near the object, nor do we wish to provide a delay line for the reference wave. Another case is that in which the object is moving. The object may simply be on an unstable base, or may be an obviously moving object. By using a LRB hologram, the reference wave experiences exactly the same shift in phase or frequency.

A third case, which has not yet been discussed fully, occurs when the object wave must pass through a turbulent wave. Goodman *et al.* (1966) showed that the effects of a turbulent medium could be cancelled partially by providing a reference wave which passes through the same region of the turbulent medium. If the paths of the object and reference waves differ appreciably, as measured in terms of the scale of the turbulence, the image quality is degraded very rapidly. One solution is to place a mirror on or near the object. The wave reflected from the mirror is the reference wave and it travels essentially the same path as the object wave. Any distortion of the phase is then equal on the object and reference wave. When the two waves are added, the result is similar to that obtained in Eq. (3). The phase of the object wave can be rewritten as

$$\phi_{od}(x, y) = \phi_o(x, y) + \phi_d(x, y), \tag{4}$$

where ϕ_{od} is the phase with distortion, ϕ_o the phase without distortion, and ϕ_d the phase change due to the phase-distorting medium. The phase of the reference is similarly rewritten with the same ϕ_d. When the difference of ϕ_{od} and ϕ_{rd} is taken, as indicated by Eq. (3), ϕ_d is removed.

A mirror need not be placed on or near the object if a portion of the object illuminating wave is focused on the object (Waters, 1972). The result is a quasi-Fourier-transform hologram having an extended reference source in the plane of the object. The image distribution is consequently related to the convolution of the object distribution and the source distribution. In general, the resolution of the image is limited by the size of the focused spot serving as a reference source. This speckle reference hologram can be used to compensate for phase distortions of the medium as well as for object motion (Cathey *et al.*, 1973). If desired, the speckle reference and LRB hologram designs could be combined so that the image of the focused spot is centered in the aperture of Fig. 1. The focused spot would increase the energy of the reference wave above that available with a LRB hologram and the iris would provide a more uniform reference wave than would a speckle reference hologram.

5.6.3   Implementation

In making a LRB hologram, a number of options are available when forming the reference wave. A small pinhole can be used or not at the focus of the image, or the wave can remain unfocused in some implementations, such as image-plane holograms. These are discussed separately below, and effects of each choice are given.

5.6.3.1   Focus on a Small Pinhole to Obtain Reference

This implementation, illustrated in Fig. 1, provides the highest quality reference wave but is of low energy level. Because the reference wave can be made to closely approximate a spherical wave, it can be used in making any configuration hologram.

5.6.3.2   Focus without an Iris or Nonfocused Reference

This implementation is more efficient in that less light is lost, but it allows phase variations to appear on the reference wave. If exactly the same variations do not appear on the hologram illuminating wave, a distorted image appears. The effect is similar to viewing an object through a piece of wavy glass such as shower glass where the phase distortion of the shower glass is equal to the difference in the phase of the reference and hologram illuminating waves.

In viewing an object through a shower glass, the effect of the glass is less for objects close to the glass. There is an analagous effect in LRB holography with an extended source. If the holographic image is close to the hologram,

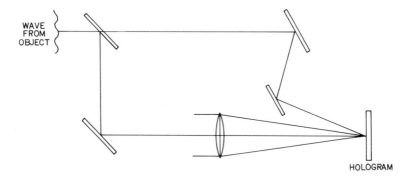

Fig. 2   One implementation for a local reference-beam image-plane hologram. The upper path provides the reference. The lens in the lower path forms an image of the object on the hologram.

5. Variation

the effects of the phase difference between the reference and hologram illuminating waves are minimized. If the image is *on* the hologram (image-plane hologram), the effect is as if an image were on a piece of shower glass; there is no distortion.

Brandt (1969) has summarized the properties of image-plane, local-reference-beam holograms. One of his holograms was made as an image-plane LRB hologram without a focused reference wave. One recording configuration is shown in Fig. 2. A nonfocused local reference beam is acceptable only in the case of image-plane holography.

REFERENCES

Brandt, G. B. (1969). *Appl. Opt.* **8,** 1421–1429.
Cathey, W. T. (1968). U.S. Patent 3415 587, December 10, 1968 (filed December 8, 1965).
Cathey, W. T. (1974). "Optical Information Processing and Holography," Section 6.3. Wiley (Interscience), New York.
Cathey, W. T., Hadwin, J. F., and Pace, J. D. (1973). *Appl. Opt.* **12,** 2683–2685.
Caulfield, H. J., Harris, J. L., and Cobb, J. G. (1967). *Proc. IEEE* **55,** 1758.
Goodman, J. W. (1968). "Introduction to Fourier Optics," Section 8.6. McGraw-Hill, New York.
Goodman, J. W., Hutley, W. H., Jr., Jackson, D. W., and Lehmann, M. (1966). *Appl. Phys. Lett.* **8,** 311–313.
Waters, J. P. (1972). *Appl. Opt.* **11,** 630–636.

Image Formation

Juris Upatnieks

6.1 IMAGE FORMATION WITH COHERENT LIGHT

6.1.1 Exact Image Formation

6.1.1.1 Thin Holograms

Exact image formation, without aberrations, size change, or distortion requires that two conditions are satisfied. One is that the wavelength for recording and viewing the hologram be the same. The second is that the direction and shape of the wavefront impinging upon the hologram be either an exact duplicate of the original reference beam or an exact conjugate of it. A conjugate wavefront is one having the same shape as the original wavefront but propagating in the opposite direction. Figure 1a–c illustrate these principles: (a) a simple recording geometry, (b) the formation of the virtual image, and (c) formation of the conjugate image. The relative positions of the point focus of reference beam in Fig. 1a and illuminating beams in Fig. 1b and c are the same relative to the plate. Besides these, several other geometrical arrangements will produce exact images if the recording is in a thin emulsion. To find these consider in Fig. 2a the impinging signal and reference wavefronts

$$a(x, y) \exp[i\phi(x, y)] + r \exp[i(bx + cx^2 + cy^2)], \tag{1}$$

where b is the carrier frequency term, $c(x^2 + y^2)$ the spherical phase term, $b = (2\pi/\lambda) \sin \theta$, and $c = \pi/(\lambda R_o)$. Both b and c are determined by the relative position of the reference beam to the hologram.

The resulting amplitude transmittance t is

$$t = 1 - k\{a^2 + r^2 + ar \exp[i(\phi - bx - cx^2 - cy^2)$$
$$+ ar \exp[i(-\phi + bx + cx^2 + cy^2)]\}. \tag{2}$$

HANDBOOK OF OPTICAL HOLOGRAPHY
Copyright © 1979 by Academic Press, Inc.
All rights of reproduction in any form reserved.
ISBN-0-12-165350-1

6. Image Formation

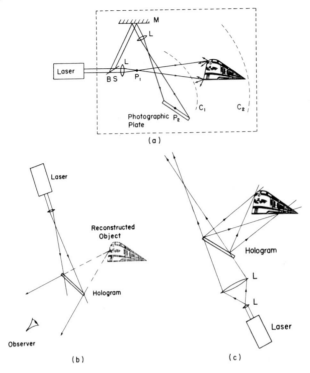

Fig. 1 Hologram recording and image formation: (a) simple recording system, (b) virtual image formation, (c) conjugate image formation.

An exact image will be formed whenever the hologram is illuminated in such a manner that either the third or fourth term in Eq. (2) will give a phase term $\exp[\pm i\phi(x, y)]$, where the positive sign signifies a virtual image and the negative sign the conjugate image. These phases can be achieved by illuminating the hologram with a beam having phases of either $\exp[i(bx + cx^2 + cy^2)]$ or $\exp[-i(bx + cx^2 + cy^2)]$. These two arrangements are illustrated in Fig. 2b and c. Furthermore, since the front and backside of a two-dimensional hologram are indistinguishable, the above wavefronts can also impinge from the opposite side of the plate with exactly the same results. This latter arrangement is illustrated in Fig. 2d and e.

Several geometrical arrangements have unique properties. If the reference beam is collimated and perpendicular to the plate, then $b = c = 0$. Illumination of the hologram with a collimated beam from either side of the hologram will produce exact virtual and conjugate images simultaneously since both $\exp[i\phi(x, y)]$ and $\exp[-i\phi(x, y)]$ terms are formed by the same illuminating beam. Therefore, exact formation of both images is possible simultaneously.

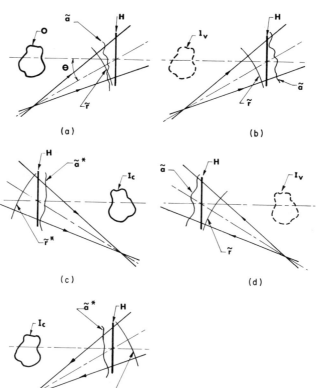

Fig. 2 Hologram recording and exact image formation with thin holograms. O is the object, I_c the conjugate image, I_v the virtual image, $\tilde{a} = a(x, y) \exp[j\phi(x, y)]$, $\tilde{a}^* = a(x, y) \exp[-j\phi(x, y)]$, $\tilde{r} = r \exp[j(bx + cx^2 + cy^2)]$, and $r^* = r \exp[-j(bx + cx^2 + cy^2)]$.

Another unique arrangement is one in which both a flat object and the reference beam are located in the same plane. For this situation each point in the object plane has a phase term of the form $\exp[i(b_x x + b_y y + cx^2 + cy^2)]$. If this is recorded with a reference beam with $b = 0$, then the resulting phase of the signal term recorded on the hologram has the phase $\exp[i(b_x x + b_y y)]$. Thus, each object point forms a grating with constant frequency in the hologram plane. When illuminated with a spherical reference wave, both virtual and conjugate images are formed at the same plane as the illuminating beam point. The location of any image point is independent of the film (hologram) location or motion. Thus, this type of hologram forms an image that remains stationary even when the hologram itself is moved. This property has been

6. Image Formation

used in some proposed motion-picture projectors that do not require a shutter since the image is stationary as the film moves at a constant rate (Bartolini *et al.*, 1970). As the film moves, one image fades out as the next comes into view.

6.1.1.2 Volume Holograms

A hologram begins to exhibit the properties of volume recording when the emulsion thickness starts to exceed $\frac{1}{8}$ of the fringe spacing (Collier *et al.*, 1971, p. 261). A hologram whose fringe spacing is 10 or more times the emulsion thickness will behave more like a thin hologram, while one in which the fringe spacing is equal to or less than the emulsion thickness will behave as a volume recording. Volume holograms exhibit several properties that are different from thin holograms:

(1) image brightness is a function of hologram orientation,
(2) only one of the images, either virtual or conjugate, is formed at a given hologram orientation,
(3) higher order images that possibly could be caused by nonlinearities of the recording process are highly attenuated or nonexistent.

The only two arrangements which form undistorted and bright images are those in which the illuminating beam duplicates the reference beam or is a conjugate to it. The arrangement shown in Fig. 2b will form the virtual image, and the arrangement shown in Fig. 2e will form the conjugate image. The arrangements shown in Fig. 2c and d will usually not provide acceptable images as the brightness would be highly attenuated.

Four other positions of the hologram can give bright but distorted images. These orientations satisfy the Bragg conditions for bright image formation over limited angular field of the image. Two of these situations are as in Fig. 3b and e except that the beams have opposite curvatures: in Fig. 3b the beam converges to the right-hand side of the plate and on axis of the beam shown, and in Fig. 3e the beam diverges from the right-hand side and is on same axis as the beam shown. Other orientations of the hologram can give bright but aberrated and/or distorted images. In general, whenever the illuminating beam is on the axis of the object or reference beam, the Bragg condition will be satisfied over a limited angular field of the image. There are eight hologram orientations that will form bright images of which only two will form aberration-free undistorted images. Characteristics of these incorrect orientations are low image resolution when projected on a screen or viewed under magnification, apparent image motion when the observer changes position, and nonuniform brightness of the image. In the latter case, the portion of the image that is bright changes as one rotates the hologram. One is tempted to choose hologram orientation entirely on the basis of image brightness, but this is

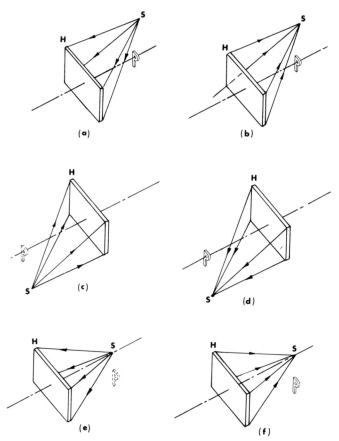

Fig. 3 Hologram recording and image formation with volume holograms: (a) recording geometry and exact virtual image formation, (b) exact conjugate image formation, (c–f) bright but distorted image formation geometries.

erroneous due to the many possibilities for aberrated image formation. The best method is to mark the plate during hologram recording and choose illumination geometry on a basis of reasoning.

Fig. 3a illustrated a hologram recording geometry where P is the object, the solid P indicates real (conjugate), and the dashed P the virtual image. Figure 3a is also the proper arrangement for exact virtual image formation and Fig. 3b for the conjugate image formation. Figures 3c and d illustrate illumination on proper axis but with conjugate (opposite) wave curvatures, and Fig. 3e and f illustrate illumination along the axis of the object beam.

229

6. Image Formation

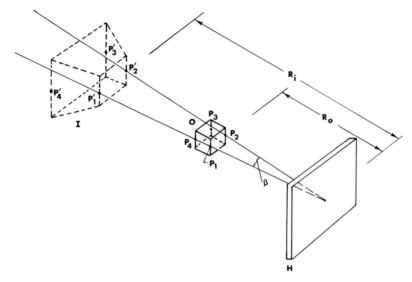

Fig. 4 Hologram image magnification: original object O subtends angle β which remains constant with magnification. Magnified image I is distorted.

6.1.2 Image Magnification

Changes in image magnification take place whenever the hologram-illuminating beam is different from the reference beam used to construct the hologram. The resultant images are located at different distances from the original position of the object relative to the hologram, aberrations are introduced, and the image is distorted. Both for direct viewing and projection, the aberrations can be negligible and apparently sharp images can be formed.

The image location is given by

$$\frac{1}{R_r} - \frac{1}{R_o} = \frac{\pm\lambda_c}{\lambda_r}\left(\frac{1}{R_c} - \frac{1}{R_i}\right), \tag{3}$$

where the plus (+) sign is for the virtual image, and the minus (−) sign is for the conjugate image, and R_r, R_o, R_c, and R_i are the reference beam point, object, illumination beam point, and image to hologram distances, respectively. For a three-dimensional object, R_o and R_i are different for each point on the object.

The effect of magnification can be discussed with the aid of Fig. 4. This figure shows two rays from the hologram to the sides of the image, a cube. The angle β subtended by these two rays remains constant regardless of the illuminating source to hologram distance as long as the wavelength remains the same. The lateral magnification M_1 is therefore simply

$$M_1 = R_i/R_o. \tag{4}$$

We should note that an object will undergo different lateral magnification for each distance R_o from the hologram, thus introducing distortion in the image. In addition, the axial magnification M_z of the image is proportional to the square of the lateral magnification M_1 (Smith, 1975):

$$M_z = M_1^2. \tag{5}$$

This causes the image to appear stretched out when $M_1 > 1$.

Change of wavelength alters the angle subtended by the image at the hologram plane of Fig. 4. If α_{o1} and α_{o2} and the angles between the object points P_1, P_2 and the reference beam, and α_{i1} and α_{i2} are the corresponding image angles, then

$$\sin \alpha_{i1} - \sin \alpha_{i2} = (\lambda_c/\lambda_r)(\sin \alpha_{o1} - \sin \alpha_{o2}) \tag{6}$$

or approximately

$$\alpha_{i1} - \alpha_{i2} \simeq (\lambda_c/\lambda_r)(\alpha_{o1} - \alpha_{o2}). \tag{7}$$

We see that the angle subtended by the image varies with wavelength of illuminating beam. Likewise the image location varies as in Eq. (3). The image is always distorted when the wavelength is changed, regardless of magnification.

6.1.3   Image Projection on a Screen

Two-dimensional images can be projected on a screen without the use of imaging lenses. Figure 5 shows an arrangement that would work fine if the hologram is recorded as in Fig. 3a. For an aberration-free image, the conjugate to the reference beam should be used for illuminating the hologram, and the beam diameter is adjusted on the basis of a compromise between depth of focus and speckle size. The depth of focus Δz is given by an equation similar to that for lenses where the f-number of the lens is replaced by the equivalent ratio R_i/w, where w is the beam diameter and R_i is the distance from hologram to screen:

$$\Delta z = 4R_i^2\lambda/w^2. \tag{8}$$

If R_i/w is very large, the illuminating beam need not be an exact duplicate of the reference beam or its conjugate because aberrations may remain small. Likewise, wavelength may be changed without obvious image degradation. If the illuminating beam is moved over the hologram, change in perspective of the object will be generated in real time and continuously.

6. Image Formation

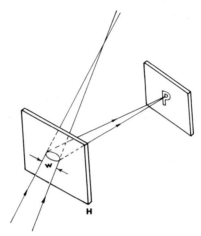

Fig. 5 Image projection on a screen using a small laser beam having diameter *w*.

The speckle visible in the projected image arises from two sources: one is the inherent speckle of the image, the other is the speckle generated by light scattered by the screen. The former can be reduced by using a larger illuminating beam diameter, while the latter can be reduced by causing the reflected light to be incoherent. Coherence of reflected light can be destroyed in numerous ways, including moving the projection screen, liquid crystal screens excited with ac voltage that cause scattering molecules to vibrate, and using fluorescent paints. The latter tend to absorb radiation and reemit incoherently at longer wavelengths.

6.2 IMAGE FORMATION WITH SEMICOHERENT LIGHT

6.2.1 Resolution with Semicoherent Light

For the purpose of estimating image resolution we shall consider only one image point. This point, from one location on the hologram plane, satisfies the equation

$$\frac{1}{\lambda_r}(\sin \alpha_o + \sin \alpha_r) = f = \frac{1}{\lambda_c}(\sin \alpha_i + \sin \alpha_c), \qquad (9)$$

where the left-hand side refers to hologram recording and the right-hand side to image formation and f is the spatial frequency. Once the hologram is recorded, λ_r, α_o, α_r, and f are constant. It is assumed that $\lambda_c \simeq \lambda_r$, $\alpha_r \simeq \alpha_o$, and $\alpha_r \simeq \alpha_c$. By differentiating the above equation and solving for the angular image resolution $d\alpha_i$, we get

$$d\alpha_i = \left| \frac{f}{\cos \alpha_i} d\lambda_c \right| + \left| \frac{\cos \alpha_c}{\cos \alpha_i} d\alpha_c \right| \tag{10}$$

in a direction perpendicular to the carrier frequency. In this equation $d\alpha_c$ is the angular width and $d\lambda_c$ is the bandwidth of the source. In a direction parallel to the carrier frequency,

$$d\alpha_i = d\alpha_c \tag{11}$$

since no dispersion or diffraction takes place. To convert these equations to special resolution ρ at the image plane we multiply $d\alpha_i$ by the hologram to image distance R_i:

$$\rho = R_i d\alpha_i = \left| \frac{R_i f}{\cos \alpha_i} d\lambda_c \right| + \left| \frac{R_i \cos \alpha_c}{\cos \alpha_i} d\alpha_c \right| \tag{12}$$

for the direction perpendicular to the carrier frequency and

$$\rho = R_i d\alpha_c \tag{13}$$

for the direction parallel to carrier frequency.

6.2.2   Compensation for Chromatic Dispersion

Chromatic dispersion can be compensated for in a variety of ways (Leith, 1976). In one version, a grating is placed after the hologram having the same carrier frequency as the hologram (Collier *et al.*, 1971, pp. 501–504), as in Fig. 6a. The reconstructed-image wavefront, after passing through this grating, is diffracted in the direction of the illuminating beam. Since the dispersion of the carrier frequency and the grating is equal and in opposite directions, all wavelengths emerge approximately in parallel and are therefore achromatic. This arrangement causes the image to be noticeably distorted if the carrier frequency is large. Another method is to predisperse the illuminating point source in such a manner that after diffraction by the hologram all wavelengths will be diffracted in the same direction, as in Fig. 6b. With this method the image is not distorted. Both of these methods compensate exactly for only one image point, with dispersion increasing with angular direction away from this point.

A common problem to both of these arrangements is that if the grating is not 100% efficient, the illuminating source will be seen superimposed on the image. This difficulty can be overcome by moving the grating in Fig. 6b away from the hologram in such a manner that the light transmitted through the grating does not fall on the hologram. This arrangement is shown in Fig. 6c.

6. Image Formation

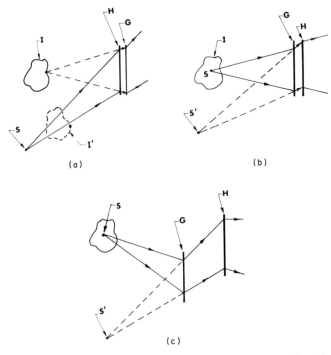

Fig. 6 Compensation for chromatic dispersion: (a) grating after the hologram, (b) grating before the hologram, (c) grating before the hologram away from the hologram.

The required spatial frequency of the grating f_g is given by

$$f_g = f\left(\frac{R_c}{R_c - R_g}\right) = \frac{R_c}{R_c - R_g}\frac{1}{\lambda_r}(\sin \alpha_o - \sin \alpha_r),\qquad(14)$$

where f is the hologram frequency. As a practical matter, the grating G should be constructed with one point source located at position S' in Fig. 6c and the other at the approximated location of the image point for which the correction is to be made.

6.2.3 Resolution of Rainbow Holograms

The "rainbow" hologram is a special kind of hologram in which parallax is eliminated in one direction to reduce coherence requirements (Benton, 1969). These holograms can be viewed with continuous spectrum lamps. The primary advantages are high efficiency and use of inexpensive lamps.

A rainbow hologram is made by a two-step process:

(1)   An ordinary hologram is recorded with geometry as in Fig. 1a.

(2)   The conjugate image is formed with a narrow strip of the first hologram illuminated, as in Figure 7.

If the narrow slit width is "b" and the image is at distance R_i from H_2, then the angular resolution is $bR_i/[R_{12}(R_{12} + R_i)]$ because of the nonmonochromatic light source and slit size, and λ/b because of diffraction. To minimize reduction in resolution, the two should be equal, and solving for the slit width and assuming $R_0 \ll R_{12}$, we get

$$b = R_{12}(\lambda/R_i)^{1/2}. \tag{15}$$

For typical constants and 100-mm depth of image, this comes to about a 2-mm width for the slit. Granularity effects decrease the image resolution further, and usually a somewhat wider slit is desirable for best image appearance, possibly two or three times greater than that given by Eq. (15).

6.3   ESTIMATION OF IMAGE BRIGHTNESS

Hologram image brightness can be calculated using the geometrical relationships shown in Fig. 8 and the following symbols (RCA, 1974): L, luminance (lm/sr m²); I, luminous intensity (lm/sr); ϕ, luminous flux (lm); E, illuminance (lm/m²); R, distance; A, projected surface area in the direction of interest; η, diffraction efficiency, defined as light intensity diffracted into the image divided

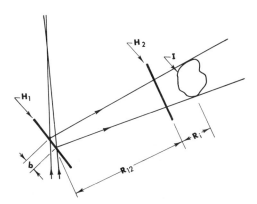

Fig. 7   Geometrical arrangement for constructing "rainbow" holograms: H_1 is original hologram of which a copy H_2 is made. H_1 is illuminated over a slit having width "b" and length of H_1.

6. Image Formation

by light intensity incident on a hologram; and ω, solid angle. The subscripts s, h, and i designate source, hologram, and image, respectively.

For a nonlaser light source, source luminance I in the direction of interest is given by

$$I = LA_s, \qquad (16)$$

and hologram illuminance E_h is

$$E_h = I/R_c^2 = LA_s/R_c^2. \qquad (17)$$

For an image having area A_i that subtends a small solid angle at the hologram, average image brightness is

$$L_i = \eta E_h R_i^2/A_i \qquad (18)$$

$$= \eta(LA_s/R_c^2)(R_i^2/A_i). \qquad (19)$$

The value of LA_s/R_c^2 is a function of the light source characteristics, and this quantity remains a constant for a given light source and desired image resolution. The quantity R_i^2/A_i is the inverse of the solid angle subtended by the image at the hologram plane.

For a laser light source, hologram illuminance E_h is

$$E_h = \phi/A_h = k(\lambda)\phi_v/A_h, \qquad (20)$$

where ϕ_v is laser output in watts, $k(\lambda)$ is the appropriate constant to convert radiant power to lumens, and k is a function of wavelength λ. The image brightness L_i for laser light sources is

$$L_i = \eta(k(\lambda)\phi_v/A_h)(R_i^2/A_i). \qquad (21)$$

For the rainbow hologram, image brightness is greatly increased because all

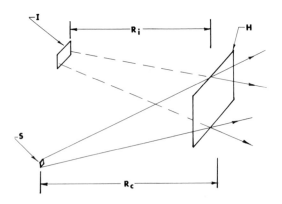

Fig. 8 Geometry for calculating image brightness: source S has projected area A_s and the image I has an overall size A_i.

236

of the light from the hologram passes through a narrow slit. If this slit has width b and is located at distance R_{12} from the hologram, as in Fig. 7, then the image brightness calculated above should be multiplied by a factor $(A_h/A_b)(R_i + R_{12})^2/R_i^2$. An increase by a factor of 100 or more can be easily achieved, and this allows the image to be viewed with low brightness light sources.

Equations (19) and (21) show several important relationships that affect image brightness:

if the image is farther from the hologram, its brightness will be higher than if it is close to the hologram;

hologram size does not affect image brightness if illuminated with nonlaser light sources but is inversely proportional to hologram size if illuminated by a laser light source;

image brightness is inversely proportional to object size.

Also the image resolution decrease due to source size and spectral bandwidth is proportional to image-to-hologram distance R_i, as can be seen from Eq. (12).

REFERENCES

Bartolini, R., W. Hannan, D. Karlsons, and M. Lurie (1970). *Appl. Opt.* **9**, 2283.

Benton, S. A. (1969). *J. Opt. Soc. Amer.* **59**, 1545A.

Collier, R. J., Burckhardt, C. B., and Lin, L. H. (1971). "Optical Holography." Academic Press, New York.

Leith, E. N. (1976). *Scientific American* **235**, No. 4, p. 80.

RCA (1974). "RCA Electro-Optics Handbook," p. 16. Tech. Ser. EOH-11. RCA, Harrison, New Jersey.

Smith, H. M. (1975). "Principles of Holography," 2nd ed., p. 172. Wiley, New York.

7

Cardinal Points and Principal Rays for Holography

Henri H. Arsenault

7.1 INTRODUCTION

Holograms may be considered as image-forming optical elements. This fact was not lost on the early workers in the field of holography, who rapidly saw the analogies between lenses, zones plates, and holograms (Gabor, 1951; Rogers, 1951).

Geometrical optics may be defined loosely as a science that describes the propagation of electromagnetic radiation under conditions where it may be assumed that the radiation propagates in straight lines. The production of images by holograms usually satisfies this condition. This allows ray-tracing techniques to be used to study holograms (Offner, 1966; Helstrom, 1966; Abramowitz and Ballantyne, 1967; Latta, 1971; Miler, 1972; Arsenault, 1975; Welford, 1975).

There is more than one pair of conjugate spaces in holography. The various pairs are shown in Table I. They are the conjugate spaces of the reconstruction source and of the direct image (type I), of the reconstruction source and the conjugate image (type II), of the direct image and the conjugate image (type III), of the object and the direct image (type IV), and of the object and the conjugate image (type V) (Miler, 1972). Types I and II, and types IV and V always go in pairs.

Conjugation types I and II are appropriate when the hologram is considered in the time sequence in which it is recorded and reconstructed, and it is the most frequently used. Types IV and V are useful when the hologram is considered as an image-forming system for the object. For this type of conjugation,

HANDBOOK OF OPTICAL HOLOGRAPHY

TABLE I

Pairs of Conjugate Spaces in Holography

Type	Conjugate spaces	Principal axis	Type	Conjugate spaces	Principal axis
I	C, I_d	RO	IV	O, I_d	RC
II	C, I^*	RO	V	O, I^*	RC
III	I_d, I^*	CI_d			

we are interested in imaginary rays emanating from the object and forming the image. For objects that are not point sources, this is usually very useful, because for conjugations of types I and II, the principal points of the hologram change with each object point. Type III conjugation is not as useful as the others and we shall not discuss it further.

A paraxial geometrical optics theory of holography was worked out by Leith *et al.* (1965), by Meier (1965, 1966, 1967), and later further developed by Miler (1972). The paraxial theory assumes that the object, reference source, and reconstruction source are in a plane perpendicular to the hologram. When this consideration is satisfied, this theory has the interesting property that it predicts the same results as the nonparaxial theory described below.

A number of authors have considered the geometrical relationships between the object and the holographic images, and attempts have been made to understand the holographic process as a combination of ordinary optical components, such as lenses and prisms or gratings (Leith *et al.*, 1965; Neumann, 1966; Champagne, 1967; Lukosz, 1968; Mandelkorn, 1973; Joeng, 1975).

The important questions of holographic aberrations and effects caused by the thickness of the emulsion are treated in Section 2.4 and will not be discussed here, except in passing.

We shall present two complementary approaches to the geometrical optics of holograms. The first is the more natural of the two, and the reader will find that it is usually faster and easier to use than the second; the second approach, however, is more similar to the geometric optics of lenses and mirrors and allows a more intuitive approach to problems in holography. In order to distinguish between the two, we shall refer to the principal points of holograms or to the cardinal points of holograms, depending upon whether we are talking about the first approach or the second, respectively.

7.2 THE HOLOGRAPHIC RAY-TRACING EQUATIONS

The geometrical properties of holograms are based on the holographic ray-tracing equations. With these equations, it is possible to trace rays through holograms. For a plane hologram, the equations are (Leith, 1965; Meier, 1965)

$$\cos \theta_i = \cos \theta_c \pm (m/\mu)(\cos \theta_o - \cos \theta_r), \tag{1}$$

where m is the scale factor between recording and reconstruction, and μ is the wavelength ratio between reconstruction and recording; θ_c, θ_r, θ_o, and θ_i are the angles between the meridian line (the intersection between the hologram and the plane containing the reconstruction source C, the reference source R, and the object O) and the reconstruction source, reference source, object, and images. (The minus sign corresponds to the direct image, and the plus sign to the conjugate image.) An equivalent form of the ray-tracing equations, given by Welford (1975), is

$$\mathbf{n} \times (\mathbf{r}_i - \mathbf{r}_c) = (m/\mu)\mathbf{n} \times (\mathbf{r}_o - \mathbf{r}_r), \tag{2}$$

where $\mathbf{n}$ is the local normal to the hologram, and $\mathbf{r}_i$, $\mathbf{r}_c$, and $\mathbf{r}_r$ are unit vectors at the hologram. This form is useful for tracing rays through holograms on curved surfaces (Welford, 1973). In this chapter, only the cases in which $m = \mu = 1$ are considered.

7.3   THE PRINCIPAL POINTS OF HOLOGRAMS

Before describing the principal points of holograms, we remark on an important point: The image is always in the same plane as the object, the reference source, and the reconstruction source. This is because at the intersection of this plane with the hologram, the hologram fringes are perpendicular to this plane, which is called the meridian plane. As a matter of fact, the hologram fringes are perpendicular to any plane containing both the object and the reference source, because the fringes are lines of constant phase in the recording step. It follows that no ray from the reconstruction source may be deviated out of the meridian plane at the hologram, because the projection of the propagation vector on a vector parallel to the fringes must remain constant and the hologram has no focusing power in a direction parallel to the fringes.

The intersection of the meridian plane (the plane containing the object O, the reference source R and the reconstruction source C) with the hologram is called the meridian line. The line passing through the reference source R and the object O is the primary axis. The intersection of the primary axis with the hologram is the primary vertex V'. The plane perpendicular to the meridian plane and containing the meridian line is called the sagittal plane. Note that this plane is usually not parallel to the hologram plane, which also contains the meridian line. The virtual object O' is a point on the primary axis on the opposite side of the hologram from the object and at a distance from the vertex equal to the distance between the object and the vertex. The virtual reference source R' is a point on the primary axis on the opposite side of the hologram from the reference source, and at a distance from the primary vertex equal to the distance between the primary vertex and the reference source. Other principal points of interest are the points symmetrical to the object and reference source, with respect to the meridian line, which bear no special names. These are the primary principal points of the hologram, and they are the principal points appropriate to type I and type II conjugation (for the direct and conjugate images, respectively) (Arsenault, 1975).

For conjugate spaces of types IV and V, other principal points are needed. The axis of the hologram is now considered to be the line passing through the reference source R and the reconstruction source C. This line is called the secondary axis. The intersection of the secondary axis with the hologram is called the secondary vertex V''. The secondary reference source R'' is a point on the secondary axis on the opposite side of the hologram from the reference source R whose distance from the secondary vertex V'' is equal to that of the reference source from the secondary vertex. The secondary reconstruction source C'' is a point on the secondary axis RC whose distance from the secondary vertex V'' is equal to that of the reconstruction source from the vertex. Some of the principal points described are shown in Fig. 1.

7.3.1 The Principal Rays for Type I Conjugation

In type I conjugation, we are interested in the conjugate spaces of the reconstruction source and the direct image. The principal rays of this section allow the determination of the position of the direct image, given the position of the reconstruction point source. This is of interest when it is necessary to

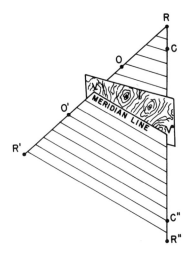

Fig. 1 Some of the principal points of a hologram.

determine the effect of changing the position of the reconstruction source, as in holographic microscopy, the study of holographic deformations caused by misplacing the reconstruction source or the hologram, the study of holographic aberrations, image formation by multielement holographic optical elements, and recording or reconstructing with extended sources.

The principal rays for the direct image in type I conjugation are

(1) a ray from the reconstruction source C through the reference source R, which is deviated at the hologram along a line through the object O;

(2) a ray from the reconstruction source through the primary vertex V' of the hologram, which is not deviated;

(3) a ray from the reconstruction source directed to a point in the meridian plane symmetrical to the object with respect to the meridian line, which is deviated at the hologram along a line through a point symmetrical to the reference source with respect to the meridian line;

(4) a ray from the reconstruction source directed through the virtual object O', which is deviated at the hologram along a line through the virtual reference source R'.

The first three rays describe the behavior of real rays, whereas the fourth does not describe the behavior of an actual ray in the system. The intersection of all those lines describes a single point that may be considered to be the ideal aberration-free image point. Except for certain special configurations, the other rays will not all pass through this point, and the actual holographic image will suffer some aberrations (Meier, 1975). Three of the principal rays for type I imaging are shown in Fig. 2.

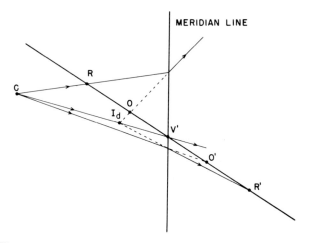

Fig. 2 Three of the principal rays for the direct image, in type I imaging.

7.3.2 The Principal Rays for Type II Conjugation

In type II conjugation, we are concerned with the conjugate image; the conjugate spaces are those of the reconstruction source C and the conjugate image I_{cp}. The principal points for the conjugate image are the same as for the direct image, and the principal rays are the following:

(1) a ray from the reconstruction source through the primary vertex V', which is not deviated;

(2) a ray from the reconstruction source along a line passing through the object, which is deviated at the hologram along a line passing through the reference source;

(3) a ray from the reconstruction source directed to a point in the meridian plane symmetrical to the reference source with respect to the meridian line, which is deviated at the hologram along a line through a point symmetrical to the object with respect to the meridian line;

(4) a ray from the reconstruction source directed along a line that passes through the virtual reference source S', which is deviated at the hologram along a line that passes through the virtual object O'.

The first three rays describe the behavior of physical rays, whereas the fourth is a fictitious ray. Figure 3 shows three of the principal rays for the conjugate image I_{cp}.

When the reference source is at infinity, and when the principal axis is perpendicular to the hologram, the principal rays for the direct and conjugate images are the same as those of a thin converging and diverging lens, respectively, when the object replaces the focal point.

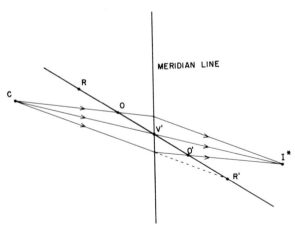

Fig. 3 Three of the principal rays for the conjugate image, in type II imaging.

7.3.3 The Principal Rays for Type IV Conjugation

The conjugate space of interest here is that of the object and the direct image. This approach is of interest when the imaging of extended objects is considered, for studying distortions caused by the holographic process, and for understanding the holographic process as an imaging process. The rays are assumed to emanate from the object and to converge to the image, although they of course emanate from the reconstruction source. The image positions determined in this manner are of course identical with those obtained by the methods described in the previous sections. For determining the image positions by graphical means, all the object points in one meridianal plane have the same principal points (except those related to the object position), whereas for type I and type II conjugation, the principal points are different for each object point, when the object is not a single point.

The principal rays for the direct image in type IV conjugation are the following:

(1) a ray from the object through the secondary vertex V'' of the hologram, which is not deviated;

(2) a ray from the object along a line through the reference source, which is deviated at the hologram along a line through the reconstruction source;

(3) a ray from the object directed to a point in the meridian plane symmetrical to the reconstruction source, with respect to the meridian line, which is deviated at the hologram along a line directed to a point symmetrical to the reference source with respect to the meridian line;

(4) a ray from the object along a line through the secondary virtual reconstruction source, which is deviated at the hologram along a line through the secondary virtual reference source.

Some of these principal rays are shown in Fig. 4.

7.3.4 The Principal Rays for Type V Conjugation

The conjugate spaces of interest in type V conjugation are the spaces of the object and of the conjugate image. The principal rays are the following:

(1) a ray from the object along a line through the reference source, which is deviated at the hologram along a line through the reconstructions source;

(2) a ray from the object along a line through the reconstruction source, which is deviated at the hologram along a line through the reference source;

(3) consider a line joining a point symmetrical to the reference source with respect to the meridian line, to a point symmetrical to the reconstruction source with respect to the meridian line; a ray from the object through the point where this line intersects the meridian line is deviated at the hologram

7. Cardinal Points and Principal Rays for Holography

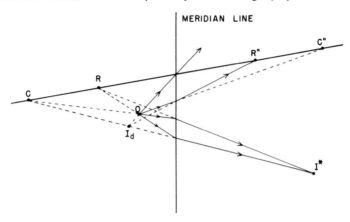

Fig. 4 Principal rays for the direct image and for the conjugate image in type IV and type V imaging.

along a line through a point symmetrical to the object with respect to the meridian line;

(4) a ray from the object through the intersection between the meridian line and a line joining the reconstruction source to the primary virtual reference source R', which is deviated at the hologram along a line through the primary virtual object.

Fortunately, only two principal rays are required to determine the location of the image, so the first two rays will usually be taken, the two others requiring the construction of an auxiliary line before the principal ray can be constructed. The first two principal rays are shown in Fig. 4.

7.4 THE CARDINAL POINTS OF HOLOGRAMS

In the previous section we have shown how certain principal points of holograms allow the rapid determination of the image positions in holography. These principal points are easy to find for any given configuration and are easy to use. However, except for certain special cases, the similarities between the principal rays that we have described and the principal rays of lenses are not sufficient to allow the holographer who is used to lens optics to transpose his knowledge of lenses to help understand a given holographic configuration easily.

In holography, it is possible to define cardinal points that are quite similar to those of lenses, namely foci and nodal points; but when this is done, it is found that the foci of the hologram are off axis, and furthermore that they are

246

different for each reconstruction point and for each object point. Only in the special case when the primary axis RO (for types I and II imaging) or the secondary axis RV (for types IV and V imaging) is perpendicular to the hologram are the foci the same for each reconstruction point (type I or type II imaging) or for each image point (type IV or type V imaging), and on an axis perpendicular to the hologram plane.

There are a number of ways that the foci can be defined, but we shall define them in such a way that the foci will be the same for any meridian plane.

This is possible only if the focal points are on the primary axis for type I and type II imaging, and on the secondary axis for type IV and type V imaging, because every meridian plane contains the axis. The focal length, measured along the primary or secondary axis, may be found by means of the conjugate relations given in Section 7.5, with the source set at infinity. This yields the focal lengths given in Table II for the various types of imaging. The conventions are the same as those that are currently used in lens optics, a positive focal length corresponds to a convergent lens, and a negative value to a divergent lens. The conventions for the source and image distances are also

TABLE II

Conjugate Relations for Holography: Polar Coordinates

Type	Principal axis	Conjugate relation	Focal length	Analog
I	RO	$\dfrac{1}{\rho_{dp}} + \dfrac{1}{\rho_c} = \dfrac{1}{F_{pd}} \dfrac{\sin \theta_c}{\sin \theta_o}$ $\theta_i = \theta_o$	$\dfrac{1}{F_{pd}} = \dfrac{1}{\rho_r} - \dfrac{1}{\rho_o}$	Lens
II	RO	$\dfrac{1}{\rho_{cp}} + \dfrac{1}{\rho_c} = \dfrac{1}{F_{pc}} \dfrac{\sin \theta_c}{\sin \theta_o}$ $\theta_i = \theta_c$	$\dfrac{1}{F_{pc}} = -\left(\dfrac{1}{\rho_r} - \dfrac{1}{\rho_o}\right)$	Lens
IV	RC	$\dfrac{1}{\rho_{ds}} + \dfrac{1}{\rho_o} = \dfrac{1}{F_{sd}} \dfrac{\sin \theta_o}{\sin \theta_c}$ $\theta_i = \theta_o$	$\dfrac{1}{F_{sd}} = \dfrac{1}{\rho_r} - \dfrac{1}{\rho_c}$	Lens
V	RC	$\dfrac{1}{\rho_{cs} \sin \theta_i} - \dfrac{1}{\rho_o \sin \theta_o}$ $= \dfrac{1}{F_{sc} \sin \theta_c}$ $\dfrac{1}{\tan \theta_i} = \dfrac{1}{\tan \theta_o} - \dfrac{2}{\tan \theta_c}$	$\dfrac{1}{F_{sc}} = -\left(\dfrac{1}{\rho_r} + \dfrac{1}{\rho_c}\right)$	Mirror (virtual object)

7.   Cardinal Points and Principal Rays for Holography

those frequently used: Object distances are positive to the left of the hologram, and image distances are positive to the right.

The nodal points for the type I and II and type IV imaging coincide with the primary vertex V' and the secondary vertex V'' of the hologram, which goes well with the idea of the hologram as a lens, but the nodal point for type V imaging is a point on the secondary axis RC at a distance from the secondary vertex V'' equal to twice the focal length. This property is reminiscent more of a mirror than of a lens and is responsible for some of the strange properties of the conjugate image, which we shall discuss in Section 7.5.2.

Considered in this manner, holograms are considered to have properties quite similar to lenses except for one thing: the principal axis is not perpendicular to the hologram plane; this is the price that must be paid if the cardinal points are not to be different for each meridian plane.

The focal planes are the loci of the points whose images are at infinity. If the conjugate relations (3) from Section 7.5 are used, the focal planes are the planes through the foci and perpendicular to the primary or secondary axis; but if the relations (4) and (5) are used, where the distances have been projected on a line perpendicular to the meridian line, then the focal planes are parallel to the hologram plane, because each focal line is parallel to a meridian line which is contained in the hologram plane.

In summary, here are the cardinal points for the various types of holograms.

For type I imaging, the principal axis is the primary axis RO, the nodal points coincide with the primary vertex and the foci F_p and F_p' are on either side of the hologram, on the principal axis at a distance from the vertex V' given in Table II. These cardinal points may be seen in Fig. 5.

For type II imaging, the principal axis and the nodal points are the same as for type I imaging, and the principal foci are the same as in type I imaging,

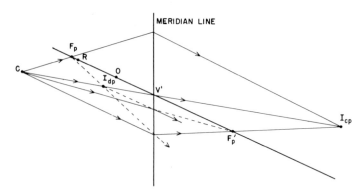

Fig. 5   The cardinal points of a hologram and the associated rays for type I and type II imaging.

248

except that the two foci are interchanged (the sign of the focal length is changed).

For type IV imaging, the principal axis is the secondary axis CR, a line through both the reference source R and the reconstruction source C, the nodal points coincide at the secondary vertex V'', and the focal points F_{sd} and F'_{sd} are on either side of the hologram at a distance from the secondary vertex V'' given by Table II (the subscript sd stands for secondary axis and direct image). These points are shown in Fig. 6.

Finally, for type V imaging, the principal axis is also the secondary axis CR; the nodal point N is on the principal axis at a distance from the secondary vertex V'' equal to twice the focal length, and both foci F_{sc} and F'_{sc} coincide on the principal axis at a distance from the vertex given in Table II. Note that the secondary vertex is not a nodal point, except when the secondary axis CR is perpendicular to the hologram.

Given the above cardinal points, it is possible to use principal rays that have properties identical to those that are commonly used for lenses and mirrors. The rays, which complement the rays described in Section 7.3, are

(1) a ray through the focus, which is deviated at the hologram along a line parallel to the principal axis;

(2) a ray through the nodal point, which is not deviated; and

(3) a ray along a line parallel to the principal axis, which is deviated at the hologram along a line through the focus.

The principal rays for the various conjugate spaces are shown in Figs. 5 and 6.

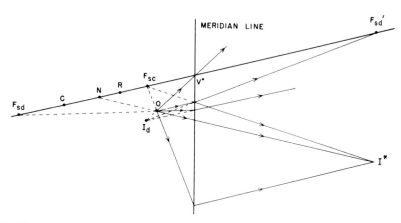

Fig. 6 The cardinal points of a hologram and the associated rays for type IV and type V imaging.

7.5 THE CONJUGATE EQUATIONS

7.5.1 The Basic Equations

The principal rays described in the previous sections allow the positions of the images to be determined by graphical means. Since these rays are not paraxial rays, the conjugate relations derived from them are not limited to the paraxial approximation but are valid in the general case to determine the ideal image positions, if they are used properly, even when the paraxial conditions no longer apply.

The conjugate equation for the direct image may be obtained from consideration of similar triangles in Fig. 5. If the sign convention used is that object, reference source, and reconstruction source distances are positive measured to the left of the meridian line and image positions are positive to the right, and that convergent lenses have positive focal lengths, the relation in polar coordinates is

$$\frac{1}{\rho_{dp}} = -\frac{1}{\rho_c} + \left(\frac{1}{\rho_r} - \frac{1}{\rho_o}\right)\frac{\sin\theta_c}{\sin\theta_o} \quad \text{and} \quad \theta_i = \theta_c, \tag{3}$$

where the distances ρ_o, ρ_r, ρ_c, and ρ_{dp} of the object O, the reference source R, the reconstruction source C, and the image I are measured from the primary vertex, and the angles are measured from the meridian line at the vertex V' in the meridian plane.

The conjugate equations may also be expressed in cartesian coordinates. If the x coordinate is taken parallel to the meridian line and the y axis perpendicular to it, and the origin is anywhere on the meridian line, the cartesian coordinates of the direct image, in terms of the object O, the reference source R, and the reconstruction source C are

$$\frac{1}{y_{dp}} = -\frac{1}{y_c} + \frac{1}{y_r} - \frac{1}{y_o}, \tag{4}$$

$$x_{dp} = \frac{x_c y}{y_c} - \left[x_r - \frac{(x_r - x_o)}{y_r - y_o}y_r\right]\left[\frac{y}{y_c} - 1\right]. \tag{5}$$

If the origin is taken on the primary vertex of the hologram, these equations yield

$$\frac{1}{y_{dp}} = \frac{-1}{y_c} + \frac{1}{y_r} - \frac{1}{y_o}, \quad x_{dp} = \frac{x_r y}{y_r}, \tag{6}$$

which is the cartesian form of the paraxial conjugate equations, except that here the equations yield the ideal image points even in the nonparaxial case. Note that the relation for y is valid for any origin taken along the meridian line. This shows that if any one or more of the object, the reconstruction source, and the reference source are moved in the meridian plane parallel to the meridian line, the distance of the image from the meridian line is not changed.

Table III shows the cartesian coordinates of the images for the various pairs of conjugate spaces in terms of the cartesian coordinates of the object O, the reference source R, and the reconstruction source C.

If the distances are measured from the primary vertex V' (type I and type II imaging) or from the secondary vertex V'' (type IV and type V imaging), the simpler expressions in the third column are obtained. If the origin is allowed to be anywhere on the meridian line, the more complicated expressions in the last column are required. The expressions required when the origin of the coordinate axes is allowed to be anywhere in the meridian plane may be easily obtained from the last column by adding to all the y distances the distance from the origin to the meridian line.

Table II gives the conjugate relations in polar coordinates for the various pairs of conjugate spaces. The focal lengths are measured along the primary

TABLE III

Conjugate Relations for Holography: Cartesian Coordinates

Type	Principal axis	Origin on vertex	Origin on meridian line
I	RO	$\dfrac{1}{y_{dp}} = -\dfrac{1}{y_c} + \dfrac{1}{y_r} - \dfrac{1}{y_o}$	$\dfrac{1}{y_{dp}} = -\dfrac{1}{y_c} + \dfrac{1}{y_r} - \dfrac{1}{y_o}$
		$x_{dp} = +\dfrac{x_c y}{y_c}$	$x_{dp} = -\dfrac{x_c y}{y_c} + \left[x_r - \dfrac{(x_r - x_o)}{y_r - y_o} y_r \right] \left[\dfrac{y}{y_c} + 1 \right]$
II	RO	$\dfrac{1}{y_{cp}} = -\dfrac{1}{y_c} - \dfrac{1}{y_r} + \dfrac{1}{y_o}$	$\dfrac{1}{y_{cp}} = -\dfrac{1}{y_c} - \dfrac{1}{y_r} + \dfrac{1}{y_o}$
		$x_{cp} = -\dfrac{x_c y}{y_c}$	$x_{cp} = +\dfrac{x_c y}{y_c} - \left[x_r - \dfrac{x_r - x_o}{y_r - y_o} y_r \right] \left[\dfrac{y}{y_o} + 1 \right]$
IV	RC	$\dfrac{1}{y_{ds}} = -\dfrac{1}{y_c} + \dfrac{1}{y_r} - \dfrac{1}{y_o}$	$\dfrac{1}{y_{ds}} = -\dfrac{1}{y_c} + \dfrac{1}{y_r} - \dfrac{1}{y_o}$
		$x_{ds} = -\dfrac{x_o y}{y_o}$	$x_{ds} = -\dfrac{x_o y}{y_o} + \left[x_r - \dfrac{x_r - x_c}{y_r - y_c} y_r \right] \left[\dfrac{y}{y_o} + 1 \right]$
V	RC	$\dfrac{1}{y_{cs}} = -\dfrac{1}{y_c} - \dfrac{1}{y_r} + \dfrac{1}{y_o}$	$\dfrac{1}{y_{cs}} = -\dfrac{1}{y_c} - \dfrac{1}{y_r} + \dfrac{1}{y_o}$
		$x_{sc} = \left(\dfrac{x_o}{y_o} - \dfrac{2x_c}{y_c} \right) y$	$x_{sc} = + \left\{ \dfrac{x_o}{y_o} - \dfrac{2x_c}{y_c} + \left[x_c - \dfrac{x_r - x_c}{y_r - y_c} y_c \right] \right.$ $\left. \times \left[\dfrac{1}{y} - \dfrac{1}{y_o} + \dfrac{2}{y_c} \right] \right\} y$

axis for type I and type II imaging, whereas they are measured along the secondary axis for type IV and type V imaging. As we have mentioned before, the distances are taken to be positive on the left of the lens for the object O, the reference source R, and the reconstruction source C, and they are taken to be positive on the right of the hologram for the images. The angles are measured from the meridian line, at the vertex and a positive focal length means a convergent effect, whereas a negative focal length means an effect similar to a divergent lens. The last column on the right compares the image-forming properties of the hologram with simple optical components. The special properties of the conjugate image of three-dimensional objects fall under type IV imaging and will be discussed further in the next section.

7.5.2 The Properties of the Conjugate Image

The conjugate image of a hologram has certain special properties, among them the phenomenon of depth reversal. The conjugate spaces appropriate to study this phenomenon is that of the object and the conjugate image (type V). Type II imaging (reconstruction source and conjugate image) is not appropriate in this case, because for an extended object, the primary vertex is different for each object point, and the conjugate relations for type II imaging do not suggest any peculiar properties.

The conjugate relations for type V imaging show that the formation of the conjugate image is akin to image formation by a mirror; but there are important differences; first, the conjugate relations show that the image is similar to that formed from a virtual object (because of the negative sign of the object distance), and that the axial symmetry around the secondary axis V'' is lost, unless the secondary axis is perpendicular to the meridian line. This can be seen on Fig. 6, and in the presence of the second term in the conjugate relation for x in type V imaging, or in the complicated relation for the angle of the image as measured from the secondary vertex in polar coordinates in Table II.

When a three-dimensional object is recorded on the hologram, the points closest to the hologram will appear closest to the hologram in the reconstructed conjugate image, seen through the hologram. If this image is on the same side of the hologram as the observer, as in Fig. 6, the result is an apparent reversal of depth, which is why this image is sometimes called a pseudoscopic image.

We now determine the conditions under which this effect will take place. We differentiate the relation for the distance from the meridian line to the conjugate image

$$\frac{1}{y_{cs}} = -\frac{1}{y_c} - \frac{1}{y_r} + \frac{1}{y_0} \tag{7}$$

relative to the object distance y_0, keeping y_c and y_r constant. The image will be pseudoscopic if

$$\partial y_{cs}/\partial y_o > 0. \tag{8}$$

This condition will be satisfied if

$$y_{cs} > 0. \tag{9}$$

For most recording and reconstruction configurations, this means that the image will be pseudoscopic only if the conjugate image is on the side of the hologram opposite to the reconstruction source, that is, if it is a real image.

7.6   APPLICATION TO HOLOGRAM TYPES

7.6.1   Gabor Holograms ($\theta_o \approx \theta_o \approx \theta_c$)

When the physical extent of the hologram contains points only in the neighborhood of the primary vertex V' of the hologram, the hologram is an in-line or Gabor hologram (Gabor, 1951). This hologram type has both the direct and the conjugate images in the same line of sight, with the result that there is some overlapping of images. In this case, the conjugate relations become

$$\frac{1}{\rho} + \frac{1}{\rho_c} = \pm \left(\frac{1}{\rho_r} - \frac{1}{\rho_o} \right). \tag{10}$$

7.6.2   Image Holograms ($\rho_o \approx 0$)

In this type of hologram, the object (often the real image from a lens) is very near the surface of the hologram, or even straddles the hologram, in the case of three-dimensional objects, whereas the reconstructed and reference sources are not usually very near the hologram; the conjugate relations yield

$$y_d \approx y_o, \tag{11}$$

which means that the image position is not very sensitive to the object position, because

$$\frac{\partial x}{\partial x_c} = \frac{y}{y_c} \approx 0. \tag{12}$$

This means that the image can be reconstructed with an extended source, while avoiding blurring of the image. The image position is not sensitive to the reconstruction source wavelength either, so that it is possible to reconstruct such holograms with a white-light extended source (Stroke, 1966).

7. Cardinal Points and Principal Rays for Holography

7.6.3 Collimated Reference Beam ($\rho_r = \infty$)

When the reference beam is collimated, the reference source R is at infinity. For each point of the object, the hologram may be considered to be a holographic lens with focal length y_0. When both the reference beam and the reconstruction beam are collimated, the direct and conjugate images are on either side of the hologram, at a distance from the meridian line equal to y_0, the distance from the object to the meridian line.

7.6.4 Lensless Fourier-Transform Holograms ($y_r = y_0$)

When the object and the reference source are at the same distance from the meridian line, the hologram is called a lensless Fourier-transform hologram (Stroke, 1965), because the recorded fringes are like the Fourier transform of the object (see Section 4.3). In this case, the primary axis is parallel to the hologram, and the image distances are

$$y_{dp} = -y_c, \tag{13}$$

$$y_{cp} = -y_c, \tag{14}$$

which means that both the direct and the conjugate images are on the same side of the hologram as the reconstruction source and at the same distance from the meridian line. Using the conjugate relations for type IV and type V imaging, the image distances from the secondary vertex are

$$x_{ds} = x_0 y_c / y_0, \tag{15}$$

$$x_{cs} = -(x_0 y_c / y_0) + 2x_c. \tag{16}$$

This means that the two images are on either side of the reconstruction source R_c, at equal distances from this source equal to

$$x'_{ds} = (x_0 y_c / y_0) - x_c, \tag{17}$$

$$x'_{cs} = -(x_0 y_c / y_0) + x_c. \tag{18}$$

This is a case in which the direct and the conjugate images are usually both virtual.

7.6.5 Fourier-Transform Holograms ($y_0 = \infty$, $y_r = \infty$)

This is probably the most important holographic configuration (see Section 4). From the viewpoint of geometry, it is a special case of the lensless Fourier-transform hologram ($y_0 = y_r$), although it is usually recorded with the object and the reference beam in the front focal plane of a lens (Vander Lugt, 1964). The conjugate relation for Fourier-transform holograms may be used, if $1/\tan \theta_0$ is substituted for x_0/y_0. If the reconstruction beam is collimated, the image is completely free from aberrations, for any angle of incidence of the reconstruction beam; the reason for this is that for each object point such a hologram must change an incident plane wave into another plane wave, which it can do without introducing aberrations. In practice, this means that accurate angular, lateral, and longitudinal positionings of the hologram are not required to avoid aberrations.

7.6.6 Random-Bias Holograms

In this type of hologram, the reference source is not a point source, but is extended in space. It is usually a diffuser (Arsenault, 1971), or in local-reference-beam holography (Caulfield *et al.*, 1967) to which this analysis is also applicable, it may be part of the object.

Using principal rays (1) and (3) for type IV imaging, described in Section 7.3.3, it may be shown from geometrical construction that a point object spreads in a direction parallel to the meridian line to an image spot having a size equal to

$$s_i = s_r y_i / y_r, \tag{19}$$

where s_i is the size of the image spread, s_r is the size of the reference source, and y_i and y_r are the distances from the meridian line of the image and reference source, respectively. But the magnification of the image is equal to y_i/y_0; therefore, the image is effectively blurred by a spread function having a width equal to $s_r y_0 / y_r$. In a typical configuration for recording holograms of three-dimensional objects, the effective spread function is about $\frac{1}{3}$ of the source size, which leads to a resolution in the image of about 0.3 mm.

REFERENCES

Abramowitz, I. A., and Ballantyne, J. M. (1967). *J. Opt. Soc. Amer.* **57**, 1522.
Arsenault, H. H. (1971). *Opt. Comm.* **4**, 267.
Arsenault, H. H. (1975). *J. Opt. Soc. Amer.* **65**, 903.
Caulfied, H. J., Harris, J. L., Hemstreet, H. W., and Cobb, J. G. (1967). *Proc. IEEE* **55**, 1758.
Champagne, E. B. (1967). *J. Opt. Soc. Amer.* **57**, 51.
Gabor, D. (1951a). *Proc. Phys. Soc. B* **64**, 244.
Gabor, D. (1951b). *Proc. Phys. Soc. B* **64**, 449.
Helstrom, C. W. (1966). *J. Opt. Soc. Amer.* **56**, 433.
Joeng, T. H. (1975). *Amer. J. Phys.* **43**, 714.
Latta, J. N. (1971). *Appl. Opt.* **10**, 2698.
Leith, E. N., Upatnieks, J., and Haines, K. A. (1965). *J. Opt. Soc. Amer.* **55**, 981.
Lukosz, W. (1968). *J. Opt. Soc. Amer.* **58**, 1084.
Mandelkorn, F. (1973). *J. Opt. Soc. Amer.* **63**, 1119.
Meier, R. W. (1965). *J. Opt. Soc. Amer.* **55**, 987.
Meier, R. W. (1966). *J. Opt. Soc. Amer.* **56**, 219.

7. Cardinal Points and Principal Rays for Holography

Meier, R. W. (1967). *J. Opt. Soc. Amer.* **57,** 895.

Miler, M. (1972). *Optica Acta* **19,** 555.

Neumann, D. B. (1966). *J. Opt. Soc. Amer.* **56,** 858.

Offner, A. (1966). *J. Opt. Soc. Amer.* **56,** 1509.

Rogers, G. L. (1951). *Proc. Roy. Soc. Edinburgh* **A63,** 14.

Stroke, G. W. (1965). *Appl. Phys. Lett.* **6,** 201.

Stroke, G. W. (1966). *Phys. Lett.* **23,** 325.

Vander Lugt, A. (1964). *IEEE Trans. Inform. Theor.* **IT-10,** 139.

Welford, W. T. (1973). *Opt. Comm.* **9,** 268.

Welford, W. T. (1975). *Opt. Comm.* **14,** 322.

Equipment and Procedures

8.1 SOLID STATE LASERS

Walter Koechner

8.1.1 Introduction

The major components of an optically pumped solid state laser oscillator are a cylindrical laser rod, a helical or linear flashlamp, a pump cavity which provides good optical coupling between the flashlamp and the laser rod, and an optical resonator, comprising a totally and a partially reflective mirror.

In order to modify the temporal, spectral, or spatial output characteristics of the oscillators, additional optical elements are usually inserted in the resonator, such as a Q-switch, an etalon, or an aperture. Auxiliary equipment of a laser oscillator includes a high-voltage power supply, energy storage capacitor, flashlamp trigger unit, and a water cooling system.

In solid state lasers, the active atoms of the laser medium are embedded in a solid host, such as a crystal or glass. The process of optical pumping consists of changing the atoms of the active material from their ground state to an excited state by means of light generated in a pump lamp and absorbed in the active material. At sufficiently high pump light intensities an inversion of the electron population in the laser material is achieved, which leads to energy storage in the upper laser level.

The optical resonator, comprising two opposing mirrors, performs the function of the feedback element. If the gain in the active material exceeds the total optical losses in the resonator, then laser output is obtained from the oscillator.

The duration of the flashlamp pulse is typically 0.5 to 1 ms long. At pump levels sufficiently above threshold the laser output follows approximately the temporal shape of the flashlamp pulse. The pulse length obtained from this

HANDBOOK OF OPTICAL HOLOGRAPHY
Copyright © 1979 by Academic Press, Inc.
All rights of reproduction in any form reserved.
ISBN-0-12-165350-1

257

conventional operation of the laser is too long for forming holograms of many objects; furthermore, it is exceedingly difficult to maintain a narrow linewidth over this time period due to heating effects in the laser host material.

With a device called a Q-switch the pulse duration can be shortened to tens of nanoseconds. A Q-switch is a fast-acting optical shutter placed inside the laser cavity. The switch remains closed and does not allow the laser to oscillate until after a period of optical pumping during which the population inversion reaches a level far above threshold. When the switch opens, the stored energy is suddenly released in the form of a very short pulse of light. The peak power of the pulse exceeds that obtainable from conventional mode oscillators by many orders of magnitude.

For further details on solid state lasers the reader is referred to Lengyel (1971) and Koechner (1976).

8.1.2 The Ruby and Nd:YAG Laser

8.1.2.1 The Ruby Laser

For holographic applications, ruby remains by far the most widely used solid state laser, primarily because of its large output energy and the wavelength of its radiation.

The rod of a ruby laser is made of synthetic sapphire Al_2O_3 which is doped with 0.05% by weight of Cr_2O_3. The substitution of a small percentage of the Al^{3+} with Cr^{3+} produces a pink-colored material. Laser action results from stimulation of Cr^{3+} ions by the pump light. The ruby laser emits red light at a wavelength of 0.6943 μm.

Typical dimensions of a ruby rod in an oscillator employed for holographic applications are 5-10 mm in diameter and 75-100 mm in length. Both end faces are polished parallel and antireflection coated. The oscillator is Q-switched either with a Pockels cell, Kerr cell, or saturable absorber.

The key element in the Pockels cell or Kerr cell is a material which becomes birefringent under the influence of an external electric field. We assume that the birefringent crystal is located between a polarizer and the rear mirror as shown in Fig. 1. The sequence of operation is as follows: During the flashlamp pulse a voltage is applied to the electrooptic cell which causes a $\lambda/4$ retardation between the x and y components of the incident beam. The incident linearly polarized light is circularly polarized after passing the Q-switch crystal. After being reflected at the mirror, the radiation passes through the electrooptic cell and undergoes another $\lambda/4$ retardation, becoming linearly polarized but at 90° to its original direction. This radiation is ejected from the laser cavity by the polarizer, thus preventing optical feedback. Toward the end of the flashlamp pulse the voltage on the cell is switched off permitting the polarizer-cell combination to pass a linearly polarized beam without loss. Oscillation within

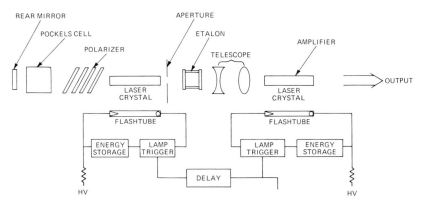

Fig. 1 Schematic diagram of a typical solid state laser oscillator–amplifier configuration employed in holography.

the cavity will build up, and after a short delay a Q-switch pulse will be emitted from the cavity.

The dye Q-switch is simply a liquid cell containing a saturable dye, such as cryptocyanine dissolved in methanol, which is placed inside the optical resonator. The dye initially absorbs the laser rod fluorescence, thus isolating the mirrors from the remainder of the resonator. As the light intensity becomes more intense the dye suddenly bleaches, the laser radiation can be reflected between both resonator mirrors, and laser oscillation occurs.

The advantages of passive dye Q-switches include low cost, simplicity of operation, and the emission of the output pulse in a narrow linewidth. However, there are a number of distinct disadvantages associated with the saturable Q-switch. The time between the triggering of the flashlamp and the emission of a Q-switched pulse is associated with a jitter which is typically of the order of 10 to 100 μs. Furthermore, with a dye Q-switch it is not possible to obtain two Q-switch pulses.

In holographic applications, very often precise timing between an event and the output pulse is required. In addition, holographic interferometry requires the generation of two Q-switched pulses. For these reasons the Pockels cell Q-switch is employed on most commercial holographic lasers.

In applications requiring greater Q-switch energy than obtainable from an oscillator, it is possible to build an oscillator–amplifier system. Ruby rods for amplifiers have diameters from 1 to 2 cm and lengths up to 20 cm. The spatial and temporal coherence are essentially preserved in the process of amplification. A discussion of ruby lasers employed in holography has been given by Koechner (1973, 1976), Wuerker and Heflinger (1971), Gregor and Davis (1969), Gregor (1971), Young and Hicks (1974), and Riley (1973).

8. Equipment and Procedures

8.1.2.2 The Nd:YAG Laser

Neodymium-doped yttrium aluminum garnet (Nd:YAG) possesses a combination of properties uniquely favorable for laser operation. In particular the cubic structure of YAG favors a narrow fluorescent linewidth, which results in high gains and low threshold for laser operation. The laser transition has a wavelength of 1.064 μm.

For holographic applications the advantages of Nd:YAG compared to ruby, namely, a more efficient operation and a high pulse repetition rate capability, are offset by two major disadvantages: A Nd:YAG laser is not capable of generating as much Q-switch energy as a ruby laser, and the output is in the infrared. In order to utilize a Nd:YAG laser, the output wavelength has to be reduced to 0.5300 μm employing a harmonic generator at the output. Frequency doubling of Nd:YAG can be accomplished by means of a temperature-controlled cesium dideuterium arsenate crystal (CD*A) or some other nonlinear crystal. Typical conversion efficiencies are of the order of 20 to 40%.

The maximum output energy obtainable from the largest frequency doubled Nd:YAG laser is about two orders of magnitude lower than the energies from large ruby lasers of comparable spatial and spectral quality. On the other hand, ruby is limited to a maximum pulse repetition rate of 1 pps, whereas a Nd:YAG laser is capable of up to 50 pps. The lower output capabilities of Nd:YAG combined with the added complexity of a harmonic generator have made Nd:YAG a not very successful contender in the field of holography. Holographic Nd:YAG lasers are discussed by Way (1975) and Bates (1973).

8.1.3 Major Design Characteristics of Holographic Lasers

Solid state lasers employed for holography are characterized by a high degree of spatial and temporal coherence. For some holographic purposes the ability to emit two pulses with a short interpulse separation is useful as well. Spatial coherence of the reference beam is desirable to obtain large, high-resolution holograms. The temporal coherence of the laser determines the depth of the object or scene from which a hologram can be made. The double pulse capability is essential for some applications in nondestructive testing. Usually two holograms, with time intervals between 1 and 1000 μs, are superimposed on the same photographic plate. Any perturbation of the test object during this time interval will show up as interference fringes on the double pulsed hologram. The double pulse technique makes it possible to apply holography to stress analysis, shock propagation, and vibration studies and to flow visualization of projectiles passing through air. See Section 10.4 for more details on this application.

8.1.3.1   Spatial Coherence

Pumping a laser rod 6 to 10 mm in diameter located in a typical laser resonator 50 to 100 cm long will cause a large number of transverse modes to oscillate simultaneously across the rod diameter. Since the oscillation frequencies of transverse modes are unrelated, the spatial coherence of the output light is very poor. The oscillator can be forced to operate in the TEM_{00} mode through insertion of an aperture of about 2 mm into the resonator. The single mode operation of the laser results in a Gaussian beam intensity profile and a uniphase wavefront.

8.1.3.2   Temporal Coherence

If a laser is operated without any axial mode-selecting elements in the cavity, then the spectral output will comprise a large number of discrete frequencies determined by the longitudinal modes. The linewidth of the laser transition limits the number of modes that have sufficient gain to oscillate. The situation is diagrammed schematically in Fig. 2, which shows the resonance frequencies of an optical resonator and the fluorescence line of the active material. Laser emission occurs at those wavelengths at which the product of the gain of the

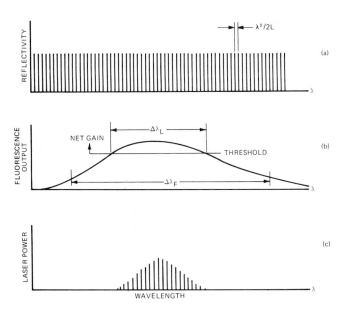

Fig. 2 Schematic diagram of the spectral output of a laser without mode selection: (a) optical resonator, (b) active material, (c) laser output, where $\Delta\lambda_F$ is the fluorescence linewidth, and $\Delta\lambda_L$ the laser linewidth.

261

8. Equipment and Procedures

laser transition and the reflectivity of the mirrors exceeds unity. In the idealized example shown, the laser would oscillate at 27 axial modes.

The wavelength separation of two adjacent longitudinal modes is given by

$$\Delta\lambda = \lambda_0^2/2L,$$

where L is the optical length of the resonator. With $L = 75$ cm and $\lambda_0 = 0.6943$ μm, one obtains $\Delta\lambda = 3 \times 10^{-7}$ μm. Depending on the pumping level for ruby and Nd:YAG, one finds a linewidth of approximately 3×10^{-5} to 5×10^{-5} μm for the laser emission in the absence of mode selection. Therefore, these lasers typically oscillate in about 100 to 150 longitudinal modes.

The temporal coherence length of a laser is strongly dependent on the number of longitudinal modes that can oscillate simultaneously. If we define the coherence length l_c as the pathlength difference for which the fringe visibility in a Michelson interferometer is reduced to $1/\sqrt{2}$, then we obtain

$$l_c = 2L/N, \qquad N \geq 2,$$

where N is the number of longitudinal modes.

The coherence length l_c of a single mode laser

$$l_c \approx 4L\sqrt{R_1}/(1 - R_1),$$

where L is the resonator length and R_1 the reflectivity of the front mirror (it is assumed that $R_2 = 1$). The single axial mode output pulse from a ruby oscillator having a cavity length of 75 cm and a front mirror reflectivity of $R_1 = 0.4$ will have a coherence length of $l_c = 5.2$ m.

Linewidth Control It is possible to discriminate against most of the axial modes by adding additional reflecting surfaces to the basic resonator. If a Fabry–Perot type reflector is inserted between the two mirrors of the resonator, it will cause a strong amplitude modulation of the closely spaced reflectivity peaks of the basic laser resonator. This will prevent most modes from reaching threshold.

The role of the resonant devices employed in interferometric mode selection is to provide high feedback for a single wavelength near the center of the fluorescence line, while at the same time discriminating against nearby wavelengths. For example, by replacing the standard dielectrically coated front mirror with a single plate resonant reflector, the number of oscillating modes can be greatly reduced.

Resonant reflectors featuring two, three, or four plates have reflectivity peaks which are much narrower as compared to a single sapphire etalon; this makes such a unit a better mode selector. The etalons fabricated from quartz or sapphire have a thickness which is typically 2 to 3 mm. This assures a sufficiently large spectral separation of the reflectivity maxima within the fluorescence curve so that lasing can occur on only one peak. In multiple plate resonators, the spacing between the etalons is 20 to 25 mm in order to achieve a narrow width of the main peak.

In an ideal resonant reflector the reflectivity as a function of wavelength shows very narrow peaks which are widely separated. Combining several mode selecting techniques, such as the use of a multiple plate resonator reflector, operation close at threshold, a saturable absorber Q-switch or a Pockels cell Q-switch with a very slow rise time, single axial mode operation of a ruby oscillator is possible.

Mode selection is considerably enhanced by operation close to threshold and by the use of a Q-switch which allows a large number of round trips in the resonator.

Longitudinal mode selection in the laser takes place while the pulse is building up from noise. During this build-up time, modes which have a higher gain or a lower loss will increase in amplitude more rapidly than the other modes. The difference in amplitude between two modes becomes larger if the number of round trips is increased. Therefore, for a given loss difference between the modes it is important for good mode selection to allow as many round trips as possible.

The development of a pulse in a dye Q-switched laser takes longer than, for example, in the case of a Pockels cell Q-switched system. However, a Pockels cell Q-switch can be operated in a manner that ensures a large build-up time by increasing the risetime or by opening the switch in two steps.

8.1.3.3   Multiple Pulse Operation

Techniques to extract multiple pulses from a ruby oscillator depend on the time separation between the pulses.

(a) Pulse Separation 1 μs to 1 ms   This time interval is the one most commonly used in double pulse holography. The output is obtained from a standard single pulse system by Q-switching the laser twice during its pump cycle. The longest pulse separation which can be achieved is determined by the length of the flashlamp pulse. The shortest time interval is determined by the switching electronics of the Pockels cell and the buildup time of the Q-switch pulse. The application of ruby lasers in doubled pulsed holography requires that the energy in the two pulses be equal. This can be achieved by adjusting the delay between the flashlamp trigger and the first Q-switch pulse, by adjusting the voltage of the Pockels cell, and by selecting the lamp input energy. By changing one or all of these parameters one can obtain equal output energies in both pulses over the time interval indicated above.

(b) Pulse Separation 20 ns to 1 ms   A technique which allows one to reduce the time interval between pulses essentially down to zero involves the utilization of a dual Q-switch oscillator. In these sytems two giant pulses are

8. Equipment and Procedures

TABLE I

Typical Performance Data of Holographic Ruby Laser Systems[a]

Type of system	Coherence length (meters)	Single pulse operation; energy (joules)	Double pulse operation[b]; energy per pulse (joules)
Oscillator	0.5–2	0.040	0.025
	5–10	0.010	0.007
Oscillator and one amplifier	0.5–2	1	0.4
	5–10	0.2	0.2
Oscillator and two amplifiers	0.5–2	10	4
	5–10	4	2

[a] Output wavelength, 0.6943 μm; pulsewidth, 15–40 ns; pulse rate (typical), 2 ppm; transferse mode, TEM_{00}.

[b] Double pulse separation 1–1000 μs.

extracted from different areas of the ruby rod. The system utilizes two separate Pockels cells with a double aperture in the cavity to select two separate TEM_{00} outputs from the rod, each one Q-switched with its own Pockels cell. The output of the laser consists of two beams separated by approximately 6 mm and separable in time from essentially 0 to 1 ms. The two output beams may

TABLE II

Typical Performance Data of Holographic Nd:YAG Lasers[a]

Type of system	Coherence length (meters)	Single pulse operation energy (joules)	Double pulse operation[b]; energy per pulse (joules)
Oscillator and two amplifiers and frequency doubler	1	0.040	0.020
Oscillator and three amplifiers and frequency doubler	1	0.150	0.070

[a] Output wavelength, 0.5300 μm; pulse width, 15–25 ns; pulse rate (typical), 5 pps; transverse mode, TEM_{00}.

[b] Double pulse separation 1–1000 μs.

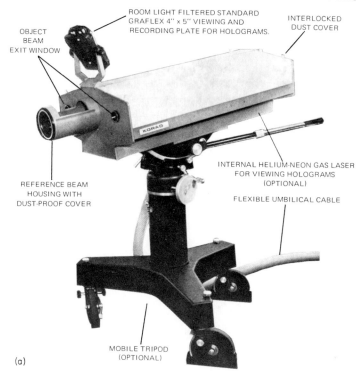

ROOM LIGHT FILTERED STANDARD
GRAFLEX 4" x 5" VIEWING AND
RECORDING PLATE FOR HOLOGRAMS.

INTERLOCKED
DUST COVER

OBJECT
BEAM
EXIT WINDOW

INTERNAL HELIUM-NEON GAS LASER
FOR VIEWING HOLOGRAMS
(OPTIONAL)

FLEXIBLE UMBILICAL CABLE

REFERENCE BEAM
HOUSING WITH
DUST-PROOF COVER

MOBILE TRIPOD
(OPTIONAL)

(a)

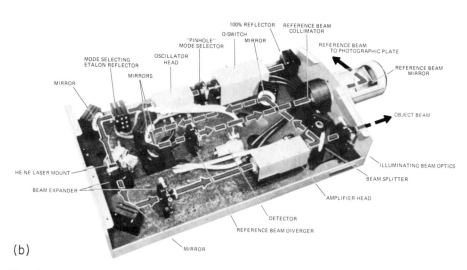

100% REFLECTOR   REFERENCE BEAM
Q-SWITCH   COLLIMATOR
"PINHOLE"   MIRROR
MODE SELECTOR

REFERENCE BEAM
TO PHOTOGRAPHIC PLATE

OSCILLATOR
HEAD

MODE SELECTING
ETALON REFLECTOR

REFERENCE BEAM
MIRROR

MIRRORS

MIRROR

OBJECT BEAM

HE-NE LASER MOUNT

ILLUMINATING BEAM OPTICS

BEAM EXPANDER

BEAM SPLITTER

AMPLIFIER HEAD

DETECTOR

REFERENCE BEAM DIVERGER

MIRROR

(b)

Fig. 3 Photograph of a holographic camera mounted on a tripod. (Korad Div., Hadron Inc., Model KHC.) (a) Operational system, (b) with covers removed.   d

be recombined by using a beamsplitter and several mirrors or prisms. Multiple cavity laser techniques are also discussed by Landry (1971).

(c) Pulse Separation 1 ms to 1 s In order to extend the pulse interval time beyond the 1-ms range, the flashlamp has to be fired twice. Since the time interval is too short for a recharge of the pulse forming network after the first pulse is issued, one usually resorts to the technique of charging two capacitor banks simultaneously. The two PFNs, decoupled from each other by means of diodes or ignitrons, are discharged at a time interval which is determined by the desired pulse separation.

(d) Pulse Separation Larger Than 1 s In these cases the system is operated in the single Q-switched mode at a repetition rate according to the required interpulse separation. Engineering aspects of holographic lasers are covered in detail by Koechner (1976).

8.1.4 Typical Performance Data

Tables I and II show typical performance data for holographic ruby and Nd:YAG laser systems. The information presented in these tables is limited to those properties of the lasers which are relevant to holography.

8.1.5 Holographic Cameras

Recording of holograms requires, besides the laser, peripheral optical components, such as beamsplitters, mirrors, filters, lenses, and a photographic plate. From a user's point of view, especially in a nonlaboratory environment, it is often desirable to have all components required to take a hologram packaged in one enclosure. In response to this requirement, several companies have developed so-called holographic cameras.

The unit shown in Fig. 3 contains a ruby laser, a detector to monitor pulse energy, optical components to manipulate the beam, and a He–Ne gas laser for reconstruction of the hologram. The object beam emerges from the left window of the enclosure, whereas the reference beam is reflected onto the photographic plate mounted on top of the unit by means of a mirror housed in the tubular structure in front of the camera.

With two amplifiers containing 15×1 cm ruby rods pumped at 6 kJ, the sytem shown in Fig. 3a is capable of producing 4 J of output in a single transverse and longitudinal mode. In the double pulsed mode two pulses with 2 J of energy and time intervals from 1 to 1000 μs are generated.

REFERENCES

Bates, H. E. (1973). *Appl. Opt.* **12,** 1172.

Gregor, E. (1971). *Proc. SPIE* **93.**

Gregor, E., and Davis, J. H. (1969). *Electro-Opt. Syst. Design* **48.**

Koechner, W. (1973). *Indust. Res.*

Koechner, W. (1976). "Solid-State Laser Engineering." Springer, New York.

Landry, M. J. (1971). *Appl. Phys. Lett.* **18,** 494.

Lengyel, B. A. (1971). "Lasers." Wiley (Interscience), New York.

Riley, L. W. (1973). *Opt. Spectra* **27.**

Way, F. C. (1975). *Proc. Electro-Opt. Syst. Design Conf., Anaheim, California.*

Wuerker, R. F., and Heflinger, L. O. (1971). *SPIE J.* **9,** 122.

Young, M., and Hicks, A. (1974). *Appl. Opt.* **13,** 2486.

8.2 GAS LASERS

N. Balasubramanian

8.2.1 Introduction

Gas lasers are the most common and widely used light sources for generating and reconstructing holograms. In fact, the emergence of holography as a practical and useful tool began with the advances in the practical manufacture of various gas laser sources. In this section, a brief discussion of the properties of gas lasers as they relate to the generation and reconstruction of holograms is given. The intent is to provide an overview of the properties of gas laser sources so as to familiarize a holographer with the available sources. The reader is referred to Bloom (1966, 1968), Sinclair and Bell (1969), and Buzzard (1976) for excellent and comprehensive reviews of the science and technology of gas lasers.

The main emphasis here is on He–Ne, argon ion, krypton ion lasers. Other gas lasers such as CO_2 lasers and He–Cd lasers are not covered since they are seldom used in holographic applications. The properties of gas lasers relevant to holography are typically determined by the resonator cavity and not by the lasing medium, except for the wavelength of laser emission. The most important property of gas lasers from the viewpoint of holographic applications is the coherence of the laser output. Gas lasers taken as a class possess coherence characteristics superior to those of any other type of lasers. Other properties of gas lasers relevant to holographic applications are the range of wavelengths of laser oscillation and the range of power output obtainable.

8.2.2 Coherence Properties of Gas Lasers

The temporal and spatial coherence of the laser source used for recording and reconstructing the hologram to a large extent determine not only the properties of the hologram produced but also the complexity of the optical system configuration that can be used for recording the hologram. Temporal coherence relates to the finite frequency bandwidth of the source and spatial coherence to its finite extent in space. In the case of a gas laser, the temporal coherence is determined by the temporal (or the longitudinal) modes of the laser cavity and the spatial (or transverse) mode of the laser cavity. The highest degree of coherence, both spatial and temporal, is obtained with a single mode operation. The exact mathematical definitions of temporal and spatial coherence of light sources and their effect on holographic recording and reconstruction are provided in Section 2.3.

The spatial mode of a laser cavity is the field distribution which repeats itself after every traverse of the cavity. Depending on the losses in the cavity the laser can oscillate in several spatial modes. In addition, the field distribution propagating within the cavity takes on a standing wave configuration determined by the separation of the cavity mirrors. Several frequencies can satisfy this resonance condition, and the possible frequencies of oscillation are referred to as temporal modes. Also, corresponding to each spatial mode there are several temporal modes.

The frequency separation between the temporal modes that exist within the resonator cavity is given by $\Delta F = C/2L$, where C is the velocity of light and L the effective cavity length. Since the gain bandwidth of the gain medium defines the frequency range over which laser oscillations can occur, the cavity length in conjunction with the gain bandwidth of the laser gain medium define the number of the temporal modes and the distance between them in frequency space. If ΔF is the width of the single temporal mode, the coherence length of such a mode is given by $L_c = C/\Delta F$. Typically, ΔF is of the order of 10^5 Hz. Hence, L_c is of the order of a kilometer. When the laser is oscillating in more than one temporal mode, the coherence length is reduced. A good mathematical treatment of the relationship between the number of temporal modes and the coherence function is given by Collier *et al.* (1971).

The expression for the coherence function can be written as

$$|\mu(L_c)| = \left| \frac{\sin(N\pi L_c/2L)}{N\sin(\pi L_c/2L)} \right| ,$$

where L_c is the optical path difference and N the number of temporal modes. It is also clear from the above expression that the degree of coherence is periodic in terms of the distance between the cavity mirrors.

From the preceding discussion, it is clear that a single temporal mode of operation of a laser is preferred for many holographic applications. Short coherence lengths of the laser source not only limit the depth of the object field that can be recorded, but also necessitate cumbersome optical system configurations for achieving equal optical path distances between interfering beams. Several techniques of internal cavity perturbation to force single temporal mode operation by the laser are usually employed to improve laser coherence length in long-cavity high-power laser systems. All of these techniques invariably introduce losses in the cavity and result in reduced laser power output. Some of the techniques used are (1) intercavity etalons (or etalon reflector end mirrors), (2) AM phase locking, and (3) FM mode coupling. The reader is referred to Bloom (1966) for details on these techniques.

The spatial mode characteristics define the spatial coherence of the laser output. A single spatial mode has the maximum spatial coherence and also gives the minimum beam divergence. In most laser resonators, the control of spatial modes is obtained by controlling the losses around the edge of the mode. For many practical purposes, a good approximation to the loss incurred in any given mode can be calculated by computing the integral of the intensity in the part of the mode that lies outside the clear region of the resonator of the laser tube. In a laser resonator, the losses around the edges are determined by the plasma tube diameter. The selection of the plasma tube diameter in conjunction with the mirror radii determine the spatial mode characteristics of a laser resonator. When a laser operates on a higher order spatial mode, each mode is spatially coherent. However, when all the modes are considered in combination, it can be treated as an equivalent extended source defined by the spot size associated with the spatial mode of the beam.

8.2.3 Coherence Characteristics of Commercial Lasers

a. He–Ne Lasers The laser beam output from most He-Ne lasers is designed to the lowest order spatial model TEM_{00}. The intensity distribution across the beam is Gaussian and the phase is uniform and the same across the wavefront. The length of the cavity of the various He-Ne lasers ranges from 25 cm to nearly 2 m. Hence, the temporal mode spacing ranges from 600 to 75 MHz. The gain bandwidth at half maximum points for a He-Ne laser operating at 6328 Å is of the order of 1500 MHz. The number of temporal modes and hence the useful coherence length depend on the length of the laser cavity.

b. Argon Ion Lasers Commercial argon ion laser cavities are designed from the viewpoint of maximum output power rather than from that of superior output beam quality. Because of the long radii mirror cavity configuration, the argon ion lasers tend to operate in higher order spatial modes. However, all commercially available ion lasers have an adjustable intracavity aperture which limits the diameter of the transverse mode. This aperture serves as a diffraction loss mechanism for the higher order spatial modes and forces the laser to operate in the fundamental TEM_{00} spatial mode. It should be noted that TEM_{00} operation of argon ion lasers usually results in about a 30% reduction of power output as compared to the case in which the intracavity aperture is fully open.

In contrast to the He-Ne laser, the argon ion laser is a multiline laser. Using broad band end mirror reflectors, the laser can oscillate simultaneously in several discrete wavelengths. All commercial argon ion lasers have the option of interchanging the end mirrors of the cavity for mirrors designed to operate at single wavelengths or the introduction of a prism assembly which permits the selection of single frequency operation. The gain bandwidth of the argon

8. Equipment and Procedures

TABLE I

Wavelength Range for Gas Lasers

He–Ne (nm)	Ar (nm)	Kr (nm)
	335.0	
		337.5
		350.7
	351.1	
		356.4
	363.8	
		413.0
	418.2	
	437.1	
	454.5	
	457.9[a]	
		461.9
	465.8	
		468.0
	472.7	
		476.2[a]
	476.5[a]	
		482.5
	488.0[a]	
	496.5[a]	
	501.7[a]	
	514.5[a]	
		520.8[a]
	528.7[a]	
		530.9[a]
		568.1[a]
611.8		
632.8[a]		
640.1		
		647.1[a]
		676.4[a]
		752.5[a]
		793.1
		799.3
		858.8
	1092.0[a]	
1150.0[a]		
3390.0[a]		

[a] Major laser lines.

ion lasers is typically about 3.5 GHz, and temporal mode spacing for a 1–2-W argon ion laser is of the order of 150 MHz. Intracavity etalons are usually used to obtain single frequency operation of the laser for any given wavelength, thus permitting extremely long coherence lengths to be achieved. When long coherence length and high laser power output are needed in holographic applications, argon ion lasers are the best choice.

c. Krypton Ion Lasers The resonator cavity characteristics of the krypton ion lasers are identical to those of the argon ion lasers. Hence, the coherence characteristics are identical to those of the argon ion lasers. Except for the wavelengths and the laser power output, there are not many differences between the two ion lasers. Krypton lasers with intracavity etalons provide high-power output and large coherence length in the red region of the visible spectrum.

8.2.4 Wavelength Range of Gas Laser Output

Between the three gas lasers, He–Ne, argon, and krypton lasers, there are as many as 40 different laser lines available. However, many of these lines are relatively weak and a very few are utilized for holographic applications. Table I shows the wavelength range of the three gas lasers under consideration. From the point of view of a holographer, familiarity with the wavelength range of the gas lasers is necessary because of their importance in determining the availability of suitable recording media and the sensitivity of the recording media. It is also important when one considers the generation of color holograms or color multiplexed holograms. Tables II–IV list the characteristics of He–Ne, argon, and krypton lasers and provide an overview of the wavelength range and relative output powers that are available for a holographer for most practical gas lasers.

8.2.5 Range of Power Outputs

Of the three types of gas lasers being considered here, the He–Ne laser is a low-power device, while the argon and krypton lasers are capable of very

TABLE II

Laser Output Power

Laser	Length (cm)	Power output
He–Ne	25–200	1–50 mW
Ar	30–200	0.01–10 W[a]
Kr	75–200	0.05–2 W[a]

[a] Multiline.

TABLE III

Output Power Range of Laser Lines

Laser line (nm)	Laser type	Output power range[a] (mW)		
		Low	Medium	High
335.00	Ar			50
337.5	Kr			25
350.7	Kr		50	500
351.1	Ar		30	400
356.4	Kr		35	450
363.8	Ar		30	500
413.0	Kr			1.2 W
418.2	Ar		3	
437.1	Ar		2	
454.5	Ar		50	800
457.9	Ar	15	150	1.5
461.9	Kr		5	
465.8	Ar		50	750
468.0	Kr		7	200
472.7	Ar		60	1.2 W
476.2	Kr		50	250
476.5	Ar	25	300	2.0 W
482.5	Kr		30	250
488.0	Ar	80	700	5.0 W
496.5	Ar	25	400	2.0 W
501.7	Ar	15	100	1.5 W
514.5	Ar	150	800	5.5 W
520.8	Kr		70	250
528.7	Ar	20	200	900
530.9	Kr		20	700
568.1	Kr		15	500
611.8	He–Ne			5
632.8	He–Ne	2	15	50
640.1	He–Ne			2
647.1	Kr	50	600	2.0 W
676.4	Kr	15	100	400
752.5	Kr		150	500
793.1	Kr		10	20
799.3	Kr		30	120
858.8	Kr			25
1092.0	Ar		100	
1150.0	He–Ne		2	10

[a] Values are approximate.

TABLE IV

Comparison of Laser Parameters

Parameter	He–Ne	Ar	Kr
Wavelength (nm)	632.8	514.5	647.1
Output power (mW)	50	100	50
Laser head length (cm)	200	70	70
Input power (W)	450	2700	2700
Optical noise (%)	1	1	1
Amplitude stability (%)	5	2	2
Beam stability (μrad/°C)	10	15	15
Tube life (hr)	6000	5000	5000
Cooling	Air	Water	Water
Approximate cost ($)	8000	10,000	10,000

high output power. The comments made relating to the power output must be considered within the context of devices having sizes that are likely to be encountered in laboratories. Table II lists the output power of the gas lasers being considered.

Many very-high-power argon and krypton ion lasers (up to 15 to 20 W) are available commercially, but because of their large size and extremely large power supply and cooling requirements, they cannot be considered suitable for holographic applications. The laser power in conjunction with the sensitivity of the recording media usually determine the time of exposure which in turn determines the susceptibility of the optical system to vibration, thermal turbulence, etc. Also, the output power determines the object field that can be recorded for reasonable exposure times.

8.2.6 Lifetime and Cost Considerations

The He–Ne lasers represent by far the most economical lasers available for holographic applications. Their lifetime usually exceeds 6000 hr and their prices range from a few hundred dollars to a few thousand dollars depending on the power output requirement. They do not usually need either special electrical power requirements or water cooling. These are the main reasons why He–Ne lasers are widely used in holographic applications.

Argon and krypton ion laser tubes have lifetimes of 5000 hr. or less and as laser systems, they are very expensive. They not only have special electrical power requirements but also need a continuous water supply to cool the plasma tube. They are the only sources available whenever color holograms need to be generated. In industrial applications where the special requirements of ion lasers do not represent an inconvenience, ion lasers are considered suitable because of their power output and long coherence lengths.

8. Equipment and Procedures

REFERENCES

Bloom, A. L. (1966). *Appl. Opt.* **5,** No. 10, 1500.
Bloom, A. L. (1968). "Gas Lasers." Wiley, New York.
Buzzard, R. J. (1976). *Opt. Eng.* **15,** No. 2, 77.
Collier, R. J., Burckhardt, C. B., and Lin, L. H. (1971). "Optical Holography." Academic Press, New York.
Sinclair, D. C., and Bell, W. E. (1969). "Gas Laser Technology." Holt, New York.

276

8.3 RECORDING MEDIA

James W. Gladden
Robert D. Leighty

8.3.1 Introduction

The future of holography is dependent upon holographic recording media. Most limitations associated with today's holographic applications can be attributed to nonoptimum recording materials. This review deals mainly with the holographic recording media which are commercially available, those "prepare-before-use" materials that have been used mainly in research environments, and materials that have demonstrated potential for achieving commercial status. Table I summarizes the major classes of holographic materials to be discussed. Subsequent sections briefly describe each of these material classes, and properties of selected materials within each class are presented in accompanying tables. Pertinent references are indicated to assist the reader in obtaining greater detail where desired.

Unfortunately, no book chapter can do justice to the broad scope and details of holographic recording media presently used or under study in the many facilities here and abroad. This summary will not deal with electrooptical devices, dry silver, amorphous semiconductors, vesicular films, diazotype films, free radical films, or the alkali halides.

8.3.2 Previous Review Articles

There are several review articles treating holographic recording materials that discuss the major classes of recording materials in general and provide some insight into their holographic parameters and photosensitive properties. No single review article will cover the broad scope and details of the field in a comprehensive manner, mainly because of the diverse nature of the field. Individual reviews will be oriented along the interests and experiences of the authors. They are valuable, within this context, and provide the reader with a general view of more than one type of recording material, not only with properties of the individual materials, but also with a contrast between materials for selected parameters. A novice interested in gaining an overview to aid in selection of a recording material for his application will usually have his

HANDBOOK OF OPTICAL HOLOGRAPHY

TABLE I

Holographic Recording Materials

Class of material	Preparation	Recording process	Processing	Readout process	Re-cyclable	Erase process	Erase time	Storage time	Repli-cation	Table No.
Photographic materials	Coating technique	Reduction to Ag-metal grains	Wet chemical	Density change	No	NA	NA	Permanent	Yes	II
Hardened di-chromated gelatin	Coating technique or chemical treatment of commercial photo-graphic materials	Photocross-linking	Wet chemical	Refractive index change	No	NA	NA	Permanent	Yes	III
Photoresists	Spin or spray coating	Formation of an organic acid, photocross-link-ing, or photopolymer-ization	Wet chemical or heated air	Surface relief	No	NA	NA	Permanent	Yes	IV
Photopolymers	Coating technique or cast-ing	Photopolymerization	None or post exposure and post heating	Refractive index change or surface relief	No	NA	NA	Permanent	?	V
Photoplastics	Evaporation and coating technique	Formation of an elec-trostatic latent image with electric field produced deformation of heated plastic	Corona charge and heat	Surface relief	Yes	Heat	~1 sec	Permanent	Yes	VI
Photochromics	Crystal wafer or disper-sion in glass or polymer films	Generally photoinduced new absorption bands	None	Density change (am-plitude) or refrac-tive index change	Yes	Actinic light and/or heat	Not known to nanoseconds	Minutes to months	Yes	VII

options narrowed to the point where he can then gainfully consult the referenced articles for details on separate materials and techniques.

Selected review articles will now be briefly outlined chronologically to provide the reader with supplemental sources and to indicate differences so that the context of this paper will be more meaningful.

8.3.2.1   Urbach and Meir (1969)

In an article entitled ''Properties and Limitations of Hologram Recording Materials,'' the authors review the role of recording materials in holographic imaging. Noise characteristics of recording media are discussed in terms of their effect on holographic recording.

8.3.2.2   Urbach (1971)

''Advances in Hologram Recording Materials'' classifies different recording materials according to their ability to form thin or thick, amplitude or phase holograms. Subclassifications according to reflection or transmission types are also treated. The article also summarizes basic performance characteristics and surveys the contemporary research. There is a discussion of materials for use in infrared and ultraviolet holography as well.

8.3.2.3   Collier, Burckhardt, and Lin (1971)

In a chapter of this text entitled ''Hologram Recording Materials,'' the authors describe the method of hologram formation, exposure and sensitivity, recording resolution, noise, recording linearity, and exposure characteristics observed with holograms. They describe the holographic properties of several recording materials.

8.3.2.4   Pennington (1971)

In the ''Handbook of Lasers,'' Pennington's chapter entitled ''Holographic Paramerers and Recording Materials'' presents many sensitivity curves, diffraction versus exposure curves, MTFs, recipes, etc.

8.3.2.5   RCA Review (**33**, No. 1, March 1972)

A topical issue of *RCA Review* entitled ''Optical Storage and Display Media'' presents a number of articles describing materials for holographic recording. The articles discuss holographic information storage, redundant holograms, recyclable holographic storage media, and the RCA Holotape. Two articles deserve further mention. Ramberg (1972), in an article entitled ''Holographic

8. Equipment and Procedures

Information Storage" surveys different holographic types, their distinguishing characteristics, and physical processes employed in preparing them. He addresses factors limiting storage capacity of plane and volume holograms and evaluates them in a semiquantitative fashion. Bordogna *et al.* (1972), in their article entitled "Recyclable Holographic Storage Media," compare performance parameters of the subject media to develop tradeoffs for their use in holographic storage and in imaging applications.

8.3.2.6 Colburn, Zech, and Ralston (1973)

A report entitled "Holographic Optical Elements" (HOE) evaluates seven materials for HOE applications through measurement of holographic sensitometric and readout parameters and investigation of their stability under differing temperature and humidity conditions.

8.3.2.7 Zech, Shareck, and Ralston (1974)

A report entitled "Holographic Recording Materials" presents detailed evaluations of a number of high-quality, dry-working recording materials for suitability to holographic data storage and optical data processing. Twelve novel recording materials are described.

8.3.2.8 Zech (1974)

A doctoral dissertation entitled "Data Storage in Volume Holograms" evaluates several photosensitive phase materials. Hologram parameter measurements described include diffraction efficiency and signal-to-noise ratio for different volume phase materials. The holographic responses of nearly ideal volume and planar phase recording materials are compared.

8.3.2.9 Kurtz and Owen (1975)

An article entitled "Holographic Recording Materials—A Review" will familiarize the reader with various options for the selection of holographic recording materials for a particular application, in addition to the materials review. The paper contains valuable tables and many references.

8.3.2.10 Bartolini, Weakliem, and Williams (1976)

This article entitled "Review and Analysis of Optical Recording Media" reviews most of the known classes of optical recording media and explains a procedure for identifying materials with potential interest for specific applications. The properties of eleven classes of recording media are summarized in tables. This article forms a substantial basis for our article.

8.3.2.11 Gladden (1978)

A report entitled ''Review of Photosensitive Materials for Holographic Recording'' builds on the preceding references, along with others, to present a deeper view of the chemistry, mechanisms, and processes associated with holographic recording materials with a view toward developing particular photosensitive materials for holographic uses. This article also forms a substantial basis for our present section.

8.3.3 Silver Halide Emulsions

The photographic emulsions have enjoyed a popularity in holographic recording not seen by nonsilver halide emulsions. There are several reasons for their importance. For example, they have very high exposure sensitivity and resolving power and a wide range of spectral sensitivities. In addition to being easily used, the photographic emulsions are versatile in that they can be used to prepare either planar or volume holograms in either amplitude or phase modes. Photographic emulsions are available on film or glass plates. The recording process is by nature photochemical, resulting in an optical density change that modulates the readout beam. Chemical processing is necessary to develop and fix the latent image after exposure. Erasure, overwriting, or recycling is not possible after fixing. Replication is usually accomplished through contact printing procedures. In this section, primarily planar amplitude holographic recordings will be treated. The production of phase holograms, volume holograms, and other modifications on the basic silver halide hologram are discussed in Section 9.1 and will be omitted here.

Commercial photographic emulsions used in high spatial frequency holographic recordings (usually greater than 300 c/mm) are listed in Table II. These emulsions will now be discussed briefly. The Kodak 649F emulsion is a spectroscopic emulsion available on either an Estar film or glass plates. The Kodak 649GH is a film containing a high resolution emulsion also used on Kodak SO-343 film and Kodak 1A and 2A plates. These high resolution emulsions are orthochromatic; the maximum spectral sensitivity in the visible is in the blue-green region. The Kodak 2A plate is two to three times less sensitive than the other emulsions in the high resolution series. It is designed to suppress Rayleigh scattering during exposure at wavelengths below 5000 Å, and it will exhibit less noise in the holographic reconstruction image than the other emulsions. The Agfa 10E56, 10E70, and 10E75 emulsions have lower resolving power than the 8E56, 8E70, and 8E75 emulsions, but they have four to seven times greater exposure sensitivity than the 8E series. The Agfa 14C70 and

TABLE II

Photographic Materials

Material	Substrate	Usable thickness (μm)	Recording wavelength range	Recording sensitivity[a] (J/cm²)	Limiting resolution (c/mm)	Reference
Kodak 649F	Estar film and plate	6	Panchromatic	$\sim 8\times 10^{-5}$	>3000	Pennington (1971)
		17				Eastman Kodak Co. (1976)
Agfa 8E70	Plate	6	Panchromatic[b]	2×10^{-5}	3000	Pennington (1971)
Agfa 8E75	Plate	6	Panchromatic[c]	2×10^{-5}	>3000	Pennington (1971) Agfa Gevaert
Kodak 131	Plate	9	Panchromatic	$\sim 2.4\times 10^{-6}$	~2500	Eastman Kodak Co. (1976)
Kodak SO-253	Estar film	9	Panchromatic	$\sim 2.4\times 10^{-6}$	~2500	Eastman Kodak Co. (1976)
Agfa 10E70	Acetate film and plate	6	Panchromatic[b]	5×10^{-6}	1500	Pennington (1971)
Agfa 10E75	Plate	6	Panchromatic[c]	5×10^{-6}	~2500	Pennington (1971) Agfa Gevaert
Kodak 649GH	Estar film	7	Orthochromatic	$\sim 9.5\times 10^{-5}$	>3000	Pennington (1971)
Kodak SO-343	Estar film (thick base)	7	Orthochromatic	$\sim 9.5\times 10^{-5}$	>3000	Eastman Kodak Co. (1976)
Kodak 1A	Plate	6	Orthochromatic	$\sim 9.5\times 10^{-5}$	>3000	Eastman Kodak Co. (1976)
Kodak 2A	Plate	6	Orthochromatic	$\sim 2.1\times 10^{-4}$	>3000	Eastman Kodak Co. (1976)
Agfa 8E56	—	6	Orthochromatic[d]	$\sim 4\times 10^{-5}$	>3000	Pennington (1971) Agfa Gevaert

Kodak 125	Plate	7	Orthochromatic	$\sim 5.0\times 10^{-6}$	>2500	Eastman Kodak Co. (1976)
Kodak SO-141	Estar film	<3	Orthochromatic	$\sim 5.0\times 10^{-6}$	~ 2500	Eastman Kodak Co. (1976)
Kodak SO-424	Minicard II film	<3	Orthochromatic	$\sim 5.0\times 10^{-6}$	~ 2500	Eastman Kodak Co. (1976)
Agfa 10E56	Acetate film and plate	6	Orthochromatic[d]	$\sim 6.0\times 10^{-6}$	~ 2500	Pennington (1971) Agfa Gevaert
Kodak 120	Plate	5	6000-7000 Å 4420 Å	$\sim 4.2\times 10^{-5}$	>3000	Eastman Kodak Co. (1976)
Kodak SO-173	Estar film	6	6000-7000 Å 4420 Å	$\sim 4.2\times 10^{-5}$	>3000	Eastman Kodak Co. (1976)
Kodak SO-285[e]	Acetate film	<4	Panchromatic	$\sim 4.8\times 10^{-6}$	~ 2500	Eastman Kodak Co. (1976)
Recordak[e] 5468 & 8465	Acetate film	3	Orthochromatic	$\sim 5.9\times 10^{-6}$	~ 2000	Eastman Kodak Co. (1976)
Kodak 3414	Estar film	<4	Panchromatic	$\sim 2.1\times 10^{-7}$	~ 1260	Eastman Kodak Co. (1976)
Kodak 5069	Acetate film	—	UV-6330 Å	$\sim 2.4\times 10^{-7}$	~ 1260	Eastman Kodak Co. (1976)
Kodak SO-410	Estar film	—	Panchromatic[b]	$\sim 0.8\times 10^{-7}$	~ 500	Eastman Kodak Co. (1976)
Agfa 14C70	Acetate film	6	Panchromatic[c]	3.0×10^{-7}	1500	Pennington (1971)
Agfa 14C75	Acetate film	6	Panchromatic[c]	3.0×10^{-7}	1500	Pennington (1971)

[a] The exposure in joules/cm² that is required to produce a density of 1.0 is listed here as the recording sensitivity. For amplitude holograms, the maximum diffraction efficiency usually is obtained at densities between 0.6 to 0.8. Thus the exposure required will be slightly less than listed.

[b] Maximum sensitivity at 6328 Å.

[c] Maximum sensitivity at 6943 Å.

[d] Maximum sensitivity at 5145 Å.

[e] These films produce a positive image (negative gamma) with conventional negative processing (i.e., develop in D-19, fix, and wash).

14C75 emulsions have still lower resolving power, but they have nearly 70 times greater exposure sensitivity than the Agfa 8E70 and 8E75 emulsions. The Kodak 120 plate and SO-173 film contain a shrinkage-resistant emulsion sensitized in the red region of the spectrum. They are used with the HeNe, krypton, and ruby lasers with spectral outputs at 6328, 6471, and 6943 Å, respectively. Kodak 131 plates and SO-253 films contain a high-speed, panchromatic emulsion that have 20 to 200 greater exposure sensitivity than the Kodak 649F materials, depending upon the spectral region being considered. The Kodak SO-285 direct positive film and the Recordak 5468 and 8468 direct positive duplicating films have negative gammas. The reversal films are reported to give a long linear amplitude transmittance as a function of exposure.

In the planar (thin), amplitude hologram, techniques were devised to reduce the effects of emulsion shrinkage on the geometric fidelity of the reconstructed image. The techniques involve overexposing and underdeveloping on an Agfa 10E70 plate and are treated by Kellie and Stevenson (1973). They observe that only a small portion of the total emulsion thickness is used for information storage, and they find that the geometric fidelity of the reconstructed image is more than adequate for photogrammetric work. Holograms are recorded on the Agfa 10E70 plates at exposures four to 16 times the recording sensitivity value listed in Table II. The plates are developed 15–40 sec (the longer developing time is used for the shorter exposure) in Kodak HRP developer at temperatures near 68°F.

Zech (1974) describes chemical processing of photographic emulsions. Obtaining optimum quality involves a number of straightforward steps in which the development is critical. In order to get constant results, the development time and temperature must be carefully controlled. The processing time is not as critical in the remaining steps, but the processing temperature of each bath must remain within a few degrees of 70°F to avoid gelatin reticulation. Zech (1974) recommends Kodak D19 and D165 and Agfa Methanol-U developers. The Kodak D19 is generally considered the best Kodak developer. D8 and HRP may be used, but they introduce staining residues in the gelatin and are not generally recommended.

8.3.4 Hardened Dichromated Gelatin

The hardened dichromated gelatin plate produces a hologram well known for its high diffraction efficiency and signal-to-noise ratio (Table III). Because of its poor shelf life the dichromated gelatin plate and the dichromated protein emulsions (e.g., egg albumen, zein, or casein) or polymer emulsions cannot be commercialized. These materials must be prepared shortly before use. The patent literature demonstrates that the use of diazos, diazo-oxides, and azides enable preparation of stable photosensitive materials with sufficient shelf life to permit commercialization, and these materials have widely replaced the use

TABLE III

Hardened Dichromated Gelatin

Material	Usable thickness (μm)	Recording process	Recording wavelength range[a] (Å)	Recording sensitivity (J/cm²)	Limiting resolution (c/mm)	Reference
Dichromated, fixed, Kodak 649F plate	12	Photocross-linking	2500–5200	5–10×10^{-2}	>3000	Meyerhofer (1972) Kosar (1965) Zech (1974)
Prepared dichromated gelatin layers	0.5–3.0	Photocross-linking	2500–5200	3×10^{-3}	>3000	Shankoff (1968)

[a] Maximum sensitivity at 3600 and <2940 Å.

of dichromated photosensitive materials in the printing industry. The same may well happen to the use of dichromated materials in holography, and perhaps the first of the improved shelf life materials used in holography is the Shipley AZ-1350 positive photoresist.

The distinguishing feature of hardened gelatin is that unlike unhardened gelatin, thin layers will not dissolve in water, although they will swell to three or four times their dry thickness. Pouradier and Burness (1966) treat hardening of gelatin and emulsions and point out that photographic emulsions are hardened during manufacture to give protection against the effects of high temperatures and humidities.

The mechanism by which hologram formation occurs in hardened dichromated gelatin is discussed in the literature. Chang (1971), Zech (1974), Meyerhofer (1972), and Close and Graube (1974) give detailed descriptions of preparations used to make the hardened dichromated gelatin plate from either commercial silver halide plates or from prepared coating solutions. Chang (1971) states that upon dehydration, the amorphous films (by placing them in dry alcohol) tend to crack more readily than the crystalline films. For further interesting information on the cross-linking of gelatin, the reader is referred to a subsection of that description in a book edited by Cox (1972). Zech (1974) gives a preparation for making the photosensitive layer from coating solutions, and Close and Graube (1974) treat preparation of a red sensitive dye layer using commercial silver halide plates. The commercial plates and the prepared gelatin plates already contain hardened gelatin. These plates are further hardened in fixer with hardener. Following washing and drying, the gelatin layer is sensitized with about 5% ammonium dichromate solution with or without ammonium nitrate and a dye sensitizer such as methylene green. Following drying, the hardened dichromated gelatin plate is exposed and the exposed plate is then washed to remove the dichromate and any other soluble compounds and dehydrated in a series of isopropyl alcohol baths. The strain produced through rapid dehydration of the gelatin layer in the alcohol brings about formation of cracks or tears in the lesser hardened, unexposed regions of the layer. Furthermore, there is formed a complex compound consisting of the isopropyl alcohol coordinated with the chromium(III) ion at the gelatin cross-linked sites as reported by Meyerhofer (1972). The strain-induced cracks are treated in some detail by Curran and Shankoff (1970). The air–gelatin interfaces denoted as cracks in the gelatin are very efficient in redirecting the light into the first order. The isopropyl alcohol chromium(III) complex entity also adds to the phase change. Consequently, diffraction efficiencies in excess of 90% for a plane wave grating and a signal-to-noise ratio of 27 dB are reported by Zech (1974). The holograms and holographic optical elements recorded on hardened dichromated gelatin have remarkable brightness and resolution. Hardened dichromated gelatin plates typically have exposure requirements of about 100 mJ/cm² and are considered to be among the least sensitive of the holographic recording materials. Underexposed or underhardened holograms exhibit a milky opacity that degrades the reconstructed image.

8.3.5 Photoresists

The photoresists form another class of photosensitive materials that produce imaged relief patterns. Upon exposure to actinic radiation they produce chemical changes in the photoresist layer that enable a solvency differentiation as a function of exposure. Development with a suitable solvent promotes dissolution of either the unexposed or exposed regions, depending upon whether the resist is negative or positive working. The resulting surface relief patterns enable preparation of reflection holograms through evaporative metal processes and multiple duplication of holograms via embossing techniques. Table IV lists many of the commercially available photoresists. Note that in the majority of cases the photosensitive material is on the order of a micrometer in thickness. The recording process is one of three types: formation of an organic acid, photocross-linking, or photopolymerization of a monomer. The recording wavelength range is UV to 5000 Å with a selection of materials with either broad band or narrow band spectral sensitivities in this range. Exposures of about 10^{-2} J/cm² are necessary to produce limiting resolutions ranging between 250 and 1500 c/mm.

A description of the preparation of holographic diffraction gratings in Shipley AZ-1350 photoresist is presented by Grime (1975). The gratings, when developed and dried, can be given a highly reflective aluminum coating by vacuum evaporative methods. Alternatively, the developed photoresist can be used uncoated as a thin phase grating of high quality. Evaluation of the Shipley AZ1350-H photoresist has been carried out by Bartolini (1974) and of the Shipley AZ1350-J photoresist by Norman and Singh (1975). Stein (1974) discusses the RCA holographic moving map display wherein three color-separated, focused-image, or image-plane holograms are angularly indexed and recorded in this photoresist. Upon developing, the material is electroplated with about 50 μm of nickel to form a master that is used to emboss full-color holograms in heated clear plastic at rapid speeds.

The class of Kodak photoresists includes five subclasses of negative resists: Kodak Photo Resists (KPR), Kodak orthoresist (KOR), Kodak metal-etched resist (KMER), Kodak thin film resist (KTFR), and Kodak microresists (KMR). Another subclass of positive resists is Kodak autopositive resist, type 3 (KAR3). Bartolini (1974) indicates a possible difficulty in the use of negative photoresist to prepare surface relief holograms which concerns the exposure needed to fix the photoresist firmly to the substrate so that the image does not become detached during development. In most negative photoresists the emulsion next to the substrate is the last ot photolyze when the actinic light enters

287

TABLE IV

Photoresists

Materials	Minimum thickness (μm)	Recording process	Recording wavelength range (Å)	Recording sensitivity (J/cm²)	Limiting resolution (c/mm)	Reference
Shipley AZ-1350 (POS)	<1.0	Formation of an organic acid	UV–5000	10^{-1} at 4416 Å	>1500	Pennington (1971), Shipley Co., Bartolini (1974)
Shipley AZ-111 (POS)	1.0	Formation of an organic acid	UV–5000	10^{-9} at 4416 Å	<1000	Pennington (1971), Shipley Co.
Micro-image Isofine (POS)	0.4–1.0	Formation of an organic acid	2850–4850[a]	~10^{-2}	1000	Clark (1975)
GAF PR-115 (POS)	~1.0	Formation of an organic acid	UV–5000	10^{-2} at 4000 Å	~1000	GAF Corp.
Kodak KAR 3 (POS)	~1.0	Formation of an organic acid	3150–4650[b]	10^{-2} at 4000 Å	>1500	Eastman Kodak Co. (1974)
Micro-image Isopoly resist (NEG)	0.8	Photocross-linking	3000–5150[b]	~10^{-2}	400	Clark (1975)
Kodak KPR (NEG)	1.0	Photocross-linking	2600–4650[c]	10^{-2} at 4000 Å	400	Clark (1975), Eastman Kodak Co. (1974), Pennington (1971)

Kodak KOR (NEG)	~1.0	Photocross-linking	2500–5500[d]	5×10^{-1} at 4880 Å	~1000	Eastman Kodak Co. (1974) Pennington (1971)
Kodak KMER (NEG)	1.0	Photocross-linking	2900–4850[b]	10^{-2} at 4000 Å	250	Clark (1975) Eastman Kodak Co. (1974) Pennington (1971)
Kodak KTFR (NEG)	0.8	Photocross-linking	2900–4850[b]	10^{-2} at 4000 Å	400	Clark (1975) Eastman Kodak Co. (1974) Pennington (1971)
Kodak micro resist 747 (NEG)	0.8	Photocross-linking	UV–5500[e]	1.5×10^{-2} at 3650 Å	400	Clark (1975) Eastman Kodak Co. (1974)
Horizons LHS7 (NEG)	0.9	Photopolymerization of monomer	UV–5500	5×10^{-3} at 4880 Å	>500	Zech et al. (1973)
Dichromated gelatin-unhardened (NEG)	0.7	Photocross-linking	2500–5200	1.4×10^{-5} at 4416 Å	>500	Meyerhofer (1971) Kosar (1965)

[a] Maximum sensitivity at 4320 Å.
[b] Maximum sensitivity at 4100 Å.
[c] Maximum sensitivity at 3200 and 4200 Å.
[d] Maximum sensitivity at 3500 and 4800 Å.
[e] Maximum sensitivity at 3650 Å.

the emulsion at the emulsion–air interface. Until photolysis occurs at the emulsion–substrate interface, the material which has not been photolyzed will simply dissolve in the developer, even though it was in an exposed area of the plate. Thus one should consider exposing negative resist through the glass or film substrate so as to better fix the resist material to the substrate.

The Horizons LHS7 photoresist is a negative dry-working material that can also be processed with liquid developers. Zech *et al.* (1974) describe this photoresist which, upon exposure to actinic light, produces changes in both the index of refraction and in surface relief because of the formation of a polymer. Following exposure, the photoresist may be processed for 90 sec in a stream of 160°C rapidly moving air. This photoresist is well suited for many holographic data storage applications since at the lower spatial frequencies the holograms compare in quality to those made with Kodak 649F plates.

Unhardened dichromated gelatin is useful as a photoresist to produce surface relief phase holograms. The unhardened gelatin may be developed out following photolysis, leaving the tanned resist behind on a suitable substrate.

8.3.6 Photopolymers

Polymerization is a chemical process by which small molecules or monomers are combined to make very large molecules or polymers. Table V lists parameters of the more common photopolymers. It will be noted that they have photosensitivities that are greater than those of the photoresists and photochromic materials but less than those of the silver halide emulsions. The holograms produced are of the phase type with either modulation in refractive indices in the bulk of the layer or modulations in the surface. Photopolymers have an advantage of completely dry and rapid processing. High resolution holograms can be produced with a range of material thicknesses and recording wavelengths. There is reason to believe that fully developed photopolymers would have good shelf life and produce images with archival properties and good geometric fidelity.

Certain photopolymers developed by Hughes Research Laboratories can be obtained commercially from Newport Research Corporation (NRC) for use in holography. The Hughes–NRC photopolymers consist of aqueous solutions of acrylamide, one of a number of dye sensitizers (e.g., methylene blue), and an initiator or "catalyst." In preparing the photopolymer for recording holograms, the monomer and dye sensitized photoinitiator are mixed just before use, because the mixture is stable for only about an hour. A few drops of the mixed polymer solution are enclosed between two glass cover plates for the holographic exposure. If no spacers are used, a film thickness between 5 and 15 μm is obtained. Exposure is accomplished in the visible spectral region and is dependent upon the dye sensitizer used. Jenny (1970) describes the use of preexposure to reduce exposure required to form a hologram, and Tomlinson

TABLE V

Photopolymers

Material	Usable thickness (μm)	Preparation	Recording wavelength range (Å)	Recording sensitivity (J/cm²)	Resolution (c/mm)	Reference
DuPont	3–150	Coating technique	3500–5500	20–30×10^{-3} (in air) 2–3×10^{-3} (in nitrogen)	3000	Booth (1972) DuPont Co.
Polymethylmethacrylate (PMMA)	mm	Cast	UV	100	5000	Tomlinson et al. (1970) Moran and Kaminow (1973)
Polymethlmethacrylate (PMMA) (Q-Doped)	mils	Cast	4880	3	5000	Laming (1971)
Hughes-Newport Research Corp.	5–15	Coating technique	Spectral sensitizers for 6328, 5145, 5300, 4880, and 4416	4–5×10^{-3}	3000	Jenny (1970)
RCA	mm	Cast	5000	1	3000	Bloom et al. (1974)
Multicomponent photopolymer systems	20–100 and 8–100 inside hollow fibers	Coating technique	3250	~1	>3000	Bartolini et al. (~1976) Tomlinson et al. (1976)

et al. (1976) indicate that any polymer formed increases the viscosity of the material, which contributes to the stability of the image. An ultraviolet sensitive fixing agent is also included in the photopolymer whose photolysis products chemically reduce the sensitizing dye to its colorless leuco form. Thus a simple postexposure to an ultraviolet source readily fixes the photopolymer.

The DuPont photopolymer material is reported by MacDonald and Hill (1973) to consist of acrylate monomers with an absorption range in the near ultraviolet (3300–3600 Å), a photoinitiator that extends the photosensitivity into the visible region (~5500 Å), and a cellulose polymer binder that forms a matrix to hold the liquid monomer. A liquid photopolymer solution in methylene chloride solvent ready for coating on film or glass has a three-month storage life if kept refrigerated. The mechanism of hologram formation in the DuPont photopolymer is explained by Colburn and Haines (1971) and they point out that the holograms show no variation in surface relief. Diffraction efficiencies aproaching 100% are reported.

Bell Laboratories have developed multicomponent photopolymer materials useful for volume phase holography which has been described by Tomlinson *et al.* (1976). In the multicomponent photopolymers two or more monomers are selected that have substantially different photochemical reaction rates and refractive indices following polymerization. During exposure, the regions with higher light intensity are polymerized to a greater degree than the adjacent regions with lower light intensity. The higher reactive monomer will be preferentially polymerized, and this produces a concentration gradient that promotes its diffusion into the polymer region. As the polymerization proceeds, the polymer rich region increases in concentration and squeezes the lower reactive matter into the region of lower light intensity. Following complete polymerization there results a modulation in chemical composition of the two materials with different molecular polarizabilities, and therefore different refractive indices. Tomlinson *et al.* (1976) report the modulation in refraction indices by this approach to be greater than that observed with other photopolymer materials.

8.3.7 Photoplastics

Photoplastics provide another class of photosensitive materials for producing phase holograms (Table VI). These materials have multilayer structures with a substrate of glass or Cronar film upon which is coated a conducting layer of doped tin or indium oxides, evaporated gold, or evaporated silver. On this is deposited a photoconductor such as polyvinylcarbazole sensitized with trinitro-9-fluorenone. A thermoplastic, Staybelite Ester 10, is deposited as the top layer. The recording technique consists of a number of steps beginning with establishing a uniform electrostatic charge on the surface of the thermoplastic with a corona discharge assembly. This charge is capacitively divided between

292

TABLE VI

Photoplastics

Material	Usable thickness (μm)	Write, erase cycles	Recording wavelength range	Recording sensitivity (J/cm^2)	Limiting resolution (c/mm)	Reference
Thermoplastic-photoconductive layers (photoplastic)	Thermoplastic layer: 0.3–1.2, photoconductive layer: 0.9–3	8000–80,000	Nearly panchromatic for PVK–TNK photoconductor	10^{-4}–10^{-5}	>4100	Credelle and Spong (1972) Goetz et al. (1972) Colburn and Tompkins (1974) Lo et al. (1975) Lo et al. (1976)

the photoconductor and the thermoplastic layers and upon subsequent exposure the photoconductor conducts imagewise in illuminated areas to discharge its voltage. However, the exposure does not cause variation in the charge on the thermoplastic; this is accomplished by recharging the surface uniformily, adding to the charge of the imaged areas. The photoplastic is then heated to the softening temperature of the thermoplastic layer allowing electrostatic forces to deform the thermoplastic surface until these forces are balanced by the surface tension of the material. Cooling the material fixes the surface relief pattern while the material takes on a frosty translucence. Reheating the thermoplastic to a higher temperature tends to restore the photoplastic film to its original state. Thus the material has a write–erase recycling capability.

Holographic recording on photoplastic materials have been studied by Cre-

TABLE VII

Photochromics

Material	Preparation	Recording processes	Usable thickness (mm)	Lower wavelength (λ) range (Å)	Sensitivity at lower λ (J/cm²)
Inorganic					
CaF_2: La, Na CaF_2: Ce, Na	Crystal wafer	Ionic, electron trap	0.1–0.8 0.3–0.9	3800–4600	Exposure is from 2 to 10 times greater than at upper λ
$SrTiO_3$: Ni, Mo, Al $CaTiO_3$: Ni, Mo	Crystal wafer	Ionic, electron trap	0.1–1.0 0.1–0.8	3300–3900	Exposure is from 2 to 10 times greater than at upper λ
$LiNbO_3$: Fe, Mn	Crystal wafer	Ionic, electron trap	5.0	UV	80
Silver halide in borosilicate glass	Silver halide crystallites in glassy matrix	Reduction to Ag metal grains	0.1–6.0	AgCl: 3200–4200[a] AgBr: 3500–5500[a] AgI: UV–6000[a]	3–15 × 10⁻³ to give o.d. = 0.1
Organic					
Salicylidene–aniline	Crystalline layer between two microslides	Photoinduced tautomerism	<20 μm	3800	UV lamp used
Stilbene	Polymer films	Photoconversion Cis–trans isomerism	?	UV–blue	0.1 for $\Delta n = 0.3$
Methylanthracene	Dimers in glassy matrix and in PMMA	Photodissociation to form monomers with conjugated π electron systems	1–2	3130	0.1 for $\Delta n = 10^{-3}$
Benzacridizinium	Dimers in transparent polymer matrix	Photodissociation to form monomers with conjugated π electron systems	1–2	3650 <2000–3800	0.1 for $\Delta n = 10^{-4}$

[a] Peak 3800–4000 Å.

delle and Spong (1972) and Goetz *et al.* (1972). They describe the unique response of the thermoplastic layer to spatial frequencies as being dependent on layer thickness. Goetz *et al.* (1972) succeeded in increasing the bandwidth by a factor of 3 by selecting a thermoplastic with low conductivity at flow temperatures. Credelle and Spong (1972) found that applying the corona charge and exposure simultaneously after heating would produce frost-free, high resolution holograms with first order diffraction efficiencies exceeding the 34% theoretical diffraction efficiency for sinusoidal thin phase gratings.

Photoplastic plates and films have photosensitivities approaching that of Kodak 649F. With limiting resolutions greater than 4100 c/mm, the photoplastic imaging process demonstrates the most promising capabilities of the electrostatic imaging processes for holographic recording. It could well be among the

Upper wavelength (λ) range (Å)	Sensitivity at upper λ (J/cm²)	Resolution (c/mm)	Storage time	Readout process	Cycle lifetime	Reference
4800–9500 Rec'd λ: 5145	2.2 to give od = 0.2	>2000	Minutes to days	Amplitude	Indefinite	Duncan (1972) Amodei (1971)
4800–9500 Rec'd λ: 5145	0.69 to give od = 0.4 2.4 to give od = 1.4	>2000	Minutes to days	Amplitude	May be cycled	Duncan (1972) Amodei (1971)
<8540	~3 to give 10% diffraction efficiency	>1000	Estimate hours	Amplitude	May be cycled	Staebler *et al.* (1973)
5300–6300	3–5 × 10⁻² to give od = 0.1	>2000	Days to months	Amplitude	Indefinite measured 3 × 10⁵ cycles with no change	Megla (1966)
4880 and 5145	0.2 to give od = 0.5 for α_2 form	>3300	Minutes, α_1 form-hours, α_2 form	Amplitude	Nonfatiguing measured 5 × 10⁴ cycles with no change	Lo (1974) Inoue and Shimizu (1971–1972)
?	Low	2000	?	Refractive index change	?	Guzik (1974)
3650	Exposure is greater than at lower λ	2000	Days	Refractive index change	Subject to fatigue	Tomlinson *et al.* (1972)
4360 <3800–5200	Exposure is greater than at lower λ	2000	Days	Amplitude and refractive index change	Subject to fatigue	Tomlinson *et al.* (1972)

more promising nonsilver imaging processes as well. Replication by embossing techniques can be accomplished for photoplastics as with the photoresists discussed previously.

8.3.8 Photochromics

Generally photochromic materials undergo a reversible color change upon exposure to light and heat. However there are materials of the photochromic class that demonstrate reversible changes in either refractive index changes or electrooptical effects, rather than color changes. Photochromism occurs in a variety of solid materials that may be organic or inorganic, in solution or in crystalline structures, as indicated in Table VII. Amodei (1971) discusses different processes in photochromic materials.

Unfortunately, the photosensitivity of the photochromics is very low; at least three orders of magnitude below that of silver halide emulsions, because the chemical reactions occur at the molecular level. For the same reason, however, photochromics are grainless with resolutions limited by the wavelength of light. Given sufficient laser power at the proper wavelengths, volume holograms can be recorded in either the darkening or bleaching mode. These holograms require no development, wet or dry, since only energy is needed for in situ recording and erasure. Inorganic photochromics have very long to indefinite cycle lifetimes, while organic materials have fatigue limitations. Holograms recorded in photochromics will have good dynamic range, but generally diffraction efficiencies will be only a few percent.

8.3.9 Transparent Electrophotographic Films

Among the more recent developments in electrophotography has been the transparent electrophotographic process TEP films. The four TEP films announced to date are Scott Graphics, Inc. P4-005 and P5-003 films, the Coulter Information Systems, Inc. KC film, and the Eastman Kodak Company SO-101 film. These materials, with their liquid developers, may offer a capability in holographic recording in the near future. Zech (1977) reports that photosensitivities approaching 10 erg/cm^2 and resolving powers as high as 1000 c/mm are projected for the KC film. It may be possible to prepare developers for phase as well as amplitude holograms in the near future.

REFERENCES

Agfa Gevaert ''Scientific Emulsions.'' Teterboro, New Jersey.
Amodei, J. J. (1971). In ''CRC Handbook of Lasers with Selected Data On Optical Technology'' (R. J. Pressley, ed.), p. 533. Chem. Rubber Co., Cleveland, Ohio.
Bartolini, R. A. (1974). Appl. Opt. **13**, 129.

Bartolini, R. A., Weakliem, H. A., and Williams, B. F. (1976). *Opt. Eng.* **15**, 99.

Bloom, A., Bartolini, R. A., and Ross, D. L. (1974). *Appl. Phys. Lett.* **24**, 612.

Booth, B. L. (1972). *Appl. Opt.* **11**, 2994.

Bordogna, J., Keneman, S. A., and Amodei, J. J. (1972). *RCA Rev.* **33**, 227.

Chang, M. (1971). *Appl. Opt.* **10**, 2551.

Clark, K. G. (1975). "Non-Silver Photographic Processes" (R. J. Cox, ed.). Academic Press, New York.

Close, D. H., and Graube, A. (1974). Holographic Lens for Pilot's Head-up Display. NTIS Rep. AD/787605.

Colburn, W. S., and Haines, K. A. (1971). *Appl. Opt.* **10**, 1636.

Colburn, W. S., and Tompkins, E. N. (1974). *Appl. Opt.* **13**, 2934.

Colburn, W. S., Zech, R. G., and Ralston, L. M. (1973). Holographic Optical Elements. Harris Electro-Optics Center of Radiation, Tech. Rep. AFAL-TR-72-409.

Collier, R. J., Burckhardt, C. B., and Lin, L. H. (1971). "Optical Holography." Academic Press, New York.

Cox, R. J., ed. (1972). "Photographic Gelatin." Academic Press, New York.

Credelle, T. L., and Spong, F. W. (1972). *RCA Rev.* **33**, 206.

Curran, R. K., and Shankoff, T. A. (1970). *Appl. Opt.* **9**, 1651.

Duncan, R. C. (1972). *RCA Rev.* **33**, 248.

DuPont Co. Wilmington, Delaware.

Eastman Kodak Co. (1974). *Kodak Tech. Bits* **13**, 7.

Eastman Kodak Co. (1976). *Kodak Tech. Bits* **4**, 5.

GAF Corp. New York, New York.

Gladden, J. W. (1978). Review of Photosensite Materials for Holographic Recording. US Army Eng. Topograph. Lab. Rep. ETL-0128.

Goetz, G., Mueller, R. K., and Shupe, D. M. (1972). IEEE Conf. on Display Devices, New York. Oct. 1972.

Grime, G. W. (1975). *In* "Non-Silver Photographic Processes" (R. J. Cox, ed.). Academic Press, New York.

Guzik, R. P. (1974). *Electro-optical Systems Design* (June), p. 22.

Inoue, E., and Shimizu, I. (1971–1972). *Graphics Arts Japan* **13**, 22.

Jenny, J. A. (1970). *J. Opt. Soc. Amer.* **60**, 1155.

Kellie, T. F., and Stevenson, W. H. (1973). Study of the Characteristics of the Holographic Stereomodel for Application in Mensuration and Mapping, US Army Eng. Topograph. Lab. Rep. ETL-CR-73-14, Pt. II.

Kosar, J. (1965). "Light Sensitive Systems: Chemistry and Application of Nonsilver Halide Photographic Processes." Wiley, New York.

Kurtz, R. L., and Owen, R. B. (1975). *Opt. Eng.* **14**, 393.

Laming, F. P. (1971). *Polymer Eng. Sci.* **11**, 421.

Lo, D. S. (1974). *Appl. Opt.* **13**, 862.

Lo, D. S., Johnson, L. H., and Honebrink, R. W. (1975). *Appl. Opt.* **14**, 820.

Lo, D. S., Johnson, L. H., and Honebrink, R. W. (1976). *SPIE Conf., San Diego, California.*

MacDonald, R. I., and Hill, K. O. (1973). Evaluation of a New Photopolymer Hologram Recording Media. NTIS Rep. N-73-32395.

Megla, G. K. (1966). *Appl. Opt.* **5**, 945.

Meyerhofer, D. (1971). *Appl. Opt.* **10**, 416.

Meyerhofer, D. (1972). *RCA Rev.* **33**, 118.

Morgan, J. M. and Kaminow, I. P. (1973). *Appl. Opt.* **12**, 1964.

Norman, S. L. and Singh, M. P. (1975). *Appl. Opt.* **14**, 818.

Pennington, K. S. (1971). *In* "CRC Handbook of Lasers with Selected Data on Optical Technology" (R. J. Pressley, ed.), p. 549. Chem. Rubber Co., Cleveland Ohio.

8. Equipment and Procedures

Pouradier, J., and Burness, D. M. (1966). *In* "The Theory of the Photographic Process" (T. H. James, ed.). Macmillian, New York.

Ramberg, E. G. (1972). *RCA Rev.* **33**, 5.

Shankoff, T. A. (1968). *Appl. Opt.* **7**, 2101.

Shipley Co. Newton, Massachusetts.

Staebler, D. L., Phillips, W., and Faöghnan, B. W. (1973). Materials for Phase Holographic Storage. NTIS Rep. AD-760343.

Stein, K. J. (1974). *Aviation Week and Space Technol.* (April); p. 50.

Tomlinson, W. J., Kaminow, I. P., Chandross, E. A., Fork, R. L., and Silfvast, W. T. (1970). *Appl. Phys. Lett.* **16**, 486.

Tomlinson, W. J., Chandross, E. A., Fork, R. L., Pryde, C. A., and Lamola, A. A. (1972). *Appl. Opt.* **11**, 533.

Tomlinson, W. J., Chandross, E. A., Weber, H. P., and Aumiller, G. D. (1976). *Appl. Opt.* **15**, 534.

Urbach, J. C. and Meier, R. W. (1969). *Appl. Opt.* **8**, 2269.

Urbach, J. C. (1971). *Proc. SPIE* **25**, 17.

Zech, R. G. (1974). Data Storage in Volume Holograms. Ph.D. Dissertation, Univ. of Michigan, Ann Arbor, Michigan.

Zech, R. G. (1977). *Internat. Conf. Electrophotography, 3rd, Washington, D.C.*

Zeth, R. G., Dwyer, J. C., Fichter, H., and Lewis, M. (1973). *Appl. Opt.* **12**, 2822.

Zech, R. G., Shareck, M. W., and Ralston, L. M. (1974). Holographic Recording Materials. NTIS Rep. AD/A002849.

8.4 HOLOGRAPHIC SYSTEMS

Robert L. Kurtz
Huang-Kuang Liu
Robert B. Owen

8.4.1 Introduction

This section has two primary purposes: (1) to provide guidance criteria for holographic component selection and (2) to offer practical guidance in designing holographic systems for specific applications.

The selection of holographic components is first governed by the intent, motivation, and budget of the experimenter. Prices for complete holographic setups range from $250 to greater than $50,000.† In this section we list the main optical components necessary for holographic operation and offer some comments on the use and restrictions of each.

8.4.1.1 Laser

Any laser used for holography must operate in the TEM_{00}, or *uniphase* mode, which produces the classical Gaussian intensity distribution. Attention must be paid to the coherence properties. The coherence length can be approximated by the cavity length for most continuous wave (cw) lasers. To facilitate path length matching it is often convenient to use something as simple as electronic lacing twine with knots as fiducial markers to measure the length of the object and reference paths. It is preferred that the laser output be linearly polarized; however, nonpolarized outputs can be used to make holograms. The necessary output power level is determined by the size of the object to be holographed, since the final requirement is on energy density in the plane of the hologram.

8.4.1.2 Shutters

Both manual and electronic shutters are available for exposure control. Manual shutters are, of course, cheaper; yet these must be used in conjunction

---

† See "A Study Guide on Holography" by Dr. Tung H. Jeong, Lake Forest College, Lake Forest, Illinois.

HANDBOOK OF OPTICAL HOLOGRAPHY
Copyright © 1979 by Academic Press, Inc.
All rights of reproduction in any form reserved.
ISBN-0-12-165350-1

8. Equipment and Procedures

with a light meter if the hologram density is to be properly controlled. Further, after the hologram arrangement has been set up and repetitive holograms are required, then perhaps a simple timed exposure with a manual shutter is sufficient. Electronic shutters are more expensive but provide much more control over exposures. They easily allow measurement of beam intensities and allow the exposure to be automatically controlled by integrating the preset energy value for the exposure.

8.4.1.3 Isolation Table

In holography the requirement for good mechanical stability during the exposure is mandatory. This is quite evident from the fact that a hologram is simply the recording of an interference pattern. If the relative phase $\Delta\phi$ between the object and reference beam changes by an amount π during the exposure, the interference fringes overlap and the interference pattern is destroyed. This value of π in relative phase is equivalent to an optical path length of $\lambda/2$ since $\Delta\phi = k\,\Delta l$ and $\Delta l = \Delta\phi/k = \pi/k$; therefore, $\Delta l = \lambda/2$. Therefore, a mechanical instability sufficient to cause a path length change of $\lambda/2$ during hologram exposure is prohibitive.

There is an extensive variety of isolation tables from sandbox type to pneumatically supported granite tables. There are two classes of vibration: high frequency combatted by "dead" materials such as wood, sand, etc., and low-frequency vibration combatted by isolation of the table from the environment by use of inner tubes, bubble packing material, and pneumatic cylinders. For general display-type holograms these cheaper sandbox-type tables can suffice. For more quantitative holography the more professional pneumatically supported surfaces are required. To test the stability of any surface, a simple interferometer can be set up on the surface and the fringes observed for magnitude and period of movement.

8.4.1.4 Mirror Components

Mirror requirements are very straightforward. The mirror surface should always be front surface type and of a good quality unmarred surface so as not to introduce additional diffraction patterns onto the wave front. The necessary adjustability of each mirror surface will be determined by the requirements of the specific experiment.

8.4.1.5 Beamsplitter

The beamsplitter may be simple glass surfaces or expensive variable density type. Some control over transmitted and reflected components is allowed with

300

simple glass by changing the angle of incidence. If sufficient control of intensities is allowed using plane glass, then the more expensive variable density type is unnecessary. If more intensity control is needed or variable intensity control is required, then circularly variable, rotating, partial front surface reflectors are commercially available. These may be automated using a "capstan" type drive and stepping motors. Birefringent prisms preceded by rotatable half-wave plates give good beam quality; yet care must be taken to reestablish proper polarization. Another useful type is made using photopolymer gratings which are themselves holographically formed.

8.4.1.6 Spatial Filter

While holograms can be made without spatial filtering, the quality is poor. A spatial filter is a necessity for good quality display and is mandatory for quantitative holography. A spatial filter consists simply of a positive lens (usually a microscope objective with a pinhole aperture at the focal point of the lens). Only light that is parallel to the optic axis of the system will pass through the pinhole and continue its divergent wavefront. Any light nonparallel to the optic axis (caused by dust, transverse laser modes, etc.) will be blocked and will not pass through the pinhole.

8.4.1.7 Lenses

Primarily, lenses are used to spread the beam out. Therefore, the major requirement here is either positive or negative lenses, usually with as short a focal length as possible to cause this spread to occur over the shortest distance. Sometimes a diffuser can accomplish this spreading for the object beam.

8.4.1.8 Film

Many forms of film or recorders are available. The first requirement is one of resolution. This is determined by the wavelength of the laser line and half the angle θ between the object and reference beam. The resolution required in lines per millimeter is given by $R = (2 \sin \theta/2)/\lambda$. Typically, for a 6328 Å He-Ne laser this value is $60 \leq R \leq 1500$ lines/mm for $2° \leq \theta/2 \leq 50°$.

Emulsions providing sufficient resolution are obtainable on rigid glass substrates and also on flexible acetate film substrates. Glass plates provide more stability, yet generally cost more. Film is cheaper and can be made stable if sandwiched between two glass plates or supported on a metal backing as in a film transport.

8.4.1.9 Film Holder

Film holders or plate holders of various descriptions and origins can be made to perform satisfactorily for most purposes. For sensitive repeatable requirements, some form of precision adjustable holders must be used. Sometimes even in situ development can be accommodated. Some holographic arrangements require fast, repetitive data taking; therefore, a 70-mm film transport is required.

It is seen that the latitude of available components is great. The choice of which to use and when to use it depends on what is to be done holographically and what the budget will allow. Sometimes ingenuity on the part of the experimenter offsets budget restrictions. Ultimately, the components decision must be the responsibility of the experimenter and be solved by his discretion.

One of the major applications of holography to date has been in the area of holographic nondestructive testing (HNDT), and the method of optical HNDT or holographic interferometry has proved to be of major use in this application. A recent and excellent text on this subject is "Holographic Nondestructive Testing" (Erf, 1974); also useful is "Optical Holography" (Collier et al., 1971), as well as Section 10.4 of this volume. The present section requires the understanding of real time, double exposure, and time averaged techniques, covered in the above references. Therefore, in this section we shall concentrate on some specific HNDT systems in an effort to offer practical guidance in the custom design of holographic systems.

Throughout this section it is assumed that the reader is familiar with the basics of holography. Because of its importance in the past and the anticipated applications for the future, the concept of holographic nondestructive testing is given major attention. General aspects of HNDT are presented first, followed by a discussion of specific systems. (See Sections 8.4.2–8.4.6.) The discussions of the specific holographic systems pertinent to HNDT are from the user's point of view. Therefore, in these discussions we stress practical rather than theoretical insight, with the exception of the section on speckle holography. Since speckle holography is a recently developed technique, a basic analytical discussion is included. The order of the techniques in Sections 8.4.2–8.4.5 has been chosen to prepare the reader for Section 8.4.6, which describes a hybrid HNDT technique that utilizes automatic data processing. This hybrid technique combines all the systems described in the preceding sections. The hybrid technique is presented not only for its own merit but, more importantly, as a concept for combining known techniques to utilize their individual advantages in the accomplishment of a given test.

Section 8.4.7 discusses the subject of holography of moving objects. The primary motivation here is to direct attention to the possible development of a holographic motion picture camera.

Section 8.4.8 presents several other holographic systems whose potential for application in the future is self-evident.

8.4.2 Description of a Composite Mobile Holographic Nondestructive Testing Technique—A Variable Sensitivity System

Because holography is an interference phenomenon, the total change in path length Δd of the object beam during the exposure must be less than $\lambda/2$, where λ is the wavelength of the light source. Because of this severe restriction, the geometry of the optical arrangement employed becomes of utmost importance to the successful recording of a moving object or particle. Therefore, attention is directed to two limiting cases of sideband geometry.

Consider the holographic arrangement of Fig. 1. This constitutes a very desirable arrangement for a stationary object because of the high energy return from the object to the film plate. For this case, the radiation propagation vector $\mathbf{k}$ ($|\mathbf{k}| = 2\pi/\lambda$) is either parallel or antiparallel to the direction of the intended object motion. Consider that during the exposure the object moves from position x_0 to a new position x_1; during the exposure time t_0 to t_1 the object translation is Δx. Then the total optical path length change is $\Delta d = 2\,\Delta x$. Using the limiting requirement for the allowed change in optical path length of a holographic arrangement, $2\,\Delta x < \lambda/2$ or $\Delta x < \lambda/4$; therefore the object cannot travel a total distance greater than $\lambda/4$ if one is to successfully record a hologram. This geometry allows the minimum object motion during the exposure.

A contrasting case is afforded by Fig. 2. The basic difference in geometry between this and the previous arrangement is the rotation of one mirror. In addition, in this case the direction of translation of the object is perpendicular to the propagation vector $\mathbf{k}$. The result is that Δx may be as large as desired without any change in the object beam path length. For this geometrical arrangement $\Delta d \equiv 0$.

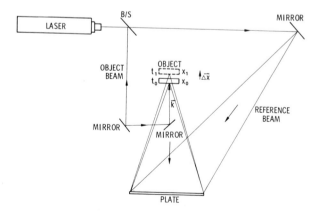

Fig. 1 Minimum motion geometry.

8. Equipment and Procedures

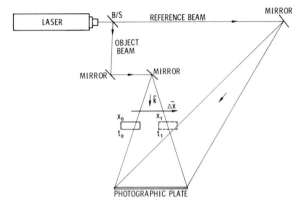

Fig. 2 Maximum motion geometry.

From the previous discussion, it can be seen that the sensitivity of an HNDT system is dependent on the geometry employed for the optical components. An HNDT system with variable sensitivity has been devised by one of the authors. It allows a large range of values in object loading. This is accomplished by the selection of a geometry which allows minor changes in optical components to control the system's sensitivity. This system will be described in some detail later, thus illustrating procedures common to most HNDT systems.

Attention is directed to configuration 1 of Fig. 3. Radiation emitted from the laser is incident on the field mirror assembly which essentially contains a spatial filter and a beamsplitter. This assembly is translatable to the left along the path ΔSb. The reflected portion of the radiation is made incident on the micrometer translatable object in a direction perpendicular to the object ($\theta = 0$). This radiation is then turned antiparallel to itself where it passes on to the film recorder. The film recorder is itself translatable to the right along the path Δf.

The radiation transmitted through the field mirror assembly is incident on mirror M, which can be translatable to the right along path Δm. From there it is turned to be incident on the film recorder and interferes with the object beam.

Configuration 2 may be traced in a similar fashion, except that the object beam makes some angle $\theta > 0$ with the perpendicular of the object.

The system is composite because one needs to only slightly manipulate three components (field mirror assembly, mirror M, and film recorder) to change from one technique of HNDT to another. It is not necessary to establish a new geometry in order to perform the various HNDT techniques. Furthermore, this adjustment facilitates the use of HNDT techniques as a field instrument since it allows control over the system sensitivity and thereby provides ease of testing of various objects. The system is mobile because all of the optical

304

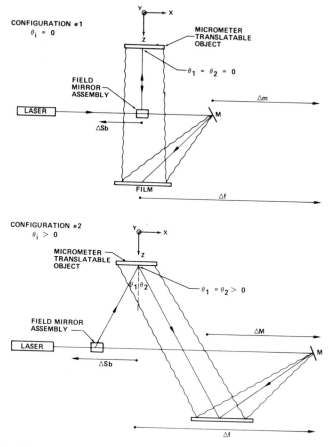

Fig. 3 Composite mobile holographic nondestructive test system.

components may be mounted on a precalibrated rigid table and locked in any position along their translatable paths. The system has variable sensitivity (which affords the composite structure) by virtue of the calibrated control over the angle θ which the object beam makes with the perpendicular of the object.

A semiquantitative theory, on fringe interpretation for the double-exposure method, derived by Liu *et al.* (1976a,b) and by Liu and Kurtz (1977) has been proved to agree quite well with experimental results. For small, general three-dimensional displacement, (D_x, D_y, D_z), the theory predicts that the fringe loci observe at an origin located at the center on the hologram will appear to be off-centered circles described by

$$A(x^2 + y^2) + Bx + Cy = (n - \tfrac{1}{2})\lambda, \tag{1}$$

305

8. Equipment and Procedures

and

$$A = \frac{1}{2} D_z \left(\frac{\cos \theta_1}{S^2} + \frac{\cos \theta_2}{H^2} \right) ,$$

$$B = D_z \left(\frac{\cos \theta_1 \sin \theta_1}{S} - \frac{\sin \theta_2 \cos \theta_2}{H} \right) + D_x \left(\frac{1}{S} + \frac{1}{H} \right) , \qquad (2)$$

$$C = D_y \left(\frac{1}{S} + \frac{1}{H} \right) ,$$

where $S = |S|$, $H = |H|$, n is an integer, and λ is the wavelength of the laser. The radius of the nth fringe circle may be written as

$$R_n = (1/2A)[B^2 + C^2 + 4A(n - \tfrac{1}{2})\lambda]^{1/2}. \qquad (3)$$

In a special case in which $\theta_1 = \theta_2 = \theta$ and $S = H$, it can be seen that in order to keep R_n the same for constant D_x and D_y, the product $D_z \cos \theta$ must be constant. In other words, if θ increases, D_z should increase as well to keep $D_z \cos \theta$ constant. This implies that the system is less sensitive at larger θ than at smaller θ for the detection of the out-of-plane displacement.

For example, if $\theta = 0$, then a specific displacement of the object along the positive z direction D_z will be sufficient to cause one fringe to be added to the fringe pattern. Yet, if $\theta = 80°$, then there must be a new displacement $D_z' = D_z/\cos 80° = 5.88 D_z$ in order to cause the addition of the same one fringe to the fringe pattern. These two values of object displacement along the positive z direction differ by approximately one-half order of magnitude. This result shows the variation of the sensitivity of this system to an out-of-plane displacement.

The composite mobile holographic technique was applied to several sandwich-structured samples of the radome of the Pershing missile system in which some specific programmed flaws had been placed. The experimental procedure for this testing consisted of the following steps.

Step 1 Load a test sample in the holder, check for pressure leaks, and place the holder properly in the holographic system.

Step 2 Obtain a reference hologram of the test object and accurately place this hologram back into position in the holographic system such that the virtual image of the reference hologram is superimposed on the real test object.

Step 3 Observe the interference pattern for continuous changes in the positive pressure loading.

Step 4 Record the fringe variations by taking photographs of the virtual image through the hologram.

These steps constitute the real-time observation and recording of the fringe pattern (real-time interferometry) and the search mode necessary to determine if the loading technique employed is adequate to locate the flaws or debonds. Finally, a double-exposure hologram is obtained as a permanent holographic record of the observed flaw or debond.

Figure 4 presents the double-exposure evidence of a Teflon disk (0.75 × 0.001 in.) placed beneath the surface and embedded in the epoxy of the sandwich structure which was a sample of the radome for the Pershing missile system. The pressure loading for this test was 0.5 psi or 1.1×10^3 dyn/cm² (Kurtz and Liu, 1974). Such a variable sensitivity HNDT system lends itself well to holographic testing outside of the laboratory. A prototype of the CMHNDT system is shown in Fig. 5.

8.4.3 Acoustooptical Holographic Nondestructive Testing

A thorough discussion of the subject of acoustical holography is beyond the scope of this section. Many complete works (e.g., Hilderbrand and Brenden, 1972; Metherell *et al.*, 1969, 1970; Sharpe, 1970) deal quite extensively with this subject. Our purpose here is to acquaint the reader with the advantages

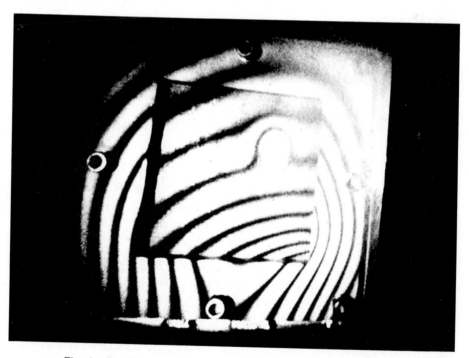

Fig. 4 Double-exposure hologram of Teflon embedded disk.

307

8. Equipment and Procedures

Fig. 5 Mobile system photograph.

and the main disadvantage of this holographic technique as it is used in the hybrid holographic system described in detail in Section 8.4.6.

8.4.3.1 Acoustical Holography

Because of the similarity of the nature of acoustics and optics, the description of acoustical holography is most similar to that of optical holography. The wavelengths of sound waves are much longer than those of optical waves. In general, this longer wavelength causes the resolution for acoustical holography to be much less than that for optical holography.

The longer wavelength, on the other hand, increases the versatility of holographic nondestructive testing because objects opaque to optical waves become transparent to the acoustic waves. This characteristic allows the test object to be interrogated throughout its volume. Data resulting from such an acoustical technique is thus a transparent three-dimensional image of the test object. This image is extremely useful in localizing the various flaw regions embedded inside the test object. Acoustical holography offers several other advantages of forming visual images of insonified objects. Specifically, these advantages are ability of real-time visualization of three-dimensional images, rapid extraction and processing of acoustical information, enormous depth of field, relative insensitivity to turbulence of the environment, capability of retrieving the information about the object from discrete sampling points, localization of defects in objects, and the capability of detecting extremely low powers that could not otherwise be detected.

8.4.3.2   Acoustooptical Holography

The ''optical'' aspect of the acoustooptical system derives from the fact that one may record the acoustical modulation of the object beam with an optical recorder. One insonifies the object with acoustic energy and allows this modulated acoustic wave to further modulate the object beam of an optical holographic system. Therefore, the modulation of interest would be recorded optically, and this would allow the use of an optical detector/readout system. Such a configuration had advantages that are described in Section 8.4.6.

There are several techniques which hold potential for this acoustooptical modulation scheme. Table I, extracted from El-Sum (1976), provides a comparison of some of these techniques.

8.4.4   Speckle Holographic Nondestructive Testing

The grainy or speckle phenomenon of laser light is caused by two inherent characteristics of lasers, the spatial coherence and the monochromaticity of the radiation. Recent development has demonstrated that this phenomenon can be applied in a variety of ways to the measurement of minute object displacements, strains, or vibrations. Basically there exist two techniques; one is called speckle beam holographic interferometry and the other is called the speckle photographic interferometry. The main advantage of the first technique is that it alleviates the stringent vibration isolation requirement in HNDT systems. The advantage of the second technique is that it results in photographic data sensitive only to in-plane components of the surface variations. The purpose of this section is to discuss the basic principles and practical limitations of these techniques and their potential applications in existing HNDT systems. In addition, three new interferometric nondestructive testing systems utilizing these techniques are presented. All the systems are based on the CMHNDT system described in Section 8.4.2 so that the merits of the original system are preserved. Experimental procedures are also outlined for the calibration and evaluation of these new systems.

8.4.4.1   Double-Exposure Speckle Photographic Interferometry

When a diffuse surface is illuminated by a continuous wave laser, the surface appears to be grainy or speckled. This is because any point in front of the illuminated surface receives diffusely scattered light of similar amplitudes but random phases from all points on the surface. The interference of these scattered coherent radiation fields produces the speckle effect.

The basic principle of the speckle phenomenon can be described as by Leendertz (1970) from Fig. 6 where S_1 and S_2 represent two diffuse surfaces

309

TABLE I

Acoustooptic Imaging Methods and Detection[a]

Imaging system	Detectors or detection technique and display	Real-time capability	Sensitivity (W/cm^2)	Frequency range (MHz)	General remarks
Liquid surface (static ripples)	Optical phase contrast or optical scanning with coherent or incoherent light	Yes	1.5×10^{-3} (normal) 10^{-5} (reported) 10^{-9} (theoretical)	0.5–10	
Bragg diffraction (direct sound-light interaction)	Coherent laser light (continuous or pulsed)	Yes	10^{-9} (theoretical)	10–100	
Deformed solids (dynamic ripples)	Laser beam scanning or electron beam scanning	Yes	10^{-3} (reported) 10^{-9} to 10^{-11} (theoretical)	100	
Image converter (Sokolov)	Scanning the back of PZT face (quartz or barium) electronically and detecting secondary emission	Yes	10^{-9} (theoretical)	Up to 0 or 20	Sealed tube very narrow angular aperture (10–20°); $3\lambda_s$–$5\lambda_s$ resolution (reported); new designs may increase aperture and frequency
Metal fiber face tube image converter (with appropriate PZT)	Scanning the back of the PZT electronically (like Sokolov tube)	Yes	10^{-9} (theoretical)	Up to 20	Improves the angular field of view of Sokolov tube
Pyroelectric face tube image converter	Pyroelectrics scanned with electron beam	Yes	10^{-3} at 3 MHz (reported)	Up to 20	Sensitivity increases with f^2 Wide frequency band (>20 MHz) Sealed tube

Method	Type		Value 1	Value 2	Remarks
Electrostatic transducers	Electric switching	Yes	10^{-8} in air 10^{-11} in water (theoretical) 10^{-3} (reported)	0.07–0.250 in air 0.3–3.5 in water	Laser beam scanning of PZT for readout has sensitivity of 10^{-4} W/cm²
Piezoelectric array with electronic focussing and scanning	Electronic	Yes	10^{-11} (theoretical) 10^{-8} (reported)	1–20 (used)	Has larger dynamic range than piezoelectrics; Has storage capability
Piezoresistive image converter	Electron beam scanning	Yes	10^{-7} (reported)	1–20 (used)	
Electroluminescent image converter	Direct conversion	Yes	10^{-6} (reported)		
Photographic and chemical methods	Direct interaction	No	~1–5 (reported)	>0.02	
Photopolymer materials	After conversion to visible or electron images	No	0.013 (reported) (with argon ion laser)		
Oil, thermoplastic and photoplastic recorders	Electron beam scanners plus optical illumination	No	0.1–1 (reported)		
Pholman cell	Direct interaction	Yes	10^{-1}–10^{-3} (reaction time 1 sec) 2.8×10^{-7} (reaction time ~60 sec)		Poor resolution, poor contrast, and limited dynamic range of 20 dB
Solid and liquid crystal display	Direct interaction	Yes	0.1–10^{-6} (reported)		Still in experimental stage
Chemical techniques: phosphor persistence changes	Direct interaction plus proper viewing system	Yes	0.05–0.1		See Berger (1969) which includes specific references; e.g., Ca–CrS stimulated by uv increases its luminescence persistence by acoustic exposure; spatial resolution of 0.2 mm reported

TABLE I (*Continued*)

Imaging system	Detectors or detection technique and display	Real-time capability	Sensitivity (W/cm²)	Frequency range (MHz)	General remarks
Extinction of luminescence			1		
Thermosensitive color changes			1		Chromotropic compound (e.g., Hg·Ag·iodide); changes color from yellow to red instantly with acoustic absorption (1 sec exposure); irreversible process
Change in photoemission			0.1 (at 5 MHz)		
Change in electrical conductivity			0.1		Semiconductor materials such as zinc and cadmium
Thermocouple and thermistor			0.1		Thermopile detects 0.1 W/cm², temperature rise of $(10^{-4})°C$
Zone plate acoustic focusing (on PZT)	Electron or optical scanning	Yes	10^{-11}		
Gabor's sonoradiography	Coherent laser beam and photo recording	No			No results reported
Acoustic tomography	PZT	No	10^{-11} (theoretical)		Mostly used in medicine
Frequency swept recording	PZT	Possible	10^{-11} (theoretical)		No results reported
Digital sampling and computer reconstruction	PZT (in water) microphone (in air)	No	10^{-11} (theoretical)		
Rutican recording devices	Light image	Yes	30 erg/cm²		Slow

[a] From El-Sum (1976).

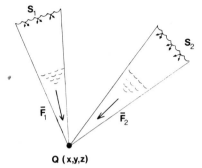

Q (x,y,z)

Fig. 6 Speckle pattern formation at $Q(x, y, z)$ due to scattering light from S_1 and S_2.

illuminated by coherent laser light. Let F_1 (x, y, z) and $F_2(x, y, z)$ be the summations of vectors representing light scattered by all the points from S_1 and S_2, respectively, which reach the point $Q(x, y, z)$. Let the resultant amplitude and phase of the light at $Q(x, y, z)$ be written as $'F_3(x, y, z)$; then

$$F_3(x, y, z) = F_1(x, y, z) + F_2(x, y, z). \tag{4}$$

When x, y, and z vary, the amplitude and phase distributions of F_1, F_2, and F_3 vary accordingly. If the phase at every point of F_1 changes by the same amount δ relative to F_2, F_3 will change in a random manner from point to point. In Fig. 6 consider an area A in the xy plane (with $z = $ const) over which δ changes equally at all points. Let the intensities at any point $Q(x, y)$ due to S_1 and S_2 be I_1 and I_2, where

$$I_1 = |F_1(x, y)|^2 \tag{5}$$

and

$$I_2 = |F_2(x, y)|^2. \tag{6}$$

If the phase angle between F_1 and F_2 is θ where $\delta = 0$, then the resultant intensity at $Q(x, y)$ may be given by

$$I_3(0) = I_1 + I_2 + 2(I_1 I_2)^{1/2} \cos \theta. \tag{7}$$

When F_1 is shifted by δ in phase with respect to F_2,

$$I_3(\delta) = I_1 + I_2 + 2(I_1 I_2)^{1/2} \cos (\theta + \delta). \tag{8}$$

Consider a photographic plate placed in the xy plane and assume one exposure for $\frac{1}{2}$-unit time is made at $\delta = 0$ and another for the second $\frac{1}{2}$-unit time is made at δ; then the total average energy received at $Q(x, y)$ per unit time is

$$I_3(\delta) = 2(I_1 + I_2) + 2(I_1 I_2)^{1/2}[\cos \theta + \cos(\theta + \delta)]. \tag{9}$$

313

8. Equipment and Procedures

If $\delta = 2n\pi$, where $n = 0, 1, 2, 3, \ldots$, then

$$I_3(2n\pi) = 2[I_1 + I_2 + 2(I_1 I_2)^{1/2} \cos \theta]. \tag{10}$$

The distribution of energy in the double-exposure record will be the same as the distribution for a single speckle pattern. It can be shown (Rayleigh, 1920; Goodman, 1965) that this distribution may be written as

$$W_1(I) = (1/I_0) \exp(I/I_0), \tag{11}$$

where $W_1(I)\, dI$ is the probability that a particular speckle has intensity I, and I_0 is a normalizing factor which represents the average intensity of the pattern.

Conversely, when $\delta = (2n + 1)\pi$ while I_1 and I_2 remain constant, from Eq. (9),

$$I_3(2n\pi + \pi) = 2(I_1 + I_2). \tag{12}$$

This result is equivalent to the effect of incoherently combining two independent speckle patterns. The resultant probability density function becomes

$$W_2(I) = 4(I/I_0^2) \exp(-2I/I_0), \tag{13}$$

where $W_2(I)\, dI$ is the new probability of a speckle having intensity I.

If $T(I)$ denotes the local intensity transmission of a particular photographic emulsion when exposed to an intensity I for unit time, then for the double exposure corresponding to $\delta = 2n\pi$,

$$T_1 = \int_0^\infty W_1(I)T(I)\, dI, \tag{14}$$

where $I = I_3(2n\pi)$ as given in Eq. (10). When $\delta = (2n + 1)\pi$, the average transmission becomes

$$T_2 = \int_0^\infty W_2(I)T(I)\, dI, \tag{15}$$

where I is given by Eq. (12).

The visibility of the speckle fringes caused by the differences of T_1 and T_2 as a result of the two exposures may be defined by

$$V = (T_1 - T_2)/(T_1 + T_2), \tag{16}$$

and the average density D of such a fringe pattern may be defined by

$$D = \log_{10}[2/(T_1 + T_2)]. \tag{17}$$

314

8.4.4.2   Speckle Reference Beam Holographic Interferometry

The technique of the speckle reference beam holographic interferometry is completely different from that of the speckle holographic interferometry described in Section 8.4.4.1. The speckle reference beam hologram is made in a system similar to the CMHNDT system discussed in Section 8.4.2 and in Kurtz (1972) and Kurtz and Liu (1974). The system consists of a coherent light source divided into two beams; one beam is used as a reference beam and the other as the object beam. The system geometry allows variable sensitivity. The unique feature of the speckle reference beam technique is that the reference beam is focused to a point on a diffuse surface. The backscattered light is collected at the film plane (Waters, 1972). The system can be illustrated by Fig. 7.

If the diffuse surface in Fig. 7 is a part of the object or is fixed to the object, then the object motion will be compensated by a change in the path length of the reference beam. Due to the compensation effect, the stringent vibration isolation requirements associated with conventional holography can be greatly reduced or even completely eliminated for some objects.

The conditions of obtaining high-quality reconstructed images with a capability of recording large areas of the object can be derived with the help of

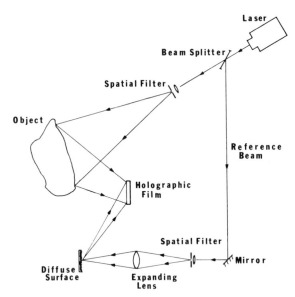

Fig. 7   Schematic diagram illustrating a typical speckle reference beam holographic system.

315

8. Equipment and Procedures

Fig. 8. Assume the object is going through random motion. The displaced object is shown with displacement components ΔX and ΔZ. The phase change at the focused reference spot is

$$\Delta\delta_0 = (2\pi\,\Delta Z/\lambda)(\cos\beta_0 - \tan\alpha_0\sin\beta_0 + \sec\alpha_0), \qquad (18)$$

where λ is the wavelength of the laser and the angular parameters are indicated in Fig. 8. The distance between the object and hologram is considered to be much larger than the random surface displacement. The phase change $\Delta\delta_n$ at a point on the illuminated portion of the object is

$$\Delta\delta_n = (2\pi/\lambda)[\Delta Z\cos\alpha_n + \cos\beta_n) - \Delta X(\sin\alpha_n - \sin\beta_n)]. \qquad (19)$$

The validity of Eq. (19) is based on the assumption that the distance from the illumination point source is much larger than the random displacement of the surface; i.e., Fraunhofer conditions are satisfied.

Since Eq. (18) has no explicit ΔX dependence, random motion in the x direction cannot be compensated. This is a limitation of the present technique. The best one can achieve for random motion compensation in this configuration is to make the two z components in Eqs. (18) and (19) equal. Therefore, maximum object motion compensation is obtained with $\alpha_0 = \alpha_n = 0$ and $\beta_n = \beta_0$.

8.4.4.3 Applications of the Techniques of Speckle Beam Holographic Interferometry and Speckle Photographic Interferometry in the HNDT System

The most important advantage of the speckle beam interferometry is that the stringent vibration isolation requirement of conventional holographic interferometry can be reduced or even eliminated for some objects. This is particularly useful for application of the holographic nondestructive test systems in realistic testing environments. The important advantage of speckle photographic interferometry is its capability to measure the in-plane component of the strain or displacement. Not only can the amplitude of one surface component be detected but also its direction. The CMHNDT system design is self-contained and quite versatile in its applications to various testing requirements. Nevertheless, it appears desirable that the advantages of laser speckle patterns be utilized so that the capability of the existing CMHNDT system will be extended.

There are many ways that speckle beam holography and speckle photography can be applied in the field of nondestructive testing; however, the following specific goals were considered in the CMHNDT applications:

(1) Preserve the present arrangement of the CMHNDT system as much as possible.

316

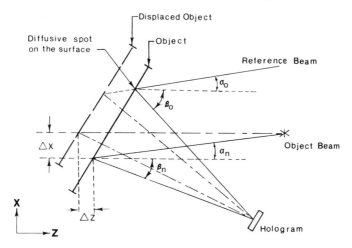

Fig. 8   Schematic diagram for the analysis of the speckle reference holographic system.

(2)   Keep the new arrangement as simple as possible.

(3)   Be able to apply or extend existing theories to the interpretation of experimental results in the new systems as much as possible.

Based on these considerations, three systems were designed. These systems are now described in detail.

(a)   Speckle Reference Beam HNDT System   The first new system proposed is shown in Fig. 9. The reference beam is created by laser light focused on a diffuse spot fixed on the edge of the object. The backscattered light, consisting of speckle patterns, distributes itself over the film plane. The spatial filter-mirror assembly can be translated, varying the angle θ_s and the sensitivity of the HNDT system. Several aspects of the system require further experimental studies.

(1)   The nature of the diffusive surface, the incidence angle of the laser beam, and the size and character of the beam focal point should all be studied quantitatively.

(2)   The optimum reference beam to object beam power ratio and exposure time need to be determined since speckle patterns are different from the uniformly expanded reference beams used in conventional holographic systems.

(b)   Symmetrical Speckle Beam Photographic and Holographic Nondestructive Test System   This new system can best be illustrated by Fig. 10. Two speckle beams are expanded by the two spatial-filter-and-mirror assemblies. The two

8. Equipment and Procedures

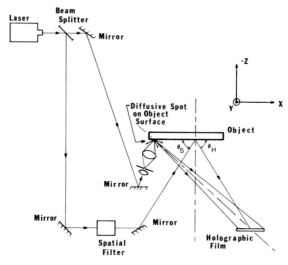

Fig. 9 Incorporation of the speckle reference beam in the CMHNDT system.

beams illuminate the object from directions symmetrical with respect to the z axis. A camera is located along the z axis in front of the object. A reference beam which is controlled by shutter B together with a spatial filter and the holographic film can be used to perform the HNDT function. If one of the speckle beams is removed by shutter C, the system becomes identical to the original CMHNDT system. Hence, the present system can be incorporated into the CMHNDT system.

The basic principles of the speckle beam photographic interferometry have been discussed previously. There are several practical considerations worth noting (Archbold *et al.*, 1970).

EFFECT OF FILM SPEED AND CAMERA APERTURE   The aperture of the imaging lens in the camera is important because it defines the speckle size and governs the exposure time necessary to record the image. If the f number of the aperture is F, the speckle size σ is given approximately by

$$\sigma = 1.2\lambda F. \tag{20}$$

For an argon laser, the wavelength λ can be chosen to be 0.514 μm. At this wavelength the resolving power of the emulsion must be better than $1600/F$ lines/mm. It was found that the fine grain Agfa–Gevaert 10E70 film (with a resolution of 2700 lines/mm) can be used as the recording material. A wide range of apertures can be used, and the small dynamic range of the 10E70 allows one to achieve high density with corresponding high fringe visibility for a small relative increase in exposure time.

318

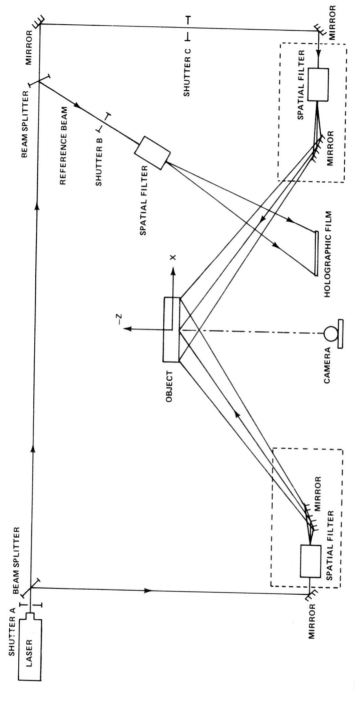

Fig. 10 Incorporation of the speckle photographic interferometry with two symmetrical beams in the CMHNDT system.

8. Equipment and Procedures

LIMITATIONS OF THE MEASURABLE RANGE OF OBJECT DISPLACEMENT It was found that the fringe visibility as defined by Eq. (16) will fall to less than one-third of its original value when the lateral displacement ΔX gives rise to an image displacement Δx equal to one speckle diameter. The relationship between Δx and ΔX is $\Delta x = \Delta X/m$, where m is the demagnification factor of the imaging lens.

Since a speckle of size σ on the image corresponds to a speckle of size $m\sigma$ on the object, the number of fringes observable is obtained by equating $m\sigma$ with ΔX in the equation

$$2\,\Delta X \sin \theta = n\lambda, \tag{21}$$

where θ is the angle between the incident speckle beam and the z axis.

Letting $n = N$ and $\sigma = 1.2\lambda F$, the result is the number of fringes observable,

$$N = 2.4mF \sin \theta. \tag{22}$$

The number of fringes observable is thus proportional to the demagnification and inversely proportional to the numerical aperture of the lens.

TECHNIQUES OF EXTRACTING DATA FROM THE SPECKLED PHOTOGRAPHS Two methods are feasible for the analysis of data from the speckle photographs (Archbold and Ennos, 1972). The first method consists of examining the recorded image point by point, using a small aperture. An alternative method of analysis is to perform a type of spatial filtering on the film record as illustrated by Fig. 11. The recorded image is first illuminated by a collimated beam of

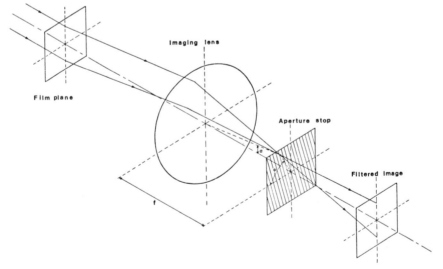

Fig. 11 Spatial filtering arrangement for analysis of double-exposure laser photographs.

light and then reimaged by means of a lens. In the focal plane of the lens, a small circular aperture stop is used to select the position of the data from the photograph. If the stop is offset from the axis by azimuth and field angles ϕ and α, the final image is formed only by light diffracted into that direction. Bright areas on the film observed through the aperture will correspond to parts of the object which have gone through a displacement, resolved in the azimuth direction ϕ, of magnitude D^* given by

$$D^* = n\lambda m/(2 \sin \alpha), \tag{23}$$

where n is the order number of the diffraction spectrum. Dark areas correspond to the half-order spectra, where n is replaced by $(n + \frac{1}{2})$ in Eq. (23).

(c) Single Speckle Beam Photographic and Holographic Nondestructive Test System If only one speckle beam is used, the symmetry of the speckle beams is destroyed, and the system shown in Fig. 9 can be simplified to that shown in Fig. 12. If the lateral surface motion is $d = [(\Delta X)^2 + (\Delta Y)^2]^{1/2}$, the corresponding image shift becomes $d = D/m$, where m is the demagnification factor. In this case, the diffraction pattern will consist of a set of Young's fringes having an angular spacing α given by

$$\sin \alpha = \lambda m/d. \tag{24}$$

The direction of the fringes will be orthogonal to the direction of the image motion.

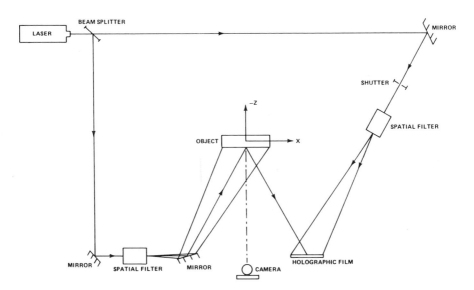

Fig. 12 Incorporation of the speckle holographic interferometry with a single speckle beam in the CMHNDT system.

8. Equipment and Procedures

The speckle size σ recorded by the film is still the same as that given by Eq. (20). Again, double-exposure holograms can be taken to calibrate the system. The procedures have already been described in the previous sections.

It is important to note that although the speckle beam interferometric techniques have many advantages, the fact that they are only sensitive to in-plane motions sets a limit to their applications for nondestructive testing, especially when the measurements of surface displacements in all directions are required. Hence it seems that it is more appropriate for the speckle techniques to serve auxiliary functions in an HNDT system.

8.4.5 Holographic Correlation

Another interesting method of HNDT is holographic correlation. Most of the work in HNDT has been done using holographic interferometry in the classical manner; that is, by generating and interpreting fringes resulting from interaction between two mutually coherent wavefronts. In this approach the areas under study are compared on a point-by-point basis. However, in holographic correlation this comparison is done as a whole, and one obtains a relative intensity which signifies the similarity of the two wavefronts being processed. The intensity is obtained by large-area integration of wavefronts and has the form of a correlation integral. The wavefronts originate from the test object which is loaded in a manner similar to other methods of HNDT.

The particular system geometry used in one of the authors' holographic correlation facilities is shown in Fig. 13. This geometry may be changed within certain constraints as required for specific studies. When a test is being made, the wavefront $h(x, y)$ backscattered off the loaded test area is operated on by the transform lens (Fig. 13), resulting in a Fourier transform, $F\{h(x, y)\}$. The object wavefront then passes through a matched Vander Lugt filter (Vander Lugt, 1964), resulting in the product $F\{h(x, y)\}F*\{s(x, y)\}$, where $s(x, y)$ is the wavefront backscattered off the unloaded test object. Construction of the Vander Lugt filter will be discussed later. This product is then operated on by the correlation lens (Fig. 13), yielding the cross correlation of $h(x, y)$ and $s(x, y)$, as can be seen from the convolution theorem (Papoulis, 1968, and Stroke, 1969). This output can be written as

$$\int\int_{-\infty}^{\infty} h(\xi, \eta)s^*(\xi - x, \eta - y) \, d\xi \, d\eta. \tag{25}$$

By placing a pinhole on the optical axis the values of x and y are made to be zero, yielding the final result

$$\int\int_{-\infty}^{\infty} h(\xi, \eta)s^*(\xi, \eta) \, d\xi \, d\eta. \tag{26}$$

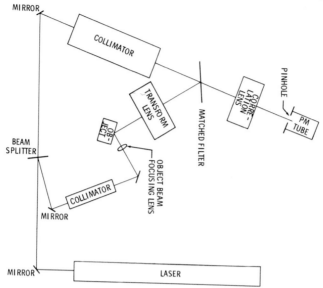

Fig. 13 Holographic correlation facility: complete system. (After H. K. Liu and R. B. Owen, 1979.)

This function reaches a maximum when $h(x, y)$ equals $s(x, y)$; that is, when the object is not loaded. The function decreases rapidly as $h(x, y)$ and $s(x, y)$ diverge. Physically the function appears as a focused point of light in the correlation plane and is detected by a photomultiplier tube and monitored by a digital voltmeter (Fig. 13).

The matched Vander Lugt filter referred to above is a Fourier transform hologram which is recorded with the system geometry shown in Fig. 12. At the hologram, $R(x, y)$ represents the complex amplitude distribution due to the reference wavefront and the distribution due to the Fourier transform of the wavefront backscattered from the unloaded test object. The intensity $I(x, y)$ at the hologram therefore can be written as

$$I(x, y) = |R(x, y) + F\{s(x, y)\}|^2$$
$$= |R(x, y)|^2 + |F\{s(x, y)\}|^2 \tag{27}$$
$$+ R^*(x, y)F\{s(x, y)\} + R(x, y)F^*\{s(x, y)\}.$$

The amplitude transmittance $T_\alpha(x, y)$ of the plate can be written as

$$T_\alpha(x, y) = T_0 + Bt[R^*(x, y)F\{s(x, y)\} + R(x, y)F^*\{s(x, y)\}], \tag{28}$$

where t is the exposure time, (E_0, T_0) the midway point of the linear section of the amplitude transmittance versus exposure curve, B the slope of the curve at (E_0, T_0), and E_0 the mean exposure. When the object is tested, the hologram is illuminated with $F\{h(x, y)\}$, which is the Fourier transform of the wavefront backscattered from the loaded test object (Fig. 14). It can be shown that in the test mode the off-axis transmitted amplitude is

$$F\{h(x, y)T(x, y)\} = F\{h(x, y)\}F^*\{s(x, y)\} \tag{29}$$

as desired. The foregoing procedure is shown in a block diagram in Fig. 15 and can be found in more detail in Espy (1974) and Goodman (1968). Variations of the technique described have been successfully used in practical HNDT applications, some of which shall be outlined in the following sections. Holographic correlation techniques are especially suited to surface inspection, and the applications that follow are generally of that nature. Correlation techniques have also been used to improve various forms of optical data processing useful in nondestructive testing (Aleksoff and Guenther, 1976; Casasent and Psaltis, 1976; Bage and Beddoes, 1976). One should note that what is being done is basically a comparison scheme in which the various methods of data processing are just the means of relating surface microstructures. One should also be aware that often the apparent surface structure is, in fact, a speckle pattern caused by the coherent nature of the object illumination; therefore, great care must be taken to have identical conditions of illumination and position for the various states being compared.

8.4.5.1 Real-Time Correlation

The correlation scheme described previously is most commonly operated in real time; that is, the hologram is recorded using an initial light pattern and is then illuminated by the pattern which is to be related to the initial one. There

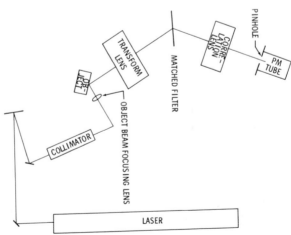

Fig. 14 Holographic correlation facility: test geometry. (After H. K. Liu and R. B. Owen, 1979.)

324

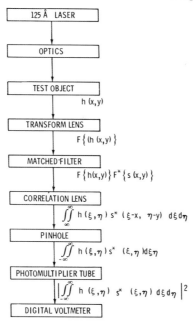

Fig. 15 Block diagram of holographic correlation facility. (After H. K. Liu and R. B. Owen, 1979.)

is therefore no delay between the illumination of a test area and the determination of its correlation value; the measurement is made in real time. We might note here that although, as will be seen in the following sections, the usefulness of holographic correlation has been demonstrated by extensive studies, its practical application has been somewhat limited by the necessity of processing a photographic plate. However, some of the newer recording materials eliminate this requirement (Kurtz and Owen, 1975). This has been effectively demonstrated by the use of real-time incoherent-to-coherent image transducers such as the PROM, EALM, Titus tube, Ruticon, and liquid crystal light valve for holographic correlation (Nisenson and Sprague, 1975; Gara, 1977). This option should greatly enhance the appeal of the correlation techniques. The possibility of automation also exists (Indebetouw *et al.*, 1976). Some practical applications of this technique are now described.

(a) Solder Joint Measurements Holographic correlation techniques have been used to test solder joints on printed circuit boards. In a study by Espy (1975) correlation data were obtained from solder joints subjected to temperature changes and to mechanical forces applied to the solder joint lead. The joints were subsequently tested destructively. It was found that the relative susceptibility of the solder joints to failure could be determined from the nondestructive holographic correlation data. In an earlier study by Jenkins

325

and McIlwain (1973), solder joints were thermally stressed and holographic correlation data taken. It was found to be possible to predict relative failure rates from these data.

(b) Fatigue Measurements Holographic correlation techniques have proved quite useful in detecting the microcracks associated with material fatigue. Extensive studies of fatigue as well as surface strain have been made by Marom (1974). Studies of changes in surface structure resulting from strain cycling using correlation techniques have also been made by Chuang (1968) and Bond (1973).

(c) Other Measurements Holographic correlation techniques have been used to monitor water pollution (Almeida and Eu, 1976) and to determine surface roughness (Léger *et al.*, 1975), particle mobility (Josefowicz and Hallett, 1975), and fluid flows (Durrani and Greated, 1975).

8.4.5.2 Delayed Correlation

In some cases (for example, where the test object is inaccessible), it may be necessary to work with a photograph or replica of the object. Correlation techniques can still be applied. A method of correlating photographs of the object in two strained conditions has been developed by Marom (1974), yielding numerical strain distributions. It is also possible to make a matched filter from one of the photographic images and use the second image as input to yield correlation peaks. This technique is amenable to automation.

8.4.6 A Hybrid Holographic Nondestructive Test System with Automatic Data Processing

No single nondestructive testing system can really satisfy all testing requirements for all objects, yet many techniques presently available are valuable in specific but limited nondestructive test situations. It is found that a combination of certain holographic techniques into one integrated unit would provide a more correlatable set of nondestructive data for a wider latitude of test objects and detectable problem categories such as flaws, debonds, and voids.

This section provides a discussion of such an integrated unit called the hybrid holographic nondestructive test (HHNDT) system. The HHNDT system integrates three holographic techniques—optical, acoustical and correlative—which were discussed in earlier sections. The specific objectives of this hybrid system are presented in Fig. 16.

Figure 17 presents a flow chart of the hybrid system. A test object can be investigated by the optical HNDT subsystem, the acoustooptical HNDT subsystem, or the correlation subsystem separately or sequentially depending on the type of test object and type of deformity sought. When the requirements

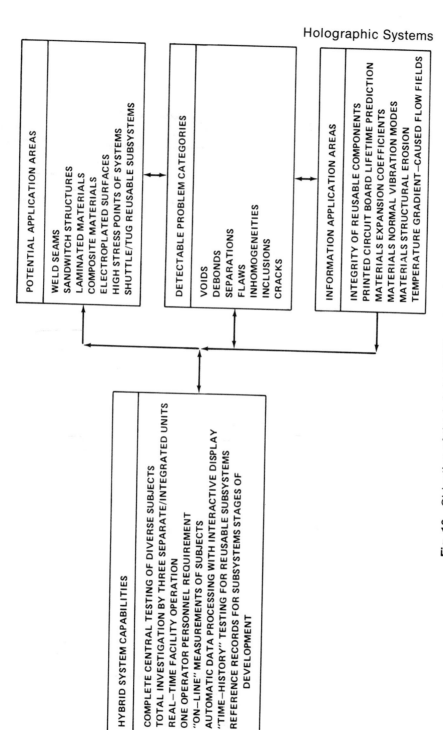

Fig. 16   Objective of the hybrid HNDT system development.

POTENTIAL APPLICATION AREAS

WELD SEAMS
SANDWITCH STRUCTURES
LAMINATED MATERIALS
COMPOSITE MATERIALS
ELECTROPLATED SURFACES
HIGH STRESS POINTS OF SYSTEMS
SHUTTLE/TUG REUSABLE SUBSYSTEMS

DETECTABLE PROBLEM CATEGORIES

VOIDS
DEBONDS
SEPARATIONS
FLAWS
INHOMOGENEITIES
INCLUSIONS
CRACKS

INFORMATION APPLICATION AREAS

INTEGRITY OF REUSABLE COMPONENTS
PRINTED CIRCUIT BOARD LIFETIME PREDICTION
MATERIALS EXPANSION COEFFICIENTS
MATERIALS NORMAL VIBRATION MODES
MATERIALS STRUCTURAL EROSION
TEMPERATURE GRADIENT–CAUSED FLOW FIELDS

HYBRID SYSTEM CAPABILITIES

COMPLETE CENTRAL TESTING OF DIVERSE SUBJECTS
TOTAL INVESTIGATION BY THREE SEPARATE/INTEGRATED UNITS
REAL–TIME FACILITY OPERATION
ONE OPERATOR PERSONNEL REQUIREMENT
"ON–LINE" MEASUREMENTS OF SUBJECTS
AUTOMATIC DATA PROCESSING WITH INTERACTIVE DISPLAY
"TIME–HISTORY" TESTING FOR REUSABLE SUBSYSTEMS
REFERENCE RECORDS FOR SUBSYSTEMS STAGES OF
        DEVELOPMENT

8. Equipment and Procedures

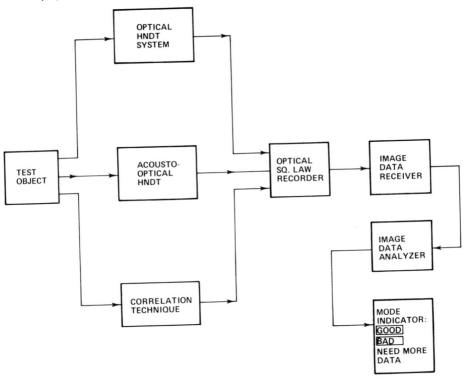

Fig. 17 Flow chart of hybrid system.

of a test necessitate the use of the optical HNDT subsystem, several subsystem operation modes can be involved, such as real-time, double-exposed, or time-averaged holography with thermal, pressure, or vibration loading of the test object. The test data will be recorded by an optical square law recorder in the form of an optical hologram. In the optical subsystem all test data will be some form of interferometric fringes superimposed on the holographic image of the test object.

When the requirements of a test necessitate the use of the acoustooptical HNDT techniques, there are several possible versions of subsystems. For example, the subsystem can operate in an immersion mode or a scanning mode (Hilderbrand and Brennen, 1972). The acoustical information will be modulated onto an optical carrier and, again, an optical wavefront (holographic or photographic) will be recorded by an optical square law detector. In all cases, with this subsystem the data to be evaluated will be an optical image of the test object revealing x, y, and z coordinates of the internal deformations.

When the requirements of a test object necessitate the use of the correlation subsystem, a Fourier transform hologram (matched Vander Lugt filter) will be recorded of the test object in an unstressed state. In all cases, with this subsystem the data to be evaluated will be an intensity correlating two wave-fronts (from the unstressed and stressed test object). This intensity is indicative of flaws or deformities in areas of interest, as was discussed earlier (Section 8.4.5).

The normal mode of operation for this hybrid system involves the sequential employment of all three of these subsystems on the test objects in the manner described later.

The optical HNDT system would be employed at the discretion of the operator to obtain interferometric fringes in proximity to the test object in the unstressed and stressed state. Examination of these fringes would determine a range of (x, y) coordinates on the test object defining areas of possible flaws or deformities. A second subsystem—the acoustooptical HNDT—now would be employed to produce an internal image of the test object, providing a display of the internal deformities seen by this system. This procedure could provide the z coordinate as further verification of the (x, y) coordinates for the flaw suspect regions found through the optical HNDT subsystem. Those regions so verified become regions, or areas of interest, to which the correlation technique can be applied as a "fine tuning" technique for the quantitative analysis of flaws or deformities. The information from these three subsystems would be recorded on a single readout device; that is, an optical square law recorder. The present format for this recorder is a 70-mm film strip; each subsystem would record its output on a single 70-mm frame.

If real-time operation is to be achieved as planned for this hybrid system, a real-time square law recording material (Kurtz and Owen, 1975) must be employed. Perhaps the best contender in the desired 70-mm format is the photopolymer presently being developed at E. I. DuPont de Nemours and Company (Booth, 1975). Already this photopolymer is capable of greater than 90% diffraction efficiency at an exposure wavelength of 5145 Å, with a spatial frequency of up to 3000 lines/mm. Further, this material satisfies the requirement of dry processing because it self-develops by means of the postexposure illumination of the photopolymer at the exposure wavelength used for recording. With such a recording material, the information from all three subsystems could be recorded and displayed in near real time.

Each of these three holographic nondestructive test subsystems will produce voluminous data which must be quantitatively analyzed before meaningful results can be obtained. This task is to be handled by an optical scanning system (image data receiver/analyzer) which includes a digital computer. The primary function of the optical scanner is to present, at high speed and in digital computer input/output format, the spatial intensity data contained in a two-dimensional scene.

This system is complicated by the existence of three different HNDT subsystems. The image data receiver must accept the following data formats: imaged information with superimposed interferometric fringe data from the optical HNDT subsystem, internal structure image information with superimposed internal flaws from the acoustooptical HNDT subsystem, and relative point intensity distributions from the correlation subsystem. The optical scanning system must analyze the data. Several new techniques are presently being formulated to properly interpret the optical subsystem fringe data, which is the most complex data format to be handled by the system.

A proposed general configuration for the optical scanning receiver/analyzer is shown in Fig. 18. The data flow shown is described briefly in the following paragraphs.

The sensor directly translates the optical data into an electrical signal. The intensity function detector translates this signal, providing input for the intensity function decoder which, in turn, generates binary signal replicas for transmission to the computer. Then, the scanning function decoder translates binary commands from the computer into the necessary analog levels, causing the scanning function driver to provide a sensor deflection field which is a precise replica of the scanning function decoder output.

The mode indicator console, complete with interactive display, is the primary system control unit. From this console the operator has automated control of one or all of the HNDT subsystems. He has automated control over the selection of the optical square law detector format.

The operator has direct access to the computer and its data bank, enabling him to interrogate the stored digitized holograms with several different techniques. If, after his analysis, the operator has not received a go/no-go condition on the test object (this would result from predetermined thresholds for flaw deformity definition), he can repeat his tests so as to acquire new data taken under different loading or stressing techniques.

Even though the hybrid system discussed here will not perform all tests on all objects, it probably has more versatility than any other single HNDT system. The integration of a digital computer-controlled scanning/analysis unit into the hybrid system to provide automatic data processing from all holographic subsystems should add speed as well as precise and objective data analysis to the overall versatility of the system.

8.4.7 Holography of Moving Objects

The objective of this section is to examine the different techniques of making holograms of moving objects and to examine the particular setups. The techniques include holographic Doppler-spread imaging (Aleksoff and Christensen, 1975), temporally modulated holography (Aleksoff, 1971), and holography of moving scenes (Neumann and Rose, 1967; Neumann, 1968; and Kurtz and Loh, 1972).

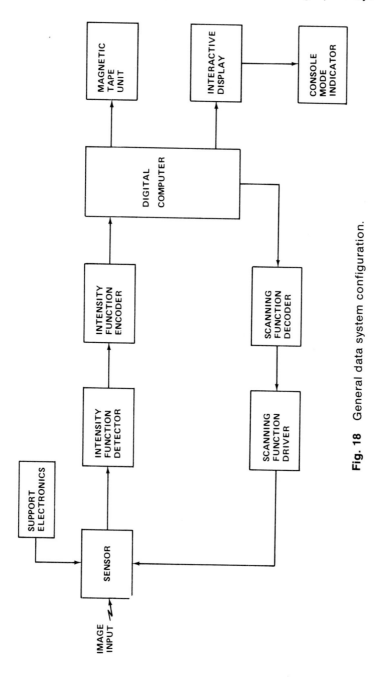

Fig. 18 General data system configuration.

8. Equipment and Procedures

8.4.7.1   Holographic Doppler-Spread Imaging

The holographic Doppler-spread imaging technique is used mainly in making holograms of rotating objects. Basically, an object is illuminated with laser light and imaged through a telescope onto the holographic film. The Doppler frequency shift associated with the rotation of the object is used as the temporal encoder. The light that is scattered from the object surface in any given direction has a special optical carrier frequency for a particular path of illumination and path of observation. Hence the reference beam has a temporal frequency shift which matches the Doppler shift in each specific direction. In other words, the temporal filtering property of the hologram changes the time-channel spread function into a spatial spread function. The width of this spatial spread function is controlled by the temporal variables. The reconstruction of the hologram can be obtained by conventional techniques.

The details of the holographic Doppler-spread imaging system are shown in Fig. 19. The object beam path traces through mirror M1, collimator 2, the object (on M2), and a unit-power concentric telescope composed of the slit S, lens L1, and lens L2. The reference beam path consists of collimator 1, mirror M2, mirror M3, lens L3, mirror M4, and lens L4. Lenses L3 and L4 also form a unit-power telescope. The object and reference beam paths are equal in length. The angle between the object beam, reference beam, and hologram plane is 15°. The slit S is located between L1 and L2 and is one focal length from each lens. The film exposure time T is controlled by the electronic shutter SH, which is synchronized with the rotation of M2 so that M2 is perpendicular to the hologram plane during the halfway portion of the exposure.

The rotation of M2 produces a linear spread of Doppler frequencies across

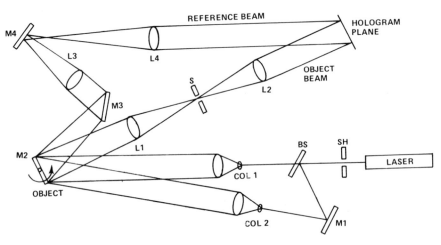

Fig. 19   Holographic Doppler-spread imaging setup.

332

the reference beam. The object is mounted on M2 so that it is rotating with M2. The speed of the mirror M2 can vary from 10^{-5} to 25 rad/sec. The lower rotation speed enables the use of lasers of lower power (approximately 10 mW) since longer exposure time is allowed in this case.

8.4.7.2 Temporally Modulated Holography

Temporally modulated holography commonly uses one of three different setups, depending on the application: making holograms of ultrasonic beams, making modulated holograms of vibrating objects, and detecting the shear wave resonance of an ADP crystal. These setups are shown in Figs. 20 through 22. Each will now be explained in more detail.

(a) Holograms of Ultrasonic Beams The making of holograms of ultrasonic beams will use a single-sideband suppressed-carrier (SSSC) modulation technique as shown in Fig. 20. This technique translates the frequency of the reference wave. A diffraction cell is used to spatially filter out the first order diffraction of the reference wave. The object wave passes through a tank of water in which the ultrasonic beam is propagating; this breaks the object wave into n diffraction orders. The diffraction cell and ultrasonic source are both connected to the same oscillator so that the system is phase-locked. The hologram plane is set at a 45° angle to the tank of water which the ultrasonic beam is propagating through. To take a hologram of the ultrasonic beam, the beam must be directed through a specific channel. The channel is made up of

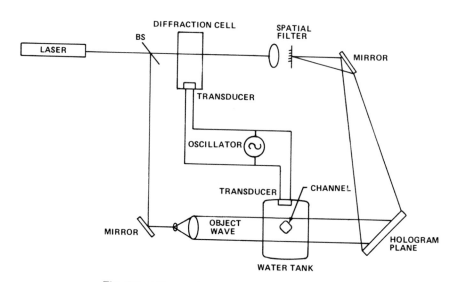

Fig. 20 Ultrasonic beam holographic setup.

333

8. Equipment and Procedures

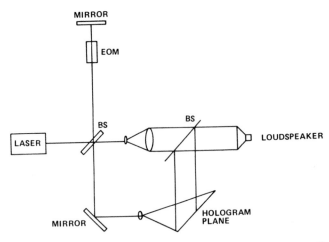

Fig. 21 Modulated holographic setup for making holograms of a vibrating loud-speaker.

two aluminum plates held apart and parallel to each other by four bolts. The channel is inclined at 45° to the beam for the purpose of concentrating the beam into a smaller area.

In the process of reconstructing the hologram, a diffuser is placed in the object wave. A virtual image can be seen under normal viewing conditions. Although a large viewing angle is used, only a small section of the ultrasonic beam which satisfies the Bragg angle restriction is seen to be bright. If the object is viewed by a telescope set at the Bragg angle, the entire ultrasonic beam can be viewed with the same brightness.

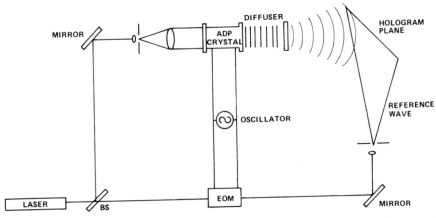

Fig. 22 Holographic setup for detecting the shear wave resonances of an ADP crystal.

(b) Holograms for Detection of Vibrations The next setup is a system that can be used to detect vibrations of an object. As an example, the specific object used in the system is a 7.6-cm-diam loudspeaker driven by a sinusoidal electric power supply, as shown in Fig. 21. When an object is illuminated and viewed along its direction of vibrations, the light reflected from the object is phase-modulated. Small-amplitude vibrations can be detected by translating the frequency of the reference wave an amount equivalent to that being applied to the loudspeaker. The reference wave is translated by an electrooptic modulator (EOM) where a sawtooth phase modulation produces an SSSC modulation.

The detection sensitivity depends on suppressing the carrier frequency of the reference wave. This problem can be handled by electronic equipment. Limitation of the system arises due to the bias buildup on the photographic emulsion. This problem can be helped to some degree by optimizing the object-to-reference intensity ratios. It was discovered that to maximize the reconstruction brightness, the object and reference intensities should be equal during the recording of the hologram.

(c) Holograms for Detection of Shear Wave Resonances The third setup as shown in Fig. 22 is used in detecting shear wave resonances of an ADP crystal. The resonances are observed by placing the crystal between two crossed linear polarizers. In this system the same signal source (oscillator) is used to keep the object wave and reference wave in phase with each other. The reference wave is passed through an electrooptic modulator which will translate the frequency of the reference wave to make it the same frequency as the object wave.

8.4.7.3 Holography of Moving Scenes

The last technique is called holography of moving scenes and has two basic setups which are used to treat cases in which the direction of the motion of the object is either known or unknown.

(a) Direction of Motion Known The setup requires that the direction of the motion be known in taking the hologram. A particular setup for the making of holograms of rotating objects is shown in Fig. 23. The object is illuminated by a uniform and collimated beam with its direction parallel to the hologram plane.

Experimental results showed that the brightness of the stationary background was greater than that of the moving scene. This is because the background is stationary; all of the illuminating light is reflected to the hologram plane. Where the object is moving, light is scattered due to its motion, and therefore not all of the light is collected at the hologram plane. The maximum speed of rotation for this experiment was 6 mm/sec.

335

8. Equipment and Procedures

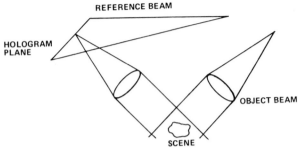

Fig. 23 Holographic setup for making holograms of moving scenes with known direction of motion.

In another system geometry (Kurtz, 1972), it was found that by using an elliptical configuration (Fig. 24) one could holographically record front surface detail from a scene moving at high velocity. Such an elliptical configuration was used to resolve front surface detail from a projectile moving at 2000 m/sec (Smigielski *et al.*, 1971). This configuration has also been used to make a three-dimensional motion picture of a moving scene whose velocity was oriented in a specific direction (Kurtz and Perry, 1973). In this experiment, however, the object was constantly illuminated using a continuous wave laser and exposure time of 1/60 sec. Here it was shown that the resolved detail of a moving object was a function of the total motion of the object during the exposure and not just a function of speed or time. Figure 25 shows this total motion dependency for a family of such ellipses.

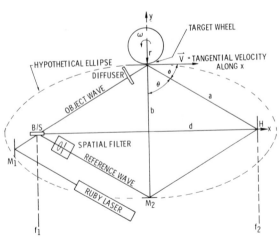

Fig. 24 Holographic setup for making holograms of moving scenes with known elliptical direction of motion.

336

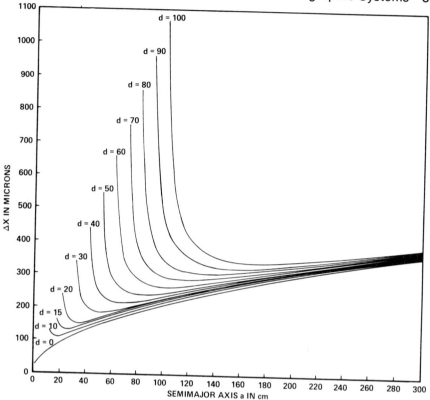

Fig. 25 Allowed travel, ΔX, for various elliptical configurations: $\Delta L = \lambda/8$.

(b) Direction of Motion Unknown The setup as shown in Fig. 26 uses a feedback-control technique. In this system the distance from the laser to the hologram point by way of the object is monitored by electronics. If the distance is changing, it should be matched by the same path length variation of the reference beam so that both fields arrive at the hologram plane with the same spatial frequency and, hence, the interference fringes will be stationary. The change in reference path length can be detected at the back of the hologram center by the use of a slit and a photomultiplier. The photodetector senses the fringe motion and sends a signal to the piezoelectric-crystal-driven mirror which then adjusts the length of the reference path. This results in a shift of the fringe-modulation envelope so that a maximum occurs in the desired direction. The correction can be achieved only at the center of the hologram by this means; blurring and loss of image brightness still occur.

The applications of these systems are numerous. For example, a holographic Doppler-spread imaging system can be very effective in increasing image

8. Equipment and Procedures

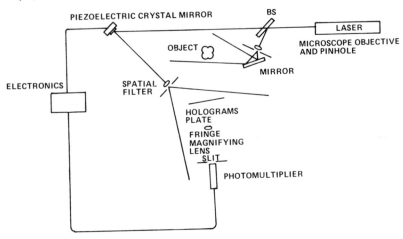

Fig. 26 Basic feedback-control holographic setup.

resolution. It can be used for imaging through diffusive media. Temporal filtering techniques are used in many cases to predict quantitatively the intensity distributions across the images of nonvibrating objects as well as vibrating objects. Finally, feedback-control techniques can be applied to make a hologram of a moving object of unknown direction of motion.

8.4.8   Other Systems

8.4.8.1   Holographic Testing of Concave and Convex Surfaces

The traditional method of testing the optical quality of concave or convex surfaces involves the comparison of a standard test surface and the surfaces to be tested through the interpretation of Newton rings. The method is usually quick and convenient, but the requirement of physical contact between the two surfaces can cause scratches and surfaces deformations. In addition, the glass surface to be tested has to be completely polished and already in its marketing stage. The defects caused by the contact testing process make the method very expensive.

In 1970, a holographic testing method was devised (Snow and Vandewarker, 1970) that involves first taking a hologram of a perfect test glass and then returning the hologram to its original position in the system. Comparison of an unknown glass to the holographic image of the perfect glass can be made. Imperfections or deformations on the surface of the tested glass will appear as real-time interference fringes.

338

A holographic configuration for testing the concave surfaces is shown in Fig. 27. In this system a telescope of appropriate power expands and collimates the beam. The collimated beam is directed through a 50/50 beamsplitter which allows half of the light to illuminate the unknown glass and the remainder to reflect to the hologram plane. This latter beam serves as the reference beam. The test procedure begins with the return of the hologram of the perfect glass to its original position. This position can be achieved by adjusting the interference patterns until a null field is reached. The hologram virtual image is then ready to be compared with any similar surface. Comparisons are allowed at any stage during the polishing and grinding of the glass. In this procedure, since the method is noncontacting, the surface of the glass under test will not be scratched.

The holographic configuration for testing a convex surface is shown in Fig. 28. Again a telescope is used to expand and collimate the beam incident on the test surface. The principle and procedure are the same as discussed in the testing of concave surfaces.

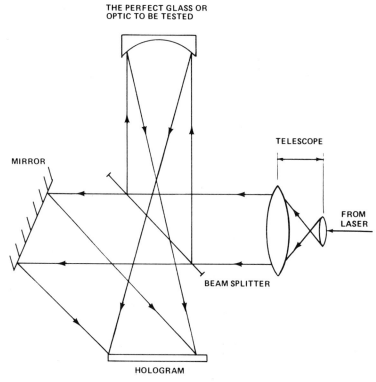

Fig. 27 System configuration used for the holographic testing of concave surfaces.

339

8. Equipment and Procedures

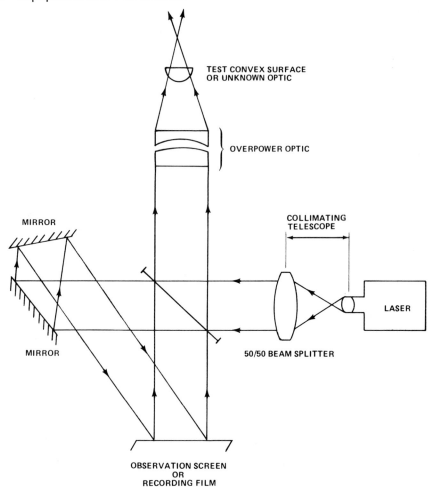

Fig. 28 System configuration used for the holographic testing of convex surfaces.

Computer-generated holograms may also be used to test optical elements. In Fig. 29 a Twyman–Greene method of testing optical surfaces has been modified by including a computer-generated hologram (MacGovern and Wyant, 1971). The computer can generate the points corresponding to the desired test surface onto a holographic plate. This is equivalent to making an original hologram of a perfect surface. The computer-generated hologram is placed in the reference beam so that its image and the image of the unknown surface can be compared.

The modified Twyman–Greene method can be aided by using a self-scanned 1024 element photodiode array and minicomputer (Bruning *et al.*, 1974) in

340

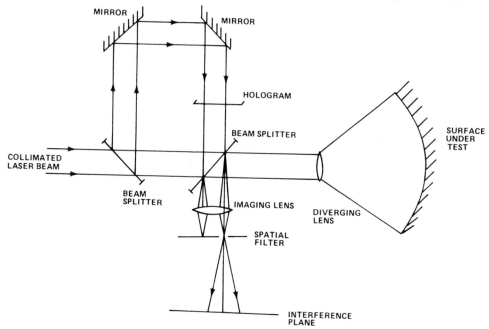

Fig. 29 A modified Twyman–Greene interferometer for the testing of a large surface. (From MacGovern and Waynt.)

measuring the phase in the interference pattern to an accuracy of $\lambda/100$. The system can be automated for routine testing; however, its construction is very elaborate. Another holographic interferometry system which enables large optics to be tested by small holograms is shown in Fig. 30. A Fresnel zone plate is illuminated by a collimated laser beam (Broder-Bursztyn and Malacara-Hernández, 1975). A photographic plate is placed in the image of the Fresnel zone plate. This image would also be a Fresnel zone plate if the surface of the test mirror were perfectly spherical. The photographic plate after development becomes a hologram. The hologram is replaced in its original position, and comparison of the surface under test and the reconstructed wavefront can be made.

All the holographic test systems described previously have two common advantages: they are noncontacting and therefore nondestructive for the optics involved, and they can be used to evaluate optical surfaces for irregularity early in the polishing stages so that manufacturing cost can be reduced.

8.4.8.2 Holographic Video Disk

Video tapes have been used by the television industry for many years. They are relatively expensive to reproduce, require a considerable amount of space,

8. Equipment and Procedures

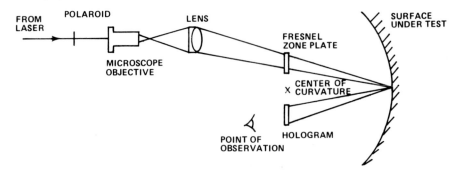

Fig. 30 Fresnel zone plate interferometer. (From Broder-Bursztyn and Malacara-Hernández.)

and are less durable and versatile than a holographic system of video storage. Hannan (1973) has demonstrated a holographic tape system that can store and read out the video information as a sequence of holograms ("Holotape system"). Each hologram is a recording of one still frame of a motion picture. Both video and audio information are stored as surface relief patterns which can be duplicated by embossing onto thermoplastic film. The object film which is used to make the master Holotape is essentially a standard 16-mm film with two adjacent frames which contain encoded color (chrominance) information and a black and white image (luminance) information of the object. A diagram of the RCA Holotape system recording apparatus is shown in Fig. 31.

Zenith (Hrbek, 1974) and Philips (Compaan and Kramer, 1973) have each advanced a disk, similar in size to a phonograph record, for video storage.

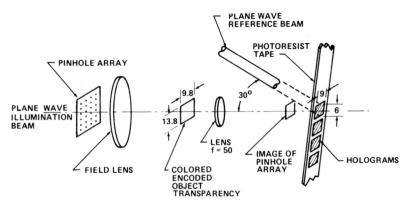

Fig. 31 A diagram of the hologram recording arrangement. The plane wave illumination and object beams are produced by beam expanders. The field lens increases the illumination efficiency of the pinhole array and controls the Fresnel imaging required for the high sampling rate. Dimensions are in millimeters. (After Hannan *et al.*, 1973.)

Both systems record the video information as a succession of short grooves or pits of variable length and repetition frequency. On the Philips VLP system, the repetition frequency and the average length of the pits determine the luminance signal, while the encoded color (chrominance) and sound signals give a modulation of the length of the pits. The Philips system uses an opaque disk with pits embossed on both sides. The disk is tracked by a small 1-mW He–Ne laser. The reflected beam is modulated by deflection of the light through diffraction at the pits. It falls on a detector and is converted into an electrical signal which is then decoded. The Zenith system is very similar to the Philips system except that the Zenith disk is transparent and the transmitted light of a 1-mW He–Ne laser is used for playback. Both sides of the disk can be used because of the small depth of field of the focused laser beam used in tracking. The focal plane of the laser light spot is merely refocused for the second side. To give good results, both the Philips system and the Zenith system must incorporate four special requirements into the video disk player:

(1) the speed of revolution of the disk must be kept constant (1 in 1000 for the Philips system),

(2) the focused laser beam must remain on the disk surface being tracked (the depth of field of the Zenith system is 3.1 μm),

(3) the beam of light must remain centered on the track even though the track may be out of round or the hole in the disk not centered or of the correct size (the player must be able to operate correctly even when the total deviation of the track from the ideal position is as much as 0.1 mm), and

(4) the complete optical system must move radially across the record at the rate at which the track advances without the aid of a continuous groove or other mechanical guide in the disk or player.

Both the Philips and Zenith optical video disks are superior to other forms of mechanical video disks because of their noncontact reproducing capability. This makes it possible to prolong the life of the pickups and disks and reduce scratching noises in reproducing signals. However, one problem with both systems is presented by the necessity for precise focusing and tracking adjustments. Although both Zenith and Philips systems seem to have solved the optical focusing and tracking problems, the high cost of the complicated servomechanisms and related player systems still remains. The RCA Holotape system, using holographic recording, avoids these tracking and optical focusing problems. However, in this system, an image frame is stored in a unit area of 7×11 mm (by 2 for colored images) as plural holograms. Also, tape-to-tape duplication is much more time-consuming and expensive than the straightforward disk-to-disk duplication technique. To incorporate the advantages of holographic storage of the RCA system (no tracking and optical focusing problems) with the high-density storage and inexpensive reproduction capabilities of the optical video disk, a technique utilizing Fourier-transformed holograms for higher density storage will be described in the following paragraphs.

(a) Random Phase Sampling Holography To make high-quality and high-storage-density Fourier-transformed holograms, the conditions of (1) high redundancy, (2) high diffraction efficiency, (3) low (speckle) noise, (4) high resolution, and (5) high fidelity of luminance tone must be satisfied. To satisfy condition (1), a hologram must be transformed exactly to its Fourier-transformed plane. Condition (2) requires that the distribution of light energy on the restricted hologram area be uniform. For condition (3), it is necessary that almost all of the light energy be contained in the restricted hologram area. Condition (4) requires that the hologram size must be as large as the bandwidth of the information. Condition (5) demands that the transfer function of the total procedure of making a hologram be constant over a wide beam intensity range.

Several methods have been tried in an attempt to satisfy these conditions and to get high-density and high-quality holograms. One of the most effective ways is the defocusing method in which a hologram is made some distance away from the exact Fourier-transformed plane. This method is useful in satisfying conditions (2)–(5), but it does not provide high redundancy. Moreover, it has the defects that the recording area is larger than the diffraction-limited size and the degree of the defocusing varies with the kinds of information. A multiple beam recording technique has been developed in order to get a good redundancy for the defocusing method, but it also enlarges the recording area.

On the other hand, the random phase shifter method has proved to be quite useful in getting high-storage-density and high-quality holograms containing digital information with diffraction-limited size. In this method, random phases are added to each beam and the beam is focused on the exact Fourier-transformed plane (Tsunoda and Takeda, 1973, 1974; Tsunoda et al., 1976).

Random phase sampling holography is an extension of the random phase shifting technique. It consists of a combination of both the shifting and the sampling method. The basic configuration of the random phase sampling method is shown in Fig. 32. Image information, sampling mesh, and the random phase shifter are closely attached and set in the optical path. The beam is passed through these devices and focused on the storage medium, which is placed on the exact Fourier-transformed plane. Next, the image information is divided into a large number of sampled portions by a sampling mesh. Then, random phases are added to each sampling point by a random phase shifter. The method is composed of the following three basic principles: (1) sampling theorem, (2) Rayleigh criterion, and (3) random walk principle. These principles are discussed as follows.

Principle 1 By utilizing a sampling technique, image information is divided

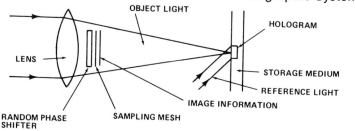

Fig. 32 Basic concept of the random phase sampling method.

into a large number of sampled portions. The large amounts of low-frequency components of the image information are eliminated on the hologram plane. The essential components of the information are spread over an area corresponding to the diameter of the sampling points. This results in information reduction.

The pitch of the sampling points is determined from the need to maintain image quality and the bandwidth of the hologram. In other words, the pitch must be equal to the required resolvable line width, which, according to the sampling theorem, is equivalent to half of the required resolvable spatial frequency.

Principle 2   The minimum size or the maximum storage density of a hologram having perfect redundancy of storage can be obtained through Fourier-transformed holography. The diameter of the hologram is related to the diameter of the sampling hole by the Rayleigh criterion. The light energy of the information is contained in the restricted hologram area. There is no speckle noise in the reconstructed image when the hologram is illuminated by a beam whose diameter is almost the same as the hologram size.

Principle 3   To get a hologram with a high diffraction efficiency and a high fidelity of luminance tone, it is necessary to distribute the light energy uniformly in the restricted hologram area. A random phase shifter can be used to eliminate keen spectra of the light energy appearing on the hologram plane through the correlations among the different beams filtering through the sampling holes.

Matsumura (1974) has reported that the use of random phase shifters also greatly reduces the macro-speckle due to scratches and dust on lenses as well as undesirable interference fringes by light reflected from surfaces. This type of speckle noise is one of the biggest noise problems in coherent imaging systems. By giving a spatially random distribution of phase to the transmitting light, high efficiency and high signal-to-noise ratio can be obtained for the reconstructed image. This is due to the fact that the diffracted light from an object illuminated through a random phase shifter is uniformly distributed at the recording medium. Therefore, the dynamic range of the recording medium can be used to its utmost.

(b) The Holographic Video Disk Player System The holographic video disk player system can be described by an example. A disk of 300 mm in diameter can have 54,000 holograms, each of which stores the luminance, chrominance, and the sound codes in a small area of approximately 1 mm in diameter. Interferences between the ghost images and the reconstructed images are eliminated by suitably choosing the incident angles of three information-bearing light waves in axially symmetric directions.

Reconstructed images are focused on the image sensors (such as solid state arrays) whose output signals are processed and displayed on a TV monitor. There is no need for precise focusing or tracking. Mechanical tolerances for adjustments lie in the region of 10 to 100 μm, which is 100 times larger than the tolerances required for time sequential recording disks.

8.4.9 Summary

We have presented one of the major applications of holography, HNDT, and have discussed it from a practical point of view. In this context we have presented and discussed several specific holographic arrangements, each having its own advantages. We have then discussed a hybrid holographic arrangement which combines the several individual holographic systems into one mobile testing unit possessing the capability for automatic data processing. The reader is invited to consider the proper hybrid combination of specific systems for his own needs. Further thought should be given at this time to the successive generations of a hybrid system incorporating integrated optics. Such a miniaturized and stable system would be useful in the inspection of space payloads.

Section 8.4.7 presents some of the successes in this area, with the possibility of ultimately realizing a true three-dimensional motion picture camera. Because of the vast potential of such a device, our hope has been to at least stimulate thought on this subject. The last section discusses other specialized holographic systems.

Finally, it should be mentioned that all of the references were carefully chosen to provide explicit detail on each concept covered in this chapter. By no means is this list totally inclusive or even complete; many other excellent works were not included because of the restriction of space.

REFERENCES

Aleksoff, C. C. (1971). *Appl. Opt.* **10,** 1329–1341.
Aleksoff, C. C., and Christensen, C. R. (1975). *Appl. Opt.* **14,** 134–141.

Aleksoff, C. C., and Guenther, B. D. (1976). *Appl. Opt.* **15**, 206–217.

Almeida, S. P., and Kim-Tzong Eu, J. (1976). *Appl. Opt.* **15**, 510–515.

Archbold, E., Burch, J. M., and Ennos, A. E. (1970). *Optica Acta* **17**, 883–898.

Archbold, E., and Ennos, A. E. (1972). *Optica Acta* **19**, 253–271.

Bage, M. J., and Beddoes, M. P. (1976). *Appl. Opt.* **15**, 2632–2634.

Berger, H. (1969). *Acoust. Holog.* **1**, 34–36.

Bond, R. L. (1973). *In* Holography and Optical Filtering, pp. 177–182, NASA SP-299.

Booth, B. (1975). *Appl. Opt.* **14**, 593–601.

Broder-Bursztyn, F., and Malacara-Hernández, D. (1975). *Appl. Opt.* **14**, 2280–2282.

Bruning, J. H., Herriott, D. R., Galalgher, J. E., Rosenfeld, D. P., White, A. D., and Brangaccio, D. J. (1974). *Appl. Opt.* **13**, 2693–2703.

Casasent, D., and Psaltis, D. (1976). *Appl. Opt.* **15**, 1795–1799.

Chuang, K. C. (1968). *Material Evaluation* **26**, 116–119.

Collier, R. J., Burckhardt, C. B., and Lin, L. H. (1971). "Optical Holography." Academic Press, New York.

Compaan, K., and Kramer, P. (1973). *Philips Tech. Rev.* **33**(7), 178–180.

Durrani, T. S., and Greated, C. A. (1975). *Appl. Opt.* **14**, 778–786.

El-Sum, H. M. A. (1976). Analytical Study of Acousto/Optical Holography, Interfacing Methods for Acoustical and Optical Holography Nondestructive Testing Research. El-Sum Consultants, Atherton, California. Rep. NASA CR-2775.

Erf, R. K. (ed.) (1974). "Holographic Nondestructive Testing." Academic Press, New York.

Espy, P. N. (1975). Testing of Printed Circuit Board Solder Joints by Optical Correlation. NASA TR R-449.

Gara, A. D. (1977). *Appl. Opt.* **16**, 149–153.

Goodman, J. W. (1965). *Proc. IEEE* **53**, 1688–1700.

Goodman, J. W. (1968). "Introduction to Fourier Optics." McGraw-Hill, New York.

Hannan, W. J., Flory, R. E., Lurie, M., and Ryan, R. J. (1973). *J. Soc. Motion Pict. TV Eng.* **82**, 905–915.

Hilderbrand, B. P., and Brenden, B. B. (1972). "An Introduction to Acoustical Holography." Plenum, New York.

Hrbek, G. W. (1974). *J. Soc. Motion Pict. TV Eng.* **83**, 580–582.

Indebetouw, G., Tschudi, T., and Herziger, G. (1976). *J. Opt. Soc. Amer.* **66**, 169–170.

Jenkins, R. W., and McIlwain, M. C. (1973). *In* Holography and Optical Filtering, pp. 183–192. NASA SP-299.

Josefowicz, J., and Hallett, F. R. (1975). *Appl. Opt.* **14**, 740–742.

Kurtz, R. L. (1971). A Holographic System That Records Front Surface Detail of a Scene Moving at High Velocity. Ph.D. Dissertation. Virginia Polytechnic Inst. and State Univ., Blacksburg, Virginia.

Kurtz, R. L. (1972). US Patent No. 3535014.

Kurtz, R. L., and Loh, H. Y. (1972). *Appl. Opt.* **11**, 1998–2003.

Kurtz, R. L., and Perry, L. M. (1973). *Appl. Opt.* **12**, 2815–2821.

Kurtz, R. L., and Liu, H. K. (1974). Holographic Nondestructive Tests Performed on Composite Samples of Ceramic-Epoxy-Fiberglass Sandwich Structure. NASA TR R-430.

Kurtz, R. L., and Owen, R. B. (1975). *Opt. Eng.* **14**, 393–401.

Leendertz, J. A. (1970). *J. Phys. E* **3**, 214–218.

Léger, D., Mathieu, E., and Perrin, J. C. (1975). *Appl. Opt.* **14**, 872–877.

Liu, H. K., and Kurtz, R. L. (1977). *Opt. Eng.* **16**, 176–186.

Liu, H. K., Kurtz, R. L., and Moore, W. W. (1976a). *SPIE* **92**, 72–86.

Liu, H. K., Commeens, E. R., Hunt, W. D., and Whitt, L. (1976b). Evaluation of a Composite Mobile Holographic Nondestructive Test System. BER Rep. No. 204-74. Univ. of Alabama.

Liu, H. K., and Owen, R. B. (1979). *Opt. Eng.* **18**.

8. Equipment and Procedures

MacGovern, A. J., and Wyant, J. C. (1971). *Appl. Opt.* **10**, 619–624.

Marom, E. (1974). *In* "Holographic Nondestructive Testing" (R. K. Erf, ed.), pp. 149–180. Academic Press, New York.

Matsumura, M. (1974). *Jpn. J. Appl. Phys.* **13**, 557–558.

Metherell, A. F., El-Sum, H. M. A., and Larmore, L. (eds.) (1969). "Acoustical Holography," Vol. 1. Plenum, New York.

Metherell, A. F., and Larmore, L. (eds.) (1970). "Acoustical Holography," Vol. 2. Plenum, New York.

Neumann, D. B. (1968). *J. Opt. Soc. Amer.* **58**, 447–454.

Neumann, D. B., and Rose, H. W. (1967). *Appl. Opt.* **6**, 1097–1104.

Nisenson, P., and Sprague, R. A. (1975). *Appl. Opt.* **14**, 2602–2606.

Papoulis, A. (1968). "Systems and Transforms with Applications in Optics." McGraw-Hill, New York.

Lord Rayleigh (1920). *Scientific Papers* **6** (Cambridge Univ. Press), pp. 565–610.

Sharpe, R. S. (ed.) (1970). "Research Techniques in Nondestructive Testing." Academic Press, New York.

Smigielski, P., Fagot, H., Stimpfling, A., and Schwab, J. (1971). *Nouv. Rev. Opt. Appl.* **2**, 587–592.

Snow, K., and Vandewarker, R. (1970). *Appl. Opt.* **9**, 822–827.

Stroke, G. W. (1969). "An Introduction to Coherent Optics and Holography." Academic Press, New York.

Tsunoda, Y., and Takeda, Y. (1973). *J. Appl. Phys.* **44**, 2422–2423.

Tsunoda, Y., and Takeda, Y. (1974). *Appl. Opt.* **13**, 2046–2051.

Tsunoda, Y., Tatsuno, K., Kataoka, K., and Takeda, Y. (1976). *Appl. Opt.* **15**, 1398–1403.

Vander Lugt, A. (1964). *IEEE Trans. Inf. Theory* **10**, 139–145.

Waters, J. P. (1972). *Appl. Opt.* **11**, 630–636.

Special Problems

9.1 PHOTOGRAPHIC MATERIALS AND THEIR HANDLING

Stephen A. Benton

9.1.1 Choice of Material

Holographic recording has been reported on a wide range of materials, including electrooptical crystals and meltable plastic films. The use of some of the most practical materials has recently been comprehensively reviewed, and among them, silver halide photographic materials still stand out because of their reliability, easy availability, high sensitivity, and generally good performance (Smith, 1977). While they do require some time for processing and do not offer an add-on or erasure capability as normally used, they continue to warrant the sobriquet of "the holographer's only friend." In this section we shall deal with techniques for getting the most out of the presently available commercial materials and will attempt to indicate some directions for future progress.

Because holographic recordings include spatial frequencies that are much higher than in normal photography, emulsions useful for holography are generally very fine grained, to the point of near transparency and of very low sensitivity. The finest grained have spherical microcrystals of silver bromide (with a little iodide and some sensitizers) about 30 nm in diameter, and require exposures of 1000–3000 erg/cm², as outlined in Table I. Other so-called "holographic" emulsions may have grains up to 100 nm or so across, requiring exposures of only 50 erg/cm (or 5 erg/cm if not bleached), although fine grained conventional materials (grain size ~500 nm, exposures $\sim\frac{1}{10}$ erg/cm may be useful in special small-reference-beam-angle applications where laser energy or exposure time is especially limited.

The most well-known holographic silver halide materials are manufactured by Eastman Kodak Company in the United States and Agfa–Gevaert in Bel-

9. Special Problems

TABLE I

Silver Halide Holographic Recording Materials

Approximate grain diameter (nm)	Manufacturer and type	Approximate exposure (ergs/cm²) for bleached holograms at [a]		
		488 nm	514 nm	633 nm
37	AG 8E56 HD	1000	700	—
50	EK 120, SO-173	16,000	16,000	1600
	EK HRP, SO-343	1000	1600	—
55	AG 8E75	8000	8000	800
58	EK 649-F	550	500	700
65	EK 125, SO-424	45	35	—
70	EK 131, SO-253	80	35	12
88	AG 10E75	—	—	25
	AG 10E56	30	40	—

[a] PAAP developed and bromine water bleached. These are not manufacturer's specifications and are subject to change without notice.

gium and are sold in the United States principally through distributors.[1,2] The range of available materials and configurations can vary widely, so readers are encouraged to request up-to-date information and recommendations from the manufacturers. Many teaching laboratories begin with 4 × 5 in. glass plates of either the Agfa–Gevaert "Holotest" 8E75 or the Eastman Kodak 120 type (antihalation coated for transmission holograms, not coated for reflection holograms) and move on as experience is gained with other types.

In principle, the wavelength range of emulsion sensitivity can be changed simply by changing the mix of sensitizers included. Unfortunately, comparable materials are often not available in both red and green/blue sensitized versions, and the choice of a laser may depend upon the current material availability. The only materials presently sensitized for red + green + blue are EK 649-F and 131/SO-253.

The same emulsion is often available coated on glass plates or plastic film. For the beginner, glass plates are probably easier to handle and have generally produced higher quality results. However, they are prone to inconsistencies and manufacturing defects, and their relatively high cost pushes routine work toward film based materials. Here the cost of fixturing, such as vacuum chucks, can be more than offset by the economies of continuous roll exposures

[1] Scientific Products Department, Agfa–Gevaert, Inc., 275 North Street, Teterboro, New Jersey 07608.

[2] Scientific Photography Markets, Department 756, Eastman Kodak Company, 343 State Street, Rochester, New York 14650.

and processing. Although polyester film is stronger and more resistant to processing solutions than acetate base, its birefringence alters the polarization of transmitted light, which can interfere with its use as a master hologram or for information processing. The anticurling and antihalation properties of film base materials also vary widely.

It is important to point out in this practical handbook that systematic differences between competitive products can be observed. Agfa–Gevaert emulsions do seem to have a significant edge in sensitivity over comparable Eastman Kodak products, up to fourfold between 8E75 and 120 plates. But it is also fair to say that they suffer from a higher incidence of cosmetic defects and a weaker customer service support system, so that their use is often accompanied by frustration. Insofar as a holographic industry exists at all in the United States, it has generally adopted domestic products for routine and reliable production. As the market grows, both aspects of this situation may change markedly, but this is the reality at present.

9.1.2 Storage

Silver halide photographic materials are markedly more stable in storage than most other recording media, but because of their small grain size, holographic materials seem to be more sensitive to storage temperature and time than conventional materials. Best results are obtained with what ought to be normal procedure: the materials are kept refrigerated at 0–7°C (32–45°F) from arrival until shortly before use. They should be allowed to warm slowly to room temperature before the protective foil or waxed paper layer is opened in order to avoid condensation, which process may take a few hours for a small box of plates, or a full day for a large roll of film. Once the moisture barrier is opened, the material should not be re-refrigerated unless it is resealed with a dessicant. Opened materials should be stored at temperatures below 27°C (80°F). Degradation is slow enough that no extreme measures need be taken, but an increase of the developed "fog" density of unexposed plates is evidence of shelf life expiration. Refrigerated materials will keep almost indefinitely if they are protected from moisture.

A process that can be associated with storage is the relieving of stress in the emulsions before use. Certain types of glass plates seem to suffer from irregular shearing of the emulsion during coating and drying. If this occurs, it can be ameliorated by a process of warming the emulsion to about 180°F, exposing to a very high humidity (~90%) for an hour or so, and they drying and cooling it to room temperature. The plates should be horizontal during this cycle to avoid restressing (Pennington and Harper, 1970). This is a fairly risky process in practice. Fog can be produced by too vigorous treatment, and inadequate hardware designs can result in condensation dripping on, and ruining, many plates. But dramatic improvements in image quality have been observed, especially in relatively soft emulsions such as AG 8E75. A similar cycling program may also be useful for drying processed plates.

9.1.3 Handling during Exposure

The slightest motion of the recording material during exposure can result in a marked loss of diffraction efficiency, or even dark bands across the hologram. Therefore, stable mounting of the material during exposure, and repositioning in the case of interferometry, is a principle requirement. Separate techniques have been developed for rigid base (glass) materials and those on flexible (plastic film) base.

9.1.3.1 Glass Plate Mounting

The first successful plate holders were of the edge-clamp design and a variety of useful and well-engineered types are commercially available. Generally, the plate is constrained against a flat surface by clamps or springs along at least two sides, and registration pins are used to locate the plate precisely. However, the plate can still vibrate in a drum-head mode, which is readily excited by sound waves, so that fairly thick plates are often needed. A serious complaint of artists is that the edges of the plates are marred by shadows. However, the central area can be left open for "live-fringe" interferometry studies.

An inexpensive, if somewhat less precise holder can be made with a simple vacuum gripper, as in Fig. 1. Here the plate is drawn firmly against the frame of the base by atmospheric pressure on one side against a lower hand-generated pressure on the back.[3] If the rubber gasket is kept clean, and perhaps moistened, no appreciable creep results. The vacuum base is positioned somewhat above center to dampen edge-ringing, and a felted bottom brace is provided for the same purpose, as well as for easy mounting.

However, capillary action has provided the key to the most versatile plate holders we have found. Here a low viscosity, high surface tension liquid is drawn between the glass plate and a flat rigid surface (Fig. 2). After draining for a few minutes, the plate is firmly and stably mounted, with little opportunity for vibration as the dynamic rigidity is roughly that of the base. In this way, reflection holograms have been made with plate glass bases under very difficult conditions. If the glass is painted black on the back, an effective antireflection means is provided for transmission holograms. We find that xylene is a very useful capillary liquid with a suitable refractive index, and its volatility reduces

---

[3] A line of hand-operated vacuum bases is sold by the General Hardware Manufacturing Company, Inc., 80 White Street, New York, New York 10013.

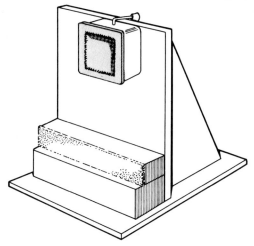

Fig. 1   Vacuum-base plate holder.

clean-up problems. Like all organic solvents, it must be handled with care and with generous ventilation, though its toxicity is rather low.

9.1.3.2   Film Base Mounting

Because of their flexibility and tendency to curl, none of the above techniques have proven useful for film base materials. The first approach is usually to sandwich the film between clear glass plates, which are clamped at the edges. However, this introduces many reflections from glass–air interfaces, which degrade the uniformity of the exposure and the cosmetic quality of the image. Filling the sandwich with refractive index matching liquid helps, but may be awkward to control.

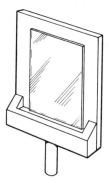

Fig. 2   Capillary action plate holder.

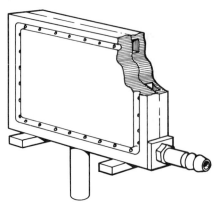

Fig. 3 Vacuum-channel film holder (cut-away view).

Rather more success has followed the use of vacuum-back holders, which draw a vacuum between the film and a flat and rigid back. Air flow can be distributed through a multiplicity of holes distributed behind the film, but the very slight dimpling over each hole is often apparent as a Bragg-angle mismatch during viewing. Instead, we use a single channel around the film periphery (Fig. 3). Its success depends on the plate being very flat indeed (as provided by a single-pass milled finish) and on waiting long enough for all the air trapped behind the film to escape. Here too, every particle of trapped dust is likely to show up as a dimple, so that the very highest quality exposures still are made on glass plates, except in the case of pulsed laser exposures with no rigidity requirements, although film flatness may remain a problem.

9.1.4 Processing

9.1.4.1 Introduction

More published attention has been devoted to the processing of holograms than any other aspect of practical holography, yet the field is still awash in poorly understood formulations and techniques. Recently, this attention has turned from bleaches toward developers, and the realization that all aspects of the process are interrelated. The fruits of this inquiry have not yet reached the commercial market, so the principles of processing chemistry will now be outlined with a few interesting examples to stimulate individual exploration.

9.1.4.2 Hardware

Holographic darkrooms tend to look very much like ordinary printing darkrooms because the processing requirements are very similar. Exotic developers, bleaches, and organic solvents may impose limitations on the container materials used, but most conventional hardware is adaptable. Special areas may be set aside for plate and film cutting and for drying. An area of particularly good ventilation, including a makeshift fume hood, is needed in many cases for bleaching and alcohol drying.

Open plastic trays are still the most flexible and economical means of processing, especially for larger sized holograms. Deep tanks are more compact and convenient for routine processing and can minimize the exposure of developers to aerial oxidation. However, tanks require much larger start-up solution quantities than trays, and special holders are needed to avoid having plates and films become stuck to the tank sides by capillarity. Odd-sized tanks are readily fabricated from acrylic plastic. Temperature sensitivity is about the same as for black and white processing, so that simple water jackets suffice for control.

9.1.4.3   Holographic Developers

Since 1964, the workhorse of holographic development has been the Eastman Kodak developer D-19 (or the comparable Agfa–Gevaert developer G3p in Europe). It is hard to say that the D-19 has any special suitability for holography, apart from being affordable and widely available, because it was optimized for black-and-white negative development. It is becoming apparent that improvements in holography will depend critically on new photochemical formulations that are optimized instead for the particularly fine-grained materials used and for the high spatial frequencies that dominate a holographic recording. However, the D-19 can work well under proper conditions and is well enough known to serve as a valuable example. The principles of developer composition have evolved through a long process of trial and error, and a serious curiosity about them can be assuaged only by a careful reading of the literature and many experiments (Mason, 1966; Kowaliski, 1972). The two types of development that arise in holography will now be outlined and illustrated with a few developer formulations of some practical interest. Much progress remains to be made, and individual investigators are still in a good position to contribute to that progress.

9.1.4.4   Types of Development

In the simplest view, the molecules of silver halide in each exposed microcrystal are separated by development into the same number of silver atoms accreted in a grain and halide ions which diffuse away. This process, called "direct" or "chemical" development, is indeed the principal mechanism in most cases and tends to proceed explosively, producing an elongated worm-like grain (Land et al., 1971). But a substantial amount of silver is often added to the grain by a process known as solution–physical development, which transfers silver atoms from nearby undeveloped microcrystals. The process begins with an etching of the microcrystal surfaces by silver halide solvents, such as sulfite, thiosulfate, or thiocyanate ions, producing mobile silver complexes. These diffuse about until a suitable precipitation site is encountered, where the developing solution can reduce them to insoluble metallic silver. Usually, the precipitation site is a nearby already developed silver grain, and the diffused silver is plated onto it in a more or less uniform way, producing spherical particles, as though the worms had been thickened. In conventional emulsions, solution–physically developed silver tends to get lost within the large multifilament grains, but in very fined-grained holographic emulsions, every silver atom counts, and effects that are ordinarily quite subtle become all too apparent. Moderate amounts of solvent (note that the process recycles solvent ions so that only small amounts are needed) markedly alter the sensitometric curve, not only by increasing the grain size, but also by uncovering additional latent image sites by surface etching, rendering more grains developable (Solman, 1966). The increase of contrast produces markedly increased diffraction efficiency in developed holograms (Smith and Callari, 1975).

Unfortunatley, silver also tends to precipitate at random within the emulsion, producing "dichroic fog" and strong milky scattering after bleaching, or on the emulsion surface, where it forms a scum. The extent of all these effects depends on the balance of development versus solvation activity in the emulsion, which is very sensitive to the developer composition and so varies with its exhaustion. Because all commercial developers contain some silver halide solvent, such as the sulfite "preservative", solution–physical development is bound to occur and to be far more noticeable in holograms, and thus to vary more markedly with developer exhaustion. This is the principal reason why conventional developers often seem so erratic and difficult to use with bleaches. After reconsidering them briefly, we will go on to some nonsolvent "direct" developers and then explore ways of exploiting the advantages of physical development, when it can be well controlled.

9.1.5 Development Recommendations

9.1.5.1 Conventional Developers

(a) **Kodak D-19** Although the manufacturer recommends 6 to 8 minute development at 20°C for most materials, we prefer 4-min development. Development time is the single most useful process variable, and termination when the desired density is reached is a very practical technique for overcoming experimental uncertainties (Biedermann and Stetson, 1969). The emulsion should be taken to a density of about 1.5 if it is to be bleached, and to about 0.8 if not. However, as the development time is extended, solution-physical development increases, causing an increase in noise, and the peak diffraction efficiency decreases. Bleached holograms come out brighter and cleaner if they are overexposed and undeveloped, with about 1-min development being a practical lower limit.

Apparently because of the exaggerated effects of solution-physical development effects in holography, we find that D-19 ages fairly quickly. Working solutions should be replaced daily, and continually fresh solutions must be used for research grade repeatability, especially when bleaches are used (Smith, 1977, p. 29).

(b) Neofin Blue Research at the University of Loughborough, England, has shown that a concentrated proprietary developer, Tetenal's Neofin Blue, works very well with very-fine-grained materials intended for bleaching (e.g., AG 8E56-HD) when the liquid concentrate is used undiluted for 5 min at 18°C (64.5°F). The kinetics are obviously quite different than in ordinary, diluted use, and the authors suggest alkali and antifoggant additions to modify the results. Addition of 0.3 g/liter benzotriazole decreases fog and increases contrast, while the further addition of 120 g/liter sodium metaborate improves results even more (Phillips and Porter, 1976).

9.1.5.2 Nonsolvent Developers

All of the vagaries of solution–physical development can be avoided if silver halide solvents are excluded from the development process. Of course, even water has some solvent effect on silver bromide, and the constituents of some emulsion types seem to as well, but a marked change occurs simply by avoiding sodium sulfite and potassium bromide. Such developers are often known as direct, chemical, or surface developers, and despite the lack of a "preservative," they seem at least as reliable as conventional types for holography.

Without sulfite protection we must turn to developing agents with photographically inert oxidation products, which tend to be rather less active than hydroquinone. One of the best known of these is ascorbic acid, as in the well-known MAA-3 developer (James and Vaneslow, 1955):

water at 40°C	0.5 liter
metol	2.5 g
d-araboascorbic acid	10
sodium carbonate (monohydrate)	55.6
water to make	1.0 liter

Develop at 20°C (68°F) for 4 min or until the desired density is obtained.

9. Special Problems

We prefer a formulation we call PAAP:

phenidone (add last)	0.5 g
l-ascorbic acid (vitamin C)	18 g
sodium hydroxide	12 g
sodium phosphate dibasic (NaH_2PO_4)	28.4 g
in water to make	1.0 liter

This is a more active developer, with a correspondingly more limited pot life because it is more vulnerable to aerial oxidation. Development times of 4 min at 68°F work well, although the densities produced are still somewhat lower than with the D-19.

Both of these developers produce clean holograms substantially free of milkiness when bleached. However, the diffraction efficiency is reduced because the same number modulation of grains produces a smaller index modulation, because the developed grains have not been "intensified" by solution-physical development. But they have produced the highest signal-to-noise-ratio results so far.

9.1.5.3 Solution–Physical Developers

In principle, the increase of silver in each grain provided by solution–physical development can lead to substantial increases of the diffraction efficiency of bleached holograms while maintaining a high signal-to-noise ratio. In practice, a certain amount of silver precipitates in fairly large randomly located chunks, which produce a milky scatter in the bleached emulsion. The ratio of correctly versus incorrectly precipitated silver varies mainly with the effectiveness of the silver halide solvent, being lowest for weak solvents such as sodium sulfite (which is why conventional developers do so poorly) and highest for strong solvents such as ammonium thiocyanate. Even with thiocyanate, the increase of scatter noise is marked, so that a minimum amount should be added to an otherwise nonsolvent developer for a slight boost of transmission hologram efficiency. Only in cases where the diffraction efficiency is marginal, such as for reflection holograms, should very much be used.

9.1.5.4 Balanced Solvation and Development

In bleach processing, about half of the emulsion microcrystals are usually exposed and direct developed. If all the undeveloped silver could be transferred successfully, a doubling of grain volume would occur, producing a fourfold increase in diffraction efficiency (and Rayleigh scattering). Such an extreme balance between solvation and development goes so far beyond normal photographic practice that we have dubbed it "IEDT," for "intra-emulsion diffusion transfer," to emphasize the essential nature of the diffusion of silver ions from unexposed to exposed areas, over distances that are small compared to the swelled emulsion thickness, but may be several fringe widths (Benton, 1974). In addition to increased image brightness, the process has other advantages for holography.

Because all the silver is developed, there is no need for a fixing bath. If the silver is bleached to produce the original emulsion constituents (principally silver bromide), the index of refraction and thickness of the emulsion are restored to their values during exposure, so that volume effects such as Bragg-angle selection are automatically and precisely compensated over a range of exposures. Thus, reflection holograms will reconstruct in the same wavelength and location that they were recorded in, an important consideration for full-color holography.

Because the lateral diffusion of the silver is limited to distances of a few fringes, the lack of emulsion thickness variation with exposure extends to signals up to a few hundred cycles per millimeter, so that no relief image appears on the emulsion surface, and scattered light is further reduced.

There is some satisfaction in pointing out that the lack of response to low spatial frequencies makes such a developer totally unsuitable for conventional photography. It seems to be the first uniquely holographic developer type!

Of course, IEDT developers share the excess scatter noise problems of solution-physical developers, and scatter reduction remains the central research problem. But quite good results can be obtained, especially in reflection holography, simply by adding ammonium thiocyanate to the PAAP developer mentioned earlier. The amount is somewhat critical, depends on the emulsion type, and may have to be adjusted for differing emulsion or chemical batches. The approach to balance as solvent is added can be followed by observing the shift of reconstruction wavelength in areas of a bleached reflection hologram that were fixed after development. Solvent should be increased until the "fixing line" practically disappears. Development time should then be cut back until it just reappears to decrease scatter and the softening effect of thiocyanate on gelatin. The exposure for the best compromise between brightness and scatter may be two to four times that for nonsolvent development. Representative ammonium thiocyanate additions to PAAP, and 633-nm exposures, for EK 120-02 are 1.0 g/liter and 3400 ergs/cm, and for AG 8E75 are 0.5 g/liter and 2100 erg/cm².

9.1.5.5   Fixer

The removal of undeveloped silver halide, called "fixing," is necessary except after IEDT development or before reversal-type bleaching. Rapid processes for teaching or interferometry sometimes use development without fixation, as do some reflection techniques, but these holograms will darken easily in room light. Solvation of the small microcrystals takes only a few

359

9. Special Problems

seconds so that ammonium thiosulfate "rapid fixers" are not necessary or recommended, as they may attack the developer silver. The hologram should be removed after double the clearing time to avoid silver attack, which may change the color of the silver and increase milkiness upon bleaching. Substantial quantities of silver are removed, so that fixer exhaustion must be monitored and may be overlooked because clearing is more difficult to observe.

Fixer F-24, nonhardening (Thomas, 1973):

water about 125°F (50°C)	500 ml
sodium thiosulfate·5H$_2$O	240 g
sodium sulfite, desiccated	10 g
sodium bisulfite	25 g
cold water to make	1.0 liter

Use for double the visible clearing time; temperature not to be higher than 68°F (20°C).

Thorough washing is necessary after fixing to assure a chemically stable hologram. Bleached holograms are especially vulnerable to residual hypo stain. Commercial hypo clearing baths can reduce the necessary wash time to a few minutes.

Satisfactory drying may require hardening of the emulsion, which usually follows fixing. We use a formaldehyde hardener, SH-1, with good results (Thomas, 1973):

water	500 ml
formaldehyde (37%)	10 ml
sodium carbonate (monohydrate)	5 g
water to make	1.0 liter

Harden for 3 min and wash thoroughly before drying.

9.1.5.6 Bleaching

(a) **Introduction** Bleaching is a general term for a variety of processes designed to produce a hologram that modulates light by retardation of the wavefront, instead of the more usual attenuation. The result is generally an impressive increase in diffraction efficiency, hence image luminance, and often a depressing decrease of signal-to-noise ratio. Because high diffraction efficiency is important to many applications of holography, a great deal of exploration of various bleach types has been reported. There are three basic types of bleaches. The first converts the developed metallic silver grains into deposits of transparent dielectric deposit having a polarizability greater than gelatin.

These salts are most commonly silver halides, and these bleaches are described as "direct" or "rehalogenizing."

"Reversal" or "complementary" bleaches dissolve the developed grains from an unfixed hologram, leaving the undeveloped silver halide grains to modulate the average refractive index. These two are "volume" bleaches, as the modulation of the average refractive index they produce is distributed through the thickness of the emulsion and produces some Bragg selection effects as well as very high diffraction efficiency. Surface-relief bleaches exploit the gelatin cross-linking effects of development or bleaching byproducts to cause a gelatin "heaving" to form during drying that follows the hologram modulation, forming a thin phase grating. A smaller such effect is also caused by the removal of emulsion constituents, usually during fixing. Because surface relief effects are substantially limited to low spatial frequencies, they are not often useful for holography, and are usually considered a source of noise where they accompany other bleaches, except in special cases (Lamberts and Kurtz, 1971).

(b) Direct Bleaches Literally dozens of bleaching procedures have been described in the literature, many bordering on modern alchemy, and most suffering from unreliability because the underlying processes are so poorly understood. Our canvass of these showed that wide variations of diffraction efficiency resulted, usually accompanied by high scatter, and darkening when exposed to light. Developer improvements have relieved some of these problems, and here we will discuss a few bleaches that help reduce the rest.

Bleaches that involve an intermediate oxidizing step almost always reduce the modulation transfer function (MTF) because of random silver diffusion, which also increases scatter, thus our preference for very direct bleaches such as chlorine and bromine water, of which the latter is the more practical. This produces such clean, bright, nondarkening holograms that the process will be described in detail. The bleach is long-lasting and easy to use, but because a small amount of free bromine is released during use, some form of fume hood is absolutely necessary. This can be as simple as a polyethylene sheet stapled to a wood frame, with an electric blower and a clothes dryer duct pipe venting to some socially and ecologically acceptable location. The same hood ought to be used for alcohol drying and other organic solvent operations and so should be part of any well-equipped holographic laboratory. Bromine water itself should be handled carefully, much as a weak acid, and a 50-g/liter sodium sulfite solution should be kept handy to neutralize any spills and to clean stains in plastic trays. Unlike bromine vapor bleach, bromine water is relatively safe and innocuous to use. Only small quantities of liquid bromine are used, which must be handled very carefully.

Bromine water is made by almost filling a large (about $\frac{1}{2}$ liter) glass-stoppered (snug fit, stopcock grease is useful) glass bottle with clean water. A small amount of bromine (about 5 cm³) is poured into the bottle to form a puddle at the bottom, and the bottle is stoppered and put aside. After about a day (sooner for colder water) the water becomes orangish with dissolved bromine, and some vapor collects at the top.

To use, decant some of the water into a glass tray. Plastic trays can be used, though the stains can only be partially controlled with sulfite solution. The bromine will begin to slowly outgas, so that the bleach should be used in a fume hood and fairly quickly, and then returned to the bottle. Be sure to leave some water in the bottle to cover the bromine, and do not pour it onto the plate, lest small drops of bromine damage the emulsion. Instead, slip the hologram evenly into the bleach and agitate gently. Clearing should begin immediately and take no more than a minute, depending on the bleach strength. After twice the clearing time, remove the hologram to a tray of water and return the bleach to its jar. The hologram should be rinsed only briefly, to leave some bromine in the emulsion, and dried in any of the usual ways. The bromine water can be reused indefinitely, needing only to be topped off with water or bromine as needed.

Bromine used in this way does not attack the gelatin, even after prolonged bleaching, and leaves a clear whitish haze as a hologram, which smells slightly of bromine for a while. The stability to light is very high as long as the hologram is kept dry. This is the best bleach to use for IEDT processed reflection holograms. The only comparable results are with a published cupric bromide plus chemically generated bromine water bleach (which also requires a fume hood!) (Lehmann et al., 1970).

Conversion instead to silver iodide is possible in a somewhat more practical bleach solution (Kido and Arai, 1976; Cross, 1976), and produces extremely light stable holograms (McMahon and Mahoney, 1970). Iodine itself is not soluble in water, but dissolves well into alcohol. A small amount of water can then be added to swell the emulsion and promote the reaction.

Direct iodine bleach:

alcohol (methyl or ethyl)	750 ml
iodine crystals	2–5 g
water to make	1000 ml

Bleaching times vary widely and are shorter with higher iodine concentration. Double the clearing time is still the rule. The bleach should be followed by an identical solution without iodine, to rinse out some of the deep yellow stain, which is a natural beginning to an alcohol drying process.

Silver iodide has a higher molecular polarizability than silver bromide (van Renesse and Bouts, 1973), so that a still higher diffraction efficiency and scattered noise level result. In fact, it seems to be very difficult indeed to get very high contrast holograms with silver iodide. This may be due in part to emulsion damage, which is particularly marked on the relatively soft emulsions, such as AG 8E75. This, and the marked iodine stain, may be alleviated by very low iodine concentrations and long bleach times. Note that some film bases may be distorted by the alcohol, which must be used with ventilation. Nevertheless, the reliability of this bleach is attested to by its use with EK SO-173 over the past several years in the production of high-quality cylindrical holographic stereograms.

A somewhat less direct but very practical rebromination bleach has recently been reported by Phillips (1979). The oxidizing agent, para-benzoquinone, does not interfere with the hardening of gelatin shells surrounding grains produced by some hydroquinone-based developers, which can reduce hologram milkiness after bleaching.

PBQ bleach:

para-benzoquinone	2	gm
boric acid	1.5	gm
potassium bromide	30	gm
distilled water	1	liter

Finely powered para-benzoquinone is very irritating to the respiratory tract and should be handled in a fume hood. The bromide concentration can be increased for very hard emulsions.

(c) Reversal Bleaches Because reversal bleaches leave behind the originally grown compact spherical silver bromide microcrystals to form an image, they are of considerable technical interest and should be part of the repertoire of every holochemist. The well-known bleaches are based on the dichromate ion, which causes hardening side effects that usually degrade image quality. We prefer a bleach based on the permanganate ion (Glafkides, 1958). The plate should be exposed and developed normally, but not fixed. After washing, use KP-4 reversal bleach:

distilled water	1.0 liter
potassium permanganate	3 g
sulfuric acid (caution)	10 cm³

The shelf life of KP-4 is erratic, but it need not be thrown out until it is clear or no longer works. As it is used, it will cast a fine and innocuous sediment, which can be filtered with cotton cloth as it is poured into the tray. Bleach for twice the cleaning time, wash 5 min, and clear the residual stains in a 50-g/liter sodium sulfite solution (ventilate!). Wash for 10 min and soak for 3 min in the stabilizer solution.

9. Special Problems

STAB-3:

methyl alcohol	880 ml
water	100 cm^3
glycerine	20 g
potassium bromide	0.12 g

The bromide serves to retard photolysis. (It may be increased as needed until crystallization appears; this concentration is for 8E75 plates). Additional desensitizing is possible with phenosafranine dye (Buschmann, 1971; see also Smith, 1977, p. 64), although the pink staining is a nuisance. A brief (1 min) methyl alcohol wash and dry should follow. Although not suitable for outdoor use, these plates are very clean, bright, and resistant to darkening.

A few drops of bromine in alcohol make a still more effective stabilizer, but it must be discarded after each use to avoid clouding during drying.

9.1.5.7 Shrinkage Effects

Generally, between 15 and 20% of the volume of an unexposed emulsion is silver halide. Because most processes involve the removal of about half that bulk, substantial shrinkage of the emulsion is expected after drying (note that the dried thickness of a gelatin layer is independent of its hardening) as well as a lowering of the average refractive index. This shrinkage tilts the diffraction planes of a transmission hologram, changing the maximum reconstruction angle from 45 to 52° perhaps, or reducing the reconstruction wavelength of a reflection hologram from 633 to about 550 nm. Shrinkage can be partially compensated by finishing the process with a water or alcohol bath containing 6% triethanolamine (Lin and LoBianco, 1967). However, this tends to introduce drying streaks and accelerates the darkening of most bleached holograms in the light. No completely satisfactory solution for shrinkage has been published, except for IEDT-type processing.

9.1.5.8 Drying

The removal of water from the emulsion layer involves a mechanical shrinkage of roughly eight times, and it is extremely difficult to achieve this without distorting the fringes through the thickness of the emulsion, creating areas where the Bragg-angle for maximum diffraction efficiency is different from that for the rest of the hologram. So severe is this problem that it is no exaggeration to say that more display holograms are lost to faulty drying than to any other single process step.

Gelatin does not dry uniformly, but by the propagation of a "drying edge," which sweeps across the emulsion. Gelatin at different depths moves laterally as the edge passes, and any impediment to the smooth transit of the edges, such as a finger print, an emulsion flaw, hardened areas, or a straggling water drop or streak, causes a change of motion and a differential fringe tip. The key to smooth drying seems to be the removal of as much water as possible before this inevitable final sweeping step.

The most common technique is a series of graded water/alcohol dehydration baths, typically 50/50, 72/25, 95/5, followed by a wash bottle rinse in super clean alcohol, and a squeegee dry (using air, rubber or vacuum squeegees). Again, smooth drying progress is the key. Very long soaks in the final bath are sometimes needed, and an extra hardening bath may help if the emulsion refuses to dry well. Above all, avoid streaks and flow-back of any kind during the squeegee step. There is, of course, a risk of scratches with rubber squeegees, although these can be cleared by index matching.

Another technique is to air dry in a very still area after a final bath of 90% alcohol, 10% water, and a small amount of wetting agent (e.g. EK Photo-Flo™). The water concentration is critical and should be monitored with a hydrometer.

Acetate-based films generally distort or curl after swelling in organic solvents, so that such dehydration techniques are harder to apply to them. Isopropyl alcohol seems to affect such films the least, but its dehydration rate is comparatively low. Ethanol denaturants can be troublesome, and the toxicity of methanol is well known, so that the choice of an alcohol is not clear cut.

9.1.6 Conclusions

In this day of prepackaged amateur photography, it is hard to imagine those early decades when every photographer struggled to coat and process his own plates with chemicals he mixed himself. Yet most areas of holographic photochemistry are now reentering that same stage of progress, where new techniques are mastered by a slow process of trial and error that science will later codify. Many will naturally balk at uncertain chemistries and unfamiliar techniques, but many rewards await those who persevere!

REFERENCES

Benton, S. A. (1974). *J. Opt. Soc. Amer.* **64**, 1393A.
Biedermann, K., and Stetson, K. A. (1969). *Photogr. Sci. Eng.* **13**, 361.
Buschmann, H. T. (1971). *Optik* **3**, 240.
Cross, L. (1976). Private communication. Multiplex Co., San Francisco, California.
Glafkides, P. (1958). "Photographic Chemistry," p. 173. Foundation Press, London.
James, T. H., and Vaneslow, W. (1955). *PSA Tech. Quart.* **2**, 135.
Kido, K., and Arai, N. (1976). Japan Kokai 76-26, 136 (issued March 1).
Kowaliski, P. (1972). "Applied Photographic Theory," Chapter 7. Wiley, New York.

9. Special Problems

Lamberts, R. L., and Kurtz, C. N. (1971). *Appl. Opt.* **10**, 1342–1347.

Land, E. H., Farney, L. C., and Morse, M. M. (1971). *Photogr. Sci. Eng.* **15**, 4.

Lehmann, M., Lauer, J. P., and Goodman, J. W. (1970). *Appl. Opt.* **9**, 1948L.

Lin, L. H., and LoBianco, C. V. (1967). *Appl. Opt.* **6**, 1255.

McMahon, D. H., and Mahoney, W. T. (1970). *Appl. Opt.* **9**, 1363.

Mason, L. F. A. (1966). "Photographic Processing Chemistry." Focal Press, New York.

Pennington, K. S., and Harper, J. S. (1970). *Appl. Opt.* **9**, 1643.

Phillips, N. J., and Porter, D. (1976). *J. Phys. E: Sci. Instr.* **9**, 631.

Phillips, N. J. (1979). Paper M3, 32nd SPSE Annual Conference, Boston, Mass. (in preparation for *Photogr. Sci. Eng.*).

Smith, H. M. (1977). "Holographic Recording Materials" (see especially Chapter 2). Springer-Verlag, Berlin and Heidelberg.

Smith, H. M., and Callari, C. A., Jr. (1975). *Photogr. Sci. Eng.* **19**, 130.

Solman, L. R. (1966). *J. Photogr. Sci.* **14**, 171.

Thomas, W., Jr. (ed.) (1973). "SPSE Handbook of Photographic Science and Engineering," p. 577. Wiley, New York.

van Renesse, R. L., and Bouts, F. A. J. (1973). *Optik* **38**, 156.

9.2 SPECKLE

H. J. Caulfield

9.2.1 Introduction

"Speckle" is the term used to describe the granular pattern all users of visible-light lasers find each time the laser light scatters from or passes through a diffuser such as paper or ground glass. Speckle is an inevitable consequence of coherence because it is simply the pattern of constructive and destructive interference throughout a region illuminated by a coherent wavefront with a "random" (irregular) phase pattern. That is, speckle is an interference pattern of an irregular wavefront. The assumption of a random (statistically describable) phase pattern is a convenient mathematical means for handling speckle.

In holography speckle is often a problem. A speckled image is not only unpleasant cosmetically but also a source of image information loss as we shall see. It is this unfortunate aspect of speckle that is treated here. Outside of holography, speckle can prove very useful in everything from mechanical strain analysis to eye testing to astronomy. Both the good and the bad aspects of speckle are covered in a recent book by Dainty (1975).

9.2.2 Describing Speckle

We begin with an explanation of speckle in very simple terms. We assume that we have a "random" phase plate illuminated by a uniform circular beam of light of diameter D and wavelength λ. By Huygen's principle, we can consider that the light arriving at some point in space subsequent to the encounter with the phase plate has contributions from every place in the illuminated aperture. Because the phases are random, we cannot predict where the interference will be constructive and where it will be destructive, but we can predict the pattern statistically. Since we expect a random interference pattern, we expect very high contrast. Of course the speckle is diffraction limited, so at a distance $L \gg D$, we expect the speckles to be randomly distributed cigar-shaped blobs of diameter

$$d_s \cong \lambda L/D \tag{1}$$

HANDBOOK OF OPTICAL HOLOGRAPHY

9. Special Problems

and of length (equivalent to depth of focus)

$$l_s \cong 4\lambda L^2/D^2. \tag{2}$$

We now state without proof some fundamental relationships. Collier *et al.* (1971) showed that the rms intensity fluctuation

$$N = [\overline{(I - \bar{I}^2}]^{1/2} = \bar{I}. \tag{3}$$

That is, the rms intensity fluctuation is equal to the mean intensity; in other words, contrast is very high.

Goldfischer (1965) showed that the autocorrelation of the speckle irradiance (power density) function is

$$\langle B(x, y)B(x + \gamma, y + \delta)\rangle$$

$$= \left[\frac{1}{2\pi h^2} \int\limits_{-\infty}^{\infty}\!\!\int P(u, v)\, du\, dv\right]^2 + \frac{1}{\delta\pi^2 h^4}|F(\gamma, \delta)|^2 \tag{4}$$

where x, y are the coordinates in output plane, u, v the coordinates in input plane, $P(u, v)$ is the irradiance at the uv plane, h the separation of uv from xy planes, and

$$F(\gamma, \delta) = \int\limits_{-\infty}^{\infty}\!\!\int P(u, v)\, \exp[2\pi i(\gamma u + \delta v)/\lambda h]\, du\, dv. \tag{5}$$

Note that except for a constant term the speckle autocorrelation has the same shape as the far-field diffraction pattern of a field of *amplitude* $P(u, v)$.

Caulfield (1971a) noted that this was true only if we assume uniformly spaced equal-area phase regions. For phase regions of shape $s(u, v)$, we obtain a total field shaped like the power spectrum of $s(u, v)$. Thus, for example, one of Caulfield's "kinoform diffusers" produced a highly anisotropic overall pattern when illuminated with an isotropic beam because the scattering cells were anisotropic. The speckles themselves were anisotropic of course because $P(u, v)$ was.

The spectral dependence of speckle is best described by George and Jain (1974). Their result requires definition of numerous terms. We go to one spatial coordinate for convenience. The light frequency ν is converted to a cyclic spatial frequency

$$\eta = (2\pi \, \Delta n/c)\nu, \tag{6}$$

where Δn is the index of refraction difference between diffuser and medium and c the speed of light. The input coordinate is u; the output coordinate is x; and the diffuser has a height $h(u)$. Let $f(h_1, h_2)$ be the joint probability of h_1

368

$= h(u)$ and $h_2 = h(u + \delta u)$. Let

$$r(\Delta x) = \int_{-\infty}^{\infty} h(x)h(x + \Delta x)\, dx. \tag{7}$$

We can write a frequency characteristic function

$$F_c(\eta_1, \eta_2) = \int\!\!\int_{-\infty}^{\infty} f(h_1, h_2) \exp[-i(\eta_1 h_1 + \eta_2 h_2)]\, dh_1\, dh_2. \tag{8}$$

We presume a space-invariant imaging system with an impulse response $z(x, \eta)$ which converts an incident field $g(x, \eta)$ to an output field

$$e(x, \eta) = g(x, \eta) * z(x, \eta), \tag{9}$$

where $*$ indicates convolution. The square-law detected signal is

$$u(x, \eta) = e(x, \eta)e^*(x, \eta).$$

We can now calculate the autocorrelation function of $u(x, \eta)$, namely,

$$R_u = \mathscr{E}[u(x + \Delta x, \eta_2)u^*(x, \eta_2)], \tag{10}$$

where $[\cdot]$ indicates the ensemble expected value. We shall need

$$R_z = \mathscr{E}[z(x + \Delta x, \eta_2)z^*(x, \eta_1)]. \tag{11}$$

George and Jain show that

$$R_u(\Delta x, \eta_1, \eta_2) = \mathscr{E}[u(x, \eta_1)]\mathscr{E}[u(x + \Delta x, \eta_2)]$$
$$+ |F_c(-\eta_1, \eta_2 : r(\gamma))R_z(\Delta x - \gamma, \eta_1, \eta_2)\, d\gamma|^2. \tag{12}$$

Let us interpret this imposing looking formula. The essence of what happens is that a source of width

$$\Delta\nu = (c/2\pi\, \Delta n)(\eta_2 - \eta_1) \tag{13}$$

has a temporal coherence length of about

$$l_c = \frac{c}{\Delta\nu} = \frac{2\pi\, \Delta n}{\eta_2 - \eta_1}. \tag{14}$$

If the diffuser roughness is small compared to l_c, we obtain coherent addition of light from all diffuser depths. If we broaden the source bandwidth $\Delta\nu$, we decrease l_c and eventually reach the point where surface roughness is large compared to l_c. In this case, speckle patterns from various depths of the diffuser are mutually incoherent and smoothing or averaging occurs.

Let us summarize what we know. The speckle-causing situation comprises a source of bandwidth $\Delta\nu$ and center frequency ν illuminating a region of size δu by δv of diffuser of characteristic optical roughness l_d in patches of roughly

$\Delta u \times \Delta v$ in size. The observation occurs in the xy plane a distance h away. What can we predict about the observed pattern? The following predictions will hold:

(1) there will be unit contrast speckles if $c/\Delta v \ll l_d$,
(2) the speckle size will be roughly

$$\delta x = ch/v\,\delta u \qquad\qquad (15a)$$

and

$$\delta y = ch/v\,\delta v, \qquad\qquad (15b)$$

(3) the pattern will cover a region of size roughly

$$\Delta x = ch/v\,\Delta u \qquad\qquad (16a)$$

and

$$\Delta y = ch/v\,\Delta v, \qquad\qquad (16b)$$

and
(4) the autocorrelation function of the speckle irradiance pattern is shaped like the Fourier transform of the irradiance pattern at the diffuser.

9.2.3 Speckle in Holograms

Speckle can enter into either of the two steps in holography: hologram formation or wavefront reconstruction. If the object for a hologram is diffuse, the object wavefront is speckled. Thus perfect recording and replay still lead to a speckled image. If the object has only weak, large scale phase variations, we call it "specular." Ideally, specular objects lead to no speckle. Actually, such imperfections as emulsion relief patterns and nonlinearities can lead to mild speckle even in these cases. However, the primary problem in holographically produced images of specular objects is "coherent noise" like "ringing" at edges of lines, or scratches and concentric rings caused by pointlike hologram defects. Diffuse illumination destroys those defects but gives us speckle in their stead. Recent work by Budhiraja and Som (1978) shows that there is a continuous transition between specular and diffuse beams and that (when possible) a compromise can be advantageous.

9.2.4 Combating Speckle

There are only two weapons in our war against speckle although there are numerous variations on each.

First, we can sometimes make the speckles small compared with the object features of interest. The speckles are essentially diffraction-limited, so if the object features are far larger, the speckle is unobjectionable. This does, however, mean that speckle is very troublesome in microscopy and other fields where maximum resolution is sought.

Second, speckle can be "averaged out." This can be done by moving diffusers (Ih and Baxter, 1978), using wavelength diversity (Geroge and Jain, 1974), varying apertures on the same hologram (Yu and Wang, 1973), using spatial diversity (Martienssen and Spiller, 1967) or time diversity (Van Lighten, 1973), and in even more complex ways (Som and Budhiraja, 1975). Each method has an occasional advantage. All of the methods degrade the image resolution below the diffraction limit for the full aperture. With the exception of rather trivial improvements seen by Caulfield (1971b) and slightly more dramatic ones observed in some of the work by Som and Budhiraja (1975), all of this work uses incoherent addition of images. Indeed, most dramatic improvements involve continuously changing patterns by Ih and Baxter (1978) and Som and Budhiraja (1975).

Probably the most popular method is the double-diffuser method of which the work done by Ih and Baxter (1978) is the latest. Very little motion is required to decorrelate speckles, so by keeping one diffuser still while rotating or translating the other, we obtain an essentially speckle-free time average.

REFERENCES

Budhiraja, C. J., and Som, S. C. (1978). *J. Opt.* **7**, 12.
Caulfield, H. J. (1971a). *Proc. SPIE* **25**, 111.
Caulfield, H. J. (1971b). *Opt. Commun.* **3**, 322.
Collier, R. J., Burckhardt, C. B., and Lin, L. L. (1971). "Optical Holography." Academic Press, New York.
Dainty, J. C. (ed.) (1975). "Laser Speckle and Related Phenomena." Springer-Verlag, Heidelberg.
George, N., and Jain, A. (1974). *Appl. Phys.* **4**, 201.
Goldfischer, L. I. (1975). *J. Opt. Soc. Amer.* **55**, 247.
Ih, C. S., and Baxter, L. A. (1978). *Appl. Opt.* **17**, 1447.
Martienssen, W., and Spiller, S. (1967). *Phys. Lett.* **24A**, 126.
Som, S. C., and Budhiraja, C. J. (1975). *Appl. Opt.* **14**, 1702.
Van Ligten, R. F. (1973). *Appl. Opt.* **12**, 255.
Yu, F. T. S., and Wang, E. Y. (1973). *Appl. Opt.* **12**, 1656.

9.3 HOLOGRAM COPYING

William T. Rhodes

Sometimes it is necessary to copy or replicate a hologram. Copies may be needed for archival purposes or commercial sale, or the original object may have been shortlived with scientific import so that copies are needed for study by others. Copying holograms is a more complicated operation than copying ordinary photographs. The difficulty results from the extremely close fringe spacing in the original hologram, which may be of the order of 1000 line pairs/ mm. Even a high-quality copy lens is incapable of imaging such fine detail, and other techniques must be employed in the replication process. One such technique is copying by reconstruction; another is by contact or by near contact copying methods under carefully controlled conditions. In the following sections we describe these two major techniques for optically copying holograms. In addition, we also briefly describe mechanical replication techniques that are used in the mass production of relief phase holograms.

9.3.1 Copying by Reconstruction

A method for copying a hologram that is conceptually straightforward is to reconstruct the hologram and use the reconstructed-image distribution as the "object" in recording a new hologram. Although this process is not copying in the strictest sense of the word, it nonetheless accomplishes the desired objective. It has the disadvantage of requiring the same interferometric stability that is required for recording an original hologram. However, it has the advantage of providing additional control over various recording parameters. For example, the ratio of object and reference illumination levels can be adjusted for optimum diffraction efficiency in the copy. Even the general nature of the reference wave can be changed. Thus a plane reference wave original might be converted to a spherical reference wave copy.

Figure 1 shows a recording geometry suitable for copying a conventional "thin" absorption hologram. Note that the unexposed holographic emulsion must be positioned so as to be illuminated only by waves from the desired "object" (in this case the virtual reconstructed image) without being illuminated by the reconstruction wave or by waves from the conjugate image. If

9. Special Problems

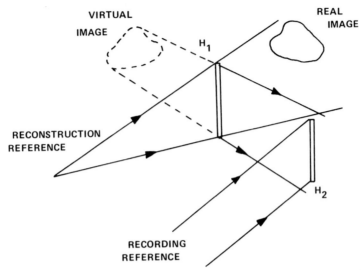

Fig. 1 Recording geometry for copying by reconstruction. H_1 is the original hologram, H_2 is the copy to be exposed. In this case, the virtual image from H_1 is being recorded. Note that the nature of the reference wave can be changed.

the original hologram exhibits sufficient Bragg selectivity, these latter distributions may have negligible amplitude, and the copy emulsion can then be positioned with greater flexibility.

An interesting and sometimes important variation on the above copying geometry is illustrated in Fig. 2. Here it is the pseudoscopic real image that serves as the object for the copy hologram. Recall that this real image distribution has an inside-out appearance when viewed: near objects are occulted by more distant ones, concave surfaces appear convex, and so forth. When the copied hologram is reconstructed, however, it is the virtual image that is now pseudoscopic in appearance; the real-image distribution, being twice pseudoscopic, has a normal appearance. The effect of viewing this nonpseudoscopic (orthoscopic) real image can be quite striking, for the observer can approach the scene as closely as he desires: there is no "window" lying between him and the image light distribution.

9.3.2 Contact Printing Methods

If the additional control over recording parameters already noted is not necessary, then contact printing replication techniques are easier to implement. They are certainly to be preferred for mass production of replica holograms.

Ideally, a contact-print hologram replica is made in the same way as a contact print from a conventional photographic negative. The original holo-

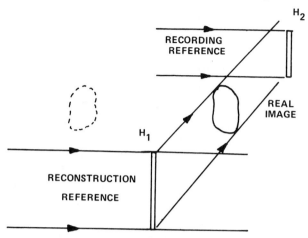

Fig. 2 Copying by reconstruction when the reconstructed real image serves as the object. Real image from copy will be nonpseudoscopic (orthoscopic).

gram is placed in close contact with a photosensitive surface—typically a sheet of satisfactorily high resolution film—the original hologram is transilluminated, and the exposed film processed as would be a conventional hologram. The nature of the light source is not critical in the ideal case so long as the illumination is uniform and the original and copy are sufficiently close together. The fringes in the (negative) copy hologram have reversed contrast. However, this reversal does not affect the appearance of the reconstructed image; it simply introduces a 180° phase change in the image light amplitude distribution with respect to the reference wavefront. Unless the image is examined interferometrically, that phase change cannot be detected.

In practice, copy holograms are extremely difficult to make by this "close-contact" method because the original and receiving emulsion must be brought microscopically close together. If diffraction effects are to be negligible, original-to-copy separations must often be of the order of a wavelength or less, and such close contact cannot generally be achieved between unpolished surfaces. Contact-copy holograms are thus nearly always of the "near-contact" type, where the separation is small—perhaps up to a millimeter or so—but need not be of the scale of wavelengths. For near-contact copies, light with a high degree of spatial and temporal coherence is required to preserve hologram fringe information in the exposure. Usually a laser source is employed. What happens with such a setup is not contact printing in the usual sense, and the copy is not a true replica of the original hologram. Rather, the copy emulsion is illuminated by the undiffracted portion of the illuminating wave and by the reconstructed-image waves; it is the interference of these waves that produces the fringe pattern preserved in the copy. The actual

375

9. Special Problems

pattern recorded depends on the nature of the original and on the separation between original and copy. We consider two important cases.

9.3.2.1 Thin Transmission Hologram Original—Double Images

If the original hologram is "thin"—by this we mean that Bragg diffraction effects are negligible—then the undiffracted wave is accompanied by two diffracted waves, one corresponding to the reconstructed real image, the other corresponding to the reconstructed virtual image. The three waves interfere in pairs to produce the total fringe pattern that exposes the copy emulsion. The fringe system resulting from interference of the two image waves is generally weak (because of low diffraction efficiency in the original hologram) compared to the other contributions and can be ignored. The two remaining fringe systems, resulting from interference of the continuation of the reference wave with the two image waves, have essentially equal amplitude and contrast.

Practically speaking, then, the copy hologram is made up of two fringe systems, whereas the original was made up of only one. On reconstruction of the copy hologram, a total of four images are reconstructed: two real, two virtual; one of each kind associated with each fringe system. The situation is illustrated in Fig. 3 for a point object and normally incident plane reference wave illumination.

An observer looking back through the copy hologram will see a double virtual image, the separation between images being twice the separation be-

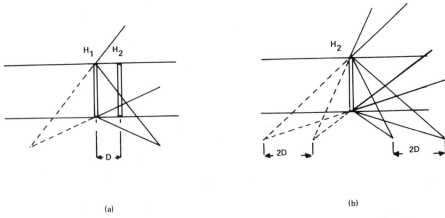

Fig. 3 Geometry for near-contact copying showing double image effect, assuming a point object: (a) recording geometry, (b) reconstruction. The two real and two virtual images reconstructed by the copy are separated by twice the distance separating H_1 and H_2 in the recording setup.

376

tween original and copy holograms. If this separation is sufficiently small, the two images merge and are seen as one. In order to assure this condition, the copy emulsion should be placed as close as possible to the original when the copy is made. Nevertheless, the separation can be large compared to a wavelength.

It should be kept in mind that the two virtual images (as well as the two real images) add on an amplitude basis. The resultant intensity pattern on the retina will thus exhibit effects of interference that depend on image separation and viewing angle. In the typical case, the original object is diffuse, and the interference patterns are little different from the speckle accompanying either of the images separately.

9.3.2.2 Copying Thick Holograms

If the original hologram is "thick," it will exhibit Bragg (angular) selectivity on reconstruction. More to the point, with a Bragg-regime hologram, it is possible, with proper orientation of the reference wave, to reconstruct a bright virtual image (or real image) without reconstructing the conjugate image. Near-contact copying without double images is thus possible in this case, even with original-to-copy separations of several millimeters. The separation should still be kept as small as possible to assure that reference and reconstructed-image waves overlap at the copy hologram. Otherwise, no fringes will be formed. Most holograms recorded in photographic emulsions exhibit a significant degree of Bragg selectivity, and the double image effect is often so minor as to be negligible. The quality of the copy is improved if an index matching liquid is used between the original and the copy to reduce reflections.

Note that in copying a Bragg-regime hologram, the original must be illuminated with an accurate replica of the original reference wave. If not—for example, if the curvature and/or direction of a reconstructing spherical reference wave is not matched to that of the original reference—both the resolution and field of view of the reconstructed image may be reduced. A change in wavelength also results in poor reconstruction of the original and, therefore, a poor copy.

If replicas of reflection holograms are to be made, these same guidelines apply: curvature, direction, and wavelength of the reconstructing wave must all be carefully matched to the original in order to produce the best possible reconstruction. This requires that the original and the copy in Fig. 3 be reversed in position, with the reconstructing wave passing through the copy emulsion before illuminating the original hologram. The fringe pattern recorded results from the interference of the illuminating wave with the back-diffracted reconstructed-image wave. The contrast of the fringe system will generally be quite low if an absorption-type reflection hologram is used as the original, because such holograms have especially low diffraction efficiency. Low effi-

377

ciency copies result. Phase-type reflection holograms, on the other hand, which are characterized by much higher diffraction efficiency, often yield excellent replicas.

9.3.3 Mass Replication of Relief Phase Holograms

The embossing method of copying relief phase holograms should also be described, for no other method is more economical in large-scale replication operations.

A relief phase hologram behaves very much like a conventional bleached-silver hologram. The diffracting phase structure, however, results from variations in hologram thickness, rather than from local changes in refractive index. A ruled transmission grating might be viewed as an extremely simple relief phase hologram.

A properly recorded relief phase hologram can be replicated by the simple process of embossing, the same process used in the mass production of phonograph records. The original hologram is usually recorded on a positive-working photoresist, such as Shipley AZ-1350. The resist is applied in a thin layer to a suitable substrate material and exposed in the same way that a conventional hologram would be. A short wavelength laser, preferably operating in the ultraviolet, must be used to match the spectral sensitivity of the resist. After exposure, the resist is developed with an etching solution. Heavily exposed areas are etched away, leaving surface corrugations in the lightly exposed regions. The result is a high-resolution relief pattern that follows the exposing fringe structure.

A replica hologram can be made directly from the hardened resist original. If a great many copies are to be made, however, it is generally better to make a nickel master from the original resist pattern. This can be done using an electroforming process similar to that used to make nickel phonograph record masters. The nickel master is used to emboss replicas in transparent vinyl sheets. Temperature and pressure must be controlled to guarantee good copies. Since the raw vinyl is quite inexpensive, the incremental cost of additional copies is low.

10

Application Areas

10.1 DIGITAL DATA STORAGE

Thomas K. Gaylord

10.1.1 Uses of Holographic Digital Data Storage

10.1.1.1 The Expanding Need

The storage and retrieval of data is of fundamental importance in almost every human endeavor. Private individuals, businesses large and small, governmental agencies, and many other institutions have always stored and retrieved information. The development of the electronic digital computer produced an immediate need for large, organized, efficient data stores (memories). The very rapid growth in computing power has been accompanied by an incessant pressure for larger, faster, and less-expensive memories for digital data storage.

A perspective view of current memory performance can be obtained by considering the memory system access time as a function of system storage capacity and cost per stored bit for various memory technologies. This is done in Figs. 1 and 2. Although the boundaries of the various memory technologies are nebulous and indeed are changing, it is still apparent that for nonholographic memories there seems to be a tradeoff between access time and both memory capacity and cost per bit; that is, fast access is associated with small, expensive memories and slow access accompanies the larger, less expensive memories. Optical holographic memories offer the possibility of overcoming these seemingly inherent tradeoffs.

10.1.1.2 Archival Storage

The need for high capacity storage may be divided into several categories depending on the accessing characteristics that are needed. Perhaps the least

379

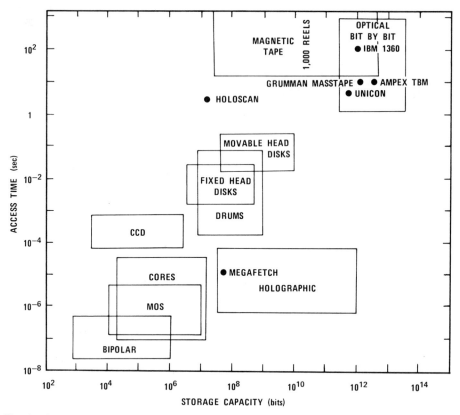

Fig. 1   A comparison of various memory types in terms of access time and storage capacity.

demanding of these categories is archival storage or record access. In this category large amounts of data need to be stored in a central memory and occasionally accessed. Examples include libraries, insurance data, medical data, seismic data, criminal data, patent records, stock market information, computer software packages, defense data, and postal data. Numerous governmental and private organizations currently have magnetic tape libraries containing over 200,000 reels of magnetic tape. Information stored in this manner is both expensive and very slowly accessible. This category of storage primarily requires a read-only memory.

10.1.1.3   Read-Mostly Storage

A second category of storage, similar to the first, is the storage of data such that the data can be occasionally changed. This requires a read-mostly mem-

380

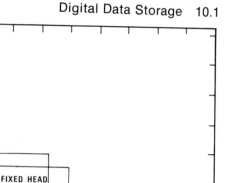
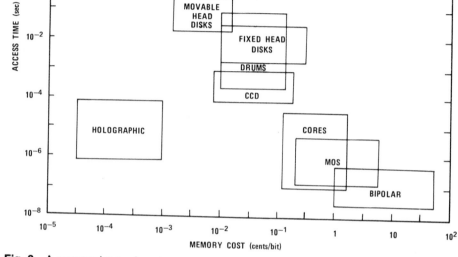

Fig. 2 A comparison of various memory types in terms of access time and cost per bit of stored data.

ory. In such a memory, data are added or changed only infrequently by computer standards. Examples of information that would be stored in this type of memory include credit data, personnel records, tax information, telephone numbers, stock market information, records of large inventories, weather data for large scale weather prediction programs, and large data bases.

10.1.1.4 Fast Recorder

A third category of storage requires high data rate recording and temporary storage. An example would be high bit rate optical communications systems. In optical communications efficient use of turbulent channels will require very high capacity, very fast, reusable mass storage for recording during temporary interruptions of these channels. Another example is data recording during a space probe flyby. Here a great amount of data is gathered during a brief period of time. If this information could be stored, it could later be transmitted at a low bit rate to minimize transmission errors in the data. These types of applications require a fast recorder.

10.1.1.5 Fast Read–Write–Erase Memory

A fourth category of storage requires rapid reading, writing, and erasing capabilities. Computer memories, both in general purpose and in specialized, dedicated processing systems are the most obvious examples of this category of storage. Present day computing systems utilize a complex hierarchy of storage devices. Some of the more important of these memories are magnetic tape, disks, drums, cores, and semiconductors. The access times, storage capacities, and costs per bit of stored data are graphically presented in Figs. 1 and 2. Modern computers use a combination of the large and slow along with small and fast memories in a hierarchical structure to realize efficient computing. The holographic optical memory, due to its potentially very high capacity and fast random access, offers the possibility of replacing a large portion of the existing memory hierarchy. Several high capacity memories using nonoptical technology have been constructed (see e.g. Houston, 1973). These include the Ampex Terabit Memory (TBM) and the Grumman Masstape System. These memory systems are shown in Fig. 1 to have access times of about 10 sec. The maximum storage capacities for these memories are 8.8×10^{11} bits for the Grumman Masstape and 2.9×10^{12} bits for the Ampex Terabit System. While this amount of storage is adequate for most applications, the long access times make these systems unusable as rapid random access memories.

10.1.2 Holographic Memory Configurations

10.1.2.1 Fundamental Design Considerations

The broad basic features of holographic memory design were established by about 1970 (see e.g. Smits and Gallaher, 1967; Anderson, 1968; LaMacchia, 1970; Rajchman, 1970a,b; Stewart and Cosentino, 1970). These fundamental features are discussed in this section. A more detailed quantitative treatment of the engineering design tradeoffs is given in Section 10.1.4. Extensive review articles on holographic memories are listed in Hill (1976), Haskal and Chen (1977), and Vander Lugt (1976).

(a) **Fourier Transform Holograms** Information should preferably be stored in holographic form as opposed to direct image or bit-by-bit storage. In the typical configuration, the hologram will be the recording of the interference pattern between the Fourier transform of the bit pattern and a plane-wave reference beam. Due to the distributed nature of the information, the storage of data in holographic form provides protection from localized loss of data due to material imperfection or surface dust. Fourier transform holograms also have the advantage of giving a position-invariant readout pattern with lateral positioning errors in the reference beam. More details on Fourier transform holograms are found in Section 4.3.

(b) Page Format Information should preferably be stored in a page-organized format as opposed to a three-dimensional isometric view. The ability of holography to provide three-dimensional views of objects is of no particular value in mass data storage. The reconstructed data will simply be in the form of two-dimensional arrays called pages.

(c) Digital Data Information should preferably be stored in a binary code as opposed to a pictorial representation. A reconstructed page of binary data would appear as a series of bright and dark spots representing the 1's and 0's of the digital data. Pictorial representations, such as a printed page, a drawing, a map or a photograph, are also usable. However, for very high information densities, constraints on the page composer and on the detector matrix favor the use of a binary code. The basic operations of writing and of readout of a digital data page are schematically illustrated in Fig. 3.

(d) Thick Phase Holograms Information should preferably be stored in the form of thick phase holograms as opposed to either thin holograms or absorption holograms. This design consideration is a result of the superior information storage capacity of phase holograms over absorption holograms and of thick

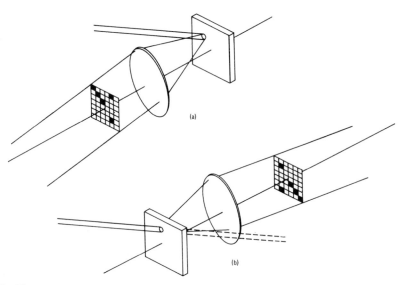

Fig. 3 The (a) writing and (b) readout of a binary data page using holographic techniques.

10. Application Areas

holograms over thin holograms. This is readily apparent from a comparison of the maximum diffraction efficiencies of these basic types of holograms.

Usually holograms are classified in terms of thickness by the parameter Q which is

$$Q = 2\pi\lambda d/nL^2, \tag{1}$$

where λ is the optical wavelength in air, d the thickness of the hologram, n its index of refraction, and L the fundamental grating spacing given by

$$L = \lambda/2 \sin \theta_a, \tag{2}$$

where θ_a is half of the angle between the recording object and reference beams in air and the wavelength is the recording wavelength (as opposed to the reading wavelength) in air. For thin holograms,

$$Q \ll 1. \tag{3}$$

For thick holograms,

$$Q \gg 1. \tag{4}$$

For a sinusoidally modulated, transmission hologram, the first-order diffraction efficiency with an input beam at the first Bragg angle is denoted by η.

For a thin absorption hologram, the diffraction efficiency is

$$\eta = \exp(-2\alpha d/\cos \theta_i) I_1^2(\alpha_1 d/\cos \theta_i), \tag{5}$$

where α is the average optical absorption, α_1 the amplitude of the sinusoidal grating absorption, θ_i half of the angle between the object and reference beams inside the medium, and I_1 the first-order modified Bessel function of the first kind. The maximum possible value for α_1 is α, and thus the maximum diffraction efficiency for a thin sinusoidal absorption hologram is

$$\eta_{max} = 4.80\%. \tag{6}$$

For a thin sinusoidal phase hologram, the diffraction efficiency is

$$\eta = J_1^2(2\pi n_1 d/\lambda \cos \theta_i), \tag{7}$$

where J_1 is the first-order ordinary Bessel function, and n_1 the amplitude of the refractive-index modulation. The maximum diffraction efficiency is the maximum value of J_1^2 and is thus

$$\eta_{max} = 33.8\%. \tag{8}$$

For a thick sinusoidal absorption hologram, the diffraction efficiency is

$$\eta = \exp(-2\alpha d/\cos \theta_i) \sinh^2(\alpha_1 d/2 \cos \theta_i). \tag{9}$$

The maximum value of the diffraction efficiency in this case is

$$\eta_{max} = 3.7\%. \tag{10}$$

384

For a thick sinusoidal phase hologram, the diffraction efficiency is

$$\eta = \sin^2(\pi n_1 d / \lambda \cos \theta_i). \tag{11}$$

Note that the argument of the $\sin^2$ function in Eq. (11) is one half of the argument of the J_1^2 function in Eq. (7). When the argument of Eq. (11) is an odd half multiple of $\pi/2$, the maximum diffraction efficiency is obtained and is

$$\eta_{max} = 100\%. \tag{12}$$

Therefore, thick phase holograms have the highest potential data storage capacity.

(e) Nonmechanical System   The optical memory system should contain no moving parts. This is necessary to achieve realistic operating speeds that are consistent with computer requirements. In addition, mechanical movements in a complex memory system will frequently reduce the reliability to an unacceptable level.

10.1.2.2 Two-Dimensional Storage System

Optical holographic memory systems may be categorized according to the thickness of the recording medium that is used in the storage and retrieval processes. A two-dimensional storage system uses thin (surface or area) holograms, whereas a three-dimensional storage system uses thick (volume) holograms.

A representative two-dimensional random-access holographic memory is shown in Fig. 4. The thin recording medium in such a system might be thermoplastic or high-resolution photographic film, for example. This configuration uses a two-dimensional storage scheme (one hologram per xy location on the recording medium) as opposed to the higher capacity, more complicated, three-dimensional storage systems discussed in Section 10.1.2.3. Apparent in Fig. 4 and basic to all optical holographic memories are an optical source, beam deflectors, a page composer, the recording material, and a detector matrix. These components are interfaced with each other using a variety of conventional optical and electronic components. The design shown in Fig. 4 is a typical configuration. Other configurations depending on the memory characteristics needed have been presented in the literature (see e.g. Aagard *et al.*, 1972; Graf and Lang, 1972; Hill, 1972; Stewart *et al.*, 1973; Vander Lugt, 1973; Kiemle, 1974; Tsukamoto *et al.*, 1974).

The operation of the composite two-dimensional holographic storage system of Fig. 4 is illustrated in Fig. 5. The writing process for recording a data page at a general xy location at the storage medium is depicted in Fig. 5a. The amplitude of the object beam at the recording material is a Fourier transform

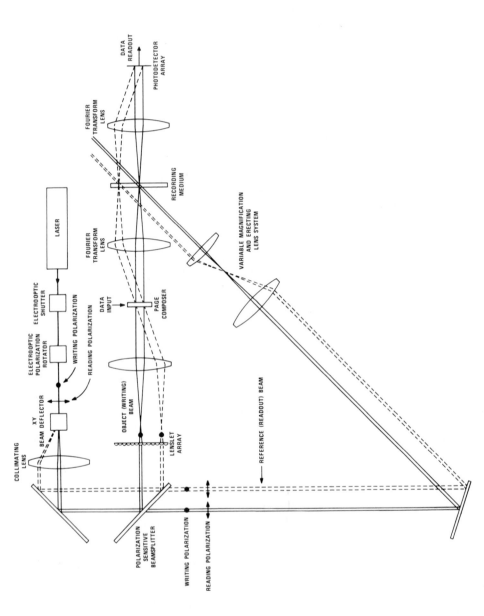

Fig. 4 A holographic optical memory system using two-dimensional (surface) storage.

(approximately) of the data page in the page composer. This amplitude pattern is interfered with the reference beam at the recording plane. The optics of the system cause the object and reference beams to intersect at the storage medium for any memory plane xy address selected (by the beam deflectors). Thus the object and reference beams automatically track each other. The readout process for reconstructing a data page is depicted in Fig. 5b. The page being read is at the same xy location as the page recorded in Fig. 5a. Now only the reference beam is present. This beam passes through the recording medium as shown in Fig. 5b. The hologram grating, however, diffracts some of the reference beam light into a complex wavefront that duplicates in amplitude, phase, and direction the original beam wavefront that was present during recording. This pattern of light spots (digital data) is incident upon and is read by the photodetector array.

10.1.2.3 Three-Dimensional Storage System

A representative three-dimensional random-access holographic memory system is shown in Fig. 6. The thick recording medium in such a system might be an electrooptic crystal or a photochromic crystal. A number of three-dimensional holographic storage systems have been designed (see, e.g., d'Auria *et al.*, 1974). These systems superpose many holograms at a single xy location inside the thick recording medium by using a different reference beam angle for each hologram. Because of their volume nature, these holograms exhibit very strong angular selectivity (Kogelnik, 1969); that is, to read a hologram, the reference beam must illuminate the hologram, within a narrow angular corridor about the Bragg angle for that hologram. Illumination outside of this angular corridor produces a rapidly decreasing intensity of reconstructed data. In addition, the thicker the hologram is, the narrower the angular corridor for reconstruction becomes (see Section 10.1.4.6). The superposition of multiple holograms at a single volume location introduces the additional problem of writing new holograms in that volume without affecting those already there. For example, when the electrooptic material lithium niobate is used as the three-dimensional storage material, this problem may be solved by the application of an external electric field (Amodei and Staebler, 1972). This greatly increases the sensitivity for writing while the sensitivity for erasure remains unchanged and at a much lower value. Thus as a new hologram is written, the other holograms at that location are only slightly erased. In addition, multiple hologram storage has been achieved in lithium niobate by applying a thermal bias (Staebler *et al.*, 1975). This has allowed the selection of the erase/write asymmetry required for multiple hologram storage. With this technique, over 500 holograms, each with more than 2.5% diffraction efficiency were recorded in 0.01% iron-doped lithium niobate. The problem of selective erasure of a single hologram among superposed holograms has been solved by writing a complementary hologram

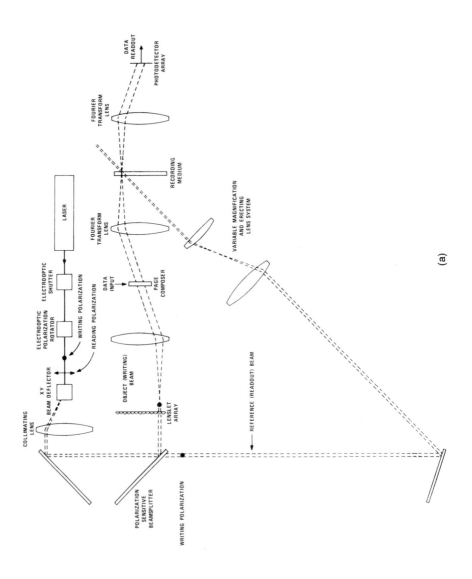

(a)

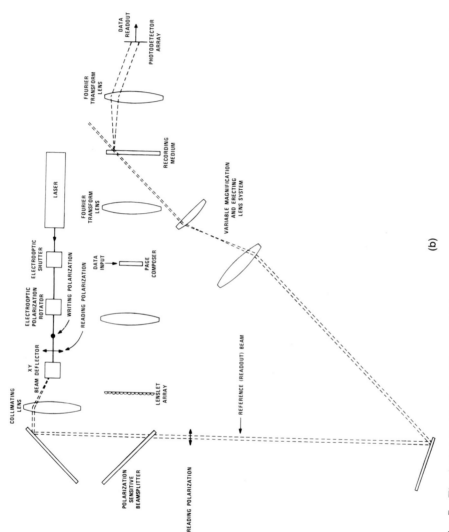

Fig. 5 The holographic (a) writing and (b) readout of a data page at a general xy location in a two-dimensional optical holographic memory.

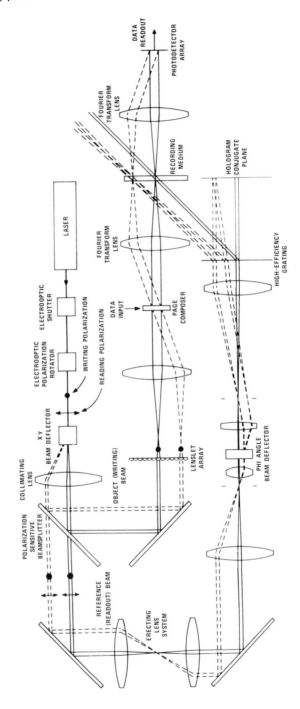

Fig. 6 A holographic optical memory system using three-dimensional (volume) storage.

in which the refractive index changes cancel with those of the original hologram (Huignard *et al.*, 1975a).

The operation of the composite three-dimensional holographic storage system of Fig. 6 is illustrated in Fig. 7. The writing process for recording a data page at a general $xy\phi$ location in the storage medium is depicted in Fig. 7a. The process proceeds as in the two-dimensional system except that an extra deflection system has been added to allow the angle of incidence of the reference beam at the recording medium to be varied. The optical system again causes the object and reference beams to intersect in the storage medium regardless of the $xy\phi$ address selected (by the beam deflectors). Automatic tracking is thus again incorporated through optical design. The readout process in a three-dimensional holographic memory is depicted in Fig. 7b. The hologram stored at the general $xy\phi$ address as shown in Fig. 7a is reconstructed in Fig. 7b. The process proceeds exactly as with the two-dimensional system case except that for each xy address there has been multiplexed numerous ϕ angular addresses.

10.1.3 Optical Memory Components

10.1.3.1 Optical Source

A laser is needed to provide the intense, collimated, coherent light required in a holographic memory system. The laser should be pulsed (possibly mode-locked) or externally gated to operate up to about 10^6 pulses/sec, each pulse of which is used for a recording or reading operation. Additionally, an average single-mode optical power of about 1 W will be needed depending on the recording medium and the writing and readout processes used. Most recording materials and photodetectors are most photosensitive in the blue–green region of the spectrum. This favors the use of an argon-ion laser as the optical source since it has a strong transition in the blue (at a wavelength of 0.488 μm) and in the green (at a wavelength of 0.5145 μm). It also can meet the requirements on frequency stability, amplitude stability, coherence length, and reliability that are needed. The disadvantages of the argon-ion gas laser are its high cost (about \$15,000) and its low efficiency of conversion of electrical power to optical power (about 0.1%). The frequency-doubled Nd:YAG laser (at a wavelength of 0.530 μm) is a promising solid state laser for holographic memories. Very high peak powers (about 10^8 W) are possible in pulsed mode operation.

Relatively short wavelengths are also desirable because the data storage density is proportional to λ^{-2} or λ^{-3} (see Section 10.1.4.1). However, in the violet and the ultraviolet, unavoidable random (Rayleigh) scattering with its intensity proportional to λ^{-4}, produces background optical noise that degrades the reconstructed data patterns.

More details on lasers are found in Sections 8.1 and 8.2 and on recording materials in Section 8.3.

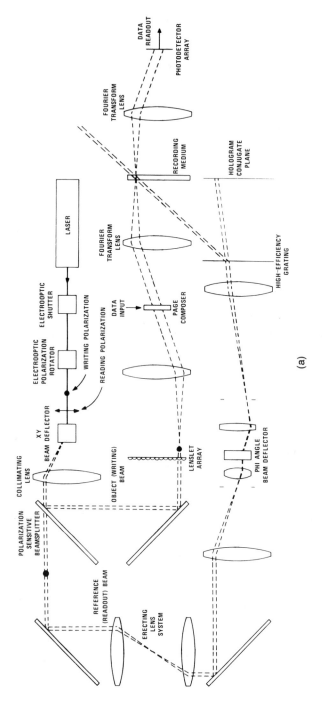

(a)

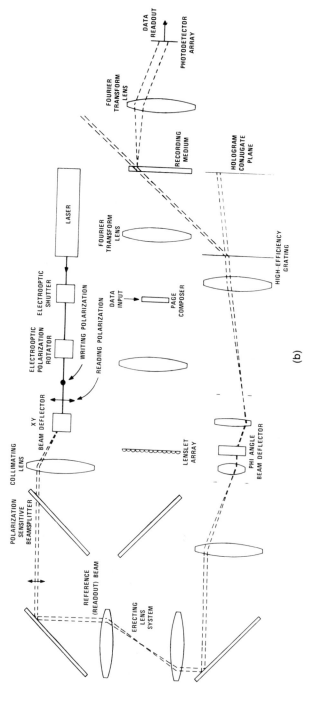

Fig. 7 The holographic (a) writing and (b) readout of a data page at a general $xy\phi$ location in a three-dimensional optical holographic memory.

10.1.3.2 Beam Deflectors

An optical memory system must utilize a number of beam deflectors to position accurately the laser beams for the reading, writing, and erasing operations. This positioning process must be both fast and accurate.

As shown in Table I, there are three basic types of deflectors: mechanical deflectors, acoustooptic deflectors, and electrooptic deflectors. A number of examples in each of these categories are also listed. The performance of a deflector may be quantified by the resolution and the random access time. Resolution may be defined as the maximum deflection angle divided by the diffraction limited angle. This ratio gives the total number of resolvable angular positions or equivalently the total number of resolvable spots M. The importance of the magnitude of M is discussed in Section 10.1.4.4.

An extensive comparison of light beam deflectors has been performed (Zook, 1974). For each of the three types of deflectors there is an engineering design tradeoff between the number of resolvable spots and the random access time. The random access time τ_a is the time required for the deflector to deflect the laser beam to a new angular position. For mechanical deflectors, the random access time is

$$\tau_a = 1/2 f_0, \qquad (13)$$

where f_0 is the resonant frequency of the mechanical system. In practice, the mechanical random access time is seldom less than 0.1 msec. Mechanical deflectors are thus too slow for fast access memory applications (which require an access time of approximately 1 μs).

In an acoustooptic deflector, a piezoelectric transducer launches an acoustic wave into the acoustooptic material. This traveling wave produces a refractive-index grating that diffracts the laser beam. Changing the acoustic drive fre-

TABLE I

Types of Beam Deflectors

Mechanical	Acoustooptic	Electrooptic
Moving iron galvonometer	Alpha-iodic acid (α-HIO$_3$)	Potassium dihydrogen phosphate (KH$_2$PO$_4$)
Moving coil galvonometer	Lead molybdate (PbMoO$_4$)	Ammonium dihydrogen phosphate (NH$_4$H$_2$PO$_4$)
Spinning polygonal mirror	Water (H$_2$O)	Lithium niobate (LiNbO$_3$)
	Tellurium dioxide (TeO$_2$)	Strontium barium niobate (Sr$_{0.75}$Ba$_{0.25}$Nb$_2$O$_6$)
	Dense flint glass (SF-8, SF-59)	Potassium tantalum niobate (KTN)
	Lithium niobate (LiNbO$_3$)	
	Gallium phosphide (GaP)	

quency changes the acoustic wavelength and thus the grating period. This causes the diffraction angle of the light beam to change and thus the device acts as a variable-angle beam deflector. The random access time is basically the transit time of the sonic wave to cross the width of the laser beam. Thus for an acoustooptic deflector

$$\tau_a = D/v_s,\tag{14}$$

where D is the aperture width and v_s is the velocity of sound in the acoustooptic medium. For water, $v_s = 1.5 \times 10^3$ m/sec. Thus for a 20-mm-aperture water deflector, τ_a is 13.3 μsec.

To increase the diffraction efficiency and to reduce the inherent beam divergence, acoustooptic deflector cells are frequently elongated in the direction of acoustic propagation. The laser beam is then focused into an elliptical spot with its major axis along the elongated cell direction. Such focusing is done with cylindrical lenses as is shown in Fig. 8. The plane containing the diffracted and the undiffracted laser beams is parallel to both the focused line and the optical axis of the lens system. The x acoustooptic deflector cell is therefore placed at the horizontally focused line, and the y acoustooptic deflector cell is placed at the vertically focused line. The aperture width D of the cell thus refers to the length of the major axis of the elliptical laser spot (line), assuming that it fills the entire elongated acoustooptic cell. Increasing the aperture width D increases the number of resolvable spots (see Section 10.1.4.4) and the diffraction efficiency, but also reduces the speed of the random access response.

Due to the velocity of the acoustic grating, the frequency of the diffracted optical beam is shifted by the Doppler frequency. Therefore, acoustooptic deflectors should be used before the object/reference beam beamsplitter so that the two beams will have the same wavelength and will thus produce stable interference fringes. Doppler shifting the frequency of only one of the two recording beams will degrade the interference pattern at the intersection of the object and reference beams.

Electrooptic beam deflectors may be either of two types: analog or digital. In the analog deflector, an electric field is applied to an electrooptic prism causing its index of refraction to change. This in turn changes the deviation

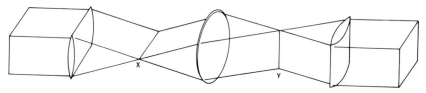

Fig. 8 The optics of a typical acoustooptic *xy* deflector system.

angle of the laser beam passing through the prism and thus produces an angular beam deflector. In practice, the electrooptic coefficients of known materials are too small to produce large angles of deflection for reasonable values of applied electric field. Multiple, cascaded prisms are thus used to increase the angle of deviation and thus the total number of resolvable spots.

The digital electrooptic deflector uses an electrooptic modulator and a birefringent prism (e.g., a calcite prism). The electrooptic modulator is used as a polarization rotator. A linearly polarized output beam has its polarization rotated by 90° when the voltage is changed by the half-wave voltage of the modulator. Upon passing through the birefringent prism, the beam is deviated into one of two directions depending on its polarization. (The two possible linearly polarized light beams experience two different indices of refraction in the birefringent material.) A cascade combination of m digital deflectors allows 2^m deflection angles. For example, a 20-stage electrooptic deflector has been constructed with 10 stages of x deflection and 10 stages of y deflection, thus producing a two-dimensional array of 1024×1024 resolvable spots (Meyer *et al.*, 1972). This device has a random access time of 0.8 μsec. The random access time τ_a for an electrooptic deflector is given by

$$\tau_a = 4CV_p^2/P_a, \tag{15}$$

where C is the deflector capacitance, V_p the maximum voltage, and P_a the available electrical driving power from the power supply. Losses through the many surfaces and high cost are significant drawbacks to these systems.

10.1.3.3   Page Composer

The data input device for the optical holographic memory is a page composer or block data composer which converts digital electrical signals directly into a two-dimensional optical array of bits. The page composer will be located in the object beam of the two-beam holographic configuration. Reconstruction of the data hologram will produce an image at the detector array that duplicates the array of 1's and 0's (bright and dark spots) generated by the page composer.

There are a number of characteristics that the page composer should possess. These requirements include:

(1)   *High frame speed*   It must be possible to change rapidly the data page in the page composer. The change time ideally must be in the microsecond range.

(2)   *High resolution*   The size of each bit in the page composer needs to be small just to fit a large number in a reasonable area. Sizes of about 100 μm would be suitable.

(3)   *Large aperture*   The total area of the page composer transverse to the laser beam needs to be large enough to accommodate the number of bits per

page desired. For many applications the bit array size should be in the range of 64 × 64 elements to 1024 × 1024 elements.

(4) *High contrast ratio*   The achievement of a high constrast ratio relaxes the subsequent requirements on the recording material and the detector matrix. A contrast ratio of 100 to 1 or greater is desirable and this has been achieved in a number of page composer type devices.

(5) *Stability*   The characteristics of page composer materials must not be degraded by exposure to high intensity light (the object beam).

(6) *Uniformity*   Material nonuniformities in the block data composer must not cause readout errors in the memory system.

(7) *Full page addressing*   All bit locations in the page composer must be able to be independently and simultaneously in either a 1 or 0 state.

A wide variety of approaches exist for the construction of page composers. A number of these approaches are listed in Table II. Obviously, a large number of physical effects and a large number of materials are potentially usable in page composers. Liquid crystal block data composers appear to be very useful. RCA has constructed a 1024 bit liquid crystal page composer (Labrunie *et al.*, 1974). A major problem with liquid crystal page composers has been their relatively slow frame speed (on the order of 100 msec). Lead lanthanum zirconate titanate (PLZT) block data composers (Roberts, 1972) also appear to be very promising. These page composers, which do not suffer from a slow frame rate, have four basic modes of operation: strain biased mode, scattering mode, edge effect mode, and differential phase mode (Drake, 1974). This last mode of operation eliminates the detrimental effects of background nonuniformities in the PLZT, but requires a double hologram exposure through the data mask. Three other promising approaches to block data composers utilize a thin, deformable, membrane mirror array (Cosentino and Stewart, 1973), the thermally induced shift in the optical absorption band edge in CdS (Hill and Schmidt, 1973), and the acoustooptic effect.

10.1.3.4   Recording Medium

Recording media are covered in detail in Section 8.3. However, since the recording material is of such central importance in the holographic memory, the characteristics needed for this particular application will be discussed.

Recording materials must possess a number of important characteristics in order to achieve the high storage capacities that have been predicted for optical memories. These requirements on the optical recording material include:

(1) *High sensitivity*   It is desirable that only a small amount of optical energy per unit area be needed to record the hologram of a data page. For a practical system an energy density of about 1.0 $\mu J/mm^2$ or less will be needed.

(2) *Large diffraction efficiency*   Diffraction efficiency is the fraction of the

10. Application Areas

TABLE II

Types of Page Composers[a]

Page composer concept	Materials	Addressing techniques
Polarization rotation by induced birefringence (electrooptic effects)	PLZT (ceramic), $Bi_4Ti_3O_{12}$, KDP, KD*P, ADP	Electrode matrix, electron beam, light beam (with photoconductor)
Phase changes by formation of surface relief pattern	Thermoplastics, photoplastics, thin metalized membranes	Electron beam, electrode matrix plus charge
Phase disturbances by piezoelectric excitation of reflecting surfaces	Mirrored piezoelectric crystals	Individual switches to an rf driver
Optical density change by induced absorption	Photochromics, cathodochromics	Light beam (uv) plus flood illumination for erase, electron beam plus flood illumination for erase
Optical scattering change by electrical agitation	Liquid crystals	Electrode matrix, light beam (with photoconductor)
Polarization rotation by magnetooptic effects	MnBi, EuO:Fe, Ni–Fe, $FeBO_3$, FeF_3	Light beam (absorption), conductor matrix
Traveling phase changes by acoustooptic interaction (Debye–Sears and Bragg effects)	Water (and other liquids), fused quartz (and other amorphous solids), $PbMoO_4$ (and other crystals)	Transverse interaction of coherent light and traveling acoustic waves
Thermally induced shift in absorption band edge	CdS, CdSe, As_2S_3	Electrode matrix for heating and heat sink substrate for cooling
Optical scattering by poled and unpoled regions of a ferroelectric	PLZT (ceramic)	Electrode matrix
Phase changes by variation of optical path length	Electrostrictive materials, PLZT (ceramic)	Electrode matrix, double hologram recording method
Reflection changes from thin, deformable membrane mirror elements	Metal films over a substrate support structure	Electrode feedthrough from transistor on back of substrate

[a] Modified version of table from Roberts (1972).

reading light (reference beam) that is diffracted into the reconstructed data beam. It must be possible to record a single hologram with a large diffraction efficiency, so that in practice many holograms may be recorded at a single location.

(3) *Erasable and rewritable* For a rapid cycle read–write–erase memory system, it must be possible to alter continuously the stored data in the memory without encountering any degradation in the material characteristics.

(4) *Long lifetime of stored information* Stored data should persist for long

periods of time before having to be refreshed. Ideally, storage should be permanent.

(5) *Nonvolatile storage* Data should remain recorded in the memory in the absence of system power.

(6) *Nondestructive readout* It should be possible to perform an essentially unlimited number of read operations without degrading or altering the stored data.

(7) *Three-dimensional storage* To achieve very high capacity storage, the information should be stored in thick (volume) holograms. Together with the requirement of high diffraction efficiency, this means that the hologram should be a thick phase (nonabsorbing) hologram.

(8) *High resolution* The storage material obviously must be capable of recording the very fine (wavelength size) variations of the interference pattern produced by the intersection of the object and reference beams.

10.1.3.5 Detector Array

An array of photodetectors is needed to convert the holographically reconstructed data pattern into an electrical signal. This photosensitive readout array would have one sensing photodiode or one sensing phototransistor and one or two switching (addressing) devices for each bit of data in the reconstructed page (see e.g. Assour and Lohman, 1969). Each sensor in the array would function as a threshold detector indicating the presence or absence of light (a binary 1 or 0). All stored holograms would be read out with the same detector matrix.

The photodetectors ideally must exhibit a high detectivity in order to differentiate between a 1 and a 0 in the presence of noise. The needed detectivity to produce a given signal-to-noise ratio is discussed quantitatively in Section 10.1.4.3.

The second basic requirement on the detector matrix is that a large defect free array be constructable with existing technology. Modern semiconductor technology has fulfilled this requirement. Bell Laboratories has constructed a silicon-diode-array camera tube that consists of 525,000 individual photodiodes on a single silicon slice. An LSI phototransistor array with 51,200 silicon phototransistors has been built (Mend *et al.*, 1970) using multilayer interconnection techniques so that any bit can be read out in about a microsecond.

A third desirable feature of the detector array is that it be able to store the incident optical energy. Since the reading format of the page will probably be by words or blocks of words, brief storage of the reconstructed bit pattern is desirable.

A fourth desirable characteristic is that the detector matrix/electronics combination allow complete random access to all words or word blocks within a reconstructed page.

399

10. Application Areas

10.1.3.6 Other Components

In addition to the laser, the beam deflectors, the page composer, the recording medium, and the detector array, numerous other optical and electronic components are needed to interface the major components. Some of the needed optical components are shown in Figs. 4–7. A number of lenses are needed. Some of these are for beam shaping and some are for Fourier transforming. Suitable high-quality lenses producing sufficiently low wavefront distortion are available. If an acoustooptic deflector is used, cylindrical lenses are also needed (see Fig. 8), and these are less available with small *f*-numbers than are spherical lenses. A number of types of polarization sensitive beamsplitters are available in addition to the simple Brewster angle beamsplitter shown in Figs. 4–7. A simple plate beamsplitter may be used together with a polarizer as shown in Fig. 9 to switch between the two-beam recording configuration and the one-beam reading configuration upon polarization rotation of the input beam. The lenslet array may be an array of individually made short-focal glass lenses, a monolithic molded array of plastic lenses, an array of graded-index optical fibers (Uchida *et al.*, 1970), or an array of holographic optical elements. For example, a 32 × 32 array of holographic binary phase zone plates has been used for the lenslet array in a holographic memory (Huignard *et al.*, 1976).

The electronic components needed depend to a large extent on the types of major components selected. For example, if acoustooptic beam deflectors are used, linear voltage controlled microwave oscillators are needed to drive the deflectors. If electrooptic beam deflectors are used, a programmable high voltage power supply is needed instead.

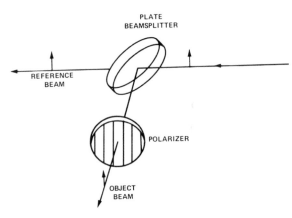

Fig. 9 An optical arrangement for switching between the two-beam writing configuration and the one-beam reading configuration upon polarization rotation of the beam.

10.1.4   Holographic Memory Capacity/Design Relationships

10.1.4.1   Theoretical Bit Capacities

The theoretical ultimate storage capacity for two-dimensional recording is one bit per λ^2 area, and for three-dimensional recording is one bit per λ^3 volume (van Heerden, 1963). This theoretical limit applies equally to directly recorded bit-by-bit storage or to holographic storage. Thus the theoretical storage density S_{2D} in two dimensions is

$$S_{2D} = 1/\lambda^2. \tag{16}$$

The theoretical storage density in three dimensions is

$$S_{3D} = n^3/\lambda^3 \tag{17}$$

for a material of refractive index n. Therefore with a material of refractive index $n = 1.5$ and for a wavelength in air of $\lambda = 0.5145 \ \mu m$, the theoretical storage density of a two-dimensional (thin) hologram is $S_{2D} = 3.78 \times 10^6$ bits/mm^2 and of a three-dimensional (thick) hologram is $S_{3D} = 2.48 \times 10^{10}$ bits/mm^3. These are theoretical values that are significantly reduced in actual practical situations.

Numerous factors limit the capacity of a holographic memory so that it is lower than what is indicated by the theoretical storage densities calculated above. While all of the limitations are not completely understood, some of them have been analyzed, and these results follow in the subsequent sections. Each degrading effect is considered in terms of the limit that it produces. If more than one effect limits a storage parameter, then obviously the lowest value of that parameter must be chosen in determining the memory capacity; that is, if there are multiple limiting effects on a parameter, then the effect that produces the greatest limitation (the dominant limitation) is the only one that needs to be considered in calculating the total capacity.

10.1.4.2   Limit on Total Bit Capacity Due to Optical Aperture Effects

The hologram diameter a at the hologram plane may be considered to be the imaging aperture for each bit at the image plane (detector array). These bits will be individually distinguishable if the Rayleigh criterion is satisfied; that is,

$$\theta_{min} = 1.22\lambda/a, \tag{18}$$

where θ_{min} is the minimum resolvable angle of the bit image at the detector plane as measured from a point at the hologram plane, a distance r away. The area occupied by a single bit at the detector plane (assuming θ_{min} is small) is

10. Application Areas

thus

$$A_{\text{bit}} = 1.49\lambda^2 r^2/a^2. \tag{19}$$

The number of bits per page, N_1, is thus

$$N_1 = A_D/A_{\text{bit}} = 0.67A_D a^2/\lambda^2 r^2, \tag{20}$$

where A_D is the area of the detector array. The number of hologram locations N_2 in the recording material is

$$N_2 = A_H/(a/\mathscr{E})^2 \tag{21}$$

where A_H is the area of the hologram array and $\mathscr{E}$ is the linear filling factor of the holograms in the recording medium. The total bit capacity C_{2D} for a two-dimensional memory is thus

$$C_{2D} = N_1 N_2 = 0.67A_H A_D \mathscr{E}^2/\lambda^2 r^2 = 0.67A_H \mathscr{E}^2/\lambda^2 F^2, \tag{22}$$

and the area storage density S_{2D} is

$$S_{2D} = 0.67\mathscr{E}^2/\lambda^2 F^2, \tag{23}$$

where F is the f-number of the detector array as measured from a point on the hologram plane. Using the example values of $\mathscr{E} = 0.70$, $\lambda = 0.5145$ μm, and $F = 4$ gives $S_{2D} = 7.75 \times 10^4$ bits/mm^2. This is greatly reduced from the theoretical value of 3.78×10^6 bits/mm^2. For a 50×50 mm hologram plane, the total capacity using the example values is $C_{2D} = 2 \times 10^8$ bits.

The storage capacity of a holographic memory can be greatly increased by using volume recording materials and by recording multiple hologram pages at each xy address on the hologram plane. If there are N_3 pages recorded at each xy location in the recording medium, then the total capacity of a three-dimensional memory is

$$C_{3D} = N_1 N_2 N_3. \tag{24}$$

If 100 pages are superposed at each xy recording address ($N_3 = 100$), then the preceding example parameters give a memory of capacity $C_{3D} = 2 \times 10^{10}$ bits. The notation that N_1 is the number of bits per page, N_2 the number of hologram xy locations, and N_3 the number of pages at each xy location will be used throughout the following discussion.

10.1.4.3 Limit on Number of Bits per Page Due to Detector Noise

The optical power diffracted to one photodiode or to one phototransistor in the detector array is

$$P_d = P_L T \eta/N_1, \tag{25}$$

where P_L is the laser power, τ the transmission coefficient for the reference beam through the optics of the system to the hologram plane, η the hologram diffraction efficiency, for the case when all detector array bit addresses are illuminated (a page of all 1's), and N_1 the number of bits per page. The detectivity D' needed to achieve a given signal-to-noise ratio SNR at a detector is

$$D' = (\text{SNR})(\Delta f)^{1/2}/P_d \quad \text{Hz}^{1/2}/\text{W}, \tag{26}$$

where Δf is the electrical bandwidth. The reciprocal of the quantity D' with $\Delta f = 1$ Hz is sometimes called the noise equivalent power. If the area of a detector element is increased, the holograms in the hologram plane can be made smaller in diameter because more diffraction spreading is now allowable. However, as the areas of the detector elements are increased, the noise in the detector also increases and the rise time increases (slower response). A larger signal-to-noise ratio and a larger bandwidth are required to correct these degradations, and thus a larger detectivity is needed. The bandwidth is the reciprocal of the fastest allowable random access time. Combining Eqs. (25) and (26), a limit on the number of bits per page is determined to be

$$N_1 = P_L \tau \eta D'/(\text{SNR})(\Delta f)^{1/2}. \tag{27}$$

For example, if $P_L = 1$ W, $\tau = 0.1$, $\eta = 0.01$, $D' = 10^{12}$ Hz$^{1/2}$/W, SNR $= 10$, and $\Delta f = 10^6$ Hz, then the number of bits per page is limited to $N_1 = 100,000$ due to detector noise.

10.1.4.4   Limit on Number of *xy* Addresses at Hologram Plane Due to Beam Deflectors

The total number of xy hologram locations at the recording medium is limited by the maximum angle of deflection Θ and the laser beam divergence $\Delta\theta$. The total number of resolvable spots in one dimension is given by

$$M = \Theta/\Delta\theta. \tag{28}$$

Typical maximum deflection angles are quite small (less than one degree). However, if the beam divergence is much smaller, a large number of resolvable spots is still obtained. These spot addresses can then be magnified optically to fill any desired size format. The divergence (full angle) of a Gaussian beam in radians due to the deflector aperture D is

$$\Delta\theta = 4\lambda/\pi D. \tag{29}$$

For example, with an aperture of $D = 10$ mm and a wavelength of $\lambda = 0.5145$ μm, the divergence angle (cone angle) is $\Delta\theta = 0.0655$ mrad or $0.00375°$.
   The total deflection angle Θ for an acoustooptic deflector, for example, is

given by

$$\Theta = \lambda \, \Delta f_s / v_s, \qquad (30)$$

where Δf_s is the maximum change in acoustic frequency possible for a given deflector and v_s is the velocity of sound in the acoustooptic material. For lead molybdate, $v_s = 3.75 \times 10^3$ m/sec. If the range of acoustic frequencies is from 50 to 150 MHz and if $\lambda = 0.5145$ μm, then $\Theta = 13.7$ mrad $= 0.786°$. For this example, the total number of resolvable spots in one dimension would be $M = 13.7$ mrad/0.0655 mrad $= 209$ spots. Two of these deflectors operating orthogonally would thus produce M^2 resolvable spots in two dimensions. Therefore

$$N_2 = \pi^2 D^2 \, \Delta f_s^2 / 16 v_s^2 \qquad (31)$$

is the limit of the number of xy hologram locations at the recording medium due to an acoustooptic xy beam deflection system. For the preceding example, $N_2 = 209 \times 209 = 43{,}681$ locations.

10.1.4.5 Limit on Number of Superposed Holograms Due to Recording Range

(a) **Lossless Refractive-Index Recording Materials** For these materials after N_{3n} hologram exposures, if at some point in the material all of the modulations add in phase to use the entire available refractive-index range Δn of the recording medium, then $N_{3n} n_m = \Delta n/2$, where n_m is the refractive-index amplitude modulation of each hologram. Thus the number of holograms at a single xy location is limited to

$$N_{3n} = \Delta n / 2 n_m \qquad (32)$$

due to refractive-index recording range. If each hologram is recorded with the reference beam in the neighborhood of θ_i (inside the medium) then, from Eq. (11),

$$n_m = [\sin^{-1}(\eta)^{1/2}] \lambda \cos \theta_i / \pi d, \qquad (33)$$

and the total number of holograms N_{3n} that can be stored at a single xy location is limited to

$$N_{3n} = d \, \Delta n / \lambda \cos \theta_i \qquad \text{for} \quad \eta = 100\%, \qquad (34)$$

$$= 4.88 d \, \Delta n / \lambda \cos \theta_i \qquad \text{for} \quad \eta = 10\%, \qquad (35)$$

$$= 15.7 d \, \Delta n / \lambda \cos \theta_i \qquad \text{for} \quad \eta = 1\%, \qquad (36)$$

$$= 49.7 d \, \Delta n / \lambda \cos \theta_i \qquad \text{for} \quad \eta = 0.1\%, \qquad (37)$$

$$= 157 d \, \Delta n / \lambda \cos \theta_i \qquad \text{for} \quad \eta = 0.01\%, \qquad (38)$$

due to the refractive-index recording range. For example, if $d = 2$ mm, $\Delta n = 10^{-3}$, $\lambda = 0.5145$ μm, $\eta = 0.01$, $\theta_o = 45°$ (outside the medium), and $n = 2.00$, then $\theta_i = \sin^{-1}[(\sin \theta_o)/n] = 20.7°$ and $N_{3n} = 65$.

(b) Purely Absorption Recording Materials For these materials after $N_{3\alpha}$ hologram exposures, if at some point in the material all of the modulations add in phase to use the entire available absorption range $\Delta\alpha$ of the recording medium, then $N_{3\alpha}\alpha_m = \Delta\alpha/2$ where α_m is the absorption modulation for each hologram. Thus the number of holograms is limited to

$$N_{3\alpha} = \Delta\alpha/2\alpha_m \tag{39}$$

due to absorption recording range. When the recording range is fully utilized, the average absorption coefficient α is equal to $\Delta\alpha/2$. Thus

$$N_{3\alpha} = \alpha/\alpha_m. \tag{40}$$

An optimum value of α exists independent of whatever value α_m has. It is (Kogelnik, 1969)

$$\alpha = (\ln 3)(\cos \theta_i)/d. \tag{41}$$

The maximum diffraction efficiency achievable for any value of α_m can be shown to be

$$\eta = (3\sqrt{3} \ln 3)^{-2}\alpha_m{}^2 d^2/\cos^2 \theta_i. \tag{42}$$

Thus combining Eqs. (40)–(42) gives the total number of holograms $N_{3\alpha}$ that can be stored at a single xy location,

$$N_{3\alpha} = 1/3(3\eta)^{1/2}. \tag{43}$$

The overall maximum diffraction efficiency that can be obtained occurs for $N_{3\alpha} = 1$ and is $\eta = 3.7\%$. In summary, the total number of holograms $N_{3\alpha}$ is limited to

$$N_{3\alpha} = 1 \qquad \text{for} \quad \eta = 3.7\%, \tag{44}$$
$$= 2 \qquad \text{for} \quad \eta = 0.93\%, \tag{45}$$
$$= 3 \qquad \text{for} \quad \eta = 0.41\%, \tag{46}$$
$$= 6.09 \qquad \text{for} \quad \eta = 0.10\%, \tag{47}$$
$$= 19.2 \qquad \text{for} \quad \eta = 0.01\%, \tag{48}$$

due to the absorption recording range. Using the same parameters as in the refractive-index recording example ($\eta = 0.01$) gives $N_{3\alpha} = 1.92$ as compared to $N_{3\alpha} = 65$ for phase holograms.

10.1.4.6 Limit on Number of Superposed Holograms Due to Angular Selectivity

The reference beam of an optical holographic memory is most conveniently restricted to a single plane of incidence (as, e.g., in the systems shown in Figs. 4–7). The angular bandwidth (the full angular width at half power) for the reference beam as measured outside of the recording material is

$$\Delta\phi = AnL/d, \tag{49}$$

where n is the average index of refraction for the material, L the grating spacing, and d the hologram thickness. The quantity A is the angular selectivity coefficient (Gaylord and Tittel, 1973).

(a) Lossless Refractive-Index Recording Materials

$$A = 0.886\text{–}0.799 \quad \text{as} \quad \eta = 0\text{–}100\%, \tag{50}$$

corresponding to increasing n_1 from 0 to $\lambda \cos \theta_i / 2d$.

(b) Purely Absorption Recording Materials

$$A = 0.886\text{–}0.895 \quad \text{as} \quad \eta = 0\%\text{–}3.7\%, \tag{51}$$

corresponding to increasing α_1 from 0 to $\ln 3 \cos \theta_i / d$. The ranges of n_1 and α_1 used (and the corresponding ranges of η listed previously) cover most practical recording situations. If the range of angles of the reference beam outside of the recording material is Φ, then the total number of holograms N_{3as} that can be stored at a single xy location is limited to

$$N_{3as} = \Phi/\Delta\phi = \Phi d/AnL \tag{52}$$

due to angular selectivity effects. For example, if the material is a refractive-index recording material and if $\Phi = 30°$, $d = 2$ mm, $\lambda = 0.5145$ μm, $\eta = 0.01$, $\theta_o = 45°$ (outside the medium), and $n = 2.00$, then $L = \lambda/2 \sin \theta_o = 364$ nm and $A = 0.886$. Thus $N_{3as} = 1624$.

10.1.4.7 Limit on Number of Superposed Holograms Due to Crosstalk Caused by Randomness of Recorded Data

In the volume superposition of data holograms at a single xy location, there will be a statistical background noise due to diffracted light from holograms other than the one being reconstructed. This is due to the random nature of the data patterns that are being recorded in holographic form. Small contributions to the total diffracted beam will be made, on the average, by at least some parts of each of the other holograms. To achieve a given optical signal-to-noise ratio (SNR) upon readout, it has been shown (Ramberg, 1972) that the number of superposed holograms at a single xy location in the material is limited to

$$N_{3C} = d/4\lambda(\text{SNR})^2 \qquad (53)$$

due to crosstalk between the recorded data pages. For example, if $d = 2$ mm, $\lambda = 0.5145$ μm, and if a signal-to-noise ratio of SNR $= 5$ is needed, then the number of pages that can be stored at each location is $N_{3C} = 39$ pages.

10.1.4.8   Limit on Number of Superposed Holograms Due to Granularity Effects in Recording Medium

In photosensitive recording materials, the microscopic light-responsive elements (e.g., dopant atoms) are, in general, randomly distributed. These elements produce the holographic recording when the individual effects of all of them are summed. From a microscopic view of the recording process it is seen that there should be at least one light-responsive element in each $(\lambda/2n)^3$ volume for each page recorded. This sets an absolute minimum concentration level of light-responsive elements. Conversely, for a given concentration c of these elements, the total number of pages that can be holographically superposed at a single location is limited to

$$N_{3g} = c\lambda^3/8n^3 \qquad (54)$$

due to granularity effects. For example, if $c = 10^{15}$ mm^{-3}, $\lambda = 0.5145$ μm, and $n = 2.0$, then $N_{3g} = 2128$ pages maximum.

10.1.5   Some Commercial Optical Memories

A possible indicator of the technological status of a device is the appearance of forms of that device as a product on the commercial market. Some milestones associated with commercial optical memories are listed in Table III.

The IBM 1360 photodigital mass storage system was introduced in 1966. This system uses electron beams to write on special 35×70 mm photographic film. Each piece of film has a capacity of 5×10^6 bits. Thirty-two films are stored in a cell and one entire memory system consists of 7000 cells of on-line storage. These 10^{12} bits are accessed for reading with a pneumatic transport system. Due to the mechanical movements, this memory system has a random access time of about 100 sec.

The Precision Instrument's 690-212 Unicon laser mass memory is a bit-by-bit high capacity optical memory with a storage capacity of 7×10^{11} bits (Gray, 1972). A focused argon-ion laser vaporizes a 1.5-μm-diameter circle of rhodium from a coated polyester data strip. A total of 13,600 data tracks separated by 7.5 μm can be recorded on each 794×121 mm data strip. This optical memory

TABLE III

Some Commercial Optical Memory Milestones

Company	Model	Utilizes laser	Utilizes holography	Nonmechanical system	Date introduced
IBM	1360 photodigital mass storage system	No	No	No	1966
Precision Instruments	690-212 Unicon laser mass memory	Yes	No	No	1971
Optical Data Systems	Holoscan	Yes	Yes	No	1972
3M Company	Megafetch data processor	Yes	Yes	Yes	1974

has a random access time of about 7 sec. The data are permanently stored and the strips are not erasable. Recorded data are virtually immune from destruction, and this provides a safety factor for this type of archival storage in addition to that available with conventional magnetic technologies.

The Optical Data Systems Holoscan read-only memory is a digital data, holographic memory (Sutherlin *et al.*, 1974). It is used for credit card and check verification. Up to 1.2×10^7 bits can be stored in holographic form on bleached film in a 35-mm cassette. A mechanical drive system allows all of the holograms to be scanned in about 2 sec. Recording of the holograms to be scanned is done with an entirely separate system. Its limited capacity and limited access time are quite adequate for its intended use.

The 3M Company Megafetch system is the first commercial nonmechanical holographic optical memory (Strehlow *et al.*, 1974). It is also a read-only memory with the data hologram plates being produced in a separate system. Each plate contains an array of 1024×1024 holograms. The system has a total capacity of 5×10^7 bits. Using a unique electron beam pumped semiconductor laser, any hologram can be accessed in 10 μsec. This system is completely computer compatible. It is capable of data rates up to 1.5×10^7 bits/ sec.

10.1.6 Parallel Digital Processing in Memory†

For continuously changeable recording media such as electrooptic crystals or undeveloped photographic film, it is possible to have a sequence of hologram exposures at a single hologram location. In this case, the orientation of the object and reference beams with respect to the hologram are unchanged. The multiple exposure process results in complex amplitude addition and subtraction in the holographic recording (Gabor *et al.*, 1965). Using a 180° phase shift

† Gaylord *et al.* (1977).

in the reference beam, this process has been used to produce coherent selective erasure in holograms that are within a stack of superposed volume holograms in self-developing lithium niobate (Huignard *et al.*, 1975a). These experimenters pointed out the equivalence of this process to the Boolean logic operation EXCLUSIVE OR. This operation is done in parallel between each bit on one data page with the corresponding bit on a second data page. There were approximately 10,000 bits/page in this demonstration, and thus 10,000 operations were being performed in parallel. They also showed the process of contrast reversal (equivalent to the logic operation COMPLEMENT). Later the Boolean logic operation OR was added to the operations that can be performed on holographically recorded digital data pages (Huignard *et al.*, 1975b).

In Boolean algebra, the combination of one similarity function (such as EXCLUSIVE OR) and one uniqueness function (such as OR) is sufficient to synthesize all other logic operations (AND, COMPLEMENT, NAND, NOR, COINCIDENCE). The definitions of these operations are summarized in Table IV. With these operations available, generalized processing also becomes possible (e.g., addition or subtraction of n-bit words). Use of only the three available basic operations already discussed (EXCLUSIVE OR, COMPLEMENT, and OR) introduces a rather high degree of complexity into general logical processing. The addition of the Boolean logic operation AND produces enormous simplifications in general. The AND operation for multiple input data pages is made possible by having multiple page composers sandwiched together in series. Each bit position is aligned with the corresponding bit positions on the other page composers. In this manner, if any page is opaque at a particular bit address, no light will be transmitted through the composite page composer at that address. After recording, the reconstructed data page will be dark at that bit location. Only for bit addresses where all individual page composers are transparent will the reconstructed data page be bright. Using dark = 0 and bright = 1, the composite page composer produces the logic operation AND upon holographic recording. Because this process can be done in a single step it may be considered to be a basic available operation in holographic processing

TABLE IV

Definitions of Boolean Logic Operations

Inputs		Outputs						
A	B	COMPLEMENT of A	EXCLUSIVE OR	OR	NOR	AND	NAND	COINCIDENCE
0	0	1	0	0	1	0	1	1
0	1	1	1	1	0	0	1	0
1	0	0	1	1	0	0	1	0
1	1	0	0	1	0	1	0	1

TABLE V

Boolean Logic Operations That Can Be Produced Between
Digital Pages *A* and *B* in a Single Step Using a Two-Beam
Holographic Configuration and a Phase Hologram Recording
Material (Dark = 0, Bright = 1)

In page composer	In recording medium	Reference beam phase change from that used for recording *B*	Operation at readout plane
A, B	—	—	AND (AB)
A	*B*	0°	OR ($A + B$)
A	*B*	180°	EXCLUSIVE OR ($A \oplus B$)
1	*B*	180°	COMPLEMENT ($\bar{B}$)

as are EXCLUSIVE OR, COMPLEMENT, and OR. The basic operations and their
one step implementations are summarized in Table V.

In all of the preceding discussion it was assumed that upon reconstruction,
no light at a bit address represented a logical zero and the presence of light
represented a logical one. Thus at the page composer, an opaque bit address
corresponds to a 0, and a transparent bit address corresponds to a 1. These
definitions, however, are arbitrary. The interpretation of the 1's and 0's may
be reversed. With the reversed definition, the logic operations performed are
changed. The logic operations for both definitions are shown in Table VI.
Even though each operation is changed, the overall computing capability is
the same in both cases.

The emphasis of the preceding discussion is on digital data processing with
holographic recording followed by read-out in a continuously changeable photosensitive medium. Some logic operations between data pages have previ-

TABLE VI

Logic Operations Achievable in
Terms of the Definition of Logical
One and Logical Zero

0 = dark, 1 = bright		0 = bright, 1 = dark
AND	↔	OR
OR	↔	AND
EXCLUSIVE OR	↔	COINCIDENCE
COMPLEMENT	↔	No change

ously been demonstrated without continuous holographic recording. For example, an EXCLUSIVE OR comparison operation may be performed using a prerecorded permanent hologram of a test page. If a matching page to be searched is in the page composer, and if the phase of the reference beam is shifted 180° from the phase used to record the test page, and if the amplitudes are equal, then a null result (dark or logical zero) may be obtained for the transmitted object wave. This principle is used in the Battelle integrated optical comparator (see e.g. Kenan *et al.*, 1976). In this lithium niobate integrated optical device, two guided waves interfere in an iron-doped photosensitive region, and a hologram is recorded and then fixed (by ion migration processes). One of the guided wavefronts originally was diffracted by a refractive-index pattern produced by a series of surface electrodes. After the test hologram has been recorded and made unchangeable (fixed), other signals may be applied to the electrode pattern. With the proper amplitude of the reference and a 180° phase shift from the phase during recording, a null output is obtained only if the input signal duplicates the signal used for the initial recording. Sensing a null at the output therefore produces a system output only when the data being searched match the prerecorded data. Another device of this type is illustrated in Fig. 10. In this system, a binary data channel is continually searched in *m*-bit word segments and these are compared using the EXCLUSIVE OR operation to *n* words prestored as a permanent Fourier transform hologram. The amplitude of the reference beam must be continually adjusted according to the transmittance of the word currently in the page composer. If a match occurs between the input word and any of the prerecorded words, then a null output occurs at all bit locations in that word in the output plane. A final cylindrical

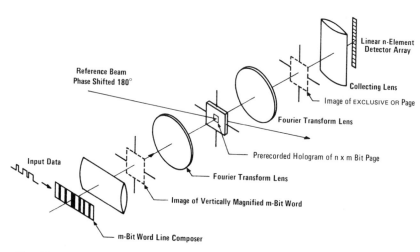

Fig. 10 Binary word detection system utilizing EXCLUSIVE OR operation.

lens collects the light output from the EXCLUSIVE OR operation and focuses it onto an n-element linear detector array. A null at any detector indicates that all bits in the corresponding word are logical zeros (dark = 0) and that a match has been found. A comparison or associative search is thus performed without recording the word or page to be searched. Likewise, however, the results of this parallel logic operation are not recorded for further processing at a later time. This latter case is necessary when a sequence of operations is required, such as in performing arithmetic operations (addition, subtraction, multiplication, division).

The various holographic processing operations have been described here in terms of digital logic and binary data pages. The same processing operations may also be performed on two-dimensional analog data pages (such as images). In this case, the same operations may be performed, but each is now linear (within the dynamic range of the material) in nature rather than binary. Analog and digital data pages may both be stored in the same electrooptic crystal.

The processing described here represents true parallel data processing. Instead of individual hardware logic gates as in electronic logic, all logic operations are performed within the memory (storage crystal) using the same optical system. Logical processing of arbitrary complexity may be synthesized in the same manner as in digital electronic design (using Karnaugh mapping technique, etc.). Although general arithmetic operations may be performed, these are likely to be of secondary importance in comparison to the parallel associative operations that are possible, such as data searching (bit-serial or bit-parallel search operations), change detection, best-fit determination, correlation, and pattern recognition.

REFERENCES

Aagard, R. L., Lee, T. C., and Chen, D. (1972). *Appl. Opt.* **11**, 2133–2139.

Amodei, J. J., and Staebler, D. L. (1972). *RCA Rev.* **33**, 71–93.

Anderson, L. K. (1968). *Bell Lab. Rec.* **46**, 318–325.

Assour, J. M., and Lohman, R. D. (1969). *RCA Rev.* **30**, 557–566.

Cosentino, L. S., and Stewart, W. C. (1973). *RCA Rev.* **34**, 45–79.

d'Auria, L., Huignard, J. P., Slezak, C., and Spitz, E. (1974). *Appl. Opt.* **13**, 808–818.

Drake, M. D. (1974). *Appl. Opt.* **13**, 347–352.

Gabor, D., Stroke, G. W., Restrick, R., Funkhouser, A., and Brumm, D. (1965). *Phys. Lett.* **18**, 116–118.

Gaylord, T. K., and Tittel, F. K. (1973). *J. Appl. Phys.* **44**, 4771–4773.

Gaylord, T. K., Magnusson, R., and Weaver, J. E. (1977). *Opt. Commun.* **20**, 365–366.

Graf, P., and Lang, M. (1972). *Appl. Opt.* **11**, 1382–1388.

Gray, E. E. (1972). *IEEE Trans. Magn.* **MAG-8**, 416–420.

Haskal, H., and Chen, D. (1977). *Laser Appl.* **3**, 135–230.

Hill, B. (1972). *Appl. Opt.* **11**, 182–191.

Hill, B., and Schmidt, K. P. (1973). *Appl. Opt.* **12**, 1193–1198.

Hill, B. (1976). *Adv. Holography* **3**, 1–251.

Huignard, J. P., Herriau, J. P., and Micheron, F. (1975a). *Appl. Phys. Lett.* **26,** 256–258.

Huignard, J. P., Herriau, J. P., Micheron, F., and Rouchon, J. M. (1975b). *IEEE J. Quantum Electron.* **QE-11,** 8D–9D.

Huignard, J. P., Micheron, F., and Spitz, E. (1976). *Optical Properties of Solids* **16,** 851–925.

Kenan, R. P., Vahey, D. W., Hartman, N. F., Wood, V. E., and Verber, C. M. (1976). *Opt. Eng.* **15,** 12–16.

Kiemle, H. (1974). *Appl. Opt.* **13,** 803–807.

Kogelnik, H. (1969). *Bell Syst. Tech. J.* **48,** 2909–2947.

Labrunie, G., Robert, J., and Borel, J. (1974). *Appl. Opt.* **13,** 1355–1358.

LaMacchia, J. T. (1970). *Laser Focus* **6,** 35–39.

Mend, W. G., McCoy, E. E., and Anders, R. A. (1970). *IEEE J. Solid-State Circuits* **SC-5,** 254–260.

Meyer, H., Riekmann, D., Schmidt, K. P., Schmidt, U. J., Rahlff, M., Schroeder, E., and Thust, W. (1972). *Appl. Opt.* **11,** 1732–1736.

Rajchman, J. A. (1970a). *J. Appl. Phys.* **41,** 1376–1383.

Rajchman, J. A. (1970b). *Appl. Opt.* **9,** 2269–2271.

Ramberg, E. G. (1972). *RCA Rev.* **33,** 5–53.

Roberts, H. N. (1972). *Appl. Opt.* **11,** 397–404.

Smits, F. M., and Gallaher, L. E. (1967). *Bell Syst. Tech. J.* **46,** 1267–1278.

Staebler, D. L., Burke, W. J., Phillips, W., and Amodei, J. J. (1975). *Appl. Phys. Lett.* **26,** 182–184.

Stewart, W. C., and Cosentino, L. S. (1970). *Appl. Opt.* **9,** 2271–2275.

Stewart, W. C., Mezrich, R. S., Cosentino, L. S., Nagle, E. M., Wendt, F. S., and Lohman, R. D. (1973). *RCA Rev.* **34,** 3–44.

Strehlow, W. H., Dennison, R. L., and Packard, J. R. (1974). *J. Opt. Soc. Amer.* **64,** 543–544.

Sutherlin, K. K., Lauer, J. P., and Olenick, R. W. (1974). *Appl. Opt.* **13,** 1345–1354.

Tsukamoto, K., Ishii, A., Ishida, A., Sumi, M., and Uchida, N. (1974). *Appl. Opt.* **13,** 869–874.

Uchida, T., Furukawa, M., Kitano, I., Koizumi, K., and Matsumura, H. (1970). *IEEE J. Quantum Electron.* **QE-6,** 606–612.

Vander Lugt, A. (1973). *Appl. Opt.* **12,** 1675–1685.

Vander Lugt, A. (1976). *In* "Optical Information Processing" (Y. E. Nesterikhin, G. W. Stone, and W. E. Kock, eds.), pp. 347–368. Plenum, New York.

van Heerden, P. J. (1963). *Appl. Opt.* **2,** 393–400.

Zook, J. D. (1974). *Appl. Opt.* **13,** 875–887.

10.2 TWO-DIMENSIONAL DISPLAYS

Burton R. Clay

10.2.1 Introduction

Recent advances in holography can be applied to projected two-dimensional image display technology with the result that several improvements over the widely used photographic film projection methods can be realized.

The major improvements obtainable are higher image brightness, greater durability of the stored image, and reduced cost of copies. Other advantages over photography offered by the use of holography are simplified processing, a very reliable retrieval means, colorimetry independent of photocompatible dyes, and reduced power input for a given brightness if negative holograms are used.

A number of techniques are described which can bring about these advantages. In some (but not all) display configurations the size of the product using holograms is slightly larger than its photographic film counterpart. This arises from the need for off-axis illumination in the nonzero-order diffraction type readers.

Because of the many kinds of holograms in existence and the associated differences in those characteristics which might be applied to the display art, a review of the properties which may benefit a display is presented in Section 10.2.3.

10.2.2 Desirable Display Characteristics

As a guide in determining appropriate parameters for application, a discussion of desirable display characteristics is presented.

10.2.2.1 Image Quality

(a) **Resolution and Brightness** For most purposes an observer would desire what he considers to be an image with good to excellent sharpness. His reaction to the image is influenced mainly by perceived edge sharpness, size (angular), brightness, contrast, surround brightness, and noise. These factors are interactive and have been difficult to quantify.

10. Application Areas

In order to establish precise trade-off criteria for the design of displays it is necessary to formulate a mathematical model of the human-display system. This has been done by Mezrich *et al*. (1977). In order to accomplish this they have developed the pertinent image descriptors and taken the necessary psychophysical measurements for obtaining data and for confirming the model. Their work is extensive and detailed and should be consulted for serious display design. The display designer must take into account the observer-display geometry, since both the size of the display and its brightness influence the perceivable resolution.

For example, they show the controlling influence of display size on contrast sensitivity as plotted in Fig. 1, and they show the influence of display brightness of threshold contrast in Fig. 2.

Consideration of these factors together with the influence of the optical system type (i.e., holographic or conventional) is necessary for the achievement of an optimal display design.

(b) Color While a display presenting only luminance information may be satisfactory for many purposes, the addition of color clearly adds to its desirability for the following reasons:

(1) For medium to high information density presentations, a given amount of information may be more quickly cognated with color coding than without

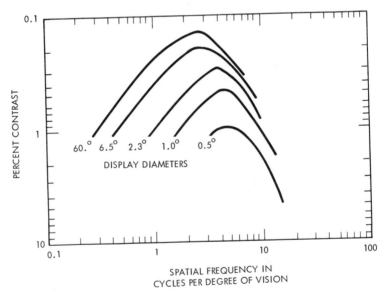

Fig. 1 Threshold contrast sensitivity of sine wave intensity functions with display diameter as the parameter. Display brightness, 34 mL; surround brightness, 3.4 mL.

416

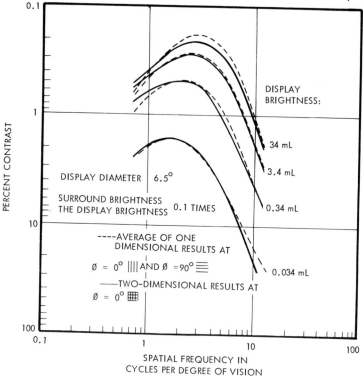

Fig. 2 Threshold contrast sensitivity for one- and two-dimensional sine wave intensity gratings with display brightness as the parameter.

it. A study by Smith (1963) has shown that for search and count tasks in which alphanumeric and small vector symbols are displayed to an observer in black and white on the one hand and with color coding using five colors on the other, the observers were able to complete their tasks more efficiently when color was employed.

(2) Fewer mistakes are made when color coding is employed for the search and count tasks just described, and significantly fewer errors were made when color coding was used.

(3) For a given error the deviation from accuracy was significantly less for search and count tasks using color encoding.

(c) Ambience   In general, displays are more useful if they can be used in environments having wide ranges of ambient light. If the viewing room is fairly dark, the limiting image contrast projected from a transparency of given density range is largely determined by internal scatter of the projector and the lens quality. However, the more severe problem is encountered in viewing rooms having high ambient brightness. The image contrast can be degraded by room light flooding the projection screen. Those portions of the image which should be black appear to be less than black because the projection screen tends to scatter room light into the viewer's visual field. By definition, the contrast is

$$C = \frac{I_{max} - I_{min}}{I_{max} + I_{min}}$$

Here I_{min} is the intensity arising from two sources: (1) the light projected from the darkest object areas together with (2) the ambient light scattered into the observer's field of view from the screen; I_{max} is the light projected on the screen from the object area of maximum transparency.

If the projector has negligible internal scatter and if the viewing room is perfectly dark, then the contrast is

$$C_d = \frac{I_{max} - 0}{I_{min} + 0} = 1$$

and the contrast of the image equals that of the object. If, on the other hand, the effective reflectivity of the screen is 0.1 and the ambient light incident on the screen is 300 fL, then the reflected light is 30 fL. Assuming the room acts as an extended source, most of this light enters the observer's field and I_{min} is 30 fL. In order to obtain a contrast of 90%, one would need an I_{max} of about 600 fL at the screen.

If the ambience is increased by say 3000 fL, typical of a shaded outdoor environment on a sunny day, the preceding contrast would be reduced to 33.3% with consequent loss of gray scale.

A desirable display must transfer a high percentage of the contrast of the object to the observer and it must do so for ambient light values as high as one expects to encounter. As will be seen, the phase hologram can be useful in achieving this end by providing the means for obtaining higher brightness. Also the screen diffuse reflectivity can be reduced by special techniques such as the use of A-R coated specular lenticular surfaces rather than the more commonly used diffuse type.

10.2.2.2 Storage Medium Merit

(a) **Information Density** The storage medium chosen may have a significant effect on various display parameters. For the most desirable general purpose display, the storage medium should be capable of storing both color and luminance information and should have good linearity over a wide dynamic range. In addition, the medium should have very low noise. In order that a large library of information occupy as little volume as possible, the medium should have high resolution. Holograms can be stored on grainless transparent media which are not amenable to photographic use. Some of these materials have very high resolution; an example of such material is Shipley AZ 1350 photoresist (Bartolini, 1974).

(b) Durability   The storage medium should be resistant to abrasion, humidity, heat, intense light, repeated flexing, and mildly corrosive atmospheric contaminants. Ideally, the image should not fade nor should the substrate transparency change with age. Where archival storage is demanded, color stability is of critical importance. Dyes used in photographic materials fade when exposed to intense light and heat. Holograms may store information simply as optical path length differences in a transparent medium—in consequence both color and luminance information are stable. Examples of such materials are given in Section 10.2.3.

(c) Cost   Summary cost must be computed from basic material cost, medium manufacturing cost, chemical development, and processing costs. Since photographic film is acceptable and in wide use, the cost of other materials will be compared to that of finished film transparencies. Holography offers some advantages in cost and these are discussed in Section 10.2.3.

10.2.2.3 Ease of Use

A desirable display is obviously one that has minimal troublesome aspects of usage.

(a) Focus Latitude   One of the great annoyances in using standard microfilm projectors, for example, is sensitivity of focal adjustment. These devices must be critically focused or serious image degradation results. Often the requisite focal tolerance appears to be incompatible with the mechanical tolerances built into the film handling mechanism. The problem arises from the necessity of obtaining an acceptably bright image from the small transparency format. If the light source brightness is increased, the image brightness increases correspondingly, but at the expense of requiring the film to modulate a greater light flux density. The modulation process depends on the absorption of light in the film. The absorption converts light to heat within the emulsion. Even though a filter may be used to remove the infrared components from the projection light source, the film temperature must rise by endothermal conversion. In the limit, the amount of light incident on the film must be less than that causing thermal decomposition. The image brightness for films which modulate light by absorption therefore is limited by magnification and lens f-number. Recourse to small f-numbers to obtain higher brightness necessarily reduces focus latitude and therefore can place demands on the system which are impractical to implement.

The obvious solution to this problem is to find a way of modulating light which does not utilize absorption. A phase hologram modulates light by diffraction, a process which does not involve the generation of heat and therefore can gate extremely large amounts of incident light flux without damage. Some of the resulting brightness increase may be traded for focal latitude by increasing the *f*-number of the projection lens, consequently the system is easier to use. This principle may be applied not only to microfilm systems but to other projection displays as well.

(b) Retrieval With display systems having large amounts of stored information, a retrieval system is often necessary. Each frame is encoded with an identifying number in the form of a machine readable code. This provides a means of generating an electrical signal for operating a search system.

The information density achievable with an array of binary bits is generally higher than with other identifying configurations. Holographic recording of this code allows maximal redundancy of the information in its allotted area so that accidental effacement of bits does not result from scratching or even from excising a substantial portion of the hologram frame number area. As will be described in Section 10.2.5, additional benefits result from the use of the appropriate hologram type.

10.2.2.4 Physical Attributes

(a) Reliability The overall reliability of a display is dependent on such factors as the life of the light source, the durability of the storage medium, and the complexity and number of its components.

A holographic readout system can offer certain reliability advantages over conventional film projection if the holograms are embossed in tough plastic film and readout with incandescent lamps rather than lasers.

(b) Size Depending on the type of hologram used, the resulting display can be the same size as a comparable film projector.

(c) Power Consumption Various types of holograms can be chosen for a display. The hologram type determines the kind and configuration of the light source used. Lasers, filtered arc lamps, and incandescent lamps have been used as readout sources; however, if it is desired to minimize the system power input, the hologram type chosen would be that allowing the use of a high pressure xenon lamp or an incandescent lamp. Section 10.2.4.2 describes a ZOD hologram technique that produces a brighter image than a conventional color slide for the same magnification of the same power input. Other circumstances indicate other directions of desirability as covered in Section 10.2.3.

10.2.3 Hologram Types Considered

10.2.3.1 Classification

Holograms may be classified according to their fringe structure: by internal morphology of the recording medium, and by the configuration of the fringes imposed by the optical recording system. Recording medium characteristics influence the former classifications and lead to the broad general categories of surface, volume, absorption (or amplitude), and phase holograms. Diffraction pattern shape and composition in the recording plane determine the latter.

(a) Diffraction Pattern Shape in Two Dimensions   FRESNEL HOLOGRAMS These holograms result from interference between a reference beam and the near-field (Fresnel transform) diffraction pattern of the object. From each object point a wavefront having a radius of curvature, determined by the object to hologram distance, is incident on the hologram. The resulting super-position of interference patterns have similarities in shape to Fresnel zone platelike sets. Because the fringe pattern for each point has a frequency gradient, the resolution of the recording medium is not optimally used and because the fringes are curved, they are not as useful for color encoding. Monochromatic sources are necessary for reconstruction.

FRAUNHOFER HOLOGRAMS   These holograms are formed from the interference between a plane reference beam and the far-field diffraction patterns of the object. (The Fourier transform hologram is a special case of the Fraunhofer in which the recording plane is located in the back focal plane of the recording lens so that the background or dc term has the coordinates 0, 0.) Since the interfering wavefronts are plane, the fringe sets are linear. This property permits optimal use of recording medium resolution and is valuable in color coding, as will be discussed in Section 10.2.4. The reconstruction source must be monochromatic.

FOCUSED IMAGE HOLOGRAMS   These holograms may be recorded with linear or curved background fringes depending on the recording system configura-tion. An image of the object is formed by a lens or by another hologram on the hologram surface—if the reference wavefront is plane, the fringes will be curved, but if the reference wavefront is made to have the same curvature as the background term from the object, the fringes will be linear. The recon-struction source need not be monochromatic; color encoding is possible and the recording medium storage density can be optimized.

(b) Fringe Shape in Three Dimensions   A further classification places any of the preceding categories in either the thin or the thick (volume) grouping. This distinction is made according to whether the modulation extends through large or small distances compared to the spatial interval of the grating. Bragg hol-ograms occupy this category. The fringe structure is such as to act as a built-in narrow band filter so that white light readout is possible. The efficiency can be very high for a particular wavelength, but image brightness obtainable from a white light source is low for color images since the filtering action of the grating is inherently narrow.

10. Application Areas

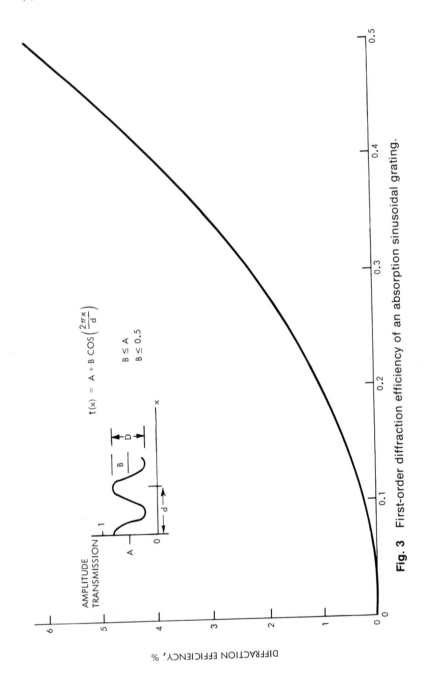

Fig. 3 First-order diffraction efficiency of an absorption sinusoidal grating.

(c) Recording Medium Effects ABSORPTION   Holograms recorded on absorption media (such as photographic emulsions) depend for their modulation on the absorption of light and are capable of a maximum first-order efficiency of less than 10%. Figure 3 shows the diffraction efficiency of an absorption hologram as a function of grating modulation depth. Since the absorption process converts the light to heat, the intensity of incident light is necessarily limited by the accompanying temperature rise in the medium. The reconstructed image brightness then is limited by the grating efficiency and the incident intensity which is a function of the thermally induced medium degradation.

PHASE   Phase holograms are essentially transparent and depend on local phase differences to modulate the incident light. This process is inherently more efficient than modulation by absorption. Figure 4 shows first-order diffraction efficiency as a function of modulation depth. Since the reconstruction beam intensity is not limited by the conversion of light to heat within the medium, the use of phase holograms enable the attainment of brighter reconstructed

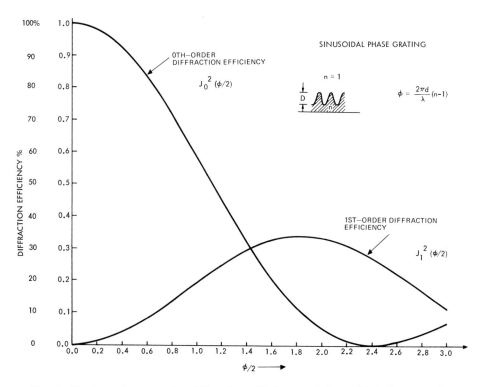

Fig. 4   First- and zero-order diffraction efficiency of sinusoidal phase grating.

423

images compared to those obtainable by conventional film, slides, or absorption holograms at the same magnification.

Diffraction in the transparent phase grating caused by periodic variations in optical path length can result from local refractive index variations, local path-length variation, or a combination of the two.

Refractive index modulation   Maximum theoretical efficiencies for thick transmission holograms as well as for thick reflection (Bragg) holograms is 100%. In practice, 100% is difficult to achieve. If the recording medium is silver halide (bleached), the granular nature of the modulation sites presents finite discontinuities resulting in some scattering and consequent small reduction in efficiency.

This effect is minimized by using very fine emulsions such as Eastman Kodak 649F and Agfa–Gevaert 10E75. The exposure values for 633-nm wavelength are approximately 1100 erg/cm² for the former and 20 erg/cm² for the latter. The limiting resolution for these materials is on the order of 2000 cycles/mm making them suitable for much holographic work. These materials after development, fixing, and bleaching not only possess refractive index changes but exhibit a small surface modulation as well. The predominate modulation effect, however, is that of index change.

Another material which exhibits refractive index modulation is dichromated gelatin. Special processing can cause an accompanying surface modulation as well (Meyerhofer, 1971).

Five kinds of materials are generally used to record volume holograms: bleached silver halide, dichromated gelatin, photopolymers, photochromics, ferroelectrics. Holograms recorded on any of these materials have limitations when applied to practical display systems. These limitations are the sensitivity, durability (Urbach, 1971), and replicability or combinations of these three.

Volume holograms do not lend themselves to replication: individual holograms must be recorded in sufficient numbers to accommodate the distribution need. This is costly. Volume holograms can be perishable depending on the materials used.

Surface modulation   The need for copies of recorded holograms is usual and is easily met by choosing a medium in which the modulation is principally or entirely a surface property.

Replication of surface modulation can be carried out by electroplating the developed hologram with nickel (after rendering the surface conductive) to a thickness of a few thousandths of an inch, then separating the hologram from the nickel, leaving an electroformed master for embossing the modulation pattern present on the nickel surface into plastic sheet stock. This process is similar to that used by the phonograph record industry in which the groove modulation is cut on a lacquer surface. This lacquer disk is rendered conductive by the deposition of silver from the solution, then transferred to a nickel plating bath for buildup of a suitable thickness of nickel. The resulting stamping master can be used to press many vinyl disks. Hologram replicas made by this process give excellent results (Clay and Gore, 1974).

Several recording materials exhibit surface modulation: silver halide emulsions, dichromated gelatin photoresist, and thermoplastic-photoconductive laminate structures. Silver halide emulsions contain modulation in both the interior of the emulsion and on the surfaces; however, this latter effect is weak and is of little benefit in embossing. Dichromated gelatin holograms generally make use of the volume effect, however; Meyerhofer (1972) has obtained holograms using the surface relief effect in a degree suitable for replication. The grating profiles appear to be changing toward square waves at lower frequencies ($\sim$300 cycles/mm). At 500 cycles/mm a grating with 26% diffraction efficiency would be formed with 100 mJ/cm^2 at λ = 441.6 nm; however, as the frequency is increased, the effective sensitivity decreases and the diffraction efficiency obtainable is reduced, i.e., 400 J/cm^2 is required to obtain 200 cycles/mm with a diffraction efficiency of only 2.4%.

Thermoplastic-photoconductor holograms exhibit surface modulation but have been shown to be difficult to plate and are therefore excluded.

A very suitable material for recording holograms for replication is photoresist. In particular, Shipley AZ 1350 positive working material has low noise, high resolution, good sensitivity, and is chemically compatible with nickel plating baths. Grain noise is absent because the photosensitive reaction is based on molecular domain interaction (rather than on silver crystal deposition). For the same reason, resolution is higher than that reported for silver halide emulsions.

Sensitivity is peaked in the blue end of the spectrum (see Fig. 5). The 441.6-nm line of an He–Cd laser is well suited as a source. At this wavelength the sensitivity is about 10 mJ/cm^2. The 488.0-nm line of an argon laser may also be used, but nearly twice the exposure is required.

Exposure to light causes the material to be more soluble in developer than the unexposed material.

Photoresists in general are designed and formulated to selectively mask areas for subsequent etching. The desired exposure slope for this use is very high (similar to a very high gamma silver emulsion).

For most holographic applications, however, the desired slope should be much less. The slope can be adjusted for specific applications by manipulating the developer composition dilution and exposure.

10.2.3.2   Selection of Hologram Type

The criteria of Section 10.2.2 demands that the hologram type chosen for the display meet the following conditions:

425

10. Application Areas

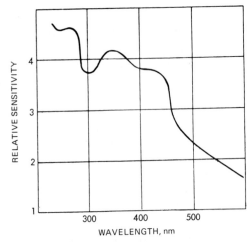

Fig. 5 The relative sensitivity of AZ1350 photoresist with wavelength.

(1)   A white light source should produce a bright image in full color by efficient broad band filtering. The focused-image hologram with linear fringes makes this possible [Section 10.2.4.1(a)]. The hologram choice may be positive, using first-order diffraction for image formation, or negative, using zero-order diffraction.

(2)   Durable replicas must be produced economically. The focused-image hologram recorded with surface modulation allows replication by embossing.

(3)   Storage density should be high—the use of high resolution photoresist enables a high spatial frequency fringe field, i.e., 2000 cycles/mm or higher to be realized.

(4)   Low cost replicas result from the embossing process (about 1% of the cost of color film copies) and from the lower processing cost compared to silver halide processing chemistry which utilizes five steps and at least four baths: develop, rinse, fix, rinse/neutralize, and dry (more steps are required for bleaching). The processing photoresist is simple; that is, only three steps and two baths are needed: develop, rinse, and dry. The total processing time for the photoresist is short. The development time is approximately 30 sec, the rinse time is about 2 sec, the drying time is about 1 sec by air squeegee.

The shorter processing time benefits result in part from the water absorption differences between the gelatinous base of silver halide and the lacquer base of photoresist; the former is hydrophilic while the latter is not.

Materials chosen for receiving the hologram impression by embossing are those that soften at temperatures far enough removed from their thermal decomposition point to avoid the need for critical control during production.

426

Film stocks such as transparent PVC and polycarbonate are excellent since the pressing temperature and pressure are not critical.†

Polyester films (such as Cronar and Estar) are almost impossible to emboss since the flow temperature is nearly the same as the thermal decomposition temperature.

While these materials have good optical properties, strength, water absorption, storage temperature limits, and abrasion resistance should be considered.

Polycarbonate film has great strength, low water absorption, good abrasion resistance and is compatible with production embossing methods. It is also storable over greater temperature extremes than PVC.

10.2.4   Surface Relief Focused-Image Hologram Applications

This versatile hologram type permits several variations in recording and readout technique. Both positive and negative color systems are achievable and alternative color encoding schemes may be used.

10.2.4.1   Positive Recording and Reconstruction

Heterochromatic color images from surface modulated holograms may be produced by the additive color process. The additive process allows a greater range of colors to be produced than that of the subtractive.

Reference to Figure 6 shows the CIE (Commission Internationale de l'Eclairage) coordinates of the spectrum locus in terms of visual response. This diagram is a computational aid in determining the specific mixture of primary colors required for producing a given color stimulus. The area bounded by the spectrum locus and the straight line joining the blue and red extremes contains all observable colors.

A common example of image synthesis using additive primaries is seen in color television. Here the primaries chosen represent an attempt to obtain the largest compass of reproducible colors (the largest area enclosed by a triangle, the vertices of which are designated by the primary coordinates of red, green, and blue). In practice, the realizable colors depend on the characteristics of available filters at the television camera and the phosphors on the receiver screen.

Holographic primaries may be chosen which are at least as close to the spectrum locus as those used in color television and if necessary may be placed on the locus. Broad band filters for each primary are used for the former while narrow band interference filters can implement the latter.

---

† PVC film can be pressed at 240°F ± 10° at 7000 psi ± 10%; polycarbonate film can be pressed at 300°F ± 10° at 12,000 psi ± 10%; pressing time should not exceed 5 sec.

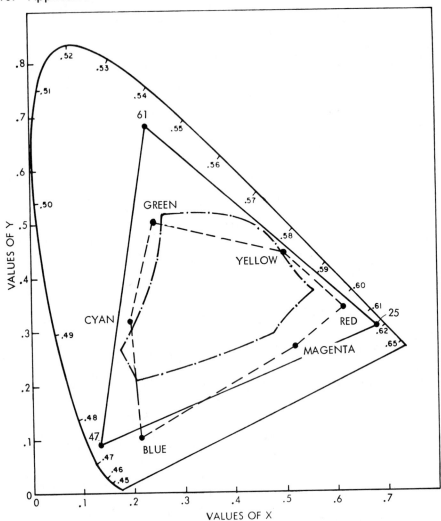

Fig. 6 The CIE color diagram showing colors obtainable by mixtures of primary colors. —, Area enclosed by Wrattan filters 47, 61, and 25; –·–, area enclosing all known printing inks; – – –, areas enclosed by colors in the zero diffraction order.

The additive color process may be implemented holographically by recording three holograms in superposition. Each of the three would contain the luminance component of the object filtered by its respective primary bandpass function. For example, three-color separation transparencies are made from the color scene on black and white film.

The three-color separations used as objects for recording the hologram are generally obtained by making a photographic film exposure of the colored scene through the appropriate set of three primary colored filters. This can introduce the film nonlinearity into the hologram recording process which results in loss of color fidelity in the final record.

Correction of this effect may be accomplished by

(1)   attempting to stay within the linear portion of the Hürter and Driffeld (H&D) curve (this can lead to loss of dynamic range),

(2)   conversion of the analog information on the film to digital by means of half-tone screening techniques (this can lead to beats or sampling noise unless the screen period is very small compared to a resolution element), or

(3)   correction by graphic arts scanners.

In the latter, the colored original is scanned optically and converted to electrical signals which are then processed electronically. Gamma control, density range expansion, or compression and other corrections can be made by programming a microcomputer controller. Owing to the versatility of such a technique, the processed data may be corrected for the intrinsic hologram nonlinearities as well. These data are then fed into the modulator of a light source which is imaged onto a photographic film and scanned in synchronism with the original. These scanners produce a set of four or more color separations for the printing industry, since four or more colored inks are used but are required to produce only three for holographic recording. Each of the three separations is recorded successively in registration on a single hologram frame.

Focused-image holograms are given the requisite linear fringe structure by the method shown in Fig. 7. The essential condition for obtaining linear, constant frequency fringes is that both the object background wave term and

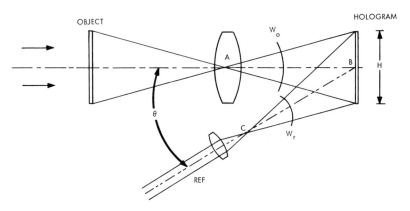

Fig. 7   Recording geometry for obtaining quasi-linear fringes with focused-image holograms.

10. Application Areas

reference wavefront have the same curvature (see Section 4.3 for more detail). The object wavefront W_o is defined by the radius AB, where A is lens second principal plane axial intercept and B is on the hologram.

The reference wavefront curvature W_r is determined by CB. When $CB = AB$, the curvatures are equal, and for $CB \gg H$ the fringes will be reasonably straight and of reasonably constant frequency for small values of θ. In practice, fringe curvatures introduced by larger H to AB ratios and the frequency gradient incurred by reference beam angles of up to 30° may be read out with excellent results for values of H of up to 20 mm or more and $AB \cong 150$ mm (Clay and Gore, 1974).

This fringe pattern has the characteristics which allow three-color separation objects to be recorded superimposed on the same frame so as to enable readout with relatively simple apparatus.

(a) Color Encoding Three methods for enabling primary color identity to be extracted from the superimposed recordings will be described; two are applicable to the positive process of primary addition, and one is suitable for the negative process of color subtraction. In the first two cases the superimposition of primary fringe fields limits the usable composite holographic efficiency to values lower than those attainable for a single field.

If three fields are superimposed, the readout intensity from each component may be estimated by making the following assumptions:

(1) each component has the same first-order diffraction efficiency;
(2) only the first-order terms are important;
(3) each of the reconstruction sources has the same intensity I.

Figure 8 shows the three grating components at G_1, G_2, and G_3 with the three primary sources I_1, I_2, and I_3 incident at θ_1, θ_2, and θ_3. Let the first- and zero-order diffraction efficiencies be η_1 and η_0, respectively, for each grating. Light diffracted into the readout axis from G_1 is ηI_1. Let $I_1 = I_2 = I_3 = 1$ for

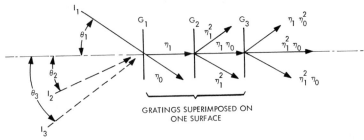

GRATINGS SUPERIMPOSED ON
ONE SURFACE

Fig. 8 Contributions from superimposed diffraction gratings using three readout beams.

simplicity. If no other component gratings are present, the image brightness will be η_1; however, when the other two are present, they each diffract light out of the system as shown and reduce the image brightness from G_1. The first-order term from G_1 acts as zero-order terms for G_2 and G_3.

The first-order term from G_1 is seen at normal incidence on G_2 where diffraction gives rise to two first-order terms (η_1^2) and the zeroth order: $\eta_0\eta_1$. Grating G_3 is now encountered where further losses occur as a result of two more first-order diffraction paths ($\eta_0\eta_1$), leaving a total of $\eta_0\eta_1$ as the primary contribution of G_1 to the image.

The other two readout sources (I_2 and I_3) incident at their prescribed angles θ_2 and θ_3 undergo the same processes; i.e., I_2 is diffracted by G_2 which produces a first-order term η_1 along the axis. This beam is at normal incidence on G_1 and G_3, each of which subtract intensity from the light finally reaching the image. Table I shows how the image is affected by various values of η_1 and the contributions of each component to the composite image. Results are plotted in Fig. 9. It can be seen that to maximize the image brightness for a three-component hologram the efficiency should be about 21%. This causes the image contribution from any primary to equal that from all three in combination; however, if a portion of the hologram consists of only two primary components, that portion of the reconstructed image will be about 10% brighter than for those image areas having one or three primary contributions. Thus red, blue, and green can produce the same intensity as white, but colors made of two primaries such as yellow, cyan, and magenta will appear very slightly brighter than white.

The preceding approximations provide a guide in determining the holographic exposure for each primary. A detailed analysis can also be made in which exact expressions are used; however, the complexity of the computa-

TABLE I

Contributions to the Image from Each Component Grating and Each Readout Source

Contribution from one primary only $I_{G_1} = I\eta_1$	Light lost by primary grating $I_{G_1} = I\eta_0$	Partial contribution from		Sum of contributions from	
		2 primary gratings $I_{G_2} = I_1\eta_0\eta_1$	3 primary gratings $I_{G_3} = I_1\eta_0^2\eta_1$	2 gratings[a] $\Sigma_2 I = 2I_{G_2}$	3 gratings[b] $\Sigma_3 I = 3I_{G_3}$
0.300	0.303	0.0909	0.027	0.1818	0.081
0.250	0.444	0.111	0.049	0.222	0.147
0.200	0.572	0.114	0.065	0.228	0.195
0.150	0.685	0.103	0.0704	0.206	0.2112
0.100	0.794	0.0794	0.063	0.159	0.189

[a] Producing secondary colors, yellow, cyan, and magenta.
[b] Producing white.

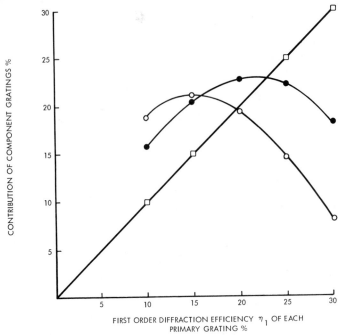

Fig. 9 Sum of contributions to the image from each superimposed primary grating and each readout source. □, single grating; ●, two-component gratings; ○, three-component gratings.

tions may not justify the additional precision obtained since various other factors demand trial and error calibration of the recording apparatus. These factors include changes in reflectivity of the mirrors with time and slight misalignments of a random nature causing intensity differences in various paths.

(b) Spatial Frequency Encoding   In spatial frequency encoding each of the primary color holograms is identified by a specific background fringe frequency. The fringe frequency for each primary is changed by changing the reference beam angle between successive color separation exposure as shown in Fig. 10. Reference beam angles θ_1, θ_2, and θ_3 are chosen for compatibility with the readout device. Figure 11 shows one of the reader configurations devised by Gale and Knop (1976).

The hologram is illuminated by the light from a pair of off-axis beams derived from a single quartz-halogen lamp, enabling a simple compact condenser optic to be used. The off-axis beams are formed by reflecting surfaces s_1 and s_2, which may be part of a prism assembly or may be reflected by plane mirrors. Since the component hologram fringes are parallel, the diffractive process

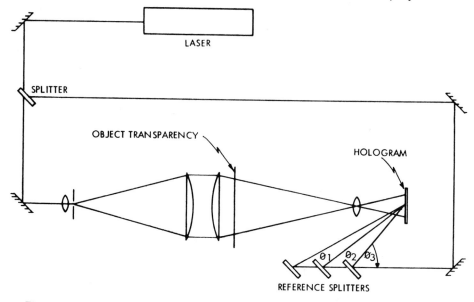

Fig. 10   Recording spatial frequency encoded focused-image holograms.

producing the colors is confined to one plane and the light source surfaces direct the light for readout in this plane; the aperture can be a slit and the lamp filament can be a line.

Collimated components of the beam of light illuminate the hologram at an angle θ. The light diffracted into the first order by the grating structure enters the projection lens which forms an image on the screen. A spatial filter (slit) in the focal plane of the lens blocks unwanted light. The rejected light consists of higher diffraction orders and first-order terms not associated with the primary color of the component hologram being read out.

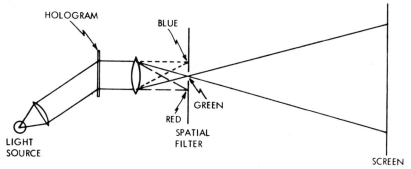

Fig. 11   Readout of spatial frequency encoded holograms showing one of the three paths.

433

One of the readout paths illuminates the blue component grating from the angle θ_B. In order for the diffraction angle to cause the blue light to enter the slit, the spatial frequency of the grating is $\nu_B = (\sin \theta_B/\lambda_B)$. The slit is wide enough to allow an appropriate passband of blue wavelengths to reach the screen.

The green color separation was recorded at θ_G producing ν_G, which at readout causes the green component of the incident white light to be diffracted along the projection axis through the center of the slit. Again the slit width dimension allows a band of wavelengths in the green region to reach the screen while the other colors are blocked. The red separation object is recorded at the smallest reference beam angle, and consequently the lowest spatial frequency fringe field is recorded so that only the red components of the white reconstruction beam reach the screen. It may be noted that for the very smallest values of slit width, the widest gamut of colors is produced in the image but at the expense of screen brightness. In practice the band pass for each primary may be rather wide, allowing good brightness and subjectively good color. As an example the band width of the primary colors used in color television are a good compromise between color saturation and brightness. If the color separation transparencies are generated by photographing the color scene successively through the Wratten filters 25, 58, and 47B, object band pass values of reasonable width are obtained. In order to obtain high color fidelity in the reconstructed image, the primary separations must undergo masking by either the method used in the printing industry to correct overall gamma or electronic scanning of the original colored scene electronically processed to produce corrected separations.

An interesting feature of this system is that one may view the image of the composite hologram in color or in black and white merely by changing the slit width. At the narrowest setting the color saturation is highest, and as the slit is opened the saturation decreases—black and white viewing is readily achieved by a wide slit setting.

(c) **Angular Encoding**   Another method of encoding the three primary color component holograms may be provided through the use of angular orientation of their fringe fields (Clay, 1972). The reference beam angle for all three components is determined by considering various system constraints, such as optimal usage of the recording medium (i.e., the spatial frequency may be chosen to lie near the shoulder of the medium modulation transfer function MTF), and convenience in placing the readout lamp assemblies in a compact structure.

By comparison, the frequency encoding technique requires fringes for all three components to be contained in one meridian, thus forcing the lamps for the readout to be contained in one plane. This, in effect, means that either one lamp (or one pair of lamps) is the maximum usable number. On the other hand, the three meridian planes used with angular encoding enable the use of a pair of lamps per plane.

The separation of the light sources for each primary allows the efficiency of the light-condenser-filter assembly to be optimized for each specific color. By contrast, optimization of the spatial frequency encoding slit yields a different width for each color, but since there can only be one slit, it must be nonoptimum for two of the three colors—thus either the brightness or the colorimetry must suffer.

Another advantage of angular encoding over frequency encoding is the relative freedom from resolution restricting interprimary fringe beats of the latter. A disadvantage of angular encoding is the added cost and complexity of using multiple readout sources.

A recording scheme for angular encoding is shown in Fig. 12. In Fig. 12a the laser beam is divided by ths splitter S into the reference and object beams. The object beam is focused by the lens L_1 to a spot at SF and spatially filtered. The divergent beam illuminates the condenser pair L_2L_3.

The plane surface of L_3 carries the object which is a primary color separation transparency. The lens L_3 converges the emergent light which after reflection from mirror M_3 enters the lens L_4 forming an image on the hologram H. The reference beam after reflection from the mirrors enters the lens L_5 which contains a pinhole filter in its focal plane. This filter is positioned so that its distance from the hologram is the same as that of the second focus of the imaging lens from the hologram.

Wavefront curvature of the reference beams is then matched to the object beam background wavefront curvature in order to obtain reasonably straight fringes. The reference beam is divided by a set of splitters A into three parts which proceed along the three primary meridian planes. The mirrors $M_4M_5M_6$ then reflect the beams toward the center of the hologram. Lenses $L_5L_6L_7$ enable filtering and provide the correct wavefront curvature for each beam.

Each of the three resulting reference beams is provided with a shutter so that only one is open during an exposure. These beams are contained in meridian planes $\phi_1\phi_2\phi_3$. The color separations for angular encoded recording are obtained as already described for frequency encoded holograms.

A very bright image may be obtained from the angular encoded phase holograms using the reader shown in Fig. 13. A pair of light sources (rather than one) is used for each primary to obtain higher image brightness. The meridian planes are the same as those shown in Fig. 12b and the reconstruction angle is the same for blue but is greater for green and still greater for red to satisfy the relation $\sin \theta = \lambda \nu$, where ν, the grating frequency, is fixed at 1100/mm by the recording wavelength of 441.6 nm and the recording angles of 30° (for all three primaries).

Additional perceived image brightness can be obtained by using a directional screen. Such a screen made with nondiffusing surfaces can be antireflection

10. Application Areas

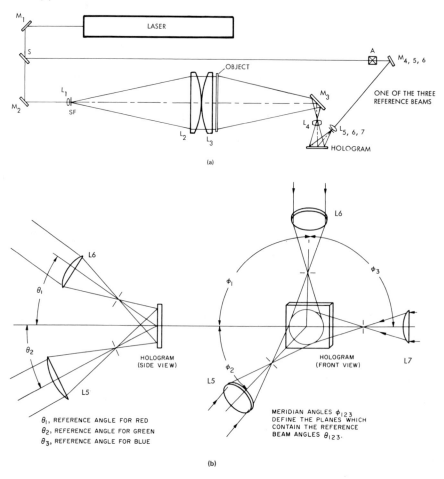

Fig. 12 Recording angularly encoded focused-image holograms: (a) with one of the three reference paths shown; (b) reference beam positions in their respective meridian planes.

coated so as to provide high contrast images even in high ambient light viewing environments.

A screen which consists of an array of lenslets bonded to the surface of a condenser lens is not only capable of low reflectivity, but also delivers a brighter image to the observer than one consisting simply of a Lambertian diffuser. The condenser lens images the projection lens aperture into the observer's plane. The observer's eye position would be confined to the image. In order to provide a large viewing area, the surface of the condenser is covered with lenslets each of which covers one picture element. Each lenslet

436

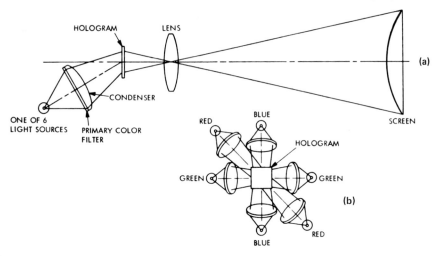

Fig. 13 (a) Readout of angularly encoded focused-image holograms. (b) Array of light sources as seen from lens.

relays its element to the observer plane. The diameter of the area covered by each lenslet image is inversely proportional to the lenslet focal length. The condenser lens provides that the image from each lenslet is coincident in the observer plane.

The ratio of the image brightness as seen by the observer with this screen over that which would have been produced by a Lambertian diffuser is called the screen gain.

The display shown produces 80 fL on the screen with holograms having an average diffraction efficiency of approximately 15% and six lamps of 60 W each, and for a screen gain of 30 produces an apparent brightness of 2400 fL. The contrast degradation by ambient light reflected from the screen for an 8000 fL ambience and an effective screen reflecticity of 1% is about 30:1, making it suitable for use outdoors on a bright day.

Resolution obtained with this display is ~200 cycles/mm at 50% MTF. The reference beam angle of 30° with the helium–cadmium laser recording source produces a fringe field frequency of 1100 cycles/mm. The hologram size is 21 × 21 mm. The magnification is 22, so that an 18 × 18 in. image could be produced; however, the projection screen is 6 in. in diameter, so that a holo-gram area of about 7-mm diameter is shown. This arrangement enables the hologram to be moved for image scanning. The resolution on the screen is about 6 cycles/mm and is limited by the projection lens. This resolution sat-isfies the criteria given in Section 10.2.2.1 for an observer at a distance of about 25 in.

The color encoding methods described each have their unique advantages and disadvantages; however, color coding need not be confined strictly to spatial frequency nor to the angular method. A combination of both may be employed for special situations where warranted (Gale and Knop, 1976). The preceding discussion has emphasized the additive color aspect of hologram image formation; however, achromatic images of even greater brightness than heterochromatic ones may be obtained by the removal of the slit stop in the frequency encoded reader and the broad band primary filters from the lamps in the angular encoded reader.

10.2.4.2 Negative Recording and Reconstruction

Images may also be synthesized from gratings utilizing the zero-order terms as information carriers. This procedure differs from the positive process already described in that, whereas the modulated grating diffracts light into the readout path of a positive system, light is diffracted out of the readout path in a negative system; that is to say, the first diffraction orders are used to reconstruct images in a positive system, whereas the zero-order diffraction terms form the image in the negative system.

(a) **Color** The optimum color primaries chosen are the standard subtractive ones used in negative color photography, i.e., yellow, cyan, and magenta.

Several advantages result from this approach:

(1) higher brightness images may be obtained for a given grating efficiency and for a given reconstruction beam intensity than for conventional first-order holograms,

(2) no off-axis readout paths are needed, so the hologram may be designed for readout in a standard unmodified slide projector, and

(3) the recording may be made with incoherent light.

Knop (1976) has demonstrated bright, high resolution color pictures using his zero-order diffraction technique in which the phase structure has a rectangular profile rather than the usual sinusoidal shape. No claim is made that this is holography, but its similarities of principle are obvious.

He shows that the $t(\lambda)$ transmission of the zero-order diffraction term for a square wave phase grating is $t(\lambda) = \cos^2(\pi a/\lambda)$, where a is the optical path difference, and that the grating transmits all light into the zero order for $\lambda = ma$, where m is an integer, and diffracts all light into the higher orders for $\lambda = (m + \frac{1}{2})a$. The color selectivity is based on the wavelength dependence of diffraction efficiency of the zero-order grating. Figure 14 shows the theoretical transmittance of three superimposed primary gratings as a function of wavelength.

The grating constant for each primary is the same; the amplitude of each of

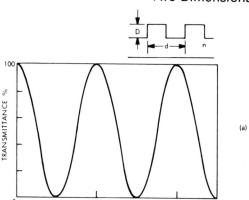

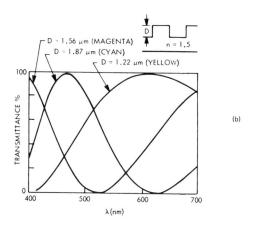

Fig. 14 Zero-order diffraction for single and three superimposed square wave phase gratings. (a) Zero-order transmittance for a square wave phase grating. (b) Zero-order transmittance in the visible for square wave grating.

the three gratings is single valued to obtain the primary color. Gray scale is obtained by screening (area duty cycle).

Three polyvinyl chloride thermoplastic sheets were embossed by electro-formed metal grating masters having amplitudes of 935 nm for cyan, 780 nm for magenta, and 610 nm for yellow. Measurements of $t(\lambda)$ indicated good conformity with theoretical values. The three separate embossed primary components are laminated in registry.

The color range obtainable with this system fits well with that occurring typically in nature and with that obtained from color films and printers' inks. Intermediate colors and gray scale can be produced by variation of the primary area. This is accomplished by screening the primary separations in the manner of halftone printing. The screen frequency must be coarser than the grating frequency, as would be expected.

To enable readout with conventional slide projectors, the grating frequency was chosen to cause the angular separation of the first and higher orders to fall outside the aperture of the readout lens.

Since the majority of slide projectors have lens f-numbers of from 4 to 2.8, a grating constant of 1.4 μm was chosen for all three gratings. Both microfiche and 35-mm formats were demonstrated with very good color fidelity, resolution, and brightness.

The primary master gratings are made by contact printing binary absorption gratings on photoresist of the correct thickness for each color. The developed profile is rectangular. Subsequent plating with nickel yields electroformed embossing masters. Next, the metal masters are coated with photoresist and exposed to a primary image. The exposed photoresist is rinsed away, but the unexposed resist remains. This is now placed in a plating bath which fills in the exposed areas but leaves grating intact in the photoresist masked areas. Thus embossing stamps for each primary component are obtained.

(b) **Negative Recording of Black and White Information**   Zero-order diffraction for producing luminance images has been described by Bestenreiner who with his collaborators (1970) recorded carrier frequency photographs in bleached silver halide emulsion by contact printing. Glenn (1959) has shown zero-order focused-image holograms in thermoplastic.

A new two-step technique has been demonstrated by Gale (1976) which produces excellent brightness contrast and resolution and includes gray scale. A sinusoidal grating profile is chosen in preference to the rectangular wave shape previously described, since the latter has a transmission minimum as a function of wavelength which is narrower than that of the former. Rectangular wave gratings are therefore more suited to color.

The transmittance of sinusoidal gratings at normal incidence is shown in Fig. 15. Curve A is for a single grating. An increase in the optical density can be obtained by the use of two crossed gratings as suggested by Ih (1973). Curve B shows the computed transmittance and density of crossed sinusoidal grating of equal amplitude.

Gale has shown that by adjusting the relative amplitudes of the gratings the neutrality of "black" can be adjusted; i.e., if the amplitudes a_1 and a_2 of the gratings are $a_1 = 790$ nm and $a_2 = 450$ nm, the computed value of zero-order color of black has the CIE coordinates $x = 0.44$, $y = 0.4$, which lies close to the color source and represents a neutral black.

A neutral gray scale cannot be obtained by simply varying the grating amplitude, therefore Gale uses the technique of halftone screening used in printing.

440

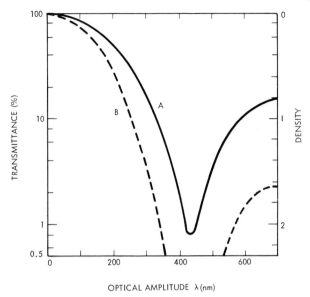

OPTICAL AMPLITUDE λ (nm)

Fig. 15 Zero-order diffraction for sinusoidal phase gratings: — single;- - -, crossed. Amplitudes are equal: $a_1 = a_2$.

He has had excellent results using grating periods of 1.4 μm together with dot patterns of 40 to 80 lines/mm (25 $\times$ 12.5 μm dot spacing) and he suggests that a further reduction by a factor of 2 should be possible. A 10 $\times$ 10 mm recorded frame projected at 20 magnification gave high-quality images when screened at 120 lines/mm. His measured resolution of unscreened images was about 300 lines/mm. The spatial interval of the fringe pattern d was chosen to fulfill the condition that $d \leq \lambda F$, where F is the projection lens f-number, thus enabling the lens to act as a higher-order spatial filter.

10.2.5   Retrieval by Holographic Means

For some applications, in which a rather large number of images are stored, a retrieval capability built into the system would be very desirable. Optical retrieval systems have been used in the past, often with less reliability than anticipated. Frame numbers recorded as binary images take little area from the information frame but are vulnerable to scratches and dust particles. This can result in loss of bits and consequently, the correct address.

For high reliability, the retrieval system must be immune to this ill. A simple way to overcome the scratch and dirt problem is to record the frame address as a highly redundant hologram.

In addition to redundancy, two other desirable characteristics of the holo-

441

gram would be

(1)   that the reconstructed image remain stationary in the direction of motion (to decrease the required circuit bandwidth) while the hologram frame
moves with respect to the readout system, and
(2)   that the hologram readout wavelength be different from the reconstruction wavelength to enable recording at the shorter wavelength demanded by
the recording medium and allow readout at the longer wavelength of a convenient source such as a gallium arsenide laser diode.

A means for doing this is shown in Fig. 16.
Let the object consist of a linear array of closely spaced points which
represent all possible binary digits of the largest frame address number. The
specific number is determined by selecting which point sources are on or off
during an exposure. This arrangement allows the use of a linear array of

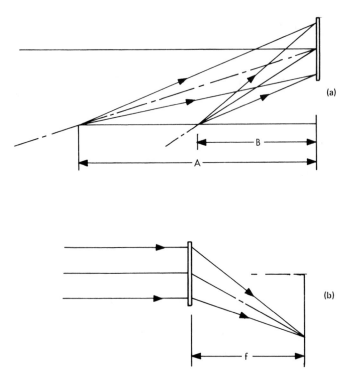

Fig. 16   Frame address (a) recording at λ_1 = 441.6 nm with (b) readout at $\lambda_2 \cong$ 1000 nm.

photodiodes such as Reticon or CCD to be placed in the reconstructed image plane to obtain the requisite signal for electronic processing.

A hologram having the zero image motion properties in the direction of motion would have a Fraunhofer fringe structure in this direction and would have a Fresnel structure in the orthogonal direction. Such a hologram requires a coherent readout source. A small gallium arsenide laser diode ($\lambda \cong 1000$ nm) is very convenient for the purpose. If Shipley AZ1350 photoresist is the recording medium and the 441.6-nm wavelength of a He–Cd laser is the recording source, then several Seidel-type aberrations are expected in the readout image.

In general, the recording of a hologram at one wavelength and playback at another introduces a number of wavefront aberrations which can be characterized as falling into the form of the Seidel aberrations of classical optics. Many applications require that these aberrations be reduced to insignificance.

A method for recording low aberration small object Fresnel holograms for readout at a different wavelength from the recording wavelength has been demonstrated by Tuft (1972) in collaboration with the author (Clay, 1972). The recording geometry parameters are calculated from considerations of the desired readout geometry and wavelength shift between recording and readout. The recording concept calls for a geometry consisting of a point source on the object and a second point source reference beam in which the points lie on a perpendicular to the hologram plane as shown in the figure. These sources are coherent and emit light at wavelength λ.

The reconstruction is made by a collimated source of λ_2 at perpendicular incidence. A detailed analysis shows that if a hologram focal length f is desired at λ_2 and if $\lambda_2 = \alpha\lambda_1$, then the distances a and b in the recording geometry are determined by the simultaneous solution of the two equations

$$\frac{1}{a} - \frac{1}{b} = \frac{1}{\alpha f} \quad \text{and} \quad \frac{1}{a^3} - \frac{1}{b^3} = \frac{1}{\alpha f^3},$$

where a is the perpendicular component of distance between the reference point source and the hologram plane and b the distance from a point on the object to the hologram plane.

Holograms recorded as prescribed above have yielded reconstructed images free from chromatically induced Seidel aberrations and have been demonstrated to operate over a wavelength shift of more than an octive. Diffraction limited reconstructions were obtained for small objects (an array of 11 dots separated by 10 dots covering a linear detector array of 6.5 mm length).

Such holograms produce images which move as the hologram moves through the readout system. To obtain the desired freedom from image motion in the orthogonal direction a cylindrical lens is introduced in the reference beam to match the reference wavefront curvature, in one plane, to that of the object beam, thus resulting in the production of linear fringes in that one meridian. Holograms which carry the information in linear fringe sets can produce a stationary readout. The spatial frequency of the linear fringes, however, varies according to a Fresnel distribution, since the cylindrical lens has zero power in a meridian 90° from the fringe direction. At readout it is also necessary to use a cylindrical lens for focusing the emerging collimated component to points in the same plane as the rays converged by the Fresnel hologram component. Thus a linear diode array may be used to readout the stationary address image, and a second array turned 90° to the first may be used to readout the position of the hologram along the y axis. A second hologram using an object consisting of a single point can be used in much the same geometry but with a single linear diode array to determine the hologram position along the x axis.

This set of two holograms provides a means of accurately positioning the image in x and y and an address number for each frame.

10.2.6 Conclusion

The use of focused-image phase holograms in display systems can display color or black and white images of higher brightness and greater resolution than comparable film systems. The cost of replicas is lower than the cost of conventional film copies since the former are embossed on transparent sheets, while the latter uses a more expensive recording structure consisting of emulsion coated on transparent sheet and must be chemically processed. The hologram replicas are more durable than color film records since the former contain the information as a material shape function, whereas the latter depends on photocompatible dyes which can degrade with the values of projection light intensity typically found in 35-mm slide projectors.

A novel retrieval and positioning system compatible with the embossing technique uses high redundancy holograms for high reliability.

REFERENCES

Bartolini, R. A. (1974). *Appl. Opt.* **13**, 129.
Bestenreiner, F., Deml, R., and Greis, U. (1970). *Optik,* **30**, 404–418.
Clay, B. R. (1972). US Patent 3695 744, Holographic Multicolor Technique.
Clay, B. R., and Gore, D. A. (1974). *Proc. SPIE* **45**, 149–155.
Clay, B. R., (1972). Holographic Multicolor Moving Map Display. Final Rep. Navy Contract N62269-72-C-0452.
Gale, M. T. (1976). *Opt. Commun.* **18**, 3 (Aug.).
Gale, M. T., and Knop, K. (1976). *Appl. Opt.* **15**, 2189–2198.
Glenn, W. E. (1959). *J. Appl. Phys.* **30**, 1870–1873.
Ih, C. S. (1973). *J. Opt. Soc. Amer.* **63**, 1282.
Knop, K. (1976). *Opt. Commun.* **18**, 3 (Aug.).
Meyerhofer, D. (1971). *Appl. Opt.* **10**, 416.

Mezrich, J., Carlson, C. R., and Cohen, R. W. (1977). Image Descriptors for Displays, RCA Labs., Princeton, New Jersey, Office of Naval Res. Rep. CR213-120-2.
Smith, S. L. (1963). J. Appl. Psyc. **47**, 6 (358–364).
Tuft, R. A. (1972). Worcester Polytechnical Inst., private communication.
Urbach, J. C. (1971). *Proc. SPIE* **25**, 27.

10.3 THREE-DIMENSIONAL DISPLAY

Matt Lehmann

10.3.1 Introduction

Three-dimensional display holography has had a very unsettled childhood, ranging from a child prodigy to a delinquent dropout. Following the initial published report by Gabor (1948), the early developments were generally directed toward improvements in microimaging. It was not until Leith and Upatnieks reported a two-beam interferometric effect in 1963 (Leith and Upatnieks, 1961; Dulberger and Wixom, 1963) followed by a report on "lensless" photography (Upatnieks and Leith, 1964) that the great rush to produce three-dimensional display holograms really started. The press envisioned everything from three-dimensional lifesize family portraits to 3D movies and television. Science was about to surpass the wildest fantasies of fiction!

Unfortunately for the orderly development of three-dimensional imaging, most of these early predictions were premature if not completely unrealizable. Some researchers, in their eagerness to obtain financing, made unwarranted, extravagant promises that they were unable to fulfill. It was no wonder that the acronym LASER was frequently translated as *L*oot *A*cquired to *S*upport *E*xpensive *R*esearch! It is understandable that 3D holography overstimulated the imagination, for one's first view of a holographic reconstruction is an exciting experience. It seemed that it was only necessary to solve a few simple technical problems to make all the science fiction dreams come true. One could envision a three-dimensional airport approach and departure display which would precisely locate all aircraft in the area as to both height and position. Even projected image traffic signs and bigger than life 3D advertising displays were seriously proposed.

We have been led to believe that a picture is much better than words and that a moving picture is better than a still photograph. Following this line of reasoning, it is easy to extrapolate to the multiple advantages of three-dimensional portrayal. This is not, however, necessarily true, as can readily be demonstrated by viewing almost any home movies! There are many times when words are more effective than pictures and frequently a good still photograph completely satisfies a requirement. This is not meant to imply that 3D imaging does not serve a useful purpose. The benefits that can be realized

447

from 3D holographic displays are real and can be exploited if the applications are intelligently evaluated and realistic solutions employed.

We should not wait for a specific need to exploit 3D holography. Many of our technological improvements have created their own need. Holography need not be an exception. Advertising has a way of convincing us that we did not know what we needed and frequently establishes the necessity for a new technique or development that at best could be classified as an expensive luxury (Boorstin, 1977). Now that holography has emerged from the laboratory and no longer necessitates a sophisticated light source such as the laser or mercury arc lamp, the way is open for creative and innovative thinking to develop 3D holographic imaging to its full potential (Leith, 1976). In reviewing the short history of holography it becomes apparent that several developments went through complex holographic procedures, becoming increasingly simplified until even the holographic method was no longer essential. In considering holographic imaging it is advisable to be aware that holography may be only the catalyst to the realization of an ultimate solution.

10.3.2 Transmission Holography

There are a number of ways to make a hologram and each procedure has its own inherent characteristics. Although the use of a laser light source is generally accepted as a requirement, holographic recordings can be made with a mercury arc or a filtered white light source. Nevertheless, the laser still remains the most efficient and certainly the most popular light source for holographic recording.

The first dramatic three-dimensional holograms were transmission holograms, and they are still the most popular for the novice and are the only way to begin a study of holographic techniques. Since coherence limitations and stability problems can be minimized and controlled, the chances of success in an initial attempt are greatly improved. There is nothing better than a successful holographic recording to maintain interest and encourage further development. It is always good policy to keep the optical setup as simple as possible. Every mirror, lens, beam divider, or additional element adds to the stability problem. The simplest system for recording a transmission hologram is therefore one in which there are no optical elements other than the laser source and an expanding lens. A spatial filter is desirable but not essential. The film or plate holder is firmly secured to the stage which is supporting the object so there can be no differential motion between the two elements (Fig. 1). If the laser source or expanding lens move slightly there will be some variation in illumination, but the fringe pattern will not be affected.

Imitation alabaster or opaque white plastic objects make the best subjects for holography since they reflect the illumination uniformly without specular reflections which may overexpose portions of the photographic plate. Metallic

Fig. 1 Simplified holographic setup.

subjects can be sprayed with dulling spray, available at most art supply stores, which serves to remove objectionable specular reflections. Painted metallic objects usually do not present any problems. Once the art of holography is mastered (and it is an art as well as a science) more difficult projects can be undertaken. If it is intended to display holograms of an art object, then a more natural illumination scheme is indicated (Fig. 2). Such an arrangement does, however, complicate the problem of stability and requires careful measurement to ensure that the coherence length of the laser source is not exceeded. Transmission holograms made in this manner can be bleached for high diffraction efficiency and, as the glass plate is almost perfectly clear, all of the requirements are satisfied for the display of an art object.

Transmission holograms can also be used to project the real image, but certain anomalies exist which must be corrected by proper shaping of the reference illumination. As the direction of the reconstructing beam through the hologram is reversed to obtain the real image, the beam shape must also be modified to the conjugate of the reference during recording. However, if the reference is a collimated plane wave, no modification is required as it is self-conjugate. Both of these requirements demand very large lenses for even a moderately sized hologram, so for economical reasons this is not a very practical solution. One way out of this dilemma is to use a very long path length (25–40 m) so that the spherical wave front is to all practical purposes a plane wave for both recording and playback. This real image is still, however, pseudoscopic (i.e., turned front to back), so this defect must be corrected by making a second hologram with the pseudoscopic real image as the object.

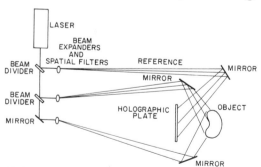

Fig. 2 Holographic geometry for double beam illumination.

10.3.3 Reflection Holography

Reflection holograms are particularly appealing because, although recorded with the coherent monochromatic light from a laser, they can be played back with an ordinary white light source. The recording process is quite simple as it only requires that the reference be incident on the opposite side of the plate from the object. This can be readily accomplished by illuminating the object with the reference beam after it has passed through the holographic plate. Obviously the object must be highly reflective as the reference beam is attenuated as it passes through the emulsion and, unless the object is highly reflective, very little object light will be available to expose the film. Film or plates with antihalation backing cannot be used as they are quite opaque to the illuminating beam. Because the fringes are formed along the bisector of the angle between the reference and object beams, the geometry for reflection holography forms the fringes roughly parallel to the emulsion surface. The fringe spacing is determined by the wavelength of the incident illumination and the sine of the angle between the reference and object

$$\text{(fringe spacing)}\, d = \lambda/(2 \sin \theta/2). \tag{1}$$

Thus when θ is approximately 180°, the fringe spacing is $\lambda/2$ (i.e., one-half wavelength of the laser source). When such finely spaced fringes are to be recorded, stability becomes an important factor. The stability problem is sufficiently critical to require that the photographic plate remain in the holder for several minutes after handling until its temperature stabilizes. The shrinkage of the emulsion for only a fraction of a degree change is sufficient to ruin the hologram.

These reflection holograms can be reconstructed with a white light source such as the sun or a spotlight. The color at which the hologram reconstructs is determined by the fringe spacing established during recording. This also poses a problem as most photographic emulsions shrink as a result of the development process so that the fringe spacing after drying is actually less than determined by the wavelength and recording geometry. A hologram recorded with the red light of a helium–neon laser will reconstruct in green or yellow. Obviously, holograms recorded with blue or green laser lines would shrink enough to reconstruct so far into the violet as to be below the level of human vision. Some photographic media shrink more than others and some, such as dichromated gelatin, actually expand during development. This makes it necessary to record at a shorter wavelength, which is fortunate since dichromated gelatin is only sensitive to the blue or shorter wavelengths. The characteristics of these photographic materials are covered in much greater detail in Section 9.1.

Reflection holography eliminates the overlapping color smearing which results when a conventional transmission hologram is reconstructed with a white light source. This spectral selectivity derives from the layered fringes. However, the image sharpness is determined by the size of the reconstructing source, so the more nearly the source resembles a point, the better the quality of the reconstructed image. This restriction can also be modified by recording the image as close as possible to the plane of the emulsion or preferably actually in the emulsion plane. This can be accomplished by projecting the image with a lens or projecting the real image from a hologram of the object. The portion of the image within the plane of the emulsion will be sharp even when reconstructed with an extended source such as a fluorescent lamp, but the portion in front or behind the emulsion will be diffuse in direct relation to their distance from the emulsion. This technique of holographic recording can be applied to either transmission or reflection holography to improve image sharpness. When this technique is used with transmission holography, color filtering must be used to minimize the color spreading as the color filtering of the multilayered fringes is only obtained with reflection holography.

10.3.4 Rainbow Holography

Where there is a problem someone always seems to find a solution. A type of holographic recording appropriately called "rainbow holography" was invented by Steven Benton of the Polaroid Corporation to solve this very problem of color smearing when using a white light source to reconstruct a transmission hologram (Benton, 1976a,b). In this type of hologram, full advantage can be taken to place the image very close to the plane of the emulsion improving sharpness and as there is no vertical parallax color smearing is minimized. To prepare such a hologram a transmission hologram is recorded in a conventional manner, except that the reference must be as close as possible to a plane wave by using a large collimating lens or long path length. Once the initial hologram is made it is masked so that only a narrow slit is available to view the virtual image. A second hologram is then made of the real image, projected through the slit hologram (Fig. 3). This second hologram has no vertical parallax as it contains only the image one would see through

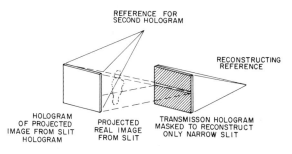

Fig. 3 Rainbow hologram recording.

the narrow slit of the original hologram. When reconstructed with white light there is a vertical color separation, but no color smearing, as each colored image is actually a separate reconstruction of the information contained only in the narrow slit. If a cylindrical lens is used to reconstruct the slit hologram and bleaching processing applied to improve diffraction efficiency, a very bright image can be seen when the hologram is viewed with a white light source. As the viewer scans vertically, the colors change but do not smear, and the image remains satisfactorily sharp and distinct. As in all holographic reconstruction, the requirement for a point source diminishes as the image is placed closer to the plane of the emulsion.

10.3.5 Multiplex Holography

Another attempt to solve the problems of white light reconstruction utilizes the narrow slit technique, but in this application the slit is vertical. This technique, currently being developed by several small companies, is described by several names such as "multiplex holography," "integrafs," and others, but is more descriptively described as a stereogram. The usual procedure is to photograph the object on a standard 35-mm black and white film with a motion picture camera. As this step of the production follows conventional photographic techniques, the object can be of any size and can move. The usual procedure is to use human figures performing some simple repeatable activity such as playing a musical instrument or dancing. The object of the 35-mm movie is placed on a rotating table and one thousand or more individual frames are exposed as the table slowly rotates. At normal movie camera rates this represents forty seconds to one minute of action that can be looped and repeated. Each separate 35-mm frame is then illuminated by a laser source and projected through a cylindrical lens on to a masked strip of film simultaneously with a shaped reference beam from the same laser making a strip hologram of the projected image. The process is repeated for each 35-mm transparency as the holographic film is moved to expose another strip. The final product is a stereographic hologram 9 in. (20 mm) wide, and several feet long (650 mm) that can be reconstructed by a vertical filament white light source. Reconstruction with white light causes some color separation from top to bottom, but otherwise gives the illusion of a three-dimensional object contained behind the curved film format. The illusion of three dimensions derives from the fact that each eye sees the object from an offset parallax. Although theoretically there is only one position for three-dimensional viewing, it is surprising to find how well human vision adapts to and corrects for rather large discrepancies.

10.3.6 Holographic Movies

So what about a complete 3D movie or television program produced holographically? Thus far there are only peripheral solutions which are more stereographic than holographic processes. A hologram is basically a diffraction grating that when properly illuminated reconstructs the image. Projection of the hologram and changing the size of this grating destroys the reconstruction process. This is because a change in grating spacing distorts the diffraction, and the projected image of a grating cannot cause light to be diffracted. Furthermore, it is necessary to look through the hologram itself to view the three-dimensional image whether it is the virtual image behind the hologram, or the projected image in front of the hologram.

A 35-mm-stripfilm holographic recording of a mechanically animated object was made by the author in 1966. This event received considerable unwarranted acclaim as the press implied that this was the forerunner of full-length holographic feature movies! This dream has not been realized, nor does there appear any probability that it will be. Nevertheless, holographic movies of this type can have a useful purpose. The study of small living objects that move in and out of the focal plane during microviewing can be enhanced by holographic methods. As holograms have no focal depth problem the individual frames of a holographic movie can be used to bring the desired object into sharp focus with a microscope. As very short focal length objectives are used, the image must be in or above the plane of the emulsion. Obviously such holographic movies must be made with very short exposures to completely stop the object motion. There are a number of pulsed laser sources currently available that will permit filming what amounts to nanosecond flash photography.

Another spin-off from the attempt to show motion in holograms was the discovery that multiple images or different aspects of the same image could be recorded sequentially on a single hologram plate. One attempt to utilize this technique was in providing images of pathological specimens. These medical specimens of unusual physical defects are kept in glass containers filled with a preservative liquid and used for study. Several such specimens were the subject of holographic storage with both aspects of the specimen jar recorded in a single holographic plate. Reconstructing the image from front to back gave the same view as examining the actual specimen. In one hologram a magnifying lens was included as part of the object. During playback this image of a lens made it possible to examine the specimen in detail as though an actual lens was present. The holographic images have all the visual advantage of the actual specimens, but are obviously much easier to store or handle than the liquid filled containers.

10.3.7 Self-Referenced Imaging

Self-referenced holography is given a detailed study in another part of this book so only that aspect that relates to 3D imaging will be discussed. Objects can be recorded by using the specular reflection from a highlight as the reference. However, the presence of specular reflection on the subject of a holographic recording usually must be avoided as each specular spot becomes a reference and the resulting multiple images ruin an otherwise good hologram. Looking back through the plate or film holder prior to loading will detect these secondary references from the object itself, or other unwanted reflections of the source.

A form of self-referenced holography was given some consideration for the photography of artificial satellites. The advantage of holographic imaging for this purpose was based on the premise that both the reference and object beams would be subjected to the same atmospheric distortion so that the fringe pattern would not be affected. A test of this theory using a piece of shower glass to simulate the atmosphere demonstrated that when the atmosphere was close to the holographic plate, the hologram recording was vastly superior to conventional photography (Fig. 4) (Goodman *et al.*, 1969).

10.3.8 Coherence and Stability

When any holographic imaging process is contemplated it is essential that the experimenter determine both the coherence length of the source and the stability of the work surface. These two factors are vital to the success of any holographic recording as the coherence length of the source determines the size of the object that can be recorded and establishes the precision to which the path length must be measured. There is a very simple procedure that will give both of these important parameters. The laser beam is directed through a beam divider and reflected back by mirrors placed equidistant from the divider. The two beams are adjusted so that the back reflections are superimposed. Expanding these two beams with a lens, the interference fringes can be easily observed and monitored (Fig. 5) (Lehmann, 1970). Any movement of the fringes indicates instability of the work surface or one of the optical elements. When one path length is increased, the fringe contrast will be reduced. When the fringes are no longer discernible, the coherence length has been exceeded. As the light beam makes a round trip to the mirrors the coherence length is twice the path length difference.

10.3.9 Light Sources for Reconstruction

Improvements in holographic recording have made it practical to reconstruct the image without an expensive light source. Initially it was necessary to use a laser source at the same wavelength and angle as the recording reference. It soon became apparent that different wavelengths could be used if consideration was given to the change in angle required to maintain the Bragg efficiency and allowance made for the change in image size and location. Light sources with a fairly narrow emission band that could be efficiently filtered, such as the

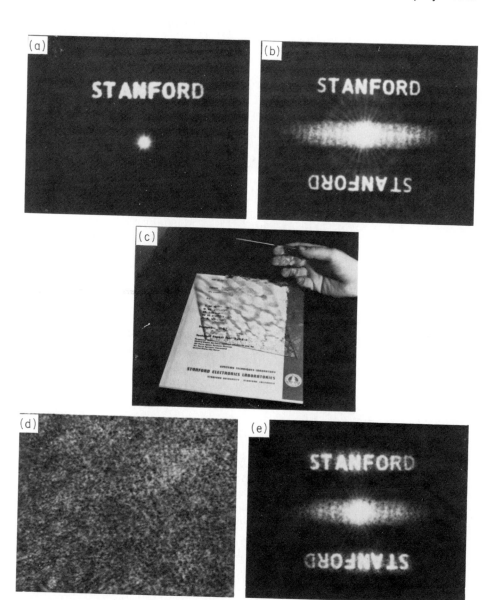

Fig. 4 Holography versus conventional photography through simulated atmosphere: (a) conventional image, (b) holographic reconstruction, (c) aberrating medium (shower glass), (d) conventional image through glass, (e) reconstruction of hologram through glass.

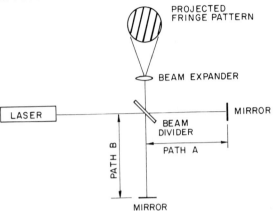

Fig. 5   Setup to determine stability and coherence length.

mercury arc, became popular for display holography. Discovery that images recorded close to the plane of the emulsion were sharply defined even when the source deviated from a point made large display holograms a reality (Mottier, 1975). Now even an ordinary vertical filament light bulb can be used for "rainbow" hologram or stereograms recorded by the multiplex process. These modifications in source requirements for display holography should certainly encourage the development and use of three-dimensional display holography as the need for an expensive laser illuminating source is no longer a limiting factor.

10.3.10   Color Holography

Satisfactory color holography still poses some problems. The very nature of holographic storage by the formation of a complex diffraction pattern in the photographic emulsion determines that different wavelengths will not be diffracted at the same angle for a given fringe spacing. A multicolor hologram, recorded with three different laser wavelengths, is actually three independent monochromatic holograms superimposed on the same plate. The thickness of the holographic fringes and the precision of the reconstructing reference angle are related by

$$\Delta\theta = \Lambda/t, \tag{2}$$

where $\Delta\theta$ is the angle change that will completely suppress reconstruction (mrad), and t the fringe depth in micrometers when the spatial frequency Λ is in cycles per millimeter.

Taking this relationship into account, it is possible to record the three holograms so that each hologram will only be reconstructed by its proper color. This use of thick holographic fringes makes it possible to reconstruct multicolor holograms without the spurious ghost images caused by one reference reconstructing more than one of the independently recorded images. Satisfactory simplified color three-dimensional holography is certainly an area that is open for development.

In a novel approach to color holography, Case (1977) of Virginia Polytechnic Institute proposes taking advantage of the selective color reconstruction of reflection holography to make color holograms with a single color laser. By changing the reference angle the fringe spacing would be varied which would cause the white light incident on the hologram to be reflected in the desired color. Thus it is possible that two reflection holograms recorded sequentially with different reference angles would reconstruct in two colors. This technique is not without problems as it is necessary to change the fringe spacing without changing the fringe orientation. Reconstruction by different wavelengths implies some differential magnification problems, and color rendition may not be realistic as the monochromatic recording source would have the same gray scale distribution of the object for both reconstruction wavelengths. However, there are some unique possibilities that could have artistic value if realism is not the sole objective.

A scheme developed at Hitachi Ltd., Japan, utilizes holographically stored images to project a full color stereogram (Tsunoda and Takeda, 1975). In the development of a system several methods were incorporated to minimize speckle and obtain a reconstructed image with luminance tones of high fidelity. The original object is recorded by angular sampling on 11 color transparencies over a 40° arc for each frame of the motion picture. Fourier transform holograms are then made on a strip of holographic film with three separate laser beams in the blue, green, and red. The color images for each transparency are recorded independently and vertically separated on the holographic film. The holograms are recorded through a random phase filter and a sampling mesh to remove vertical parallax, minimize speckle, and select the desirable spatial frequencies to be recorded in the hologram. The 33 holograms (11 of each color) are projected by color laser beams on a lenticular screen backed by a diffuser (Fig. 6). The lenses of the screen run vertically in order to be unilaterally direction-selective so that an illusion of a three-dimensional image is obtained when viewed from the proper position in front of the screen. The screen is large enough to allow several viewers to be in position for the images to appear in proper perspective. This method appears to have some advantages over "multiplex" stereograms as true color images can be seen and the information is stored on a strip of 35-mm film for projection viewing, rather than requiring viewing through a large film format.

457

10. Application Areas

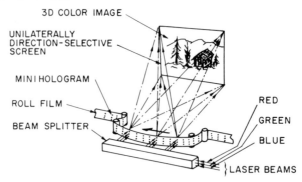

Fig. 6 Projection of composite color holographic stereogram.

10.3.11 Holographic Display Systems

There is always the decision as to whether a virtual or projected image display system should be used. Shortly after television became commonplace, science fiction envisioned projected images of the actors so that the viewer could enter into the scene and become personally involved in the plot. When it was discovered that a holographic image could be projected, it looked as though this fiction might be realized. There are too many unsolved problems, however, both with projection holography and the transmission of a hologram to suggest that the science fiction dream will ever be realized.

Both virtual and projected image holography have their applications, and several dramatic displays have combined the two in a single hologram. The purpose of the display will dictate the mode to be used. Usually valuable art objects are displayed behind glass, so it is no offense to realism to use a virtual image as it will appear within the glass enclosure much as the actual object would be displayed. A projected image would, for this purpose, be an unreal display mode. The dramatic positioning of the image in space in front of the hologram would not be consistent with display procedure for valuable artifacts. The same would be true of holographic storage in the plane of the emulsion for, although resolution is markedly improved, it is not a realistic portrayal of a solid object.

10.3.12 Decorative Holography

The charm of holography is in the realistic appearance of the image. This quality is being exploited in the creation of what can best be described as holographic jewelry. One of the initial defects in silver halide emulsion storage was the dark gray or black appearance, but this distraction was soon eliminated by bleaching chemistry which made possible holographic imaging in an almost clear glass plate. The earlier bleaching efforts were not completely satisfactory since the silver halide tended to discolor when exposed to light for any period of time. A bleaching and clearing technique was developed to minimize this defect and gave excellent results unless the hologram was exposed for several weeks to direct sunlight (Lehmann *et al.*, 1970). An emulsion without photosensitive silver halide was the obvious solution, so various photopolymers were tried to eliminate the problem of discoloration. Possibly the best results to date have been had with dichromated gelatin emulsions which have all the advantages of silver halide, but leave no sensitized products after development processing. The dichromated gelatin has very low sensitivity which limits the use of the material to comparatively small holograms in which high energy densities can keep exposure times to reasonable values. These small holograms are very satisfactory for jewelry such as pendants, pins, and rings. The jewelry holograms are made to be viewed with diffuse reflected illuminations; so are reflection holograms of flat objects, such as coins, or image plane holograms of projected images so that the information is close to or in the plane of the emulsion. A procedure has been reported that makes it possible to emboss the hologram fringes on plastic, making the duplication of originals very economical (Benton, 1976a).

A form of holographic display that has not been explored in depth is its use in architectural fenestration. One can imagine the impact of holographic "stained glass" windows. Holographic gratings and images can be designed and oriented in a window format that would cause the colors to change with the sun angle. It should be possible to arrange the individual holographic elements so that the composite images would change both in shape and color as the sun angle varied during the day. This architectural application of holography should be far more dramatic than the suggested use of holographic three-dimensional murals to produce the illusion of an outdoor window. The author has made holographic grating "wind bells" which were more colorful than the prism type commercially available.

10.3.13 Applications of 3D Display

There have been several attempts to use holographic methods for the display of art objects. Unfortunately, the museums and art commissions have not been convinced of the advantages of holographic displays, except as novelties. However, this use of holography would permit a much wider exposure of rare and valuable artifacts without exposing the originals to theft and vandalism. Many of the art objects are monochromatic, being of gold, jade, ivory, or similar precious materials so there is no need for multicolor holography to assure realism. Also valuable art objects are almost always displayed in a glass enclosed case, so the use of bleached holographic recordings would be as realistic as though the actual object were present.

Several years ago Emmett Leith and Juris Upatnieks of the Environmental

10. Application Areas

Research Institute of Michigan made a series of full circle holographic displays of rare and valuable musical instruments from a collection in the University of Michigan School of Music, demonstrating the value of holographic display for objects that would not generally be viewable (Fig. 7). Another remarkable projected image hologram was made by Tribillon and Fournier (1977) of the Laboratoire de Physique Générale et Optique, in Besancon, France. This hologram of the Venus de Milo at the Louvre was made on a 1 × 1.5 m photographic plate with the projected image the full size of the original statue (2.18 m) (Fig. 8). These two examples clearly illustrate that holographic display of art objects are a practical way to expand an audience for appreciation of fine art. This application could be the most productive use of 3D display holography.

10.3.14 The Future of Display Holography

Three-dimensional display holography must come out of the laboratory and into the hands of the artists if it is to have a productive future. This does not

Fig. 7 Holographic reconstruction of seventeenth century guitar. (Courtesy Juris Upatnieks, Environmental Research Institute of Michigan.)

Fig. 8   J. M. Fournier holding the 1 × 1½ m hologram of the Venus de Milo. (Courtesy J. M. Fournier and G. Tribillon, Laboratoire de Physique Générale et Optique, Besancon, France.)

mean that the science community will cease to contribute, but the new ideas and better ways to exploit the technique will come from the art oriented students and practitioners. The scientist is satisfied when he has solved a problem, published the results, and been recognized by his colleagues. The artist must have his creation appreciated and enjoyed by the public, and it is in this environment that 3D holography will grow and prosper.

It is always hazardous to predict future events, but holographic display techniques have developed far enough to predict that 3D holography will be utilized for the secure display of art objects and will itself become an art form.

10. Application Areas

REFERENCES

Benton, S. A. (1976a). White light transmission/reflection holographic imaging, *Proc. ICO Conf., Jerusalem, August 23-26.*

Benton, S. A. (1976b). Three dimensional holographic displays, *Proc. Electro-optics/Laser Conf., New York, September 14-16,* pp. 481-485.

Boorstin, D. J. (1977). Tomorrow: the Republic of Technology, *Time* (Jan. 17), 36-38.

Case, S. K. (April 1977). Production of a Multicolor Hologram with a Single Color Laser, private communication.

Dulberger, L. H., and Wixom, C. (1963). *Electron.* **36,** 44.

Gabor, D. (1948). *Nature* **161,** 777-778.

Goodman, J. W., Lehmann, M., Jackson, D. W., and Knotts, J. (1969). *Appl. Opt.* **8,** 1561-1586.

Lehmann, M. (1970). "Holography, Technique and Practice." Focal Press, London.

Lehmann, M., Lauer, J. P., and Goodman, J. W. (1970). *Appl. Opt.* **9,** 1948.

Leith, E. N. (1976). *Scientific American* (Oct.) 80-95.

Leith, E. N., and Upatnieks, J. (1961). *J. Opt. Soc. Amer.* **51,** 1469.

Mottier, F. (Sept. 1975). 90 × 60 cm Display Holograms made at Brown Boveri, Switzerland, private communication.

Tribillon, G., and Fournier, J. M. (April 1977). Large Sized Holographic Interferometry in Real Image, private communication.

Tsunoda, Y., and Takeda, Y. (1975). *IEEE Trans. Electron. Devices* **Ed-22,** 784-788.

Upatnieks, J., and Leith, E. N. (1964). *J. Opt. Soc. Amer.* **54,** 579-580.

10.4 HOLOGRAPHIC INTERFEROMETRY

Gerald B. Brandt

10.4.1 Comparison of Holographic and Classical Interferometry

All of the classical interferometers which have been developed for the measurement of optical path changes, either by transmission through or by reflection from an optical quality element, have their analogs in holographic interferometry. Classical interferometers are characterized not so much by their arrangement of optical components, since this can vary considerably depending upon the application, as by the fact that the wavefronts which are being compared interferometrically are nearly plane or spherical waves with relatively small phase deviations from an ideal wavefront. As a consequence, optical components used in assembling a classical interferometer must be fabricated with a high degree of precision so that they do not contribute spurious fringes to the resulting pattern. In contrast, holography makes possible the reconstruction of wavefronts with arbitrary variation of phase across the wavefront, thus opening up the areas of application of interferometry to components of lower optical quality. With holography, interferometry can be done on diffusing components, which cannot be studied at all with classical methods. Since wavefronts are compared, not recorded, in classical interferometers, classical interferometry takes place in "real time," thus requiring great stability from the optical components in the interferometer, and to a certain extent, an equivalent stability of the phenomenon under study. By comparison, the wavefronts to be compared are stored in a holographic interferometer. Thus the additional dimension of time is made available to the experimenter. The time variable adds an important part of hologram interferometry which leads to a number of important new applications, particularly in the area of vibration studies.

Classical interferometers have been discussed extensively in the literature [Tolansky (1973) gives a good summary of these, also Tanner (1965)]. Wavefront-dividing interferometers such as the Mach–Zehnder and Michelson interferometers utilize beamsplitters to separate a plane wave into two paths, one of which passes through the region under test while the other acts as a reference. A second beamsplitter (or in the case of the Michelson, a second

10. Application Areas

pass through the beamsplitter) recombines the beams leading to constructive and destructive interference across the wavefront. These interference fringes represent contours of constant path difference across the wavefront. Holographically, these interferometers are realized by replacing the beamsplitter function with a hologram. (see Fig. 1 for two such realizations.) An interferometer is formed by recording a hologram of a given arrangement of optical elements (Horman, 1965; Heflinger *et al.*, 1966; Tanner, 1966). Reconstruction of the hologram at the place where it was formed provides the reference wavefront (namely that exiting from the optical system in the undisturbed state) from which interferometric measurement can be made of subsequent

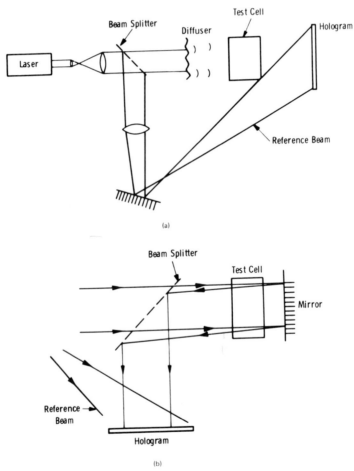

Fig. 1 Holographic analogs of classical interferometers: (a) single-pass Mach–Zehnder type, (b) double-pass Michelson type.

changes in the system. As a result, fundamentally, any hologram is an interferometer, providing the original wavefronts used to form it are available to interfere with the reconstruction.

Other classical interferometers use beamsplitters or polarizing elements to shear the wavefront in a plane perpendicular to the wave propagation direction. Shearing interferometers compare different portions of the same wavefront and thus are sensitive to wavefront phase changes across the wave rather than the absolute value of the phase at a given point. Holographically, a shearing interferometer is realized in a single hologram (Bryngdahl, 1968) or by making two holograms of the wavefront (see Section 10.4.4.3 for further discussion of this technique). In either event, whether the shear is introduced before or after the exposure of the hologram, the information is identical to that obtained from a nonholographic interferometer.

Real time holographic interferometers share the stability requirements of their classical analogs. Unique to holography is the ability to record multiple images on the same hologram which, when reconstructed, interfere as if they were independent wavefronts. This ability to do double exposure interferometry with holograms makes it possible to relax the stability criterion for interferometry to a time scale on the order of the exposure time. It is the ability to record wavefronts at different times combined with the fact that arbitrary wavefronts can be used in interferometry (Haines and Hildebrand, 1966a,b) that makes the holographic approach much more versatile than the classical arrangements. Even holograms of moving objects contain useful information about the motion, an observation which has revolutionized vibration studies (Powell and Stetson, 1965).

10.4.1.1 Diffuse and Other Uniquely Holographic Interferometers

Although the early holographic interferometers (for flow studies, at least) were conceived along the lines of their classical counterparts, it was soon realized that diffusing elements could be included in the optical system with no loss of interferometric information (Brown et al., 1969). This result follows directly from a modification of the basic hologram equations to include, upon reconstruction, the presence of a fourth wavefront propagating along the direction of the original object beam, and thus propagating collinear with the reconstructed "true image" beam. For an off-axis hologram, the expression for the amplitude of a wave reconstructed by a reconstructing beam of the form $c \exp(i\phi_c)$ is given by

$$V \exp i\phi_v = \gamma car \exp i(\phi_a - \phi_r + \phi_c), \tag{1}$$

where the proportionality constant γ depends on the hologram recording parameters, and the subscripts a and r refer to the original object and reference

465

waves. In Eq. (1) the amplitude terms c, a, and r are functions of the lateral coordinates of the hologram x and y, as are the phase terms ϕ_a, ϕ_r, and ϕ_c. When the original object beam is present, as is the case with real time holographic interferometry, an additional term of the form $a' \exp i(\phi_a + \Delta\phi[x,y])$ is added to Eq. (1) to represent the modified wave which differs in phase across the wavefront by an amount described by the term $\Delta\phi(x,y)$. Reconstruction with simultaneous inclusion of the modified object wave produces an intensity variation along the direction of the reconstructed beam equal to the square of the sum of the two waves. Following the same procedure used to derive the original hologram formulas, we find that the intensity of the pattern comprising the two waves is

$$I = (\gamma car)^2 + a'^2 + \gamma cara' [\exp i(\Delta\phi_a + \phi_r - \phi_c) \tag{2}$$
$$+ \exp - i(\Delta\phi_a + \phi_r - \phi_c)].$$

Equation (2), in turn, can be reduced to the form of a cosine distribution of intensity given by

$$I- = (\gamma car)^2 + a'^2 + 2(\gamma cara') \cos(\Delta\phi_a + \phi_r - \phi_c), \tag{3}$$

which is a cosine variation in intensity with optical phase difference $\Delta\phi_a$, whose fractional modulation depth is given by

$$F_{mod} = 2cara' /[(\gamma car)^2 + a'^2]. \tag{4}$$

Equation (3) shows that the intensity of the output of any real time holographic interferometer contains a term which is a function of the phase differences introduced in any beam of the hologram between the recording and reconstructing steps in addition to those variations in intensity which result from the intensities of the beams themselves. In fact, the distinction between object, reference, and reconstructing beams is irrelevant in holographic interferometry since a phase variation $[\Delta\phi(x,y)]$ in any one of them produces the same interference pattern when the holographic and object wavefronts are superimposed.

To the extent that the hologram is able to reconstruct one of the original wavefronts used to make it, it becomes a perfect interferometric element, since any variation in amplitude or phase across the wavefront aperture is recreated accurately in the reconstruction. A practical consideration is modulation depth, indicated in Eq. (4), since it determines how clearly the fringe location can be seen and measured. Modulation depth (and hence fringe contrast) is greatest when the amplitude of the reconstructed image (proportional to (γcar), is equal to that of the illuminated image which is transmitted through the hologram (a'). As a practical matter, fringe contrast in real time holographic interferometry is adjusted to a maximum by filtering the object illumination and reconstructing beams so that the reconstruction and object have the same brightness. Since most holograms have a diffraction efficiency less than 10%

(i.e., the quantity $\gamma a r$ in Eq. (1) is <0.1), this operation involves reducing the illumination of the object and increasing that of the reconstructing beam during the analysis. Clearly, good reconstruction efficiency is helpful in real time studies.

When double exposure or multiple hologram techniques are used, a similar analysis can be used to predict the intensity of the resulting holographic reconstruction. In these cases, the diffraction efficiency of both hologram images is the same, and as a result, the contrast of the fringes in double exposure (a double plate) interferograms is nearly equal to 100%. Since fringe contrast is independent of the hologram recording parameters and overall reconstruction efficiency in double exposure holography, this technique is often much easier to apply than real time interferometry.

Unlike classical interferometers, interference fringes are formed in holographic interferometers even though the wavefronts in the object beams may have very complicated spatial phase variation. As a result, holographic interferometry makes it possible to study, with interferometric precision, diffusely reflecting or transmitting objects which simply cannot be accommodated in classical interferometers. This versatility is only one of the unique advantages of the holographic approach.

10.4.1.2 Time as a Dimension in Holographic Interferometry

Classical interferometers are basically real time instruments since the various optical elements forming the instruments are fixed and the only recording takes place at the output of the experiment. Since the first observation of holographic interferometry in the time average mode (Powell and Stetson, 1965), several alternatives have developed to encompass double exposure, stroboscopic, and single or multiple pulse modes of operation (Aleksoff, 1971; Mayer, 1969; Aaidel' et al., 1969; Neumann et al., 1970; Vilkomerson, 1976). When a hologram is exposed, then processed in place, or replaced exactly after processing, it acts as a combination of beamsplitter and wavefront combiner, and since it directly compares two wavefronts in real time, it is closely analogous to a classical interferometer. What holographic interferometry offers is the fourth dimension of time.

Double exposures of a test scene on a single plate are reconstructed as independent wavefronts and thus a single hologram can operate as a complete interferometer upon reconstruction. Multiple exposures of a hologram have the same effect as double exposures, providing that they are synchronized with the changes under study. For example, stroboscobic holographic interferometry synchronized with the vibration period of a test object can provide contour fringes of peak displacement in a given mode of vibration when the strobe period and phase are chosen so that the exposures are effectively taken at the peak and null of the vibration cycle. Multiple exposures with varying phase have the effect equivalent to multiple beam interferometry in that various contributions are summed at various phase levels and the result is the squared average of their sum. In this instance, the intensity of the resulting fringe pattern is a function of the average phase change seen by the hologram during its exposure. When these phase changes are random and uncorrelated, no hologram results. Correlated phase changes such as those produced by sinusoidal or linear motion of the object during the exposure produce patterns which may be predicted (Vilkomerson, 1976; Lurie, 1968). Generally then, a hologram reconstruction is a function of, and may be used as a measure of, the temporal coherence of the light used to make the hologram.

10.4.1.3 Light Source Requirements

Because hologram reconstruction brightness depends strongly on the coherence of the light used to form the hologram, single mode, high stability lasers are preferred for interferometry, just as they are preferred for pictoral holography. However, when a coherent light source of sufficient power is not available to the experimenter, interferometry can be performed with a less ideal source, provided that path-length compensation and wavefront matching are accomplished in the optical apparatus (Brooks *et al.*, 1966). Since no laser is perfectly coherent, some matching must be done in any experiment; techniques to accomplish this matching will be discussed in the sections describing individual experiments.

Slowly varying phenomena or periodically vibrating phenomena are studied with continuous lasers. The most popular of these is the He–Ne laser which is available in power ranges from a fraction of a milliwatt to 100 mW. Where studies of larger objects dictate a higher power output, the argon ion laser can produce as much as several watts, single line, single mode. Operating multimode an argon laser can produce 10 W or more in the visible region of the spectrum. For repetitive phenomena a continuous laser can be shuttered in a variety of ways or a repetitively pulsed laser can be used. Argon lasers are available with pulse lengths on the order of 20 μsec and peak powers of 5 W operating at pulse repetition frequencies (PRFs) up to 20 kHz. For many experiments this time duration and energy is sufficient. Interferometry of large objects moving at high velocities must be done with pulsed ruby laser systems. A typical "holographic" ruby laser will have an output energy per pulse of 30 mJ and a pulse duration of 20 nsec. Amplifier stages can be used to increase the energy to several joules, however, large ruby laser systems are neither cheap nor simple.

In summary, the light source requirements for holographic interferometry are the same as those for holography, namely sufficient energy to illuminate

468

the object so that the hologram is exposed properly, and sufficient coherence so that the hologram can be formed. In reality the experimental object will set additional requirements on the light source power and time duration. Ultimately, considerations of cost may determine the degree to which an ideal light source can be approximated for any given task.

10.4.2   Interferometers for Transparent Media

Generally, the holographic arrangements for the study of transparent media for flow, heat transfer, etc., and for testing of optical components are similar to their classical analogs in arrangement. Reconstruction methods of the holograms can differ, depending upon the type of hologram made, whether it is diffuse or not. Interpretation of holographic interferograms is the same as that for classical interferometry, however, the relative simplicity of obtaining multiple images or data points can make the holographic technique much more powerful than the classical one in practice. Although much of the work in holography has been done in a laboratory environment where vibration, air currents, and optical stability can easily be controlled, by utilizing certain practical techniques, similar experiments can be performed successfully in the relatively hostile environment of the field.

10.4.2.1   Geometrical Arrangements and Considerations

The most common arrangements for transparent media are the holographic analogs of the single-pass Mach–Zehnder and two-pass Michelson interferometers (see Fig. 1). In these arrangements, the reference beam merely takes the place of one leg of the classical interferometer. Since the recording and wavefront comparison processes are done holographically, a great deal of versatility is possible with the arrangement of the optical components. The nature of the experiment will usually determine whether single or multiple pass is used. In the case of highly refracting media or strong turbulence, where a ray path deviates strongly from a straight line, a single pass arrangement is preferable. Interpretation of the fringes from a single pass interferometer is simpler than the interpretation when the rays pass twice through the medium, and in addition, nonplanar wavefronts can be used as the test beam once there is no requirement for the rays to retrace their path.

In some experiments, the physical arrangement of the test cell may make it impossible to bring the test beam back to the hologram without passing through the cell a second time. In this case the double-pass Michelson analog is indicated, since only a single mirror need be placed behind the test cell. This mirror can, with care, be mounted independently of the rest of the holographic apparatus. Double-pass interferometry doubles the sensitivity of the interferometer which may be of advantage when the phenomena produce small optical phase shift and thus small refractive effects. Meaningful interpretation of multiple-pass interferograms requires that the test rays retrace their paths while returning through the cell; to accomplish this the optical system must retrofocus; i.e., either a collimated test beam with a plane mirror behind the cell must be used, or a spherical test wave with a curved mirror whose curvature matches that of the wave at the back of the cell must be used.

Except when a collimated test beam is needed, lenses and focusing mirrors whose diameter equals that of the test cell are not usually necessary—an advantage in cost when the cell is large. Unless the hologram plate is the same size or larger than the test cell, lenses will be needed to reduce the test beam to the hologram diameter. Simple lenses can be used for this purpose and they need be of only sufficient optical quality to assure adequate image quality of the final interference pattern. The fact that optical components in holographic interferometers need not possess interferometric quality leads to considerable cost savings, particularly when large apertures are involved.

If the laser used to make the holograms were perfectly coherent, then no effort would have to be made to compensate optical path length differences between object and reference beams. Unfortunately, all lasers have a finite broadening of their line which can be translated into a coherence length L which represents (roughly) the maximum difference in optical path between the reference and test beams, which produces stationary, high contrast fringes. If a laser line frequency spectrum has a width Δf, then the coherence length L is related to Δf by (Collier *et al.*, 1971, p. 27)

$$L = c/\Delta f, \tag{5}$$

where c is the speed of light in the medium surrounding the experiment. For lasers, the maximum Δf over which oscillation is obtained is determined by the natural line width of the line (or lines) in the laser medium. For example, for an argon laser Δf is on the order of three gigahertz, yielding coherence lengths on the order of one centimeter of less when there is no mode selection to reduce the number of oscillating modes within the natural linewidth. Often various methods of mode selection can be used to reduce Δf and thus increase L. Often the laser power loss introduced by these methods cannot be tolerated and the low coherence length must be accepted. In such cases great care must be taken to assure that the total optical lengths of the reference and optical paths are identical. Even with highly coherent lasers, path length differences should be minimized. Path length compensation can be accomplished in the reference path by using small mirrors to fold the laser beam before it is diverged to fill the hologram aperture.

Symmetry of the optical elements is important since each portion of the wavefront will not possess the same phase as every other portion, unless a perfectly single mode laser is used. When multimode lasers are used, to achieve higher power it is important to match the reference and object wavefronts spatially at the hologram so that fringe contrast remains constant. Early pulsed laser holography depended upon accurate wavefront matching, and several arrangements were developed to assure that the waves arrived at the hologram plane with the same portions overlapping (Brooks *et al.*, 1966). When arranging the optics to accomplish path length compensation and wavefront matching, care must be taken to retain the same orientation of polarization of the two beams, since orthogonally polarized beams do not interfere to form a hologram.

After the laser beam is split into the components which will form the reference and test beams, relative motion between the beams may cause optical path length differences which show in the final interferogram as spurious fringes. Relative motion of the beams must be kept to a minimum either by extremely rigid mechanical construction of the optical elements or by monitoring the motion so that exposure is made only when such differences are small. Motion of the laser beam relative to the first beamsplitter does not affect the optical path difference between the object and reference beams, and thus rigidity between the laser and the first beamsplitter is not necessary (an important consideration when the laser is too large to be included in the same structure as the rest of the optics). When motion of the input beam relative to the interferometer must be tolerated, the relative positions of equivalent rays in the reference and object paths will change unless wavefront matching has been accomplished. Interferometer structures of certain symmetry automatically accomplish wavefront matching and thus nullify the effects of such relative motion (Mollenauer and Tomlinson, 1977). Whenever both beams of the interferometer undergo the same number of reflections in a given plane, the motion induced in the two paths of the structure due to motion of the input ray will be the same magnitude and direction. Such arrangements should be used in holographic experiments where input beam motion is present.

Since wavefronts with arbitrary spatial variation of phase can be used in holographic interferometry, a diffuser may be placed behind the test cell to illuminate it with rays from a large number of directions. In this case, the hologram contains optical path information for many passes through the cell, and reconstructions through different portions of the hologram can be used to obtain information about the three-dimensional nature of the disturbance within the cell. A typical example of one such hologram is the double exposure hologram of a lamp shown in Fig. 2. In this type of hologram, the extended diffusing source behind the study object extends the range of angles throughout the object which are recorded on the hologram by transmitting a number of rays through the test object from each point on the diffuser. Often with diffusely illuminated double exposure holograms, the fringes are located in the space near the diffuser where they may be easily photographed for further analysis. However, when strong refractive gradients are present in the test region, the fringe localization is highly dependent upon the nature of the refractive-index changes in the object and various portions of the fringe field

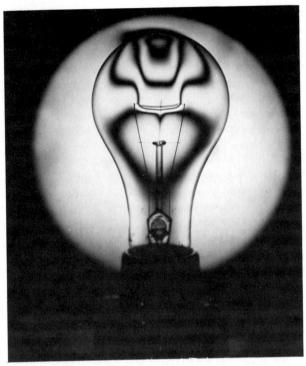

Fig. 2 Diffuse holographic interferogram of a 100-W incadescent lamp operating at partial power. One exposure was made with the lamp off and the other made with the power on. These interference fringes are caused by the heated gas within the envelope.

can appear in focus at different points in space (Vest, 1975). Nevertheless, the advantages of three-dimensional information about the optical path differences introduced by the object overcome any disadvantage presented by fringe location difficulties.

When proper attention is paid to the geometrical aspects of experimental design, successful interferometric studies can be performed in relatively hostile environments (Trolinger, 1975; Wuerker, 1976). The fullest utilization of the holographic technique usually involves a combination of diffuse and nondiffuse holograms.

10.4.2.2 Reconstruction Methods

In real time holographic interferometry there is no choice but to use the original reference beam for reconstruction; however, when double exposure holograms are used, there is a greater choice of reconstruction methods. In

472

interferometry as in other applications of holography, the perfect reconstruction takes place only when an exact replica of the reference beam is used. Diffuse holograms of transparent objects require monochromatic illumination for satisfactory reconstruction since the broad spatial frequency bandwidth occupied by the diffuse hologram causes image blur when extended polychromatic sources are used.

Reconstructing holograms of nondiffuse object beams, such as those which result in the holographic realizations of classical interferometers, can be accomplished with white light sources, however. In such holograms, ray angle differences are small and the information is contained in a form which occupies a relatively small bandwidth around the hologram carrier spatial frequency.

In this instance, as in the case of focused, image plane holograms (Brandt, 1969), spatial and temporal coherence of the reconstructing light source are unimportant since the fringe information is located in the plane of the hologram. Each portion of the hologram acts as a plane grating element which has a one-to-one correspondence with a portion of the object. Upon reconstruction, the contrast of the image is simply proportional to the hologram fringe contrast at that point. Since fringe contrast is a function of the optical path differences introduced between the exposures of the hologram, the interferometric information is contained on the hologram plane itself. Illumination of a double exposure image plane hologram by a white light source produces an image colored by diffraction wherever hologram fringe contrast is high and produces a dark image at those portions of low contrast. Thus reconstruction of image plane holograms is relatively simple to obtain with white light sources. Location of the interferometric information in the image plane prevents not only blur of the fringes due to diffraction but also image blur caused by finite source size. An example of one such reconstruction is shown in Fig. 3.

The increased versatility of using a white light source is not gained without some penalty, however. Upon reconstruction, a hologram will tend to produce an image at a distance Z_1 from the hologram plane given by the holographic formula

$$1/Z_1 = 1/Z_c + 1/Z_o - 1/Z_r, \tag{6}$$

where Z_c, Z_o, and Z_r represent the reconstructing, object, and reference beam distances from the hologram, respectively. This formula implies that with the general case of the reconstructing source on the same side of the hologram as the original reference beam, the object will originate as a virtual image behind the hologram. As illustrated in Fig. 4, this gives rise to a limited field of view of the hologram, since only a small angular portion of the hologram is effectively illuminated at any one time. This situation can be changed, however, by arranging the optics (Brandt et al., 1976) either during construction of the hologram or during its reconstruction so that a real image of the object is produced. Now placing an imaging system such as the eye or a camera lens

Fig. 3 Reconstruction from an image plane hologram made of flow in an aerodynamic cascade. Flow is from top to bottom. Discontinuities in the fringe pattern in the downstream side are caused by shock waves. The dark lines running through the blades are reference marks. This was reconstructed in white light, so the background between fringes was colored.

at the location of the real image point allows for reconstruction over the whole field of the hologram. As the viewing optical system is moved in a plane perpendicular to the hologram fringes, the color of the reconstruction changes due to diffraction of the hologram grating. Field of view, on the other hand, does not change as long as the viewing system lies within the reconstructed cone of light from the hologram.

When a laser is used to reconstruct holograms of transparent media (Tanner, 1966) in the manner shown in Fig. 4, additional information can be extracted from the hologram. Schlieren information about refractive-index gradients can

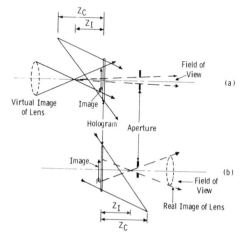

Fig. 4 Reconstruction geometry for image plane holograms of transparent objects. The image is visible through a cone of light whose angular extent is determined by the parameters used to make and reconstruct the hologram.

be obtained by placing a knife edge or other stop in the position of the focal spot of the reconstructed object beam. Imaging of those rays which pass by the knife edge gives information about refractive-index gradients recorded on the hologram.

Shadowgraph information about the refractive-index second spatial derivatives can also be retrieved from holograms of this type by direct photography of the reconstructed beam as it comes through the hologram. Regions of very strong refractive gradient, such as areas around shock waves, will appear dark on the hologram reconstruction because the light rays passing through that area have been refracted out by the strong refractive curvature. This phenomenon is useful for determining the location of such regions, but within them interferometry is impossible since no rays pass directly through them.

10.4.2.3 Data Reduction

Whether real time or double exposure mode of operation is used, the basic data which interferometry of transparent objects provides is the integral along a ray path from the source to the observing plane of the optical path difference (OPD) between the two hologram exposures. In the case of real time operation, the integral represents differences between the sample condition at the time the hologram was made and the present time. This expression is

$$\text{OPD} = \frac{2\pi}{\lambda} \left[\int_s^0 n(p)\, dp \bigg|_{t_1} - \int_s^0 n(p)\, dp \bigg|_{t_2} \right], \qquad (7)$$

475

where p indicates the coordinate along the path, $n(p)$ the refractive index along that path, and t_1 and t_2 the times of the exposures. Basically the data reduction problem involves inverting Eq. (7) to obtain the values of refractive index $n(p)$ along the path. In general this inversion cannot be accomplished except in certain cases of reduced symmetry (Abel inversion problem). Fortunately, many problems of technical interest can be reduced to experimental arrangements which have either cylindrical or planar symmetry for which the Abel inversion problem becomes relatively simple. For example, in the case of flow measurements in cascades, turbine blade sections of uniform thickness are constrained between parallel windows. Assuming that the influence of the flow at the window boundary is negligible, the inversion of Eq. (7) is trivial since these assumptions imply that the OPD at any point is simply equal to the product of the refractive index at that point and the distance between the windows. Since each OPD interval of one optical wavelength produces a change in the fringe from dark to light and vice versa, the interferogram represents a contour map of the flow refractive-index field.

From the refractive-index field, information about temperature, pressure, or polarizability of the fluid can be obtained by using physical relationships appropriate to the experiment at hand. Discussion of these relationships is beyond the scope of this section.

10.4.2.4 Removing Ambiguities in Fringe Interpretation

As in classical interferometry, the use of finite fringe interferograms is useful in holographic interferometry for removing the ambiguity which results from the inability to distinguish between positive and negative phase shifts. For interferograms of steady state phenomena this ambiguity can be removed by tilting one wavefront a known amount in a known direction between the exposures of a double exposure hologram. This will produce a background of straight fringes which can be used to indicate the sense of optical path variations in the sample. Positive and negative deviations of OPD will produce displacements in the fringes in opposite directions; an absolute determination of increase or decrease in optical phase can be made if the sense of the background pattern is known. This can be determined by knowledge of the actual angle of tilt between exposures, or more frequently, from knowledge of the phenomenon under study. A change θ in the angle of the object beam between two exposures of the interferogram produces fringes whose spacing D is given by

$$D = n\lambda/(2 \sin \theta/2) \qquad (8)$$

along a direction perpendicular to apex of the tilt. This background pattern is added to that created by a refractive-index change Δn which for cell thickness *t* gives rise to an OPD of

$$OPD_{\Delta n} = 2 \, \Delta n \, t/\lambda. \tag{9}$$

Equation (9) can be considered a perturbation term on the pattern given by Eq. (8). Thus from deviations of the straight fringes of (8) the relative sense of OPD variations from (9) can be established. In many experiments, the lower refractive-index portions of the interferogram are obvious since the positions of lower pressure or temperature are set by the nature of the experiment. In most nonlaboratory experiments, particularly those using pulsed laser holography, this method of interpretation is likely to be more practical than adding background fringes would be.

10.4.2.5   Fringe Position Measurement

Although a great deal of qualitative information can be gained by visual inspection of interferograms, quantitative utilization of the information requires measurement of fringe position and determination of fringe order over a significant area of the interferogram. Fringe order determination is best accomplished by human interpretation utilizing knowledge of the phenomenon under study. Coordinates of each fringe can be measured manually with rulers or optical comparators, but such a process becomes very tedious to an operator without some additional help in mechanizing the process. One solution to the problem of obtaining quantative data from interferograms involves partial mechanization of the measurement process in which the advantages of the human operator in ability to interpret fringe order and position are combined with the ability of a machine to measure and record coordinates accurately. One realization of such a device is shown in Fig. 5 in which an optical system is used to image the hologram reconstruction on a television monitor. A precision *xy* table driven by stepping motors and appropriate counting electronics moves the hologram so that a cross hair is optically superimposed on the fringe whose position is to be measured. The operator moves the table with an *xy* control to the fringe location and indicates the fringe order at the time of reading by pressing a signal button. This process activates a minicomputer which then records the position and fringe order for that data point. In this way data can be acquired rapidly for a large number of fringe points. Automatic realization of a fringe interpretation system for specific applications is of course possible (Erf, 1974). However, the complexity of a system for general interpretation of interferograms is so great that a more economical solution lies in the use of automation of the measurement process while leaving the interpretation to a human operator.

10. Application Areas

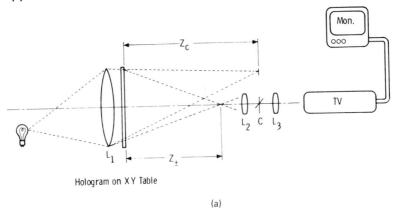

Hologram on XY Table

(a)

Fig. 5 (a) Schematic of a semiautomatic device for data reduction of fringes from holograms. (b) A photo of the completed device.

10.4.2.6   Practical Considerations for Field Operation

Although many interesting and useful experiments can be done with holographic interferometry in the laboratory environment, study of flow, heat transfer, shock and vibration of actual components requires that experiments be done in less than ideal circumstances. The problems presented to the holographer concerning ambient light (which would expose the hologram), dirt (which ruins optical components), and relative stability of the optical components forming the holographic apparatus are solved by proper design of the apparatus, which once constructed may be glorified with the title "holocamera." Mechanical shutters and filters provide adequate protection from stray light, and sound mechanical design assures rigidity of the optics. Generally it is good practice to contain reference and object beams within opaque, air tight tubes or bellows enclosures wherever possible both to reduce the effects of turbulence on the interferograms and to provide a degree of safety protection from inadvertent exposure of personnel to the laser beam.

Apart from the fact that the field operation takes place in a noisy, vibration prone environment, the phenomena under study are usually of short duration whether single shot or repetitive in nature. This need to study short duration events dictates the use of a pulsed laser, of which the Q-switched ruby laser is the most highly developed and most commonly used (Brooks et al., 1966). Pulse duration is nominally 20 nsec, short enough to stop the action of subsonic phenomena. For transparent objects, the output from a single mode oscillator in the range 10–50 mJ is adequate to expose even the slower holographic plates since very little light is lost in transmission optical arrangements, even when diffusers are used. Double exposure interferograms may be made by exposing the test chamber in its quiescent state, then in the test state, using the single pulse mode of the laser.

Rapidly changing phenomena, such as those found in transonic turbulent flow in turbines, are more easily studied using the double pulse mode of the laser. Pulse intervals in this mode are set by the electronics in the laser control system and may range from 40 nsec to several seconds between pulses depending upon the laser type. When multiple pulses are necessary at time intervals close to that of the pulse length, multiple lasers must be used. With either single or double pulsed operation of a laser some means for triggering the laser in coincidence with the phenomenon must be provided. In the case of rotating machinery, a magnetic inductive or optical pickup attached to the rotating parts can provide pulses to drive selective circuits which will fire the laser at the desired time. Where electrical signals induce the test, as for example a spark induced shock wave, timing can be performed entirely electronically as part of the experiment. When pulse intervals on the order of a few or tens of nanoseconds are needed from multiple lasers, physical optical path differences can be introduced between the lasers and the test cell to affect fixed delays between pulses. Since light travels 1 ft in 1 nsec in air, delays up to a few hundred nanoseconds are possible in large rooms. When different paths are used in the experiment, arrangement for compensation of the reference beam has to be made so that the reference and object beam pulses overlap in time when they reach the hologram plane.

Even when quasi-static phenomena are studied, such as cascade flow tests or heat transfer studies, synchronization of the laser may be necessary to assure that the holocamera is aligned with the test cell. For example, when the single-ended holographic equivalent of a Michelson interferometer was used in cascade tests (Brandt *et al.*, 1976), the mirror which was placed behind the test cell was often mounted on a support separate from the rest of the optics forming the camera. Any relative motion between the mirror and optics will cause a background pattern to be formed on the hologram. In one experiment (Brandt *et al.*, 1976) in order to eliminate or control this pattern, it was necessary to monitor the relative orientation of the mirror relative to the rest of the optics. A low power He–Ne laser beam transmitted through the test cell to the mirror where it was returned to a pinhole, photodiode arrangement on the optical table containing the rest of the optics. A simple electronic coincidence circuit guaranteed that the laser could only be fired when the probe laser beam returned through the pinhole. Thus the mirror alignment relative to the optics was guaranteed for the two exposures.

The best optical quality of lens and window components is not required for the successful utilization of holograpic interfereometry, since it is basically a differential technique for comparing arbitrary wavefronts. Nonetheless, in experiments in which there is considerable lateral motion or thermal distortion of the window material between exposures, high-quality window material may be required to avoid false interference patterns caused by the optical inhomogeneities of the window. When schlieren or shadowgraph information is needed, stria and other imperfections in the optics will be reproduced in the interferogram along with the refractive gradients associated with the test phenomenon. Finally, if the optics are not of sufficiently good optical quality that the test cell can be imaged clearly, then obviously, the patterns which are produced by the phenomenon cannot be associated with a specific location in the cell.

10.4.2.7 Summary of Applications to Transparent Objects

In those applications where field operation of holographic systems are needed for visualizing transparent media, satisfactory results can be obtained, provided that careful consideration is given to the components of the experiment. Once optical access to the test region is achieved by using appropriate windows and mirrors, proper design of the mounting and enclosure of the optical elements ensures stability and immunity to stray light and physical contamination. Even with a well-designed arrangement for the optical components, careful consideration has to be given to the timing and synchronization of the exposures with the phenomenon. By taking these elements into consideration it is possible to perform interferometric tests under extremely hostile conditions, ranging from steam turbine cascade tests to jet engine and rocket combustion studies.

10.4.3   Interferometry of Three-Dimensional Diffuse Objects

As we have seen, holographic interferometry is very versatile and useful for the study of transparent media since it extends the capabilities of classical interferometry. However, in the study of three-dimensional diffuse objects, holographic interferometry has had a truly revolutionary effect because it enables measurements to be made which simply are not possible by classical means. Not only are measurements possible on surfaces whose roughness would make them totally unsuitable for conventional optical study, but the technique is applicable to a surface whose depth prevents precision measurement due to focusing depth of field limitations of ordinary optics. Holographic interferometry captures the time behavior of the process as well, adding a unique dimension to vibration and strain studies (Sampson, 1970; Sollid, 1975; Robertson, 1976; Erf, 1974). Fortunately, the techniques for utilizing this approach are no more difficult than those for conventional holography. Principal differences for holographic interferometry experiments involve sample excitation and light source timing. Interpretation of interferograms, like those for transparent media, can be useful either qualitatively or quantitatively. In the latter instance, advanced data reduction methods are necessary for optimum results.

10.4.3.1   Geometrical Arrangements for Three-Dimensional Objects

Holographic interferometry does not require any different geometrical arrangements of optical elements than those used in pictorial holography. Primary considerations are the ability to make high quality holograms of the object in its quiescent state and its typical state of static or vibrational strain. A typical interferometer experiment will consist of, as in the case for any hologram, a means for splitting the laser beam into a portion which illuminates the object and a portion which acts as a reference. Generally, since the objects are not highly reflecting, the fact that most of the object light does not reach the hologram due to absorption and scattering losses at the object dictates that the majority of light be sent to the object (as much as 90%) so that as little

481

light as possible is lost by attenuating the reference beam to an appropriate level.

Since, in addition to light lost by scattering, the experimenter is usually trying to study as large an object as possible, consistent with the light available from the laser, those arrangements which return a maximum of light from the object are preferred. Thus one tries to place the object as close to the hologram plane as possible to minimize $1/R^2$ intensity loss, and one tries to arrange the illumination beam so that little is wasted by illuminating anything but the object under study. Uniformity of illumination is important too, particularly for time-averaged experiments, since fringe contrast decreases as the vibration amplitude increases. Visibility is improved if those portions of the object with highest amplitude of motion are given more illumination than the stationary portions. It is difficult to illuminate a large area uniformly with the gaussian intensity distribution from most lasers. The drop-off in intensity toward the edges of a gaussian beam can be compensated for in part by using a lens with a large amount of spherical aberration following the pinhole-spaced filter in the object beam (Ennos, 1977). A short focal length condenser lens used with its highest power surface towards the pinhole is quite effective in smoothing gaussian illumination for this purpose.

Conservation of the light once it strikes the object is important too. To this end, it is beneficial to prepare the surface of the object to ensure that a maximum amount of the incident light is returned to the hologram. Painting the surface is very helpful in this respect when possible. One very successful treatment involves a layer of metallic aluminum paint covered with a second layer of flat white paint. This combination appears to possess a good balance between specular and diffuse reflectance and does not appear to depolarize the laser light as much as does a purely diffuse surface. It is important to keep in mind that depolarization of the object beam is as important a consideration as reflectivity, since the hologram records only that component of the object beam which has the same polarization as the reference beam.

Polarization effects must also be considered when measuring beam ratios with a photometer. That component of the object beam whose polarization is orthogonal to that of the reference beam will contribute only to the background exposure of the hologram and not to its image reconstruction. In experiments where the object is highly diffusing a check of the amount of depolarization introduced by the object is a worthwhile precaution. When the amount of depolarization is large, the reference to object beam intensity ratio should be increased to compensate.

Interferometry of three-dimensional objects can be performed using the image plane technique in order to gain the advantages of white light recon-struction. Unless the object is a relatively flat one, good white light reconstruc-tion will take place only over those portions of the image near the focal plane.

When the image on the hologram is demagnified, the light gathering ability of the lens is useful in decreasing exposure time.

It is even more important to check for equal reference and object beam path lengths for three-dimensional objects than it is for transparent ones since the depth of the object can easily place some of it out of the coherence length of the source. Since the objects have depth, compensation can only be a compromise, and thus that point on the object which is compensated exactly should be chosen to maximize the overall contrast of the pattern. For example, in time averaged interferometry the fringe contrast is lowest in those regions of the hologram in which the vibration amplitude is highest. Whenever possible, the experiment should be arranged so that those portions have zero path length error relative to the reference beam.

Sometimes it is difficult to provide for a long enough reference path to compensate for the distance of the object from the hologram. In this case and in the case in which the object cannot be held fixed relative to the hologram optics, the reference beam can be derived from either a portion of the object itself, or more simply, from a mirror fastened firmly to the object. This arrangement has the advantages that path length compensation is accomplished automatically and that phase disturbances in the atmosphere between the object and hologram are experienced nearly equally by the two beams and thus do not have much effect on the reconstruction. When interpreting the interference patterns from such holograms, the fact that the reference beam originates from a nonstationary portion of the object must be kept in mind.

Although any optical system which is usable for pictorial holography can be used for interferometry, in setting up the experiment care must be given to produce uniform illumination and exact path length compensation, particularly in those portions of the object where fringe contrast will be lowered. Quantitative interpretation of the fringe pattern may also dictate the geometrical arrangement of the illumination source, reference source, and object locations relative to the hologram, since each interference fringe represents changes in total optical path. It may be easier to arrange the taking geometry of the hologram so that complex corrections need not be applied to the final data.

10.4.3.2   Test Object Preparation

Apart from the surface preparation described earlier to assure adequate light return to the hologram, the most important sample preparation step involves mounting of the object, both so that it does not influence other optical components during the test and so that its excitation is realistic in amplitude and is what is expected in direction. Care given to the mechanical design of the mechanical mounting system can mean the difference between success and failure of the experiment, since unwanted translations or tilts of the object during the test can make interpretation of the final interference patterns impossible. For example, when vibration tests are made of the fundamental and lower frequency modes of cantilevered structures such as turbine blades, the mode patterns and frequencies depend strongly on the rigidity of the base mounting structure. In order to obtain realistic data on these low frequency modes, the blade roots must be sealed in their mounting block, which in turn is welded to a massive table. For smaller structures, less extreme methods will suffice, however, it is always better to err on the side of making the mounting arrangements more rigid and stronger than intuition would suggest is necessary.

Before embarking on a series of tests, some preliminary checking of the system using real time holography is very helpful. With a hologram of the system which has been replaced in its holder, the rigidity of the mountings for all of the components can be verified by successively tapping or preloading components and determining whether this process causes any permanent change in the fringe pattern. At the same time, any hysteresis in the exciting mechanism can be checked and any unexpected thermal or mechanical strains introduced by the exciter can be discovered. Once the experiment had been checked with a real time hologram, tests can continue with confidence that no spurious fringes are generated by the apparatus.

10.4.3.3 Exciting the Test Object

When interferometry is being used to study an object under its actual operating conditions, no particular problem arises in excitation. Such a direct measurement has the advantage that the interference patterns generated represent directly the static or vibratory strains experienced by the object under what one hopes are typical conditions. The disadvantage of such a study is that interpretation of the strain patterns is difficult since rarely will actual operation result in a "pure" mode of vibratory or static motion. As a result, laboratory tests in which known exciting forces are applied to the test sample are often useful in characterizing the mechanical behavior of real structures or their models. When holographic interferometry is used for flaw detection in composite structures, the mode of excitation must be designed so that the flawed areas are excited relatively more than the sound areas. In all cases, thought must be given to the mode of excitation in order to ensure the correct amplitude of excitation as well as the appropriateness of its mode.

Static strain experiments with holographic interfereometry are difficult because small known deflections must be applied in order to keep the number of fringes in the interferogram to a manageable number. Mechanical devices such as micrometers exhibit backlash and hysteresis effects which are of the same order of magnitude as the measured strain. Contact points have a tendency to wander so it is rare that a micrometer or other screw arrangement applies its force accurately along the axis desired. Excitation methods which do not involve moving joints or bearings are much preferred. One effective means of providing static force to an object is to utilize thermal expansion induced by localized heating of a portion of the support to provide the motion. Thermal excitation may be applied directly to the object by means of lamps, resistance wire, or flames. If the electrical characteristics of the object are suitable, it may be heated by passing a current through it, its own resistance providing the source of self-heating. This can be useful in flaw detection where void or delaminated regions would be selectively heated. The principal disadvantage of thermal stressing is its lack of spatial selectivity. Its principal advantages lie in simplicity and wide applicability.

Hydraulic forces are useful in cases where the object contains closed cavities which can readily be pressurized or evacuated. With most forms of static test, some preloading will be necessary to set the object in its fixturing. Once hysteresis is removed from the system, relatively large strains can be studied by successive applications of small strain moments while recording each increment with double exposure holography.

Excitation of vibration in structures is easier than excitation of static strain, since a variety of noncontacting methods are available. Airborne acoustic excitation is the simplest form and often appears as an unwanted exciting mechanism of poorly designed holographic experiments. When single frequency excitation is desired, as in mode structure studies, a loudspeaker can be used as an exciter. Public address system speakers are preferred since single frequency excitation by an oscillator places power demands on the speaker which often cannot be tolerated by home-type speaker systems. The advantage of acoustic excitation is that it is nonselective regarding the position of drive on the object. This is an advantage in exciting complex vibration modes when the position of the best driving points (antinodes) is not apparent. It is a disadvantage when the rest of the holographic apparatus is excited at the same time as the object.

Magnetic drivers are useful for driving at audio frequencies from about 50 to 20,000 Hz. One convenient type of driver is the electromagnetic pickup used in machinery to measure RPM by counting pulses generated by gear teeth passing the pickup (see Fig. 6). These units consist of a magnetic rod typically about $\frac{1}{8}$ in. in diameter around which a coil is wrapped; typically they present an electrical impedance on the order of 10 to 100 Ω so they are easily matched to an audio power amplifier. Although they are intended as inductive proximity pickups, they work well as inductive noncontacting drivers for ferromagnetic objects. Larger units will handle several watts of power without burning out, a level of excitation which is adequate at audio frequencies for many structures. To use these pickups as drivers they are mounted in close proximity to the magnetic surface, typically 0.010 in. from the surface. Drive effectiveness is increased with smaller air gaps, but care must be taken to avoid the situation

Fig. 6 Some typical magnetic drivers used in vibration studies.

in which the permanently magnetized core of the driver pulls the object into solid contact. Nonmagnetic objects can also be driven magnetically by bonding a ferromagnetic strip to them. When a single driver does not provide sufficient amplitude, several units can be driven in parallel from the same amplifier. Drivers can be placed at different locations on the structure to emphasize different modes. In general, driver placement is governed by the consideration that a given mode of vibration is most easily excited at its antinode with the largest amplitude of vibration. With complicated mode shapes, real-time holography may have to be used to determine the most suitable position for the driver. By using separate drivers connected to separate sources, multiple modes may be excited simultaneously.

At low frequencies the magnetic pickups lose effectiveness due to the reduction of the magnetic forces, and at high frequencies, inductance increases the driver impedance; however, in the range of audio frequencies, the electromagnetic pickup is a convenient driver whose presence does not influence the vibration characteristics of the structure under test. Magnetic shakers and shaker-tables, which are used in conventional vibration testing, produce displacements which are much larger than those needed for holographic interferometry. In addition, they require mechanical contact to the object, thus measurements represent the characteristics of the object-shaker system rather than the object itself. Nonetheless, such drivers may have utility in certain tests. Where a broad spectrum of excitation frequencies is needed, as in flaw detection, impulse excitation is often used. The ''driver'' in this case is a hammer or projectile.

In all of these types of experiments, real time mode of holographic interferometry is useful in verifying the operation of the apparatus before a time averaged or double exposure hologram is made. Proper level of excitation can be verified as well as proper placement of the driver. Undesirable motion of the rest of the holographic apparatus can be monitored at the same time. For vibration studies, a particularly useful experimental procedure combines acoustic driving with real time holographic interferometry to scan the mode spectrum. The nonspecific nature of the acoustic drive guarantees that no vibration mode will be missed due to misplacement of the driver during the frequency scan. With stroboscopic vibration holography, real time monitoring is a necessity for setting the phase of the strobe relative to the vibration cycle. A real time hologram, where it can be used, should always precede more elaborate testing; even though the real time hologram may not be a perfect zero-fringe match, much about the vibration tests can be learned by its use.

10.4.3.4   Time Modulation Methods

Real time holographic interferometry using continuous, stable lasers represents one extreme of the time scale available to the experimenter while pulsed laser holography represents the other extreme. While it is obvious that pairs of holograms made with short enough pulses to stop the object motion can be compared interferometrically in just the same way as double exposure holograms of statically deformed objects, it is not obvious what type of information is obtained from a hologram made with a continuous source of an object which is moving during the exposure.

Motion of the object during the hologram exposure produces interference fringes on the hologram whose nature is determined by the functional form of the motion (Powell and Stetson, 1965; Lurie, 1968; Lurie and Zambuto, 1968). For sinusoidal motion the analytical relationship between peak vibration amplitude and fringe visibility was worked out by Powell and Stetson (1965) whose analysis and verifying experiments originated the field of vibration studies with holography. A later paper which discusses this along with other aspects of hologram interferometry may be useful (Brown et al., 1969). When an object is vibrating sinusoidally with amplitude a, it gives rise to an excursion in optical phase, $\phi = 4\pi a/\lambda$, when the illumination and viewing directions are parallel to the direction of motion. In general the geometrical paths must be

10. Application Areas

considered, so for a ray incident on the surface at angle θ_i and scattering toward the hologram at·an angle θ_0, the phase function ϕ is

$$\phi = (2\pi a/\lambda)(\cos \theta_i + \cos \theta_0). \tag{10}$$

In this case, the hologram reconstruction is crossed by fringes whose intensity I is of the form

$$I = I_0 J_0^2(\phi), \tag{11}$$

where $J_0(\phi)$ is the zero-order Bessel function of argument ϕ. Since the value of $J_0(0)$ is equal to 1, the brightest fringe on the hologram occurs for those portions of the object which are stationary, i.e., the nodal lines of vibration. For larger amplitudes, the Bessel function oscillates periodically with decreasing peak amplitude. Each fringe maximum and minimum corresponds to an equivalent oscillation in the Bessel function. For large ϕ, these oscillations represent nearly uniform increments of vibration amplitude. Time averaged holography is particularly useful in locating nodal lines but may be used for quantitative strain measurements as well. One example of a time averaged hologram is shown in Fig. 7, a hologram of a high order vibration mode of a large turbine blade group. In this photograph the nodal lines quite clearly show as bright fringes. Making time averaged holograms is no more difficult than making a single exposure hologram, and the information obtained is both useful in itself and complementary to that obtained with other methods.

Time averaged holography, the term customarily applied to the process of making a hologram of a moving object under continuous illumination, is really only a special case of "temporally modulated holography," a subject which has been treated in depth by Aleksoff (1971). Equation (11) describes the intensity modulation of the hologram reconstruction when the object beam is sinusoidally phase modulated by scattering from a vibrating object and the reference beam is unmodulated. Phase modulation of a light beam [or any other sinusoidal wave for that matter (ITT, 1968)] at a frequency ω produces *temporal* sidebands on the beam at frequencies of ω, 2ω, 3ω, . . . whose amplitudes are proportional to the Bessel functions of order 1, 2, 3, . . . , n, namely, $J_1(\phi)$, $J_2(\phi)$, $J_3(\phi)$, ... (Argument ϕ is related to the peak phase shift introduced.) Meanwhile, since the light in these frequency shifted sidebands must be taken from the incident, unshifted light beam, the zero order or nonfrequency shifted beam must also vary in amplitude with phase of modulation, and in fact, its amplitude varies as the zero-order Bessel function $J_0(\phi)$. Thus the object beam scattered from a sinusoidally vibrating object contains a portion of the beam which is unshifted in frequency relative to the original illuminating beam and other portions of the beam at multiples of the phase modulating frequency, whose relative amplitudes are a function of ϕ which in turn is related to the amplitude of displacement of the vibration through Eq. (10).

Fig. 7 Time averaged hologram of a five-blade 31-in.-long turbine blade group oscillating at a high order mode at 653 Hz. This type of hologram is particularly useful for illustrating nodal lines.

A hologram acts as a heterodyne receiver tuned to the temporal frequency of the reference beam, since only those components of the light beam which are at the reference beam frequency remain stationary in the hologram. Since those components at ω, 2ω, 3ω, . . . form fringes which move across the hologram plane due to the frequency shift, those components are "washed out" during the exposure process, leaving only those fringes which are formed by the unshifted object beam. Since the amplitude of this wave is proportional to $J_0(\phi)$, the intensity of the hologram reconstruction for those portions of the hologram with phase modulation is proportional to $J_0^2(\phi)$, thus leading us back to Eq. (11). Temporally modulated holography extends the heterodyne received to the detection of higher frequency sidebands in the object beam. The principle of operation of temporally modulated holography remains the same

489

as that for time averaged holography; those components of the object beam whose frequencies exactly match that of the reference beam will be recorded and other components will not.

Phase modulation of the reference beam offers the opportunity to retrieve information which is contained in the frequency-shifted portion of the object beam. When the reference beam used to make a time averaged hologram has been shifted by the vibration frequency of the object, the maximum brightness of the reconstruction will appear at the maximum of the first-order Bessel function $J_1(\phi)$ rather than at the maximum of $J_0(\phi)$, as is the case for the time averaged hologram which records the zero order. The effect of such a shift is to move the position of maximum brightness on the hologram from a position of a true nodal line of zero motion to one corresponding to that motion which produces a phase shift needed to maximize the first Bessel function. When the reference beam is frequency shifted to match the second harmonic of the modulation frequency, then the hologram reconstruction brightness will follow the functional form of the second-order Bessel function. While a direct frequency shift of the reference beam relative to the frequency of the object beam is appealing in theory, in practice it is nearly impossible to accomplish; thus a more realistic form of modulation of the reference beam involves sinusoidal phase modulation at the same frequency as the object excitation. Analysis in the general case where reference and object phases are not identical is complicated, nevertheless, the cases where the reference beam is either exactly in phase with the object motion, or is 180° out of phase with it, provide very useful information, since in the first case, the fringe order which corresponds to that equivalent of the reference beam is enhanced, and in the second case, the same fringe is reduced in intensity by destructive interference. Thus, to first order, phase modulation of the reference beam can be used to study the relative phases of nodes and antinodes of vibration. In addition, by shifting the amount of phase modulation along with the reference beam amplitude, qualitative information about the phase of the motion of various portions of the object can be obtained, even if quantitive information is difficult to obtain (Aleksoff, 1971). Modulation of the reference beam can increase the range of effective utilization of time averaged holography by providing high degrees of phase shift which interact with those components associated with large amplitude. Time modulation is effective in increasing the sensitivity at the subfringe level, too, since if the reference beam modulation and phase are adjusted properly, the overall motion can be cancelled, leaving the smaller variations to show as increases in brightness which can easily be detected against the dark background. In this manner a small fraction of a wavelength of motion can be detected (Vilkomerson, 1976). In a certain sense then, time modulation of the reference beam is of very general utility. Its application ranges from time averaged holography in which only one beam is modulated by motion of the object, to those in which the phase or frequency of the reference beam is modulated to enhance certain portions of the motion. In the following section we see that the application of temporal modulation extends beyond phase modulation of the light used to form the hologram to the arena of amplitude modulation. Pulsed modulation of the exposure of the hologram can convert the information from a relatively long exposure to that which is equivalent to two or more relatively short ones.

10.4.3.5 Stroboscopic Techniques

Stroboscopic holography has been treated as a special case of time variation of the illumination function (Aleksoff, 1976), but in its most simply used realization it is, practically speaking, equivalent to double exposure holography. Usually, stroboscopic techniques are applied to holography in the same manner as in normal stroboscopy, namely the illumination is pulsed at the frequency of the vibration of the object with a pulse length short enough that the motion of the object is stopped during the pulse. Stroboscopic techniques are very useful in real time vibration holography studies to determine the relative phases of motion in complex modes of a structure. When the strobe frequency is slightly different from the vibration frequency in a real time interferometer, the interference fringes move in and out towards their maximum position at a frequency equal to the difference frequency. Real time stroboscopy is useful in a survey or catalog of modes but usually some permanent record is desired, and in this case, double exposure stroboscopic holograms are a very useful addition to a time averaged study.

Double exposure strobed holograms are made by adjusting the strobe frequency so that it is identical to that of the periodic motion under study and then adjusting the phase of the strobe so that the motion is stopped at the desired point. Since the strobed pulse interval cannot be made arbitrarily short without increasing the hologram exposure time excessively, the best exposures result when the pulse length is moderate (such as 10–15% of the interpulse interval) and the phase is chosen so that the pulses are synchronized with the maximum amplitude of the vibration. An example of a double exposure strobed hologram appears in Fig. 8 which shows the same mode as the time averaged hologram in Fig. 7. Here, since only the peak of the motion is compared with the stationary state, each interference fringe corresponds directly to an equal increment of vibration amplitude and each fringe order has equivalent contrast and brightness. In mode studies where quantitative measurements of the motion are derived, this type of hologram is the most convenient to use since no correction need be made for the Bessel Function relationship observed in time averaged holograms. A typical strobed holography experiment is shown in Fig. 9.

Stroboscopic modulation of the laser beam can be achieved with mechanical chopping for frequencies up to 10 kHz or so. By focusing the beam to a small

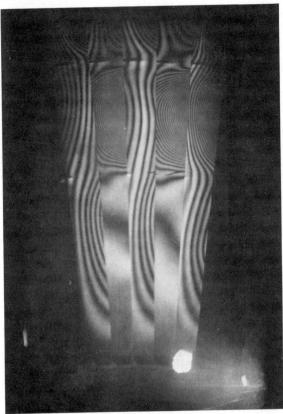

Fig. 8 Double exposure strobed hologram of the 31-in. blade group oscillating in the same mode as Fig. 7. Here the contour intervals and contrast of this hologram were made with the blades stationary and the other was made with the strobe adjusted so that the laser was shuttered on at the peak of the vibration cycle.

pinhole such frequencies can be reached with moderate sized chopping wheels at reasonable wheel velocities. Such choppers are difficult to regulate in speed and, more particularly, in phase relative to the driving signal. Electronic modulation means are more convenient. Electrooptical modulators can be used, but they require high drive voltages and a secondary polarizer. Losses of light are usually high with such modulators. Acoustooptic modulators utilizing Bragg diffraction of the laser light have become quite popular and relatively inexpensive. They may be used in the straight through mode in which light is removed from the beam, or they may be used to send light into a single Bragg diffraction order at an angle to the straight through beam. The first mode suffers from the limitation that contrast ratio between on and off

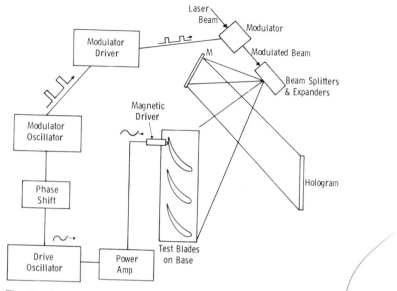

Fig. 9 Schematic of a stroboscopic holography experiment.

states is limited to perhaps 10 to 1, but it has the advantage that light losses and beam shifting are minimal. Bragg mode offers good contrast but requires realignment of the optics between the on and off states. The principal disadvantage of the straight through scheme can be alleviated by using more than one pass through the cell with spatial filtering introduced in between passes (Brandt *et al.* 1977).

An interesting extension of the stroboscopic technique involves time modulation of the interpulse interval to enhance certain fringe orders by selective sharpening (Mottier, 1970). In this method, multiple pulses are applied to the modulator during each cycle of the vibration motion. These pulses, which are generated electronically, are timed so that constructive interference from each pulse gives rise to bright fringes on the hologram reconstruction at fringe intervals equal to the number of stroboscopic pulses per vibration cycle. With this method interferometric studies can be extended accurately to higher amplitudes than is possible with unmodulated techniques.

10.4.3.6 Multiply Exposed Holograms

Double or multiple exposed holograms using Q-switched lasers are often as useful for front lighted diffuse objects as they are for flow studies. Here the principal problem is appropriate synchronization. When the phenomenon occurs as a direct result of an electrical trigger, signals are directly available.

Choice of the interval between pulses will depend on the expected motion during the pulse interval. Interpretation in pulsed holograms may be more complicated than that in strobed holograms since there is no particular knowledge of the degree to which the two exposures will repeat with the time interval used. Nevertheless, the double exposure pulsed method is very useful in studies of random and pseudorandom motion.

10.4.3.7 Data Acquisition from Diffuse Holograms

Qualitative information about diffuse objects (Briers, 1976) is often all that is necessary, and this can be obtained by visual observation of the hologram reconstructions or their photographs. A great deal of useful information is revealed by the position of nodal lines and antinodal regions of the object. Thus a survey application of holographic interferometry, such as a study of vibration mode shapes as a function of frequency, can be accomplished satisfactorily by recording photographs of hologram reconstructions of each mode to produce a final catalog.

If the hologram is of good quality, photography of the reconstructed image does not present any extraordinary difficulties, since the hologram acts as a window on the object through which it and its superimposed fringe information can be photographed. In practice some difficulties may be experienced. Usually since as much light is needed on the hologram as possible (Aaidel' *et al.*, 1969), the object will be placed as close to the hologram as illumination and mechanical constraints make convenient. Satisfactory photography of the reconstructed image can be accomplished only by using optics appropriate for "macro" or close-up photography. In addition, care must be taken to avoid image fogging from light scattered from the hologram plate. Furthermore, unwanted glare from the reference beam can cause a problem if it is allowed to strike the lens.

Fringes from interferograms are usually not located in a single plane (Vest, 1975; Molin and Stetson, 1970a,b; Steel, 1970), so usually in an effort to obtain as much depth of field in the fringe system as possible, the photographic lens is stopped down to small aperture. Beyond a certain point, this process is self-defeating since speckle size is increased and the information desired in fine fringes is lost as smaller lens apertures are used. This effect is often difficult to discover visually since the human observer tends to average these effects when watching the hologram and generally will ignore the effect of speckle. Image plane holograms have a minimum of such problems since the depth of the object plane is inherently limited by the hologram process.

When only qualitative data is needed, ordinary photographic techniques give adequate accuracy in irradiance and geometrical dimensions. When quantitative measurements of fringe irradiance or accurate position are needed, it is best to avoid an intermediate photographic step, and measurements should be made directly from the holographic reconstruction. In this manner, the geometrical distortions and irradiance errors introduced by the photographic step can be avoided.

10.4.3.8 Data Interpretation and Reduction

(a) Geometrical Effects In double exposure or real time holographic interferometry, each fringe represents the location of the locus of constant optical path length difference between the two exposures. Except when the illumination and viewing directions are parallel to the direction of motion of the object, a geometrical correction must be made to fringe location in order to achieve the actual amplitude of motion of the surface. If the angle between the object illumination ray at a given point of the surface and the direction of motion is θ_i and the angle between the motion vector and angle of view is θ_v, then each fringe order represents an optical path change which must be multiplied by a factor which is equal to

$$F = (\pi/\lambda)(\cos \theta_i + \cos \theta_v). \tag{12}$$

When the illumination and viewing points are at close enough distances from the object that the angles in Eq. (12) vary significantly across the view of the object, interpretation of the equation becomes difficult. A simple means of interpreting fringe position in such cases is to recognize that Eq. (12) is that of an ellipse whose foci are the point of origin of the reference and viewing points. This variation has been formalized in the "holodiagram" (Abramson, 1974, 1975, 1976), comprising a series of ellipsoids and their orthogonal hyperbolic functions, which maps those portions of space along which given components of the motion give rise to the same interference patterns. Simply, any coordinate of motion along a direction parallel to an ellipse whose foci are the reference and observing points produces no change in the fringe pattern, while those components perpendicular to that direction give rise to a maximum change in fringe location. Thus ellipses in the holodiagram represent loci of minimum change of optical path length change and their orthogonal hyperbolas are the directions along which maximum sensitivity to motion is observed. In all cases, optical path length changes are measured in terms of the physical path change Δx and the index of refraction change Δn and represent the integral of

$$OP = \int_{source}^{obj} n(x) \, dx + \int_{obj}^{holog} n(x) \, dx, \tag{13}$$

when the reference beam remains fixed relative to the hologram.

(b) Interpretation of Time Effects in Holograms As already discussed, for time averaged holograms the interpretation of fringe amplitude is related to

495

the optical path differences through the functional form of a Bessel function whose argument contains the optical path as a variable. In the case where the motion of the object is purely sinusoidal, the contrast of the fringes forming the hologram decreases with increasing amplitude, and the contrast is given by the well-known equation derived first by Powell and Stetson (1965) (see Section 10.4.3.4). In those instances when the object is not vibrating in a single pure mode, modifications of the pattern appearance will result from the superposition of the various vibratory components. In the case of time averaged holograms, secondary nodal and fringe patterns characteristic of another mode of the object will be superimposed on the primary pattern. When the two modes are harmonically related, interference effects can be observed between the patterns. In general, however, the motion is not harmonic with the drive and tends to merely degrade the contrast of the interference fringe pattern. When stroboscopic methods are used to study vibration, additional modes tend to superimpose patterns on the fringes which enhance or degrade certain portions of the fringe pattern, giving a mottled look to the interferogram. Additionally, when the strobe frequency is detuned from the drive frequency in real time holography, the fringe motion takes on a jerky appearance instead of the sinusoidal motion at the difference frequency which is observed when only a pure mode is being excited. This is a useful check on mode purity when a series of holograms is being made of the vibration spectrum of an object. In general any motion which is not the pure mode under excitation will decrease the contrast of either time averaged or stroboscopic interferograms. In the case where the motion is at the frequency of another mode of vibration, the changes in pattern will correspond to a vibration pattern for that mode.

When the pure mode spectrum of an object is under study, every effort must be made to excite the object in only one mode at a time by adjusting the driver position and assuring driver linearity. (With magnetic drive, saturation of the driver or the material being driven can easily cause nonharmonic drive of unwanted modes.) For those objects, such as musical instruments or magnetic devices, where harmonic excitation is present, time averaged holograms will usually indicate nodal regions, even under complex excitation, and stroboscopic holograms can often be used to sort out some of the vibratory components. These cases are very difficult to interpret, and often little more than mapping of the positions of nodes and antinodes can be accomplished by such studies.

(c) **Data Smoothing** Quantitative measurements from interferograms require careful determination of the positions of the interference fringes, whether the desire is to map a flow or a strain field. In all such cases, improved fringe location on the hologram, whether achieved by multiple exposures or by enhancement of the original interferogram improves accuracy of the final result. In the area of using vibration or static double exposure holograms for

496

strain measurements, fringe accuracy, and more importantly, variations in spacing between various fringe pairs are critical factors. Since holograms measure only amplitudes of motion, strain determination must utilize the spatial derivatives of the amplitude data acquired from the fringe location, and thus noise in the fringe data becomes a particualrly severe problem in such studies. Data smoothing can be helpful in such cases, provided that the functions are chosen with care (Taylor and Brandt, 1972; Brandt and Taylor, 1972). In one study of smoothing of interference fringe locations for strain analysis, spline functions (cubic representatives of a bending of a thin flexible spline) were found to be an advantageous choice of smoothing functions (Brandt and Taylor, 1972), principally because these functions inherently retain the smoothness of their higher derivatives, a necessary condition if meaningful strain measurements are to be made. In the smoothing of holographic data, which is inherently a discrete representation of a continuous variation in the surface of the test object, care must be taken not to choose functions which vary more rapidly than the spatial variation of the phenomenon itself. Successful smoothing methods will provide for reduction of the number of fitted points and adjustment of their position to maximize the "smoothness" of the function which has been fit. Unless this is done, the final function may bear little relation to the actual behavior of the test data.

10.4.4 Special Techniques and Tricks

10.4.4.1 Techniques for Real Time Interferometry

Zero-fringe operation of real time holographic interferometry is more difficult than double exposure or time averaged studies principally because changes in the holographic plate and/or mechanical apparatus holding the optics and the object are very difficult to avoid. It is in this area that sound kinematic design (Whitehead, 1934) of plate holders, optical mounts and object constraints truly pays off in improved experimental results. The basic principle involved is that it is necessary to provide the absolute minimum number of mechanical constraints on an object to prevent motion of each particualr degree of freedom. For example, all hologram plate holders, whether used for interferometry or not, should utilize kinematic design to minimize distortion of the plate during exposure. Only three support pins are necessary to orient a rectangular plate in a plane in both position and angle. In the same way, only three points are required to establish the position of that plane; thus a kinematic plate holder need have only six support points in order to assure accurate alignment of the hologram plate. Fingers are needed to hold the plate against the pads and to provide frictional forces to retain the plate against its orientation pins, but the forces here need not be very great. A plate holder designed along kinematic principles will not warp the plate and can be used to replace a hologram after it has been exposed with sufficient accuracy that if nothing else has changed in the interim, a zero fringe condition will be maintained over the whole view of the hologram. In this case, a liquid cell may be used to contain the hologram during the experiment. Such a cell eliminates shrinkage effects but requires a kinematic holder for the plate. The ultimate in achieving zero fringe operation results when all of the processing is done in the cell itself without removing the hologram. Automatic machines are available for performing all the chemical processes in situ, however, they are expensive.

For the majority of work, immersion of the hologram is not necessary and real time holography can be performed quite satisfactorily with a dry kinematic holder using the holographic plate edges as references and using manual replacement of the hologram once it has been processed. In such a system, since the edges of the plate are used as reference surfaces, critical work may require that the edges of the plate be dressed lightly with an abrasive stone, particularly if the plate has been chipped in shipment or has been cut in the laboratory. Finally, the necessity for extreme rigidity cannot be overemphasized. Particularly in the case of vibration studies where acoustic excitation is present, the plate itself may vibrate, even though it is kinematically mounted. In such cases, more satisfactory results may be achieved by smaller plates and kinematic holders than large ones. A good kinematic holder for 4 × 5 in. plates without a liquid gate gives very satisfactory performance for the majority of interferometric applications, and the construction of such a holder need not be elaborate provided that sufficient attention is given to rigidity.

10.4.4.2 Use of Holographic Films in Interferometry

Probably the majority of experimental holographic arrangements utilize glass plates rather than film as the recording medium in spite of the greater cost of the medium, primarily because it is relatively easy to mount a plate rigidly. If cost per exposure is an important factor when multiple experiments are to be made sequentially or when very large hologram areas are involved, film is the preferred medium. Smaller sized films, such as 35-mm film, can often be accommodated adequately in standard, quality camera mounts without film motion or creep causing problems with the holography. Sheet films with relatively thick substrates can be used to make holograms in conventional film holders up to 4 × 5 in. provided that the film is mounted securely and carefully and that adequate time is allowed to elapse for the film to "settle" into place. Even at 4 × 5 in. the usual edge-clamped film holder may give trouble, and for larger sizes it is totally inadequate. For all sizes the best mount for film is a vacuum or pressure mount which applies support over the whole film surface during exposure. If properly designed, a vacuum holder is preferred over a pressure holder since no window is needed between the hologram and object.

When pulsed lasers are used, there is usually no need for special film mounting precautions since film motion is negligible during the pulse interval.

10.4.4.3   Double Exposure Methods for A Posteriori Fringe Compensation

One disadvantage of double exposure holographic interferometry is the inability to alter the background fringe pattern to determine the relative sense of motion of various portions of the hologram once the hologram has been made. This disadvantage can be overcome by recording the two exposures separately in such a manner that the two reconstructed images can be manipulated separately after reconstruction. The simplest of these methods, called "sandwich holography" (Abramson, 1974, 1975, 1976; Bjelkhagen, 1977), requires no more change from the conventional apparatus than a modification of the plateholder so that two plates can be exposed simultaneously. Although any close spacing of the plates can be used, the most convenient arrangement is one in which the plates are pressed together with their emulsion sides in contact. An exposure is made with a pair of plates of the object in its first state followed by an exposure of the second state with a second pair of plates. Following processing and drying, the front plate (on the object side of the hologram) is aligned with the back plate of the other pair, emulsion sides in contact, and the new pair of holograms is reconstructed with a replica of the original reference beam. Since each image is reconstructed separately, the fringe pattern caused by changes in the two states can be manipulated by translating and rotating the holograms relative to one another. For example, tilting one hologram relative to the other is equivalent to inducing a tilt in the object between exposures, and similarly, rotation of the holograms is equivalent to rotation of the object between exposures. Suitable manipualtion of these hologram pairs can be used to resolve phase ambiguities in the fringe patterns by observing changes in the direction of known background fringes.

More complicated holographic arrangements can be used to accomplish the same manipulations on a single hologram. When the two exposures are made using reference beams separated in angle by an amount sufficient to prevent image overlap of secondary images upon reconstruction, the images reconstructed by two reconstructing beams can be manipulated relative to one another by making changes in the reconstruction geometry (Tsuruta *et al.*, 1968). This form of interferometry in which different spatial frequency carrier fringes are used is only one of several types of spatial multiplexing which can be utilized. Spatially complimentary photographic masks can also be used to record the two holograms, and color or polarization masks can be used on reconstruction to illuminate the appropriate areas of the hologram with separate reference beams. However, of the methods which provide separate recording and reconstruction of the two images, the sandwich method offers the greatest simplicity, both in the formation and postreconstruction manipulation stages.

10.4.4.4 Increasing Fringe Sharpness

Although photographs of holographic fringe patterns may not give a precise density replica of the fringe intensity variation, fundamentally, the fringe resolution is determined by the irradiance distribution which is of the form $J^2(\phi)$ for time averaged holograms and of the form $\sin^2(\phi)$ for double exposure holograms. Ability to observe density differences in the fringe structure sets the basic ability to see spatial fringe shifts and thus sets the limit on displacement resolution. Fringe position measurements become critical when strain measurements are made from holograms because such measurements depend upon differences in fringe position (derivatives of the amplitude function) and thus small errors in fringe position quickly magnify in the strain calculations. While in principle multiple wavelength holographic techniques could be used to sharpen double exposure or vibration fringes in much the same manner as classical multibeam inteferometry, the difficulties of arranging such an experiment make a system based on more conventional approaches attractive.

One approach for photoelectronically determining fringe location differences uses a dual reference beam double exposure hologram arrangement in which one of the reconstructing beams is frequency offset from the other by a convenient acoustic frequency (Dandliker *et al.*, 1976). This frequency offset modulates the interference fringe position on the hologram reconstruction which, in turn, can be detected photoelectrically by placing a detector in the image plane of a lens which projects the reconstruction. At a point in the reconstructed image from a double exposure hologram at which the optical phases in the initial and final states of displacement are $\phi_1(x,y)$ and $\phi_2(x,y)$, respectively, the phase of the alternating signal will be proportional to

$$\phi_1(x,y) - \phi_2(x,y) + w(t) = \Phi + w(t), \tag{14}$$

where $w(t)$ is the instantaneous phase of the modulating signal at time t. A second detector placed (say) at $x + \Delta x$, and y will detect a similar signal whose phase in time is

$$\phi(x + \Delta x) - \phi_2 (x + \Delta x) + w(t) = \Phi' + w(t), \tag{15}$$

which has the same time variation as the first signal but a phase difference $\Phi - \Phi' = \Delta \Phi$ equal to the phase difference in the interference pattern at x and $x + \Delta x$. Thus temporal phase differences between two detected signals Δx apart are proportional to the spatial phase differences in the interference patterns at those points and relate directly to the slope of the displacement between those points.

By converting the spatial modulation on a double exposure hologram to a time modulation, fringe location can be specified by a phase measurement (which can be done quite accurately) relative to that of any other point on the reconstruction. In particular, a pair of detectors mounted a fixed distance apart can be used to measure the displacement slope directly, an important result since direct slope measurement increases the accuracy of strain measurements.

10.4.5   Conclusion

Intrinsically, holographic interferometry provides the same type of information that can be obtained from conventional interferograms, namely, a measure in terms of the optical wavelength used of changes in optical path introduced during the experiment. Its utility greatly exceeds that of conventional methods in most practical situations for a veriety of reasons. When conventional interferometers are applicable, holograpic interferometers can be used at apertures greatly exceeding those possible with reasonably priced high quality optics. When random or diffuse wavefronts are involved, holographic interferometers work where other methods would not give usable signals. Holography adds the fourth dimension, time, to interferometry through the variety of single and multiple exposure techniques which are available to the holographer. Finally, the versatility of the technique and its relative simplicity make available to anyone with access to a laser a very powerful sophisticated tool for measurement wherever changes in the experiment can be converted to differences in optical path.

REFERENCES

Aaidel', A. N., Malkhasyn, L. G., Markova, G. V., and Ostrovskii, Yu. I. (1969). *Sov. Phys.- Tech. Phys.* **13**, 1470–1473.

Abramson, N. (1974). *Appl. Opt.* **13**, 2019–2025.

Abramson, N. (1975). *Appl. Opt.* **14**, 981–984.

Abramson, N. (1976). *Appl. Opt.* **15**, 200–205.

Aleksoff, C. C. (1971). *Appl. Opt.* **10**, 1329–1341.

Bjelkhagen, H. (1977). *Appl. Opt.* **16**, 1727–1731.

Brandt, G. B. (1969). *Appl. Opt.* **8**, 1421–1429.

Brandt, G. B., and Taylor, L. H. (1972). Holographic Strain Analysis Using Spline Functions, *Proc. Symp. Eng. Appl. Holography, SPIE, Redondo Beach, California,* February.

Brandt, G. B., Rozelle, P. F., and Patel, B. R. (1976). *In* "The Engineering Uses of Coherent Optics" (E. R. Robertson, ed.), pp. 577–591. Cambridge Univ. Press, London and New York.

Brandt, G. B., Gottlieb, M., and Conroy, J. J. (1977). *J. Opt. Soc. Amer.* **67**, 1269–1273.

Briers, J. D. (1976). *Opt. Quantum Electron.* **8**, 469–501.

Brooks, R. E., Heflinger, L. O., and Wuerker, R. F. (1966). *IEEE J. Quantum Electron.* **QE2**, 275–279.

Brown, G. M., Grant, R. M., and Stroke, G. W. (1969). *J. Acoust. Soc. Amer.* **45**, 1166–1179.

Bryngdahl, O. (1968). *J. Opt. Soc. Amer.* **58**, 865–871.

10. Application Areas

Collier, R. J., Burckhardt, C. B., and Lin, L. W. (1971). "Optical Holography." Academic Press, New York.

Dandliker, R., Eliasson, B., Ineichen, B., and Mottier, F. M. (1976). *In* "The Engineering Uses of Coherent Optics" (E. R. Robertson, ed.). Cambridge Univ. Press, London and New York.

Ennos, A. (1977). National Physical Laboratory, Teddington, private communication.

Erf, R. K., ed. (1974). "Holographic Nondestructive Testing." Academic Press, New York.

Haines, K. A., and Hildebrand, B. P. (1966a). *IEEE Trans Instrum. Meas.* **IM-15,** 149-161,

Haines, K. A., and Hildebrand, B. P. (1966b). *Appl. Opt.* **5,** 595-602.

Heflinger, L. O., Wuerker, R. F., and Brooks, R. E. (1966). *J. Appl. Phys.* **37,** 642-649.

Horman, M. H. (1965). *Appl. Opt.* **4,** 333-336.

ITT (1968). "Reference Data for Radio Engineers," 5th ed., pp. 21-27. ITT, New York.

Lurie, M. (1968). *J. Opt. Soc. Amer.* **58,** 614-619.

Lurie, M., and Zambuto, J. (1968). *Appl. Opt.* **7,** 2323-2325.

Mayer, G. M. (1969). *J. Appl. Phys.* **40,** 2863-2866.

Molin, N., and Stetson, K. A. (1970a). *Optik* **31,** 157-177.

Molin, N., and Stetson, K. A. (1970b). *Optik* **31,** 281-291.

Mollenauer, L. F., and Tomlinson, W. J. (1977). *Appl. Opt.* **16,** 555-557.

Mottier, F. M. (1970). *In* "Applications de l'Holographie" (J. Ch. Vienot *et al.*, eds). Besancon, France.

Neumann, D. B., Jacobson, C. F., and Brown, G. M. (1970). *Appl. Opt.* **9,** 1357-1362.

Powell, R. L., and Stetson, K. A. (1965). *J. Opt. Soc. Amer.* **55,** 1593-1598.

Robertson, E. R., ed. (1976). "The Engineering Uses of Coherent Optics." Cambridge Univ. Press, London and New York.

Sampson, R. C. (1970). *Exp. Mech.* **10,** 313-320.

Sollid, J. E. (1975). *Opt. Eng.* **14,** 460-469.

Steel, W. H. (1970). *Optica Acta* **17,** 873-881.

Tanner, L. H. (1965). *J. Sci. Instrum.* **42,** 834-837.

Tanner, L. H. (1966). *J. Sci. Instrum.* **43,** 81-83.

Taylor, L. H., and Brandt, G. B. (1972). *Exp. Mech.* **12,** 543-548.

Tolansky, S. (1973). "An Introduction to Interferometry," 2nd ed. Wiley, New York.

Trolinger, J. D. (1975). *Opt. Eng.* **14,** 470-481.

Tsuruta, T., Shiotake, N., and Itoh, Y. (1968). *Japan. J. Appl. Phys.* **7,** 1092-1100.

Vest, C. M. (1975). *Appl. Opt.* **14,** 1601-1606.

Vilkomerson, D. (1976). *Appl. Phys. Lett.* **29,** 183-185.

Whitehead, T. N. (1934). "The Design and Use of Instrument and Accurate Mechanism; Underlying Principles." Macmillan, New York.

Wuerker, R. F. (1976). *In* "The Engineering Uses of Coherent Optics" (E. R. Robertson, ed.), pp. 517-540. Cambridge Univ. Press, London and New York.

David Casasent

10.5.1 Introduction

Pattern and character recognition are two of the most widely pursued application areas of optical data processing. The purpose of a pattern or character recognition system is to determine the presence (and usually the location of) a reference pattern in an input image. In character recognition, a fixed bank or set of reference functions from a large font or class are usually used, and the purpose of the system is to determine which member of this class of reference functions is present in the input (and usually its location in the input scene). The key operation in all optical pattern and character recognition systems is the correlation of an input and reference function or two input functions. This discussion will thus concentrate on optical correlator systems. In this section we sketch the architecture of the major optical pattern recognition correlators, include the salient equations describing each, tabulate the resolution requirements for the filter plane material for each, and discuss practical engineering considerations such as stability, positioning requirements, and implementation using real time spatial light modulators.

Since many optical processors are special-purpose systems, several specific applications are included. In many of these systems, advantage is taken of the ability to control the format of the input data. A particularly attractive new adaptation of this approach involving the use of space-variant optical correlators is also included. Both image and signal correlator systems are described as well as systems in which the two functions to be correlated are both on-line rather than one being a fixed reference function. In all systems, the reference function can be changed (at increased system complexity) as the application warrants. Other adjuncts to effect practical optical pattern recognition systems involve the use of electronic pre- and postprocessing, hybrid systems (Casasent and Sterling, 1975), and multiple matched filters.

In the most widely used system, the correlation is performed by multiplying the Fourier transforms of the input and reference functions. In such a case, the reference pattern is stored as the conjugate Fourier transform of the reference function. Since this reference pattern is placed in the Fourier plane, it is a spatial filter. This stored function is complex and is thus similar to a

503

hologram (see Chapter 1). However, its purpose is to determine if the input pattern (or part of it) matches the reference function (not to produce an aesthetically pleasing output image, as in holography). We thus refer to this complex Fourier plane reference pattern as a matched spatial filter (MSF) (See Section 4.3), since it is said to be matched to the desired key object or pattern to be searched for. The more classic descriptions of MSF theory consider a signal in noise. For our pattern recognition purposes, the signal is the desired key object or reference function being searched for, and the noise is any other object present in the input or those parts of the input that are not the key object.

The synthesis of the matched spatial filter and the optimization of its parameters are of vital and practical concern. The highlights of this aspect are presented in Section 10.5.15. Particular attention will be given to the often ignored but practical problem of performing pattern recognition when the input is a degraded version of the reference function. The correlation loss encountered and a discussion of how the matched spatial filter synthesis parameters can be chosen to enhance such correlations are included. For generality, we still refer to this filter function as an MSF, although the degree to which it is perfectly matched is clearly different in this case.

In Section 10.5.17, we include general remarks on the differences between pattern and character recognition and define the particular issues that we see as problems in each case. General remarks on future directions for work in optical pattern and character recognition conclude the presentation.

From the many optical correlator architectures possible, 13 have been selected for extended discussion. In all systems, a consistent notation is used, in which the input, transform, and output correlation planes are P_1, P_2, and P_3 with spatial coordinates (x_1, y_1), (x_2, y_2), and (x_3, y_3), respectively. The spatial frequency coordinates of the transform plane are denoted by $(u, v) = (x_2/\lambda f_1, y_2/\lambda f_1)$, where λ is the wavelength of the light used and f_1 the focal length of the Fourier transform lens (we use f_2 to represent the focal length of the second Fourier transform lens). Lower case variables describe spatial functions, and the corresponding upper case variable denotes the Fourier transform of this function.

10.5.2 Frequency Plane Correlator

The first and the most widely used optical correlator is the frequency plane correlator (Vander Lugt, 1964) or matched spatial filtering system. A practical laboratory version of this system is shown in Fig. 1. To correlate $g(x_1, y_1)$ and $h(x_1, y_1)$, we first form a matched spatial filter $H^*(u, v)$ of $h(x_1, y_1)$ at plane P_2. This is done by placing $h(x_1, y_1)$ at the input plane P_1 and recording the interference of its transform $H(u, v)$ (formed at plane P_2 by lens L_1) and a plane wave reference beam. The variable attenuator (VA) facilitates com-

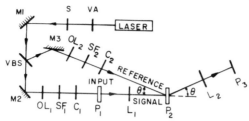

Fig. 1 Schematic representation of a frequency plane correlator. Key: VA, variable attenuator; S, shutter; M, mirror; VBS, variable beam splitter; OL, objective lens; SF, pinhole spatial filter; C, collimating lens; L_1 and L_2, Fourier transform lenses.

ponent alignment and visual inspection of the light distributions in the various planes. The variable beamsplitter (VBS) enables the beam balance ratio (intensity of the reference to signal beams) to be controlled.

The folded path from the VBS to mirror M_3 along the reference beam path is used to maintain equal reference and signal beam path lengths. This also facilitates setting the reference beam angle at the 15° angle value we normally use. This choice of reference beam angle simplifies placement of the components on the table. The resultant fringe pattern is at a high spatial frequency when a 15° angle is used. Film usually has adequate resolution to record this pattern. However, if the resolution of the P_2 material is not adequate, a smaller reference beam angle is essential. A beamsplitter and more involved optics (or longer focal length lenses) are thus needed to combine two wavefronts at a small angle. The lens pinhole/spatial filter systems (OL, SF) in each beam path help to remove noise from the laser beam. They are placed after the mirrors to reduce the effect of mirror imperfections. In our system, we choose the collimation optics to produce very uniform plane waves with less than 5% spatial intensity variation. Such measures are often not necessary, however.

When the matched spatial filter has been produced and is repositioned at plane P_2, we rotate it by 180°. This causes the correlation to appear at +15° into the table to the path of the signal beam rather than at −15° along the extended reference beam path. This is a simple expedient that allows use of a narrower table. During the correlation process, the reference beam is blocked and a transparency with amplitude transmittance $g(x_1, y_1)$ is placed in the input plane. With the matched filter plate rotated and repositioned at plane P_2, the correlation wave emerges from plane P_2 at 15°. The second Fourier transform lens L_2 is placed one focal length from plane P_2 along this 15° path and not coaxial with the signal beam as it is usually shown. This greatly reduces the required aperture and field angle for this lens.

Let us now consider the salient equations describing this system. First, consider the matched spatial filter synthesis step. When a transparency with amplitude transmittance $h(x_1, y_1)$ is placed at the input plane P_1, the amplitude of the light distribution incident on plane P_2 along the signal beam path is $U_2(x_2, y_2) = H(u, v)$, where $H(u, v)$ is the complex Fourier transform of $h(x_1, y_1)$ and is

$$H(u, v) = (1/i\lambda f_1) \int \int h(x, y) \exp[-2\pi i(ux + vy)]\, dx\, dy.$$

For simplicity, we assume all transforms to be exact and thus ignore constant amplitude transmission factors and the $-i/\lambda f_1$ constant in front of the transform. The spatial frequency coordinates (u, v) of plane P_2 are related to the physical distance coordinates (x_2, y_2) of this plane by $x_2 = f_1 \lambda u$ and $y_2 = f_1 \lambda v$, where λ is the wavelength of the light used and f_1 the focal length of the Fourier transform lens L_1.

The light amplitude distribution at plane P_2 due to the reference beam of uniform amplitude r_0 at an angle θ to the signal beam is $U_r = r_0 \exp(-i2\pi\alpha x_2)$, where $\alpha = (\sin\theta)/\lambda$ is the spatial frequency associated with the off axis reference wave assumed to be at an angle θ to the x_2 axis of plane P_2.

The amplitude transmittance of the energy sensitive material (usually photographic film) placed at plane P_2 is assumed to be proportional to the irradiance of the light distribution incident on it. (This is the $\gamma = -2$ condition in Sections 2.4 and 2.6.) When the signal pattern U_2 and the reference beam U_r are interferred at plane P_2, the subsequent transmittance of plane P_2 is

$$t(x_2, y_2) = |U_2 + U_r|^2$$
$$= r_0^2 + |H|^2 + r_0 H \exp(i2\pi\alpha x_2) + r_0 H^* \exp(-i2\pi\alpha x_2). \tag{1}$$

With the reference beam blocked, the matched filter described by $t(x_2, y_2)$ is placed at plane P_2 and $g(x_1, y_1)$ placed at the input plane P_1, the light amplitude distribution incident on plane P_2 is $G(u, v)$ and that leaving plane P_2 is $G(u, v)t(x_2, y_2)$. With plane P_2 in the front focal plane of a second Fourier transform lens L_2 of focal length f_2, the pattern in the back focal plane of this lens is the Fourier transform of $G \cdot t$ or

$$U_3(x_3, y_3) = r_0^2 g\, \delta(x, y) + [h \star h \star g]\, \delta(x, y) + r_0[h * g * \delta(x_3 + \alpha\lambda f_2, y_3)]$$
$$+ r_0[g \star h * \delta(x_3 + \alpha\lambda f_2, y_3)], \tag{2}$$

where $*$ denotes convolution and $\star$ denotes correlation.

From Eq. (2), we see that the first two terms appear centered on-axis in this output plane; the third term is the convolution $h * g$ and emerges from plane P_2 at an angle $+\theta$ and appears centered at $(0, +\alpha\lambda f_2)$ in the output plane; the last term containing the desired correlation $g \star h$ emerges from plane P_2 at an angle $-\theta$ and appears centered at $(0, -\alpha\lambda f_2)$ in the output plane. In practice, the matched filter plate at plane P_2 is rotated by 180° when it is reinserted in the system, and lens L_2 in Fig. 1 is placed off-axis at $+\theta$ to intercept this last term. The pattern located on-axis in the back focal plane of lens L_2 is the desired correlation $g \star h$. The various regions of the input plane are mapped about the origin of this output correlation plane. The location of a bright spot of light in this plane is proportional to the location of the reference function h in the input function g. The irradiance of this peak of light is indicative of the degree of match between h and g. It is customary to define the (x_3, y_3) coordinates of plane P_3 in the inverted directions to simplify notation.

Because the system is linear, multiple occurrences of h in g produce multiple correlation peaks in plane P_3. A vivid demonstration of this is shown in Fig. 2. Figure 2a shows the input function g (a paragraph of text with six occurrences of the word RADAR); the reference function h used was a matched filter of the word RADAR. The pattern in the output correlation plane P_3 of Fig. 1 is shown in Fig. 2b. A thresholded version shown in Fig. 2c demonstrates our point. The locations of the six correlation peaks are proportional to the six locations of the word RADAR in the input, as a comparison of Figs. 2a and 2c will show. The scaling factor between the input and correlation planes is f_2/f_1 (the ratio of the focal lengths of the two transform lenses). The many cross correlations visible in Fig. 2b are indicative of one problem that occurs in optical word or character recognition. This will be discussed further in Section 10.5.17.

If the reference function h corresponds to a much larger scene or ground area than the ground area covered by g, the correlation peak will occur on-axis in plane P_3 if g is placed in the same location in the input plane at which it occurs in h. Any departure in g from this position will cause the correlation peak in plane P_3 to be displaced. If g is placed on axis in the input plane, the location of the correlation peak in plane P_3 will be proportional (by a factor f_2/f_1) to the location of f in g.

Let us now consider a most important feature of this or any optical correlator, the resolution requirements of the Fourier transform plane material (Casasent and Furman, 1977a). We consider one-dimensional (1D) functions only for simplicity, denote the widths of the functions g and h by W_g and W_h, respectively, and consider three cases: (a) $W_g \gg W_h$, (b) $W_h \gg W_g$, and (c) $W_h = W_g$. We first consider the resolution required at plane P_2 to just record the matched spatial filter. The impulse response of the matched spatial filter, found by substituting $\delta(x, y)$ for the input function $g(x, y)$ in Eq. (2), is

$$U_3(x_3, y_3) = r_0^2 \, \delta(x_3, y_3) + h \star h + r_0 h * \delta(x_3 \pm \alpha\lambda f_2). \qquad (3)$$

The location of these impulse response terms in the correlation plane and their size for each of the three cases are given in Table I. Similar parameters for the cross-correlation case in Eq. (2) are given in Table II. For the general case, the width of the impulse response pattern is $4W_h$ (where we require $\alpha\lambda f_2 = 1.5W_h$ to just separate the terms in the impulse response). The resolution required at plane P_2 in units such as lines per millimeter thus appears to be $\frac{1}{4}W_h$ and the space bandwidth that the Fourier transform plane recording

10. Application Areas

THE DEVELOPMENT OF RADAR DURING WORLD
WAR II BROUGHT RADAR FROM A LABORATORY
CONCEPT TO A MATURE DISCIPLINE IN JUST
A FEW SHORT YEARS. SINCE 1945 RADAR
TECHNOLOGY HAS BECOME SO SOPHISTICATED
THAT THE BASIC RECTANGULAR PULSE RADAR
SIGNAL IS NO LONGER SUFFICIENT IN THE
DESIGN OF MANY NEW RADAR SYSTEMS.
MORE COMPLEX RADAR SIGNALS MUST BE
TAILORED TO SPECIFIC REQUIREMENTS.

(a)

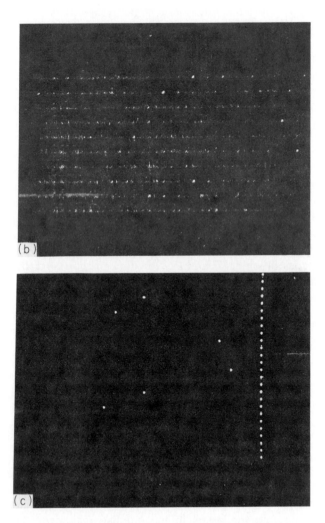

Fig. 2 Example of optical character recognition using the frequency plane correlator of Fig. 1. (a) Input image, (b) output optical correlation plane pattern, (c) thresholded version of (b). (Casasent and Sterling, 1975.)

TABLE I

Location and Size of Terms in the
Impulse Response of the Matched
Spatial Filter for the Frequency
Plane Correlator

Term	Location	Width
r_0	Origin	0
$h \star h$	Origin	$2W_h$
$r_0 h$	$x_3 = -\alpha\lambda f_2$	W_h
$r_0 h$	$x_3 = +\alpha\lambda f_2$	W_h

material must support to be $4W_h/\lambda f_1$. However, when we consider the requirement to separate the cross-correlation term from the other terms in the complete correlation plane pattern (Table II), we find that we require $\alpha\lambda f_2 \geq \frac{1}{2}(2W_g + 3W_h)$. Using this $\alpha\lambda f_2$ value (e.g., $\alpha\lambda f_2 = 5W/2$ for the $W_h = W_g = W$ case) in the impulse response calculation, we find the width of the impulse response to be $(2\alpha\lambda f_2 + W_h)$. The resolution (in units such as lines per millimeter) and space bandwidth (in units of lines or resolvable points) requirements for three cases for the frequency plane correlator are summarized in Table III.

We next consider the fringe visibility of the pattern in plane P_2. This will prove useful as background for the discussion on matched spatial filter parameter optimization (Section 10.5.15) and to facilitate analysis of other correlator architectures. If we rewrite Eq. (1) with the magnitude and phase of H explicitly denoted by $H(u, v) = |H(u, v)| \exp(j\phi)$, we can describe the exposure of plane P_2 by

$$E(x_2, y_2) = (r_0^2 + |H|^2)T + 2r_0|H|T\cos[2\pi\alpha x_2 + \phi(x_2, y_2)], \qquad (4)$$

where T is the exposure time. If we define the bias exposure as $E_B = r_0^2 T$, the average exposure as $E_0 = (r_0^2 + |H|^2)T$, and the beam balance ratio as K

TABLE II

Location and Size of Terms in the
Correlation Pattern for the Frequency Plane
Correlator

Term[a]	Location	Width
$r_0 g$	Origin	W_g
$h \star h * g$	Origin	$2W_h + W_g$
$h * g$	$x_3 = -\alpha\lambda f_2$	$W_h + w_g$
$h \star g$	$x_3 = +\alpha\lambda f_2$	$W_h + W_g$

[a] h = filter; g = input.

509

TABLE III

Resolution Requirements for the Frequency Plane Correlator

Case[a]	Resolution	Space bandwidth
$W_h \gg W_g$	$\frac{1}{4} W_h$	$4 W_h / \lambda f_1$
$W_g \gg W_h$	$\frac{1}{2} W_g$	$2 W_g / \lambda f_1$
$W_g = W_h$	$\frac{1}{6} W$	$6 W / \lambda f_1$

[a] h = filter, g = input, W = spatial width.

$= r_0^2 / |H|^2$, we can rewrite Eq. (4) as

$$E(x_2, y_2) = E_B[1 + 1/K + (2/\sqrt{K}) \cos \psi], \qquad (5)$$

where $\psi = 2\pi \alpha x_2 + \arg(H)$. The only points we wish to make at present are that K varies spatially since $|H|$ does and thus so does the fringe modulation, and that the modulation depends on K, E_B, and the t-E (transmittance versus exposure) curve for the recording material used at plane P_2. The choice of optimum matched spatial filter parameters and their use in reducing correlation degradation are discussed at length in Sections 10.5.15 and 10.5.16.

As a final topic of practical concern, consider the positional and vibrational tolerances of this correlator. A 1D analysis is again used for simplicity. We consider the autocorrelation of g,

$$\rho(x) = \int G(u) G^*(u) \exp(-i2\pi u x) \, du, \qquad (6)$$

where G is the 1D Fourier transform of g,

$$G(u) = \int g(x') \exp(-i2\pi u x') \, dx'.$$

If the fitter $G^*(u)$ is displaced by Δu, we obtain

$$\rho(x) = \int g(x') g^*(x + x') \exp[i2\pi \, \Delta u(x + x')] \, dx'. \qquad (7)$$

From this, we find the phase distortion due to the displacement Δu of the filter to be $|\phi| \leq \pi W x \, \Delta u$, where W is the width of the input object and the width of the correlation is assumed to be $\pm W/2$. To maintain interference, we require $\phi \leq \pi/4$. Using $u = x_2 / \lambda f_1$, we find the allowed positional tolerance of the filter to be

$$\Delta x_2 = \lambda f_1 / 4W, \qquad (8)$$

with $\lambda = 633$ nm, $f_1 = 600$ mm, and $W = 35$ mm, and we find a required filter positioning tolerance of $\Delta x_2 = 2.7 \ \mu m$.

Space does not permit the luxury of a detailed presentation such as this one for each of the 13 correlator architectures to be discussed. We have included these various correlator topics for this one optical correlator architecture since it is the most used topology and since such a presentation demonstrates the analysis methods and procedures that can be used to determine such items for the other correlator architectures.

10.5.3 Scaling Correlator

With a slight modification to the basic frequency plane correlator (i.e., placing the input P_1 behind the Fourier transform lens L_1) as shown in Fig. 3, the scale of the transform occurring in plane P_2 can be changed by varying d and d_2 (Vander Lugt, 1966). With $g(x_1, y_1)$ placed at plane P_1, the light amplitude distribution incident on plane P_2 is

$$U(x_1, y_1) = (f_1/i\lambda d^2)\, G(u, v)\, \exp[ik(x_2{}^2 + y_2{}^2)/2d], \qquad (9)$$

where $k = 2\pi/\lambda$. From this, we see that the scale of the transform can be varied by varying the distance d and hence the position of the input plane P_1. In use, a matched spatial filter $G^*(u, v)$ of $g(x_1, y_1)$ is formed at plane P_2 by interfering the distribution given by Eq. (2) and a plane wave. A new function $g_1(x_1, y_1)$, a scaled version of $g(x_1, y_1)$, is then placed in the input plane P_1 with plane P_1 now a distance $D = md$ from plane P_2, where m is the scale factor between the two functions. As the distance D is varied, the scale of the transform of g_1 is changed until the transforms of g and g_1 match, at which point a bright maximum correlation peak occurs at plane P_3.

In practice, the focusing and transform conditions are satisfied by fixing the distance from planes P_1 to P_3 at $3d$ and maintaining $D \ll 2d$. This correlator can thus accommodate only about a 20% scale search, since with $m = 1.2$ the correlation plane is out of focus by 30 mm or 20% for a 762-mm focal length transform lens (Casasent and Furman, 1977c). We have found that a lens and input plane mount with z-axis control facilitates use of this correlator. Correlators that permit a scale search are always needed since it is difficult in practice to obtain two images of the same scene without at least some scale change present between them. The placement of the input behind the transform

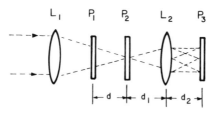

Fig. 3 Schematic representation of the scaling correlator, P_1 is the input plane, P_2 the Fourier transform plane, P_3 the output correlation plane, and L_1 and L_2 the Fourier transform lenses.

511

10. Application Areas

lens as in this correlator also somewhat reduces the requirements on the Fourier transform lens.

Another approach to scale invariant correlation involves the use of multiple matched spatial filters at plane P_2; such systems are discussed briefly in Section 10.5.13. A novel and most promising approach to a scale invariant correlator is the use of a space-variant optical processor based on the Mellin transform. This correlator is discussed in Section 10.5.10.

10.5.4 Joint Transform Correlator[†]

The schematic representation of the joint transform correlator is shown in Fig. 4. Here the spatial versions of the two functions to be correlated are placed side by side in the input plane P_1. Each function is assumed to be of width b and the center-to-center spacing of the functions is $2b$, as shown. The amplitude transmittance of plane P_1 can be described as

$$U_1(x_1, y_1) = g(x_1, y_1 - b) + h(x_1, y_1 + b). \qquad (10)$$

The amplitude distribution of the light incident on plane P_2 is the Fourier transform of this distribution,

$$G(u, v)\exp(-i2\pi vb) + H(u, v)\exp(+i2\pi vb). \qquad (11)$$

The modulus squared of this light distribution is recorded at plane P_2, and we assume the subsequent amplitude transmittance of plane P_2 to be

$$t_2(u, v) = |G|^2 + |H|^2 + GH^*\exp(-i4\pi vb) + G^*H\exp(+i4\pi vb). \qquad (12)$$

This pattern is recorded by the write light λ_1 in Fig. 4. Plane P_2 is now illuminated with a plane wave read beam (shown as λ_2 in Fig. 4) reflected from the beamsplitter (BS). Lens L_2 then forms the Fourier transform of $t_2(u, v)$ at plane P_3, where the light distribution is

$$U_3(x_3, y_3) = g \star g + h \star h + g \star h * \delta(x_3, y_3 + 2b)$$

$$+ h \star g * \delta(x_3, y_3 - 2b). \qquad (13)$$

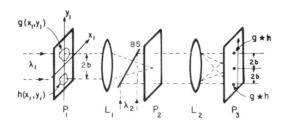

Fig. 4 Schematic representation of the joint transform correlator.

[†] Rao (1967).

512

Plane P_3 thus contains the desired correlation of g and h centered at $(0, \pm 2 f_2 b / f_1)$.

We will discuss this correlator at some length since it has many practical features and advantages. To realize the full advantages of optical processing, real time spatial light modulators (Casasent, 1977) are essential at plane P_1 (and often at P_2 of Fig. 1 also). The plane P_2 media (for the Fig. 1 system) need be real time and reusable only if the reference function is changed quite frequently. In the system shown in Fig. 4, the plane P_2 material must be real time and reusable. Most of these candidate real time devices are optically addressed and operate in reflection. Thus the beamsplitter (used for readout) will normally be placed on the right side of plane P_2. But more important, many real time devices require separate wavelengths of write and read light (λ_1 and λ_2 in Fig. 4). If the matched spatial filter in the frequency plane correlator of Fig. 1 is written at λ_1, and then reading and correlation are performed at λ_2, a scale change and resolution loss in the reference data will be produced. Furthermore, many candidate real time and reusable materials for use at plane P_2 do not exhibit storage. This makes them useless in the frequency plane correlator but causes no problems in the joint transform correlator. The positional tolerances of this system are also far less than those of the frequency plane correlator since plane P_2 is now illuminated with a plane wave. This system is also preferable in applications in which both functions to be correlated are on-line or vary each frame, or when the reference function does not remain fixed for many cycles. Thus, for a number of practical reasons, we find this system preferable to the classic frequency plane correlator, even though it requires a real time and reusable device at both the input P_1 and filter P_2 plane.

Now consider the resolution and space bandwidth requirements of the material used in the transform plane P_2 (Casasent and Furman, 1977a). (If one input plane material or device is used, it must have three times the space bandwidth of the input for the frequency plane correlator. However, two input planes and two lenses L_1 can be used.) The input plane device requirements are then the same for both the frequency plane and joint transform correlators. We proceed as in Section 10.5.2 to analyze the resolution and space bandwidth requirements for plane P_2. The impulse response of the plane P_2 pattern is the same as the output correlation pattern, since plane P_2 is illuminated with a plane wave during correlation. The width and location of the terms in the impulse response (or correlation pattern) are given in Table IV for this system and follow directly from Eq. (13). The full width of the impulse response for the three cases: (a) $W_h \gg W_g$, (b) $W_g \gg W_h$, and (c) $W_g = W_h = W$ are $4W_h$, $4W_g$, and $6W$, respectively, from which the resolution and space bandwidth requirements for plane P_2 given in Table V follow directly. Comparing these data to those of Table III, we see that the resolution requirements of both correlators are the same except for the case $W_g \gg W_h$. In this case (analogous

10. Application Areas

TABLE IV

Location and Size of the Terms in the
Impulse Response or Correlation Pattern
for the Joint Transform Correlator

Term	Location	Size
$g * g$	Origin	$2W_g$
$h * h$	Origin	$2W_h$
$g \star h$	$y_3 = -2b$	$W_g + W_h$
$h \star g$	$y_3 = +2b$	$W_g + W_h$

to a filter of a single character or object in a large scene correlated against an input paragraph or large area scene), the resolution requirements of the joint transform correlator are two times larger and the frequency plane correlator is preferable.

Another advantage of the joint transform correlator is that it produces full visibility interference fringes and thus full modulation of all spatial frequency components of g and h. Thus, it does not require (nor does it allow) the matched spatial filter parameter control (Section 10.5.15) possible and required in the frequency plane correlator. To demonstrate this, we rewrite Eq. (12) in terms of the actual coordinates (x_2, y_2) of plane P_2 as

$$t_2(x_2, y_2) = |G|^2 + |H|^2 + 2|G||H|\cos[4\pi y_2 b/\lambda f_1 + \phi(x_2, y_2)], \quad (14)$$

where $\phi = \arg(GH^*) = \arg G - \arg H$. For $g = h$, we obtain

$$t_2(x_2, y_2) = 2|G|^2[1 + \cos(4\pi y_2 b/\lambda f_1)]. \quad (15)$$

Comparing this to Eq. (5) for the frequency plane correlator, we find that the joint transform correlator produces 100% fringe contrast at all spatial frequencies. In some instances, such a feature is desirable, and this system is preferable to the frequency plane correlator and in one sense is optimum.

TABLE V

Resolution and Space Bandwidth
Required for the Transform Plane P_2
Material in the Joint Transform
Correlator

Case	Resolution	Space bandwidth
$W_h \gg W_g$	$\frac{1}{4}W_h$	$4W_h/\lambda f_1$
$W_g \gg W_h$	$\frac{1}{4}W_g$	$4W_g/\lambda f_1$
$W_h = W_g$	$\frac{1}{6}W$	$6W/\lambda f_1$

Since the fringe intensity at plane P_2 is a haversine, the exposure incident on plane P_2 varies from 0 to the peak of the sine wave and all modulation must come from the input light. For modest exposures, we will be on the knee of the t–E curve where there is little gain. In this case, a noncoherent bias exposure for plane P_2 is needed.

10.5.5 Multichannel 1D Correlator†

The correlators discussed previously are specifically intended to be used in two-dimensional (2D) image pattern recognition. Here we discuss a multichannel 1D correlator of use in signal processing of coded waveforms and in the synthesis of ambiguity functions. All of the prior correlators can easily be converted to multichannel 1D correlators by replacing each Fourier transform lens by a cylindrical/spherical lens combination. The present correlator (shown in Fig. 5) differs from these 1D correlators by use of a spherical lens for L_2. We present it as an example of the many correlator architectures possible, as an example of signal correlation, and as a demonstration of the use of input data format control.

The combination of lenses L_{c1} and L_{s1} images vertically and transforms horizontally. At plane P_2 we find each of the N horizontal signal lines at plane P_1 imaged onto N lines in plane P_2 and the Fourier transform of each of these N signals displayed horizontally on the respective line in plane P_2. We consider the case in which all N signals are the same coded waveform $s(x_1)$. When its multichannel 1D transform and an off-axis reference beam are interfered at plane P_2, we find the subsequent transmittance of plane P_2 to be

$$U_2(x_2, y_2) = \sum_{n=1}^{N} S^*(u, y_2 - nd_2) \exp(-i2\pi\alpha x_2) \exp(-i2\pi u x_r). \quad (16)$$

The separation d_2 between the N channels in plane P_2 is related to the separation d_1 between the N channels in the input plane P_1 by $d_2 = (f_{1s}/f_{1c})d_1$, where f_{1s} and f_{1c} are the focal lengths of the spherical and cylindrical lenses. The $\exp(-i2\pi\alpha x_2)$ factor is due to the off-axis reference beam, and the

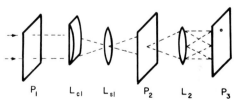

$$P_1 \qquad L_{c1} \quad L_{s1} \qquad P_2 \qquad L_2 \qquad P_3$$

Fig. 5 Schematic representation of a multichannel 1D correlator.

† Casasent and Klimas (1978).

515

$\exp(-i2\pi u x_r)$ factor is due to the horizontal reference starting location $x_1 = x_r$ chosen for the input signal code.

With this multichannel reference matched spatial filter in place at plane P_2, a multichannel input function is recorded at plane P_1 with a different starting location $x_1 = x_s$, where $\Delta x_r = x_s - x_r$, for the signal on line $n = 1$ and with an incremental shift Δx_ϕ in the starting location of the input signal from line to line. We describe the transmittance of plane P_1 for this case as

$$t_1(x_1, y_1) = \sum_{n=1}^{N} s(x_1 - x_s + n\,\Delta x_\phi, y_1 - nd_1). \tag{17}$$

This is representative of a recording of the time histories of the received signals from N elements of a phased array radar on N lines. The relative change in the starting location $\Delta x_r = x_s - x_r$ of the signal is proportional to the fine range of the target in a range bin (hence the choice of the subscript r). The differential shift Δx_ϕ in the starting location of the signal between successive lines is proportional to the azimuth angle ϕ of the target (hence the choice of the subscript ϕ).

The light amplitude distribution incident on plane P_2 is the 1D horizontal transform of Eq. (17) or

$$U_s(x_2, y_2) = \sum_{n=1}^{N} s(u, y_2 - nd_2) \exp(i2\pi u x_s) \exp(i2\pi un\,\Delta x_\phi). \tag{18}$$

The light amplitude distribution transmitted by plane P_2 is the product of the light incident on P_2 described by Eq. (18) and the stored reference pattern described by Eq. (16) or

$$U_2'(x_2, y_2) = U_s U_2 = \sum_{n=1}^{N} S(u, y_2 - nd_2)S^*(u, y_2 - nd_2) \exp(-i2\pi\alpha x_2)$$

$$\times \exp(i2\pi u \Delta x_r) \exp(i2\pi un\,\Delta x_\phi). \tag{19}$$

Spherical lens L_2 forms the 2D Fourier transform of U_2' at plane P_3, where we obtain

$$U_3(x_3, y_3) = [s(x) \star s(x)] * \delta(x_3 - \alpha\lambda f_2 - \Delta x_r, y_3 - \Delta x_\phi). \tag{20}$$

From Eq. (20), we see that the output plane P_3 pattern consists of the autocorrelation $s \star s$ of the coded waveform (thereby retaining the signal-to-noise and signal-to-clutter advantages of various waveform codes). Its horizontal position (relative to the $x_3 = \alpha\lambda f_2$ reference center of the correlation plane) is proportional to the fine range Δx_r of the target (i.e., to the positional difference between the starting location x_s of the input signal and the starting position x_r of the coded reference waveform that defines the zero range gate reference). The vertical position of the correlation peak is proportional to the

516

differential shift Δx_ϕ of the input signal from line-to-line and hence to the vertical phase slope across the input signals and is thus proportional to the azimuth angle ϕ of the target. In a multitarget environment, multiple correlation peaks occur at the range/azimuth coordinates of each target. The system is directly adapted to processing pulsed Doppler and pulse burst radar data. In this case, successive returned pulses are written on successive lines in plane P_1 and the output display is a range/Doppler plot or the ambiguity function of the coded radar waveform chosen.

10.5.6 Long Coded Waveform Correlator†

In many advanced applications, long pseudorandom coded waveforms are transmitted. The length of these codes can easily exceed several thousand bits and thus exceed the linear resolution of an input spatial light modulator. In such cases, the problem is to detect the presence of the code and its starting location in the input bit stream sequence even if the code might extend over parts of many lines in the input plane. An optical signal correlator for use in such cases is now described. The basic correlator architecture is the same as that of Fig. 1. However, the input data format and the format chosen for the matched filter function are unique.

To best convey the concept, we consider the detection of a 63-bit pseudorandom binary code 101110011000010101010001111101010010111010000-111011010010101000. For this relatively simple 63-bit code, we would use a 9 × 7 element spatial light modulator on which the received signal is recorded in raster form. If the received signal pattern for the code started at the actual starting bit of the code, the input raster recorded pattern would appear as shown in Fig. 6a. A simple holographic matched filter of this input pattern would successfully detect the presence of this input code. However, if the received signal originated at bit 42 of the input sequence, the received input signal pattern would appear as shown in the indicated portion of Fig. 6b. The matched spatial filter made from the input pattern in Fig. 6a would clearly not be effective in recognizing this latter input pattern.

The conventional matched filter correlation for recognition of an M-bit code at any starting location would require an $M \times M$ bit reference mask. In this new scheme, we use the mask shown in Fig. 6b. This mask is generated by selecting any starting bit location (for the mask shown we chose bit 33) and recording the next $2m$ bits of the code on the first line of the mask. If the input is $m \times n = 9 \times 7$ as here, we record on line 2 of the mask the $2m$ bits of the code beginning at bit 42 ($= 33 + m$ in a cyclic fashion, with the code repeated when we exceed 63). This pattern is continued for $2n$ lines until the

† Casasent and Kessler (1976).

```
1 0 1 1 1 0 0 1 1

0 0 0 0 1 0 1 0 1

0 1 0 0 0 1 1 1 1

1 0 1 0 1 0 0 1 0

1 1 1 0 1 0 0 0 0

1 1 1 0 1 1 0 1 0

0 1 0 1 0 1 0 0 0
```

(a)

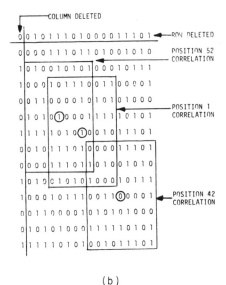

(b)

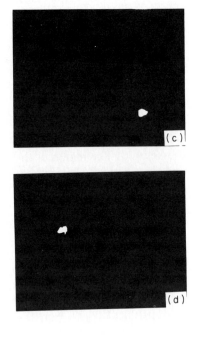

(c)

(d)

Fig. 6 Long coded waveform correlator for 63-bit code. (a) Input format for bit-1 starting position, (b) matched spatial filter plane mask, (c) and (d) output correlation plane patterns for bit-42 and bit-52 starting positions of the code, respectively. (Casasent and Kessler, 1976.)

$2m \times 2n$ pattern of Fig. 6b is completed. The first row and column of the mask are then deleted.

By comparing Figs. 6a and b and the code used, we see that each of possible $m \times n = 9 \times 7$ bit input patterns (corresponding to any starting bit choice) is contained in Fig. 6b. We denote the corresponding 9×7 bit blocks for starting bit locations 1, 42, and 52 specifically on Fig. 6b to demonstrate the point. The advantage of this type of matched filter is that only a $(2m - 1) \times (2n - 1)$ bit mask is needed. A holographic matched filter of this mask is made and

518

placed at plane P_2, and the input pattern is placed at plane P_1 of Fig. 1. The output plane P_3 correlation pattern contains a bright correlation peak. From its location, the starting bit location and presence of the chosen code can be found. Typical output correlation plane patterns for codes starting at bit locations 42 and 52 in the sample 63-bit coded sequence are shown in Figs. 6c and d.

10.5.7   Image Plane Correlator

Before the discovery of the matched spatial filter, the most widely used optical pattern recognition system was the image plane correlator of Fig. 7a. In this system, the spatial domain representations of the two functions to be correlated are placed in planes P_{1a} and P_{1b}. Lenses L_1 and L_2 image plane P_{1a} onto plane P_{1b}. The light amplitude distribution emerging from plane P_{1b} is

$$U_1(x_1) = g(x_1)h(x_1), \tag{21}$$

where one-dimensional functions have again been used for simplicity. Lens L_3 forms the Fourier transform of $U_1(x_1)$ at plane P_3, where we obtain

$$U_3(x_3) = \int g(x_1)h(x_1) \exp(-i2\pi u x_1) \, dx_1. \tag{22}$$

If the input function $f(x_1)$ is now displaced horizontally from $x_1 = x_a$ to x_b and we look only at the point $x_3 = u = 0$ in plane P_3, we obtain

$$U_3(x_3) = \int_{x_a}^{x_b} g(x_1 + x')h(x) \, dx_1 \tag{23}$$

which is the desired correlation of g and h.

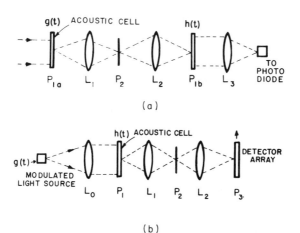

(a)

(b)

Fig. 7 Schematic representations of acoustooptic correlators: (a) spatial integrating correlator, (b) a time integrating correlator.

10. Application Areas

A myriad of practical problems complicate this correlation method. Chiefly, one of the inputs must be accurately displaced physically, and the time history of the output at $u = 0$ recorded in synchronization with the movement of plane P_{1a} to provide the desired correlation $g \star h$ at the output. This system is thus restricted to the correlation of 1D signals. Displacement of plane P_{1a} both horizontally and vertically is possible and would provide a time history output at $u = 0$ in plane P_3 that is equivalent to the 2D correlation. Such a scheme is of course a very complex mechanical effort, but it is possible.

Since the signals recorded at planes P_{1a} and P_{1b} cannot be negative, they must be recorded on a bias level. This bias level is removed by a dc stop in the Fourier transform plane P_2. Although this correlator has been used many times with input data on reels of film at plane P_{1a} and a synchronized output film drive at plane P_3, the mechanical motion required reduces the speed and accuracy of the system. Because this is basically an imaging system, its positioning requirements and coherence requirements are far lower than those of the frequency plane correlator. This basic topology is chiefly of interest because it can and has been successively implemented using acoustooptic cells at plane P_1 to produce easily and accurately the required displacement of one signal with respect to the other. We discuss this and other acoustooptic correlators in the next section. They offer the advantages of high bandwidth and speed, but are restricted to use with 1D signals only.

10.5.8 Acoustooptic Correlators†

When pattern recognition is applied to 1D signals and when the signal bandwidth is the important parameter, acoustooptic correlators are the most promising pattern recognition systems. We consider the two basic types of acoustooptic correlators. Numerous versions of each are under research and development and are documented elsewhere (*Opt. Eng.*, 1977).

The acoustooptic spatially integrating correlator is shown in Fig. 7a. Because of the phase modulation mechanism of the acoustooptic cell, a slightly different analysis method is used here. The input received signal $g(t)$ is fed to the input acoustic device at plane P_{1a}. The phase modulated output wave is described by

$$U_{1a}(x_{1a}) = \exp i[\omega t + K g(x_{1a} + v_a t)] \tag{24}$$

where K is a constant and v_a is the velocity of the acoustic wave in the plane P_{1a} material. With low phase modulation, we can rewrite Eq. (24) as

$$U_{1a}(x_{1a}) = [1 + iK g(x_{1a} + v_a t)] \exp(i\omega t). \tag{25}$$

† Flores and Hecht (1977).

520

The Fourier transform of U_{1a} formed at plane P_2 is

$$U_2(x_2) = [\delta(0) + iKG(u) \exp(i2\pi uv_a t)] \exp(i\omega t). \tag{26}$$

The zero spatial frequency term in Eq. (26) at $x_2 = 0$ is blocked at plane P_2 and the filtered pattern reimaged onto plane P_{1b}, where the reference signal $h(t)$ is stored as a spatial mask with transmittance $[1 + h(x_{1b})]$. The light amplitude distribution leaving plane P_{1b} is

$$U_{1b}(x_{1b}) = [1 + h(x_{1b})][iKg(x_{1a} + v_a t)] \exp(i\omega t). \tag{27}$$

Lens L_3 collects all of the zero-order transmitted light and focuses it onto a detector at plane P_3 whose time integrated output amplitude over the aperture time length T is the zero-order term in the transform of Eq. (27) or

$$U_3(t)\bigg|_{x_3=0} = K^2 \left| \int_0^{v_a T} g(x_{1a} + v_a t) h(x_{1b}) \, dx \right|^2, \tag{28}$$

which is the square of the correlation function.

Variations of this correlator using multiple reference masks, cylindrical/spherical optics, a single acoustic cell with two transducers, a reflective acoustic correlator, and acoustic correlators for chirp waveforms which are self-focusing have all been reported. The reference signal can be made variable rather than fixed by using a second acoustic cell at plane P_{1b} with the time reversed reference signal injected into the bottom of the cell.

The second type of acoustic correlator is a version of a noncoherent correlator (Section 10.5.14) and is shown schematically in Fig. 7b. The input signal $g(t)$ is added to a bias level B_1 and used to modulate the intensity I of the light source as

$$I(t) = B_1 + g(t). \tag{29}$$

The reference signal $h(t)$ is added to a bias level B_2 and used to amplitude modulate the carrier signal to the acoustooptic cell in plane P_1. With plane P_1 imaged onto the output plane P_3 and a Schlieren slit at plane P_2 passing only the first-order beam, the intensity detected at plane P_3 by a detector array is

$$I_3(x_3 t) = [B_1 + g(t)][B_2 + h(x_1 + v_a t)]. \tag{30}$$

Integration of the detector array output results in an output intensity at detector n after ac coupling, given by

$$I_3(x_3)_n = \int_0^T g(t) h(x + v_a t) \, dt, \tag{31}$$

which is the desired correlation.

As noted earlier, acoustooptic correlators are the choice when high bandwidth signals (200–400 MHz) must be correlated. Their major disadvantage is a limited time-bandwidth of 1000–2000 and their 1D nature.

10. Application Areas

10.5.9 Cross Path Correlator†

A coded waveform signal processor that easily produces the ambiguity function $\chi(v, \tau)$ of a coded waveform can be produced from a crossed-input pattern in which the function $g(t)$ whose ambiguity function is desired is rotated by $+45°$ and by $-45°$ and the two patterns superimposed. The input pattern can then be described by

$$t_1(x_1, y_1) = g\left(\frac{t + \tau}{\sqrt{2}}\right) g\left(\frac{t - \tau}{\sqrt{2}}\right). \tag{32}$$

Its 1D horizontal transform with respect to t is formed by a cylindrical/spherical lens system. The output plane pattern is

$$I(v, \tau) = 2|\chi(\sqrt{2}\,v, \sqrt{2}\,\tau)|^2, \tag{33}$$

which is proportional to the modulus squared of the ambiguity function of the coded waveform.

This system has been realized in real time using a pair of crossed acoustooptic lines as the input transducer. The input signals can now be fed to the acoustooptic lines at high frequency, 2D Fourier transformed, and the proper output cross-terms filtered. A 1D Fourier transform of this filtered pattern then yields the ambiguity function in range τ and Doppler v space.

10.5.10 Mellin Transform Correlator‡

The conventional optical processor is space invariant and thus the performance of all Fourier transform based optical pattern recognition systems will be degraded by scale changes present between the input and reference functions. Digital processors handle such practical problems by use of sophisticated software algorithms and have thus been more widely accepted than their optical correlator counterparts which require an exact match between the input and reference function if a severe loss in the peak intensity I_p or signal-to-noise ratio SNR of the output correlation is to be avoided. A recent optical approach to this problem is a space variant correlator in which a coordinate transformation is applied to the input data and this coordinate transformed data used as the input to a conventional space variant correlator.

In the particular system used to produce a scale-invariant correlator, the coordinates of the $g(x, y)$ input function are log scaled and $g(e^x, e^y)$ is used as the input to the correlator. The Fourier transform of $g(e^x, e^y)$ is the Mellin transform of $g(x, y)$. The required log scaling can be realized by analog log modules in the deflection system of the input device (spatial light modulator

---

† Said and Cooper (1973).
‡ Casasent and Psaltis (1976a, 1977).

522

or closed circuit television) or by computer-generated holograms. The system used in this or similar (Section 10.5.11) space variant correlators is shown in functional form in Fig. 8. It consists of a conventional correlator as in Fig. 1, with an input coordinate transformation preprocessing step. For the scale invariant Mellin transform, this transformation is $(x, y) \rightarrow (\xi, \eta) = (\ln x, \ln y)$. To describe this correlator mathematically, we use the development of Section 10.5.2. We concentrate only on the correlation term of concern and use one-dimensional functions for simplicity only. We record $h(e^x)$ at the input plane P_1. The pattern at plane P_2 is the Fourier transform of $h(e^x)$ or the Mellin transform $M_h(u)$ of $h(x)$. When this pattern and an off-axis plane wave are interferred at plane P_2, the term of interest in the resultant transmission of plane P_2 is

$$t_2(x_2, y_2) = M_h^*(u) \exp(-i2\pi\alpha x_2). \tag{34}$$

With this function recorded at plane P_2, consider an input function $g(x) = h(ax)$ (i.e., a scaled version of $h(x)$ with a scale factor "a"). At plane P_1 we record the log-coordinate scaled version $g(e^x)$ of this function. The light distribution incident on plane P_2 is then $M_g(u)$. The Mellin transforms M_g and M_h are related by

$$M_g(u) = a^{-i2\pi u} M_h(u) \exp(-i2\pi\alpha x_2)$$

$$= M_h(u) \exp(-i2\pi u \ln a) \exp(-i2\pi\alpha x_2), \tag{35}$$

from which we see that $|M_g| = |M_h|$ or that the Mellin transform is scale invariant. The light amplitude distribution leaving plane P_2 is now

$$M_g(u)M_h^*(u) = M_h(u)M_h^*(u) \exp(-i2\pi u \ln a) \exp(-i2\pi\alpha x_2). \tag{36}$$

Lens L_2 forms the Fourier transform of this pattern. The light amplitude distribution in plane P_3 is

$$U(x_3) = h \star h * \delta(x_3 + \alpha\lambda f_2 + \ln a). \tag{37}$$

From Eq. (37), we see that the amplitude of the output cross correlation of two functions that differ in scale is equal to the autocorrelation of the function

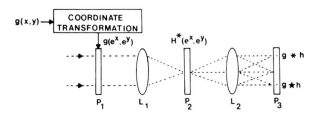

Fig. 8   Schematic representation of a space-variant correlator.

523

and hence that there is no loss in I_p or SNR and that this Mellin transform correlator is truly scale-invariant. From Eq. (37), we see that the location of the correlation peak is displaced from the normal $x_3 = -\alpha\lambda f_2$ location by ln a and thus that the scale change between the input and reference function can be found from the location of the correlation peak. This analysis and system are directly extendable to the 2D case, in which we find the horizontal and vertical displacement of the correlation peak to be proportional respectively to the horizontal and vertical scale differences between the input and reference functions.

A particularly unique application of this space variant correlator is in Doppler signal processing. Since a Doppler shift is equivalent to a scale change in the input signal and since the location of the output correlation peak is proportional to this scale change and hence the Doppler shift between the input and reference signal, novel approaches to Doppler signal processing are possible using a Mellin transform system. To demonstrate this application, the use of input data format control as a vital adjunct to optical pattern recognition, and to demonstrate another new correlator architecture, we consider the multichannel 1D joint transform correlator of Fig. 9a. The log-coordinate scaled reference coded waveform $h(e^x)$ is replicated on N lines on the left side of the input plane and the log-coordinate scaled versions $g_n(e^x) = h(e^{nax})$ of n Doppler versions $g_n(x) = h(nax)$ of it are recorded on N lines in the right-half of the input plane. If the physical length of each signal is b and the two patterns are recorded with a center-to-center spacing of $2b$, we describe the transmittance of plane P_1 by

$$t_1(x_1, y_1) = \sum_{n=1}^{N} [h(e^{x_1} - b, y_1 - nd_1) + h(e^{nax_1} + b, y_1 - nd_1)]. \quad (38)$$

The pattern recorded at plane P_2 by the cylindrical/spherical lens set L_{1s} and L_{1c} is the 1D horizontal Fourier transform of t_1. The subsequent transmittance of plane P_2 is

$$t_2(x_2, y_2) = \sum_{n=1}^{n} M_h(u, y_2 - nd_2)M_h^*(u, y_2 - nd_2)$$

$$\times \exp(-i2\pi un \ln a) \exp(-i2\pi u2b). \quad (39)$$

The pattern at plane P_3 is the 1D transform of Eq. (39) or

$$U_3(x_3, y_3) = \sum_{n=1}^{n} h \star h * \delta(x_3 + 2b + n \ln a, y_3 - nd_3), \quad (40)$$

from which the output plane pattern is seen to consist of N peaks of light on N lines. Each peak of light is equal to the autocorrelation of h (so that there is no loss in SNR or I_p) and the horizontal coordinate of each peak of light is proportional to the scale change "a" (and hence the Doppler difference)

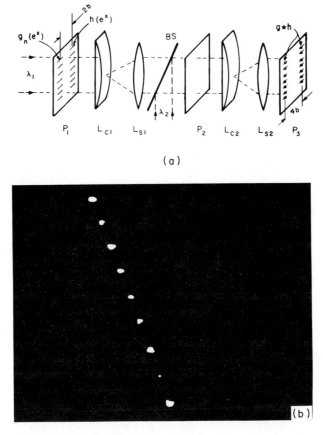

(a)

Fig. 9 Doppler processor using Mellin transforms. (a) Schematic representation of a 1D joint transform correlator, (b) output correlation of Doppler signals. (Casasent and Kraus, 1976.)

between the input and reference signal on the line in question. The output correlation plane pattern for $N = 9$ signals, formatted as described above is shown in Fig. 9b. The input signals $g(x)$ had a linear Doppler change of 100% from the top to bottom line, and hence the horizontal displacements of the correlation peaks in Fig. 9b track out the log a portion of the log curve from $a = 0$ to 1.

10.5.11   Rotation Invariant Correlator[†]

A second type of space-variant correlator using the same (coordinate transformation followed by a conventional space-invariant correlator) basic archi-

[†] Casasent and Psaltis (1976b).

tecture of Fig. 8 can be used to achieve rotation invariant correlation. The input coordinate transformation is now a polar transformation $(x, y) \rightarrow (\xi, \eta) = (\rho, \theta)$. We consider the input function to be $g(\rho, \theta)$ and the reference function to be a rotated version of it, $g'(\rho, \theta)$. The 1D transform (in θ) of g' is

$$G'(\rho, \omega_\theta) = G_1(\rho, \omega_\theta) \exp(-i2\pi\omega_\theta\theta_0)$$

$$+ G_2(\rho, \omega_\theta) \exp[i2\pi\omega_\theta(2\pi - \theta_0)], \quad (41)$$

where θ_0 is the angle through which the input is rotated, g_2 the portion of the unrotated input subtended by the polar angle from 0 to $2\pi - \theta_0$, g_1 the portion lying between $\theta = 2\pi - \theta_0$ and 2π, and G_1 and G_2 the 1D transforms of g_1 and g_2.

With g' placed at plane P_1 and the holographic matched filter G^* of g stored at plane P_2, the light distribution leaving plane P_2 is

$$G'G^* = G_1G_1^* \exp(-i2\pi\omega_\theta\theta_0) + G_2G_2^* \exp[i2\pi\omega_\theta(2\pi - \theta_0)], \quad (42)$$

The transform of Eq. (42) produced by lens L_2 at plane P_3 is

$$U(x_3, \theta') = g_1 \star g_1 * \delta(x_3, \theta' + \theta_0)$$

$$+ g_2 \star g_2 * \delta(x_3, \theta' - 2\pi + \theta_0). \quad (43)$$

From Eq. (43) we see that the output plane P_3 contains two correlation peaks separated by 2π, that the sum of their intensities equals the autocorrelation of g, and that the location of the correlation peak is proportional to the rotational difference θ_0 between the input and reference functions. Thus this correlator has an output SNR and I_p equal to that of the autocorrelation and is thus rotation invariant, and the rotation angle θ_0 can be found from the location of the output correlation peak.

10.5.12   Transposed Correlator[†]

The transposed processor is the first of the multiple matched spatial filter correlators we consider. The optical system of Fig. 1 is again used. However, now we place an array of multiple spatial reference functions $\{h_i\}$ side by side in the input plane P_1, and at plane P_2 we place a matched spatial filter G^* of the input function (i.e., the role of the input and filter functions are reversed). The transmittance of plane P_1 is described by

$$t_1(x_1, y_1) = \sum_{m,n} h_{mn}(x_1 - mW_{xp}, y_1 - nW_{yp}), \quad (44)$$

where an $m \times n$ array of functions h_{mn} separated from each other by W_{xp} and

---

† Vander Lugt and Rotz (1970).

W_{yp} are assumed. The light amplitude distribution incident on plane P_2 is

$$U_2(u, v) = \sum_{m,n} H_{mn}(u, v) \exp(-i2\pi um W_{xp} - i2\pi vn W_{yp}), \tag{45}$$

and the light amplitude distribution transmitted through plane P_2 is $G^*(u, v)U_2(u, v)$. The transform of this product appears centered at $(0, -\alpha\lambda f_2)$ in plane P_3 and is

$$H_{mn} \star g * \delta(x_3 + m W_{xp}, y_3 + n W_{yp}), \tag{46}$$

where unit magnification ($f_2 = f_1$) is assumed.

The location of the correlation peak in the output plane denotes which of the $m \times n$ reference functions h_{mn} is present in the input function g. The principal advantage of this correlator architecture is that common features of the $m \times n$ reference functions are suppressed and thus cross-correlation intensity reduced. The obvious disadvantages of this scheme are an increased input plane size with the more severe lens requirements associated with it, the need to record the reference functions as spatial (rather than frequency) functions spatially multiplexed, and the need to make a new matched spatial filter G^* of each new input function g.

10.5.13   Multiple Matched Spatial Filter Correlators†

A wide variety of multiple hologram storage schemes are discussed in Section 5.2 and we will not attempt a complete survey here. Rather, we note that only some of these schemes are applicable for use in storing multiple matched spatial filters for optical pattern or character recognition. If the matched spatial filter is formed from a composite input array of M functions

$$h = \sum_{i=1}^{M} h_i * \delta(x_1 - x_i, y_1 - y_i), \tag{47}$$

as before, and g is placed in the input plane, the output correlation plane pattern centered at $(x_3, y_3) = (-\alpha\lambda f_2, 0)$ consists of M correlation patterns:

$$U_3(x_3, y_3) = \sum_{i=1}^{M} g \star h_i \, \delta(x_3 - x_i, y_3 - y_i), \tag{48}$$

where we assume $f_2 = f_1$ in all cases.

If the full size of h is $W_{hx}W_{hy}$ and the size of each h_i is $W_{hxi}W_{hyi}$, we find that the columns and rows of h must be separated by at least $W_{gx} + W_{hxi}$ and $W_{gy} + W_{hyi}$, respectively. Assuming $W_{hx} \gg W_{gx}$ and $W_{hy} \gg W_{gy}$, we find that we require $\alpha\lambda f_2 \geq 3 W_{gx}/2$, and for optimum packing, $W_{hy} = 4W_{hx}$. This results in a required plane P_2 resolution of $1/(2\sqrt{2}\, W_{hx})$ and a plane P_2

---

† Burckhardt (1967).

bandwidth of $B = 2\sqrt{2}\,W_{hx}/\lambda\,f_1$. The maximum number of matched spatial filters that can be stored is $(B\lambda\,f_1)^2/2W_{gx}W_{gy}$ for $W_g \gg W_{hi}$ and $(B\lambda\,f_1)^2/2W_{hix}W_{hiy}$ for $W_{hi} = W_g$.

The multiple matched spatial filter can be made either by a single exposure (coherent method) or M multiple exposures (noncoherent method). However, the efficiency η and hence I_p and SNR of the correlation peak decrease as the number of multiple exposures increases. As a consolation, we find that we can store over three times the number of multiple filters on a plane P_2 material of given resolution using a multiple exposure method.

An averaged filter (Vienot *et al.*, 1973) can be produced by forming the matched spatial filter from an array (1D function assumed for simplicity) of M input functions described by

$$h(x_1) = \sum_{i=1}^{M} [h_i(x_1 - i\Delta_1) + \delta(x_1 - i\,\Delta_1 - \Delta_2)]. \tag{49}$$

When the magnitude squared of $h(x_1)$ is recorded, its impulse response will contain a term that is the desired average of the signal set $\{h_i\}$. Its holographic interference pattern is then the desired ''average'' filter. The recording resolution for this case is the same as that required for a single input image h_i. Separation of the desired terms in the transform is one major problem in this scheme. In addition, coherent recording of this filter in one exposure requires use of a separate impulse response for each element.

Intermodulation terms that arise in coherent (single exposure) multiple filter recording are difficult to mask out and are the source of the reduced (factor of 3) number of matched spatial filters possible in coherent versus noncoherent recording. Thus most researchers prefer the multiple exposure (or noncoherent) filter synthesis method and synthesize the filter by changing the reference beam angle or shifting the location of each reference function between exposures to encode each function by frequency multiplexing.

A final multiple matched spatial filter approach (Groh, 1970) involves forming an array of Fourier transforms of the input with a combination lens and multiple point source hologram and filtering these Fourier transforms with an array of matched spatial filters. If the same reference point location is used for more than one matched spatial filter, averaged filtering results, but interference fringes occur as the filter outputs overlap. The spatial rather than frequency multiplexing in the transform plane of this system and the associated lens system requirements are the major problems in this approach.

10.5.14 Noncoherent Correlators†

We have described various correlators which use holographic data recording to achieve correlation. A common element in all of these systems is the use

---

† Armitage and Lohman (1965).

of coherent light and its interference properties to record complex (amplitude and phase) data patterns. Although the subject of this volume is holography, we feel that at least some mention should be made of noncoherent correlators for pattern recognition if only in the interest of completeness.

In the first noncoherent correlator to be discussed (Armitage and Lohman, 1965), a bank of N spatially separated spatial reference functions are illuminated with monochromatic light and their spatially separated intensity only Fourier transforms $|S_i|^2$ recorded through a diffuser. This multiple reference pattern is then imaged onto an output plane with a transparency of the unknown input function serving as the aperture function of the system. The correlation plane pattern consists of N correlation images. The origin of each of these N correlation plane images is given by

$$U(x_3, y_3) = \int |S_i|^2 |S|^2 \, du \, dv, \qquad (50)$$

where the integral is over all space frequencies. These central correlation regions are dark for cross correlations and bright only for the autocorrelations.

A second noncoherent correlator topology that is most attractive and worth noting uses a light emitting diode (whose output intensity is modulated by a 1D input signal) as the single source (Monahan *et al.*, 1977). This light source is imaged by a condensor lens onto the entrance aperture of an imaging lens behind which a transparency of a reference library of N 1D signals is placed and focused into the output plane by an imaging lens. The input signal is described by

$$g(t) = B + Ks(t/a) + n(t), \qquad (51)$$

where B is the bias level, s the signal, n the background noise, and where K and "a" enable scale and frequency changes to be made in the input signal pattern. We denote the transmittance of the ith reference channel by

$$f_i(x) = B_i + K_i r_i(x/a_i). \qquad (52)$$

The light leaving the mask is then $g(t) f_i(x)$.

Between the mask and imaging lens a mirror is placed that rocks back and forth causing the $g(t) f_i(x)$ image to scan the output plane at a velocity v. The resultant light intensity incident on a vidicon placed in the output plane is

$$k_i(x, t) = g(t) f_i(x - vt - \phi), \qquad (53)$$

where ϕ is the phase of the mirror's scan. The vidicon integrates this light intensity over a single scan time T. This integral contains the correlation of the signal s and the N reference functions r_i. By varying v until $v = a_i/a$, a scale search of the signals can be achieved. By use of special binary masks, discrete cosine and Walsh transforms of the input signal are possible. The output vidicon can also be replaced by a linear or planar CCD array and noncoherent matrix multiplication achieved.

10.5.15 Matched Spatial Filter Parameter Optimization†

We now consider the synthesis of the matched spatial filters (MSFs) required in the frequency plane correlator. Specifically, we consider the selection of three MSF parameters: the bias exposure E_B, beam balance ratio K, and the spatial frequency band f^* in which K is set. As goodness criteria for the optimum correlation, we use the peak intensity I_p and signal-to-noise ratio SNR of the output correlation. In Eq. (5) we found the exposure incident on the MSF plate to be

$$E(x_2, y_2) = E_B[1 + 1/K + (2/\sqrt{K})\cos\psi], \tag{5}$$

where $E_B = r_0^2 T$, T is the exposure time, $E_0 = (r_0^2 + |H|^2)T$ is the average exposure, $K = r_0^2/|H|^2$ is the beam balance ratio, and $\psi = 2\pi\alpha x_2 + \arg(H)$. The subsequent transmittance of the MSF is a function of E_B, α, and the t–E curve of the film. The transmittance of the MSF can be described by

$$t = t_0 + (m/2)\cos\psi + A = t_0 + 2d\cos\psi + A, \tag{54}$$

where t_0 is the average transmittance, A denotes higher order terms, m is the peak-to-peak ac amplitude transmittance swing or the modulation of the cosine wave, and $d = \sqrt{\eta}$ is the amplitude diffraction efficiency. The peak intensity of the autocorrelation of h then becomes

$$I_p = \left|\iint H(x_1, y_1)d(x_1, y_1)\, dx_1, dy_1\right|^2.$$

Let us now analyze these results. $H(u, v)$ is a function of the spatial frequency and thus so are K, m, and d; m and d also depend on the exact t–E transfer curve of the film and the E_B and K values chosen. Since d and K vary with spatial frequency, the spatial frequency f^* at which K is measured must be provided. Such data are rarely included in any pattern recognition demonstration. Once $t(E)$ is known and E_B chosen, $d(K)$ can be found. Since K varies with spatial frequency, so does d, and it can be determined once the spatial frequency response $G(u, v)$ of $g(x_1, y_1)$ is known.

The fringe visibility $V = 2\sqrt{K}/(K + 1)$ versus η curves for most MSF materials are available, and from them the $d = \sqrt{\eta}$ versus K curves at various average exposures and the $\sqrt{\eta}$ versus E curves can be found. The linear portion of a constant E_0 curve is generally used. Linear recording requires $E < 2E_B$ (where E_B corresponds to $t = 0.5$) which corresponds to $K \geq 5.8$. This

---

† Casasent and Furman (1977b).

is consistent with the use of large K values $\geqslant 10$ for linear data recording. For $K \leq 0.17$ saturation occurs and for $0.17 \leq K \leq 5.8$ clipping occurs in MSF synthesis; K is normally chosen to be 1, and the question is at which f^* value to set $K = 1$. From these brief remarks, we see that MSF synthesis differs from normal holographic recording in which the desired output is a high quality image not a high quality correlation.

To best demonstrate the effects of these MSF parameters, the results of several experiments are described. For linear recording, an average exposure $t_0 \simeq 70\%$ and large K values are optimum. From extensive correlation experiments, graphs of I_p and SNR versus E_B were obtained for a variety of inputs, films, and lenses. The peak E_B value for all cases occurred within 10% of the same E_B value. A factor of 100 loss in I_p was observed for even a 50% decrease in E_B, and a 2:1 slope was found for the I_p versus SNR curve, indicating that noise increases with departures from optimum E_B and that proper E_B selection is critical. Since small changes in t_0 result in large E_B changes, E_B is the more sensitive parameter and the more realistic one to control than t_0.

If different image font data (e.g., rural, urban, and structured) are used, their spectra will be different with more energy in the higher (lower) frequency components of the transform of the urban (rural) imagery. We thus expect the optimum f^* value to be higher for the urban than the rural image. The typical variation of I_p versus f^* for a rural (A), urban (B), and structured (C) image shown in Fig. 10 demonstrate these expected trends. Since the spatial frequency content of an image generally decreases with increasing spatial frequency so does K. MSFs with low K values thus correspond to imagery whose high spatial frequency content is emphasized. In general low K values improve discrimination (but reduce I_p). In Fig. 10 we verify these data and provide a measure of how low a K value to select and the I_p and SNR loss expected for a given choice.

The I_p correlation value that results is proportional to the square of the area of the image (for approximately equally dense imagery) and the SNR correlation value obtained is proportional to the area of the image. The data of Fig. 10 and other correlation experiments have confirmed these expectations.

When multiple MSFs are used, the area of the frequency plane for each MSF must be restricted. To see the effects of this, we graphed I_p and SNR versus the size of an aperture placed in the MSF plane for various input fonts and f^* choices. We found that I_p and SNR for the MSFs formed with higher f^* values dropped as soon as the MSF aperture was reduced, whereas for those MSFs formed with lower f^* values, far lower I_p and SNR loss rates resulted until the aperture size was reduced to the f^* value of the MSF.

In these and other cases (Section 10.5.16), we find lower f^* values to be preferable to reduce the rate of change of I_p and SNR, and in practical situations it is preferable to accept a lower I_p and SNR autocorrelation value

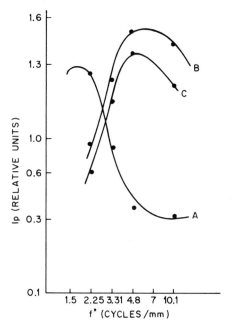

Fig. 10 I_p of the autocorrelation of a rural (*A*), urban (*B*), and structured (*C*) image versus the spatial frequency band f^* in which the beam balance ratio $K = 1$. (Casasent and Furman, 1977b.)

(by choosing lower f^* values than those that would produce maximum I_p and SNR values).

10.5.16   Sources of Correlation Degradation†

The extent of and the methods to overcome various practical sources of correlation degradation are now considered. All results reported here were obtained on aerial imagery. A scale change "*a*" between the input and reference function is an obvious source of correlation I_p and SNR loss that we have previously noted. I_p has been shown to decrease due to a scale change as $(1 - a)^4$ for a 2D image with a more severe rate of loss for higher space bandwidth data that for lower space bandwidth imagery. This has been experimentally verified for the case of a small apertured input and an MSF made from a large area reference function (case AF) and for the autocorrelation of the large area image (case FF). I_p was lower for case AF (since I_p is proportional to the square of the input area), but no appreciable loss in I_p was found

† Casasent and Furman (1977c).

until a 1% scale change occurred. For case FF, a severe 10-dB loss in I_p resulted from the same 1% scale change between the input and reference functions. The scaling correlator (Section 10.5.3) was used in these experiments.

The effects of rotational misalignments between the input and reference function have already been noted as a source of correlation error. To experimentally determine the magnitude of this error and the effect of different size input apertures, different input space bandwidths and f^* settings an MSF of a large ground area image was made and correlated against the full input (case FF) and three apertured regions of it that were predominantly rural, urban, and structural imagery (cases AF, BF, and CF, respectively). The results are shown in Fig. 11. The SNR loss as the rotation angle θ (between the input and reference) was increased was far more severe for case FF with a 20-dB loss in SNR for a 1.7° rotation (compared to only a 3-dB loss for a 2.5° rotation for case AF). The rate of decrease is clearly proportional to the input space bandwidth. The conclusion reached is that one should not use more input space bandwidth than needed or the system will require severe tolerances on rotation, scale, etc. When the analogous I_p versus θ curve is overlaid on this SNR versus θ curve, we find that both overlap indicating that noise is constant and the additional data stored in the MSF beyond that present in the input do not produce additional noise in the form of cross correlations.

The positional tolerance on the MSFs was also experimentally measured for these data and a negligible I_p and SNR loss encountered for 50-μm displace-

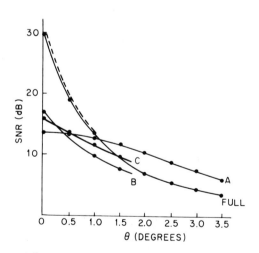

Fig. 11 SNR versus θ for rural, urban, and structured portions of a large area input correlated against an MSF made from the full area input (curves *A, B,* and *C,* respectively) and for the autocorrelation of the large area input (curve FULL). (Casasent and Furman, 1977b.)

ments of the MSF for any apertured input case, whereas a severe 20-dB SNR loss was encountered for a 50-μm displacement for case FF.

Considerable control of I_p and SNR versus θ can be realized by proper selection of the spatial frequency band f^* in which $K = 1$. In Fig. 12, SNR versus θ is shown for the autocorrelation of an urban image with $K = 1$ set at a low (band B), medium (band C), and high (band H) spatial frequency f^*. The effects are obvious. For curve H, f^* is high, the high spatial frequency image data are emphasized and a severe loss of SNR with θ results. For curve B, f^* is low, and the high frequency data that increases SNR is less emphasized and the reward is obvious. There is essentially no SNR loss with θ out to 0.12°, whereas a 25-dB loss results from curve H.

Similar tests of SNR and I_p variations with θ and scale have been performed for character recognition (Vander Lugt, 1965). We discuss specific differences between patterns and character recognition in the next section.

10.5.17   Pattern and Character Recognition

As the data in Sections 10.5.15 and 10.5.16 (all of which were taken on aerial imagery) indicate, the major problem in pattern recognition is obtaining a correlation peak. Because of the detailed pattern and structural content of an image, erroneous cross correlations do not seem to occur. However, in character recognition (as the example in Fig. 2 indicates), all inputs and all portions

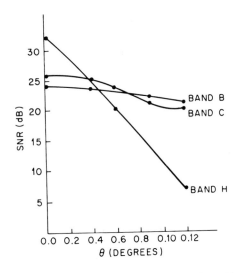

Fig. 12   SNR versus θ for the autocorrelation of an urban image with $K = 1$ at three spatial frequencies (f^* for band B is less than for band C which is less than for band H). (Casasent and Furman, 1977b.)

of the input consist of patterns or characters that are members of the general set to which the reference or key object belongs. Thus character recognition is generally characterized both by a multiple filter bank of somewhat similar patterns and likewise by the presence of strong erroneous cross correlations whose presence and detection must be suppressed. These cross correlations arise because the general shape of all letters are predominantly similar. The structure of all letters consists of vertical, horizontal, diagonal, and spiral lines of relatively fixed width. These all contribute to the strong erroneous cross correlations present in character recognition.

One solution to optical character recognition that seems most attractive is to apply digital algorithms to these output optical character recognition patterns. In this approach, one does not attempt to solve the entire problem by either optical correlation or by digital processing, but rather one uses the proper hybrid optical/digital processor that combines the best features of both approaches. An alternate approach that is most attractive is the use of optical word recognition rather than optical character recognition (Harris Corporation, 1976). Data indicate that the cross correlations decrease rapidly as the length of the key word or phrase searched for increases. The variance of the data likewise decreases appreciably for the optical word recognition case. This is in agreement with the reason for the lack of erroneous cross correlations in pattern recognition, where the space bandwidth, intensity, and textural variations in the optical input data are far higher.

The use of and need for a multiple MSF filter bank in some character recognition cases greatly complicates the resultant system and the MSF synthesis procedure and places severe requirements on the material used to record the multiple MSFs.

10.5.18   Summary, Conclusions, and Future Work

Thirteen different optical image and signal correlators have been described and their advantages and disadvantages discussed. The frequency plane correlator remains the most used, but we feel that the joint transform correlator is the most promising if multiple matched spatial filters are not required. Acoustooptic correlators are essential in high bandwidth signal processing applications.

If optical pattern or character recognition is to succeed, two system components are vital: input and matched spatial filter plane spatial light modulators. Considerable work and system integration is needed, with particular attention devoted to high optical and cosmetic quality devices. It is generally agreed that a hybrid optical/digital system with digital preprocessing and/or postprocessing will be the final system topology that will prevail. Much work is still needed here, especially in the area of digital algorithms appropriate for use on optically processed data.

10. Application Areas

A combined acoustooptic correlator whose output is recorded on a 2D spatial light modulator is an attractive system approach that promises to combine the high bandwidth advantages of acoustooptics and the high space bandwidth and 2D nature of an optical processor.

Space variant systems such as the Mellin transform must be pursued beyond initial conceptual designs and their applicability assessed in producing a correlator that is invariant to other expected distortions between the input and reference function. Advances in this area have recently been made and these systems applied to the correlation of nonvertical imagery with both image scale and tilt angle differences present (Casasent and Furman, 1977d). As a final note, we feel that if optical pattern recognition is to succeed, we must address real cases when distortion differences exist between the input and reference imagery.

REFERENCES

Armitage, J., and Lohmann, A. (1965). *Appl. Opt.* **4**, 464.
Burckhardt, C. (1967). *Appl. Opt.* **6**, 1359–66.
Casasent, D. (1977). *Proc. IEEE* **65**, 143–157.
Casasent, D., and Furman, A. (1977a). *Appl. Opt.* **16**, 285–286.
Casasent, D., and Furman, A. (1977b). *Appl. Opt.* **16**, 1662–1669.
Casasent, D., and Furman, A. (1977c). *Appl. Opt.* **16**, 1652–1661.
Casasent, D., and Furman, A. (1977d). *Appl. Opt.* **16**, 1955–1959.
Casasent, D., and Kessler, R. (1976). *Opt. Commun.* **17**, 242–244.
Casasent, D., and Klimas, E. (1978). *Appl. Opt.* **17**, 2058 (1978).
Casasent, D., and Kraus, M. (1976). *Opt. Commun.* **19**, 212–216.
Casasent, D., and Psaltis, D. (1976a). *Opt. Commun.* **17**, 59–63.
Casasent, D., and Psaltis, D. (1976b). *Appl. Opt.* **15**, 1795–1799.
Casasent, D., and Psaltis, D. (1977). *Proc. IEEE* **65**, 77–84.
Casasent, D., and Sterling, W. (1975). *IEEE Trans.* **C-24**, 348–358.
Flores, L., and Hecht, D. (1977). *SPIE J.* **118**, 182–192.
Groh, G. (1970). *Opt. Commun.* **1**, 454–456.
Harris Corp. (1976). Final Rep. on Contract 30602-75-C-0073, for RADC, April.
Monahan, M., Bromley, K., and Bocker, R. (1977). *Proc. IEEE* **65**, 121–129.
Opt. Engr. (1977). Special Issue on Acousto Optics (July).
Rao, J. (1967). *J. Opt. Soc. Amer.* **57**, 798.
Said, R. A. K., and Cooper, D. C. (1973). *Proc. Inst. Elec. Eng.* **120**, 423.
Vander Lugt, A. (1964). *IEEE Trans. Inform. Theory* **IT-10**, 139–145.
Vander Lugt, A. (1966). *Appl. Opt.* **5**, 1760–1765.
Vander Lugt, A., Rotz, F., and Klooster, A. (1965). "Optical and Electro-Optical Information Processing," pp. 125–141. MIT Press, Cambridge, Massachusetts.
Vander Lugt, A., and Rotz, F. (1970). *Appl. Opt.* **9**, 215.
Vienot, J., *et al.* (1973) *Appl. Opt.* **12**, 950–960.

10.6 IMAGE PROCESSING

Sing H. Lee

10.6.1 Introduction

Image processing means, in a broad sense, the manipulation of multidimensional signals, which are functions of several variables. Examples of multidimensional signals include reconnaissance photographs, medical x-ray pictures, television images, electronmicrographs, radar and sonar maps, and seismic data. The purpose of processing usually falls into one of the following categories: image enhancement, information extraction, efficient coding, pattern recognition, and computer graphics. We shall try to illustrate how the principles of holography and coherent optics can be applied to achieve various purposes of image processing. Some of these principles are based on linear, space-invariant processing, while others are based on nonlinear or space-variant processing.

10.6.2 Linear Processing of Images

Linear processing means the processed (output) image is linearly related to the original image. Examples of linear processing operations are bandpass filtering, subtraction, convolution, and correlation. Since lenses can conveniently Fourier transform images under coherent illumination, image enhancement by bandpass or high-pass filtering is easily achieved (Iwasa, 1976; Aldrich *et al.*, 1973; Ansley, 1969). In this section we shall, therefore, only describe and comment on spatial filtering and other more sophisticated methods (e.g., coherent optical feedback method).

10.6.2.1 Image Deblurring with Inverse Filters

Enhanced or deblurred images can be obtained from photographs which have been blurred either by accident (motion, imperfect focus, turbulence, etc.) or deliberately (e.g., when "coded" in view of special image processing or synthesis applications). Let the blurring or point spread function be $h(x, y)$ and the blurred image $g(x', y')$ described mathematically by

$$g(x', y') = \int \int f(x, y) h(x' - x, y' - y) \, dx \, dy, \qquad (1)$$

10. Application Areas

the enhanced image is obtained by performing a deconvolution operation on the blurred image. To carry out the deconvolution operation by spatial filtering, the filter function of $1/H(p, q)$ is needed because the Fourier transform of Eq. (1) is

$$G(p, q) = F(p, q)H(p, q), \tag{2}$$

and

$$F(p, q) = G(p, q)/H(p, q), \tag{3}$$

where $F(p, q)$, $G(p, q)$, and $H(p, q)$ are the Fourier transforms of $f(x, y)$, $g(x, y)$, and $h(x, y)$, respectively. To synthesize the filter function of $1/H(p, q)$, Stroke and Zech (1967) suggested the combined use of two filters, one with the amplitude transmittance of H^* and another with $1/(HH^*)$ as in Fig. 1. The H^* filter is made by the Vander Lugt technique of interfering an oblique plane wave with the Fourier spectra of $h(x, y)$. The $1/(HH^*)$ filter is obtained by careful photographic recording of the Fourier spectra of $h(x, y)$ with a gamma of (-2). The two filters are then sandwiched together and accurately aligned so that when they are illuminated, the product of their amplitude transmittances $(H^*)(1/HH^*)$ is obtained for the desired filter function $(1/H)$. Figure 2 shows a large pinhole x-ray photograph of the sun enhanced by this method (Stroke, 1970). Figure 3 shows another result on enhancing a scanning electron micrograph (Stroke and Halioua, 1971).

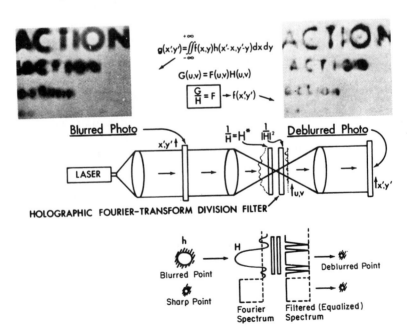

Fig. 1 Optical image deblurring with inverse filter (Stroke and Zech, 1967).

538

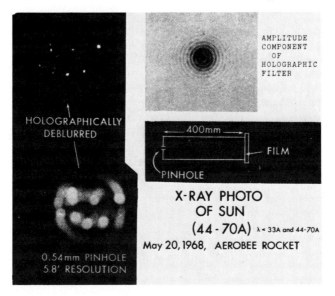

Fig. 2 The large-pinhole x-ray photograph of the sun, taken in May 1968, was deblurred by Stroke and Zech in the summer of 1969 (Stroke, 1970).

10.6.2.2 Image Coding and Decoding with Computer-Generated Spatial Filters

Image coding and decoding is an interesting and important field. If an image $f(x, y)$ is to be coded into $g(x, y)$ by spatial filtering, the filter function needed is $G(p, q)/F(p, q)$. Because a division of *two* complex functions is

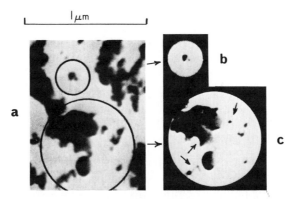

Fig. 3 (a) Original SEM micrograph, under optimum conditions (200-Å resolution, 50,000× magnification, 25 kV). (b) and (c) Holographically sharpened images showing resolution enhanced to better than 70 Å and correspondingly increased contrast (Stroke and Halioua, 1971).

539

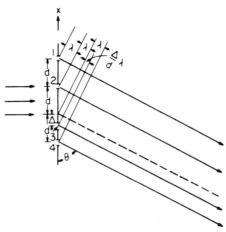

Fig. 4 Computer-generated spatial filters making use of the detour phase principle to synthesize complex filter functions (Lohmann and Paris, 1968).

involved, the filter function is more easily synthesized by computer than by ordinary holographic methods.

The computer-generated hologram for an arbitrary complex filter function was first invented by Lohmann and his co-workers (Brown and Lohmann, 1966; Lohmann and Paris, 1968). To generate this type of hologram, the complex field is first sampled. The complex field at each sampled point is represented by a slot with size proportional to amplitude and lateral displacement (from the sampled location) proportional to the phase. The computer-generated hologram can be considered as a diffraction grating with purposely introduced defects. The desired complex field is obtained by means of the detour phase effect at one of the diffraction orders (Fig. 4). Such a hologram filter for converting the letter "G" into the sign "+" is shown in Fig. 5. The experimental result employing this filter was obtained by Lohmann and his co-workers (1967) as shown in Fig. 6.

10.6.2.3 Image Subtraction and Differentiation with Gratings

To observe the changes between two scenes or to evaluate the rate of change in information within a scene, image subtraction and differeniation operations are useful. Holographic and computer techniques have been applied to generate spatial filters for these operations (Lohmann and Paris, 1968; Bromley *et al.*, 1971). Simple gratings or the superposition of a few of them have also been shown to be valuable in achieving the same objectives (Lee *et al.*, 1970;

Fig. 5   The binary spatial filter for the conversion G → + (Lohmann *et al.*, 1967).

Yao and Lee, 1971). Since gratings are generally available or easily producable, the theory of their filtering operations is discussed here.

For image subtraction, a sinusoidal grating whose maximum transmittance is displaced by a quarter of a fringe from the optical axis will provide results of optical subtraction in the central portion of the output plane as shown in Fig. 7, when the two nonoverlapped input images are symmetrically positioned at distances of $\pm(p\lambda f/2\pi)$ from the optical axis, where $(p/2\pi)$, λ, and f are the grating frequency, optical wavelength, and lens focal length, respectively. To analyze this process, let the transmittance function of sinusoidal grating be

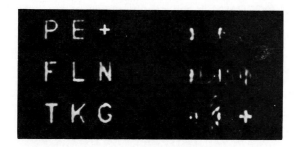

Fig. 6   The image of the code translator process. At left the zeroth diffraction of the grating-like spatial filter, reproducing the object. At the right the filter output with + in place of G (Lohmann *et al.*, 1967).

10. Application Areas

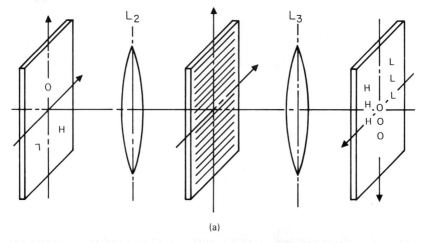

(a)

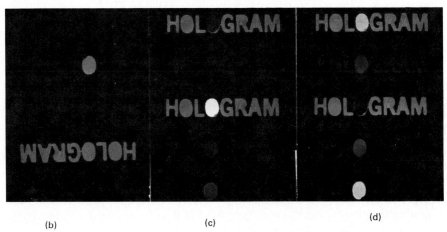

(b)   (c)   (d)

Fig. 7 Complex amplitude addition and subtraction with gratings. (a) The optical system. (b) The two input pattern functions. (c) The pattern function "0" is added onto that of "HOLOGRAM" in the central region of the output plane. (d) The pattern function "0" is subtracted from that of "HOLOGRAM" in the central region of the output plane (Lee *et al.*, 1970).

represented by

$$H(p, q) = \tfrac{1}{2}\{1 + \exp[i(px + \pi/2)] + \exp[-i(px + \pi/2)]\}, \qquad (4)$$

and the light incident on the grating be

$$G_i(p, q) = G_1(p, q)e^{-ipx} + G_2(p, q)e^{ipx}, \qquad (5)$$

where $G_1(p, q)$ and $G_2(p, q)$ are the Fourier transforms of the two input functions g_1 and g_2 in the upper half and lower half of the input plane,

542

respectively. The light amplitude behind the grating is then

$$G_0(p, q) = G_i(p, q)H(p, q)$$

$$= \tfrac{1}{2}[G_1(p, q)e^{i\pi/2} + G_2(p, q)e^{-i\pi/2}] + 4 \text{ other terms,} \quad (6a)$$

and we obtain for the output

$$g_0(x, y) = (i/2)[g_1(x, y) - g_2(x, y)] + 4 \text{ other terms.} \quad (6b)$$

For image differentiation, two sinusoidal gratings of slightly different frequencies will be necessary, with the maximum transmittance of one displaced by half a fringe with respect to the other. The transfer function and the impulse response of this composite grating are

$$H(p, q) = 1 + \cos(px) - \cos[(p + \epsilon)x],$$

$$h(x, y) = \tfrac{1}{2}[\delta(x + 2\pi/p, y) - \delta\{x + 2\pi/(p + \epsilon), y\}] + 3 \text{ other terms.} \quad (7)$$

An input image processed by the composite grating will yield at one of the diffraction orders the differentiation output (Fig. 8) because

$$g_0(x, y) \propto \lim_{\epsilon \to 0} (1/\epsilon)[g(x, y) * \{\delta(x + 2\pi/p, y)$$

$$- \delta(x + 2\pi/(p + \epsilon), y)\}]$$

$$= \lim_{\epsilon \to 0} (1/\epsilon)[g(x + 2\pi/p, y) - g(x + 2\pi/(p + \epsilon), y)]$$

$$= \partial g/\partial x, \quad (8)$$

where $*$ denotes the correlation operation.

10.6.2.4   Contrast Control with Coherent Optical Feedback

The visual quality of an image is to a large extent dependent on the contrast or the relative intensities of the information-bearing portions of an image and the everpresent background. The contrast of a (developed) photographic transparency, in certain instances, needs to be altered. For example, the contrast of aerial photographs often needs to be reduced, whereas the contrast of x-ray photographs requires an increase. Coherent optical systems with feedback (Fig. 9) can be utilized to control contrast when the input transparency is made to modulate the light multiply reflected between the feedback mirrors before escaping from the feedback system (Jablonowski and Lee, 1975; Lee et al., 1976). Depending on the mirror separation, image contrast can be enhanced or reduced as the results of constructive or destructive interferences between multiple reflections.

The output amplitudes from the coherent feedback systems of Figs. 9a and

Fig. 8 Optical differentiation with composite gratings. (a) The object pattern. (b) Experimental results of $\partial g/\partial x$. (c) Experimental results for $\partial g/\partial x + \partial g/\partial y$. (d) Experimental results for $\partial^2 g/\partial x^2 + \partial^2 g/\partial y^2$ (Yao and Lee, 1971).

9b can easily be derived to give Eqs. (9a) and (9b), respectively:

$$\frac{a_0(x, y)}{a_i} = t_i(x, y)t_m^2[1 + r_m^2 t_i(x, y)e^{i\phi} + r_m^4 t_i^2(x, y)e^{i2\phi} + \cdots]$$

$$= t_i(x, y)t_m^2/\{1 - r_m^2 t_i(x, y)e^{i\phi}\}, \tag{9a}$$

$$\frac{a_0(x, y)}{a_i} = t_i(x, y)t_m^2/\{1 - r_m^2 t_i^2(x, y)e^{i\phi}\}, \tag{9b}$$

where a_i is the light amplitude of input illumination, $t_i(x, y)$ the amplitude transmittance of the original image, r_m, t_m the mirror amplitude reflectance and transmittance, and $e^{i\phi}$ the phase delay of light traveling between mirrors which is dependent on mirror separation. There is a difference in the denominators of Eqs. (9a) and (9b) because the light reflected between mirrors experience once or twice modulations by $t_i(x, y)$, respectively, during each round trip. The corresponding output intensities are

$$T_c(x, y) = \left|\frac{a_0(x, y)}{a_i}\right|^2 = \frac{T_i(x, y)T_m^2}{1 + R_m^2 T_i(x, y) - 2R_m t_i(x, y) \cos \phi}, \tag{10a}$$

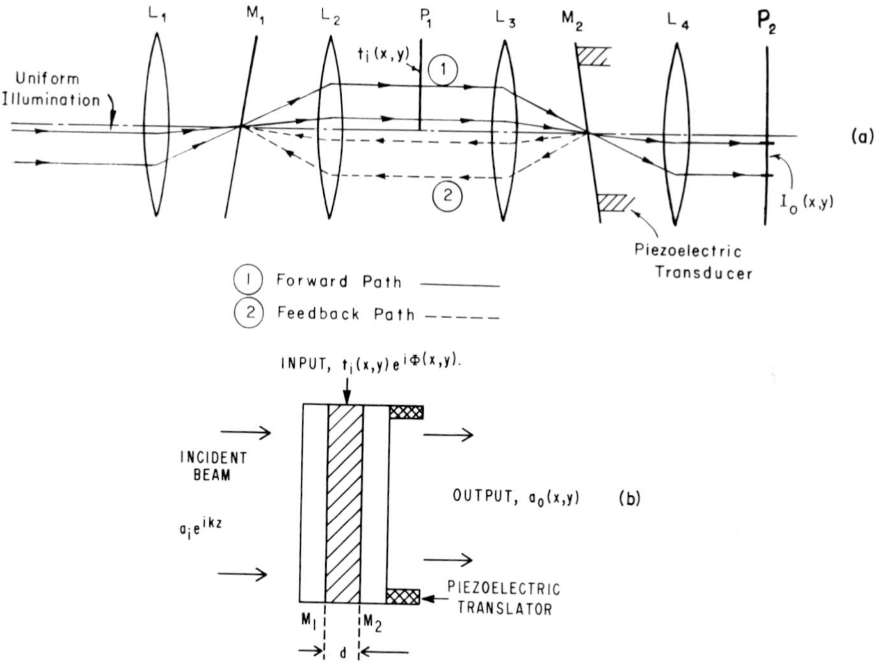

Fig. 9 Contrast control with coherent optical feedback. (a) A coherent feedback system with lens and plane mirror (Jablonowski and Lee, 1975). (b) The plane parallel mirror system (Lee *et al.*, 1976).

$$T_c(x, y) = \frac{T_i(x, y) T_m^2}{1 + R_m^2 T_i^2(x, y) - 2R_m T_i(x, y) \cos \phi},$$ (10b)

where $T_i(x, y)$, T_m, R_m are intensity transmittances and reflectances. Therefore, by controlling the mirror separation, which affects $e^{i\phi}$, the output intensities will show various contrast (see Fig. 10a and b). Experimental results are illustrated in Fig. 11.

Beside contrast control, coherent optical feedback systems can also be useful for deblurring images and solving partial differential equations (Jablonowski and Lee, 1975; Cederquist and Lee, 1977); beside image coding and decoding, computer-generated spatial filters can also be applied to deblurring and differentiating images (Lohmann and Paris, 1968). Hence, the discussion in this section on linear processing is not intended to be exhaustive, but rather to exemplify that there are a number of coherent optical techniques suitable for a variety of image processing tasks.

10.6.3 Nonlinear Processing of Images

Logarithm, quantization, intensity level slicing, thresholding, and analog-to-digital conversion are a few interesting and important nonlinear processing

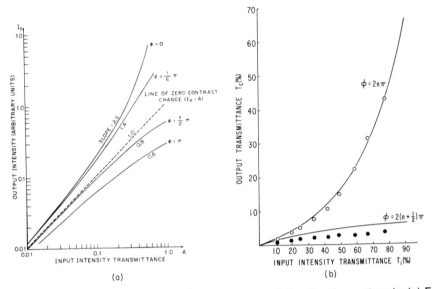

Fig. 10 The output versus input transfer characteristics for the system in (a) Fig. 9a, $R_1 = R_2 = 0.9$; (b) Fig. 9b, $R_1 = 0.65$, $R_2 = 0.54$.

operations which have been successfully demonstrated with coherent optics. For implementing these nonlinear operations, several schemes are presently available. They are the halftone-screen process, theta modulation, and nonlinear devices with feedback. The principles of operation of these schemes are now summarized.

10.6.3.1 Halftone Screen Process

When a slowly varying object function $g_i(x, y)$ is contact printed onto a high contrast film through a halftone screen, an image consisting of a dot array results. The size of the dots is dependent upon both $g_i(x, y)$ and the dot profile of the halftone screen as illustrated in Fig. 12 (Kato and Goodman, 1975). By properly controlling the dot profiles of the halftone screens, the size of the dots in the halftone image will be nonlinearly related to $g_i(x, y)$. Then, upon low-pass filtering either by eye or with an optical processor, the halftone image will yield a filtered image $g_0(x, y)$ which is nonlinearly related to $g_i(x, y)$ in a monotonic manner.

Experimentally, this principle has been verified for the logarithmic transformation by modulating or coding the input image with a logarithmic contact screen, which was obtained by making a contact negative duplicate of a Kodak Gray Contact Screen (100 lines/in., elliptical dot) on Kodak Contrast Process Ortho film (Kato and Goodman, 1975). In Fig. 13 an input image which is the product of two input components is converted by logarithmic transformation into the sum of the two input components. In this illustration the two input components are gray tone gratings oriented perpendicular to each other. It is noted that when the transmittance through the two crossed component gratings are recorded through the logarithmic contact screen, the resultant coded image yields a spectrum which is the superposition of the two spectra from the two component gratings, with each component grating yielding spectral contents along one spectral axis only. On the other hand, when the same transmittance through the two crossed gratings are recorded linearly without using the log-

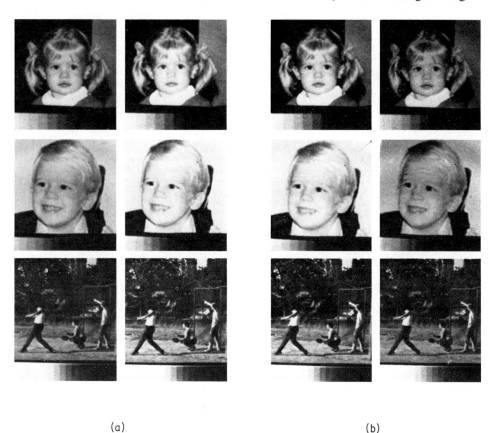

(a)   (b)

Fig. 11   (a) Experimental results of contrast enhancement with the system shown in Fig. 9a. Left column: references. Right column: output from feedback (Jablonowski and Lee, 1975). (b) Experimental results of contrast reduction with the system shown in Fig. 9a. Left column: references. Right column: output from feedback (Jablonowski and Lee, 1975). (c) Experimental results of contrast control with the system shown in Fig. 9b. Picture in the center is the reference. Contrast increases in clockwise direction (Lee *et al.*, 1976).

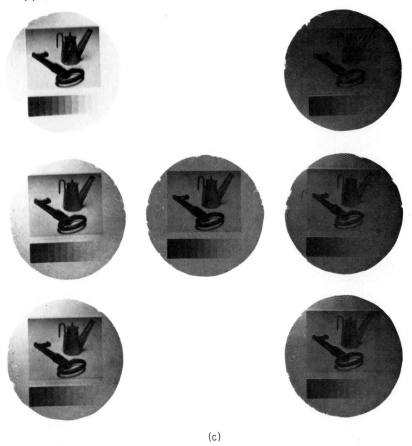

(c)

Fig. 11 Continued

arithmic screen, the resultant image yields a spectrum which is the convolution of the two spectra from the two component gratings, thus yielding the cross modulation spectral components off-axis.

To achieve nonmonotonic nonlinear effects, higher diffraction orders from the halftone image must be selected, instead of low-pass filtering (Sawchuk and Dashiell, 1975; Lohmann and Strand, 1975). To understand the concepts involved, consider the halftone image to consist of many localized regions. In every localized region we have a simple rectangular grating whose grating width w is dependent on the transmittance of the original input in that region. The diffraction from the rectangular grating to higher orders will be nonmonotonically dependent on the grating width, though the dependence is monotonic at the zero order, as shown in Fig. 14. Since the width of the grating in each localized area in the halftone image is controlled by the transmittance of the

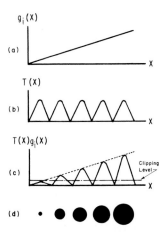

Fig. 12   The halftone screen process. (a) Continuous-tone light distribution input. (b) Transmission characteristics of the halftone screen. (c) Light distribution falling on hard clipping film. (d) Halftone image recorded on hard clipping film (Kato and Goodman, 1975).

original input in the same area, the diffraction to higher orders will also be nonmonotonically dependent on the input. Experimentally, the concepts have been verified for the level slicing operation (Fig. 15) (Sawchuk and Dashiell, 1975), in isophote production (Fig. 16), and analog-to-digital conversion (Fig. 17) (Lohmann and Strand, 1975; Strand, 1975, 1976).

10.6.3.2   Theta Modulation Technique

The original object $g_i(x, y)$ is converted into a modulated signal $g_m(x, y)$ with a local grating angle θ, which is proportional to the amplitude distribution in the object (Armitage and Lohmann, 1965)

$$\theta(x, y) = K g_i(x, y), \qquad K = \pi/\max g_i. \tag{11}$$

An example of the modulation scheme is illustrated in Figs. 18a and 18b. When the modulated signal is illuminated with a collimated, coherent beam in a coherent optical processing system, light is diffracted into various angles in the Fourier plane. In fact, the light from all elemental gratings in $g_m(x, y)$ oriented in the same angle, which correspond to all image elements of the same intensity in $g_i(x, y)$, is diffracted into one angle in the Fourier plane (Fig. 18c). Now, if a filter is placed in this Fourier plane whose transmission function $T(\theta)$ is a nonlinear function of the azimuth angle θ, the output image amplitude $g_0(x, y)$ will be nonlinearly related to $g_i(x, y)$. An example is given in Fig. 18d. The following three examples of $T(\theta)$ will further illustrate the versatility of this nonlinear processing technique.

549

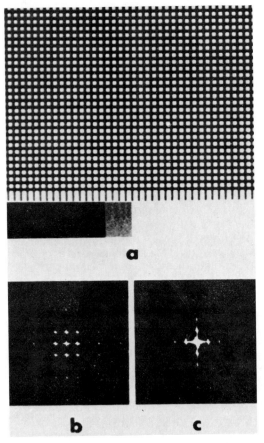

Fig. 13 The effect of the logarithmic transformation on the Fourier spectrum. (a) The original pattern of two multiplied gratings perpendicular to each other. The dynamic range is from 0 to 2 in density. (b) Normal spectrum of the linearly copied crossed gratings, with intermodulation. (c) Spectrum of the logarithmically transformed crossed grating obtained using the logarithmic contact screen (Kato and Goodman, 1975).

Example 1 If the filter $T(\theta)$ is a slit oriented at one angle θ, the output image will be an equiamplitude or equidensity line image, i.e., in the image $g_0(x, y)$ appear sharp lines, representing the contour for one amplitude value in $g_i(x, y)$. If the filter $T(\theta)$ consists of multislits, instead of a single slit, oriented at equiangular spacings, the output image becomes a contour map of equiamplitudes.

Example 2 If the filter $T(\theta)$ is zero for $0 < \theta < \theta_0$ and unity for $\theta_0 < \theta < 2\pi$,

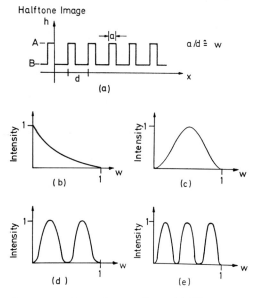

Fig. 14 Diffraction from a rectangular grating. (a) The rectangular grating. (b) Zeroth diffraction order. (c) First diffraction order. (d) Second diffraction order. (e) Third diffraction order (Lohmann and Strand, 1975).

the threshold operation results with the elimination in the output image of any regions of amplitudes or intensities below a certain value corresponding to θ_0.

Example 3 If the filter $T(\theta)$ has a logarithmic transmittance dependence on θ, the output will be logarithmically related to the original object.

Though the theta modulation technique is versatile, one should be cautioned that it presently suffers from the lack of a practical method for accomplishing the coding operation required to convert $g_i(x, y)$ into $g_m(x, y)$ of good resolution. A promising coding method under investigation, however, will be offered: it is one of the few versions of the combined scanning laser beam and Mach–Zehnder interferometer system,† since laser beam scanners capable of scanning 1000 resolvable spots or better are commercially available and Mach–Zehnder interferometers can easily provide more than 10 grating lines within each scan spot of typical size 15 to 20 μm. One version of the combined system is shown in Fig. 19.

† The investigation on the combined scanning laser beam and Mach=Zehnder interferometer systems for θ-modulation is currently jointly conducted by G. Dial and S. H. Lee at UCSD.

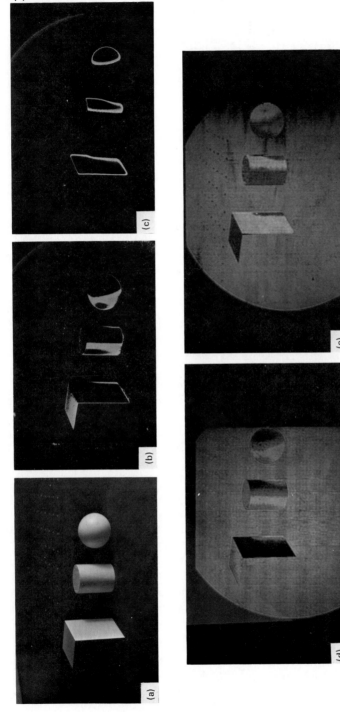

Fig. 15 Nonmonotonic nonlinear processing with halftone screen. (a) Original photograph of geometrical figures to be processed. (b) Level sliced at one setting. (c) Level sliced at another setting. (d) Quantified to three levels. (e) Notch filtered (Sawchuk and Dashiell, 1975).

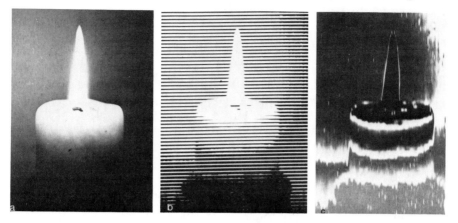

Fig. 16 Isophot results. (a) Original image. (b) Halftone image. (c) Isophots (Strand, 1975, 1976).

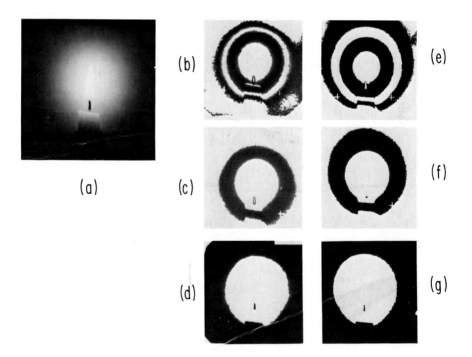

Fig. 17 Results of optical bit-plane generation compared to digital electronic bit-plane generation. (a) Original image. (b) First bit-plane optically generated. (c) Second bit-plane optically generated. (d) Third bit-plane optically generated. (e) First bit plane electronically generated. (f) Second bit-plane electronically generated. (g) Third bit-plane electronically generated (Strand, 1975, 1976).

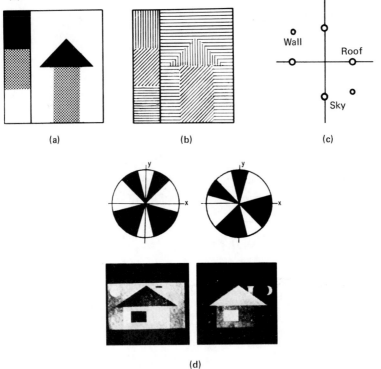

Fig. 18 Principle of theta modulation. (a) Object with gray ladder. (b) Same object in theta modulated form. (c) Diffraction pattern of theta modulated object. (d) Results of applying theta modulation principle to modulate one object, which is then demodulated with two different masks (Armitage and Lohmann, 1965).

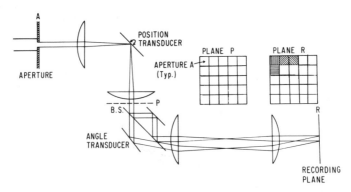

Fig. 19 One version of combining a laser beam scanner with a Mach–Zehnder interferometer to produce an array of small gratings of controlled frequencies and orientations. Both the laser beam scanner and the Mach–Zehnder interferometer are controlled by a microprocessor. (See footnote on p. 551.)

554

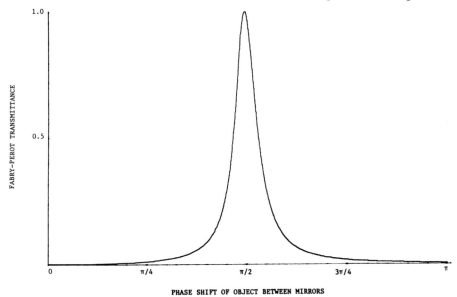

Fig. 20 Transmittance of high finesse Fabry–Perot interferometer. R_m = 95% (Bartholomew, 1978).

10.6.3.3   Nonlinear Devices

Intensity level slicing can be performed on images if their intensity variations are recorded as phase variations on a transparent medium located between the mirrors of a high finesse Fabry–Perot interferometer (Lee *et al.*, 1976; Bartholomew, 1978).‡ The transmittance $T_p(x, y)$ of the interferometer with a phase variation $\Phi(x, y)$ recorded on the transparent medium is

$$T_p(x, y) = T_m^2/[1 + R_m^2 - 2R_m \cos\{\phi + 2\,\Phi(x, y)\}]. \tag{12}$$

For high finesse, large values of R_m are chosen. Then the device acts as a narrow band filter, transmitting light only in those areas of the image where $\Phi(x, y) + (\phi/2) = n\pi$ (Fig. 20). If the phase variation $\Phi(x, y)$ is recorded as a monotonic function of the input intensity with a range of less than π, different values of Φ can be selected by a piezoelectric translator which controls the mirror spacing and ϕ. With R_m = 95%, the full width at half maximum of T_p is about 0.1 rad; approximately 30 values of Φ or 30 gray levels of an image can be resolved within one free spectral range of interferometer scanning.

‡ B. Bartholomew is the principal investigator on the nonlinear processing scheme described in this section.

Fig. 21 Optical analog-to-digital conversion of an image with eight gray levels (Bartholomew, 1978).

This device can also be used to compute different functions of the original intensity distribution. For example, suppose the square root of the original image intensity distribution is desired and the image is recorded such that Φ is proportional to the intensity. The output image is constructed by incrementing the mirror separation to select the various values of Φ. For each value of Φ, the intensity of the incident beam is made equal to the square root of the intensity of the original. The intensity distribution in the output will thus be the square root of the intensity distribution in the input.

As another example, analog-to-digital conversion of an image can be achieved (Fig. 21). The least significant bit plane of an eight gray level image is generated by turning the laser on when levels 1, 3, 5, and 7 are selected. It is on for levels 2, 3, 6, and 7 for the next most significant bit plane and it is on for levels 4, 5, 6, and 7 for the most significant bit plane. Thus, the gray level of the roof in Fig. 21a, which is level 4, is converted to white, black, black (100) in the three output bit planes (Fig. 21b).

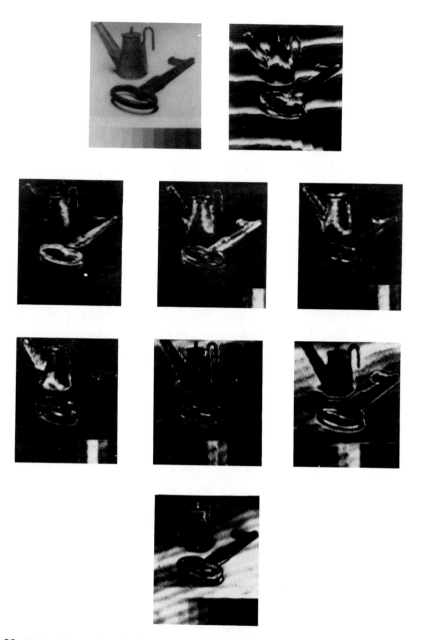

Fig. 22 Intensity level selection with bleached photographic plate. Top left: original image; top right: output of the system with the bleached plate between the mirrors. The mirrors have been tilted slightly to emphasize the phase shifts produced by the bleaching process. The remaining pictures show intensity level selection with the mirrors parallel. The individual levels were selected by moving one mirror with a piezoelectric translator (Lee *et al.*, 1976).

Experimentally, intensity level selection was demonstrated with a bleached image recorded on a high resolution photographic plate. Results are shown in Fig. 22. Research to demonstrate other nonlinear processing operations and to replace the bleached photographic plate with real time electrooptic materials is in progress.

10.6.4 Space-Variant Operations on Images

In Sections 10.6.2 and 10.6.3 linear and nonlinear processing operations were discussed. Most of these operations are space-invariant in the sense that all points within the input field are identically affected by the operations. In this section we shall discuss space-variant operations which generally affect different input points differently.

An example of space-variant operation is geometrical transformation by which $g_1(x, y)$ is transformed into $g_2(p, q)$. Using computer-generated holograms in the optical system shown in Fig. 23a, Bryngdahl has obtained the experimental results shown in Fig. 23b (Bryngdahl, 1974a,b). The principle behind the design of the computer-generated hologram for geometrical transformation can be understood by considering it as a generalized grating in which

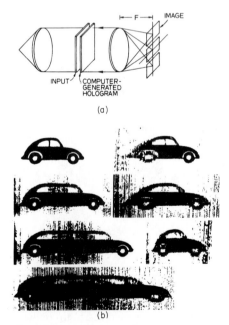

Fig. 23 (a) Optical method for realizing a coordinate transformation with a computer-generated hologram. (b) Experimental results of (a) (Bryngdahl, 1974a).

the grating frequency varies as a function of the coordinate (x, y). The local spatial frequency of the grating (ν_x, ν_y) at location (x, y) will diffract light to location (p, q) in the focal plane of the lens:

$$p = (\lambda f)\nu_x, \qquad q = (\lambda f)\nu_y, \tag{13}$$

where λ is the wavelength of light and f is the focal length of lens. (Geometrical transformations can, of course, be accomplished also by means of a nonlinear beam-scanning device which writes the data onto a coherent optical light valve with storage capability (Casasent and Psaltis, 1976). But this method sacrifices the fully parallel processing advantages normally associated with optical processing, which in some cases is acceptable.)

Other optical methods for implementing various space-variant operations exist, e.g., holographic multiplexing techniques (Deen *et al.*, 1975) and matrix multiplication techniques (Heinz *et al.*, 1970). Descriptions and present status of these methods can be found in the excellent review article by Goodman (1979). Though interest in space-variant processing is fairly recent, further progress can be expected in the future in view of the increased attention such problems are receiving now.

10.6.5 Concluding Remarks

We have tried to illustrate how the principles of coherent optics and holography can perform various image processing tasks, although no attempt was made to summarily review the vast amount of work reported in the literature of image processing. Some of the principles of coherent optics and holography are based on linear, space-invariant processing, while others are based on nonlinear or space-variant processing. Certainly, the processing techniques can be combined to solve more complex image processing problems. For example, linear processing techniques described in Section 10.6.2 may be combined with coordinate transformation techniques mentioned in Section 10.6.4 to solve scale-invariant pattern recognition problems (Casasent and Psaltis, 1976), to restore images degraded by certain aberrations (Sawchuk and Peyrovian, 1975), or to restore motionblurred, distorted imagery (Sawchuk, 1973).

Besides processing images with optical techniques, there are also many important digital techniques (Huang, 1975; Andrews, 1970). Generally speaking, optical methods offer the advantage of capacity, the ability to handle images of high information content. Digital methods offer the advantage of versatility, the ability to readily program a combination of many mathematical operations. Continued development in optical–digital interface devices will probably lead to the fruitful result of hybrid processing in which the advantages of both optical and digital methods are realized.

ACKNOWLEDGMENTS

The support of the National Science Foundation in the area of optical image processing is gratefully acknowledged. Partial support has also come from the Air Force Office of Scientific Research and National Aeronautics and Space Administration at Goddard Space Flight Center.

REFERENCES

Aldrich, R. E., Krol, F. T., and Simmons, W. A. (1973). **IEEE Trans.** *Electron Devices* **ED-20,** 1015.

Andrews, H. C. (1970). "Computer Techniques in Image Processing." Academic Press, New York.

Ansley, D. A. (1969). *Electro-Optical System Design* pp. 26–34. (July/Aug.).

Armitage, J. D., and Lohmann, A. W. (1965). *Appl. Opt.* **4,** 399–403.

Bartholomew, B. (1978). Ph.D. thesis, Univ. of California at San Diego.

Bromley, K., Monahan, M. A., Bryant, J. F., and Thompson, B. K. (1971). *Appl. Opt.* **10,** 174–181.

Brown, B. R., and Lohmann, A. W. (1966). *Appl. Opt.* **5,** 967–969.

Bryngdahl, O. (1974a). *Opt. Commun.* **10,** 164–168.

Bryngdahl, O. (1974b). *J. Opt. Soc. Amer.* **64,** 1092–1099.

Casasent, D., and Psaltis, D. (1976). *Opt. Eng.* **15,** 258–261.

Cederquist, J., and Lee, S. H. (1977). A Confocal Fabry-Perot for the Solution of Partial Differential Equations. *Proc. Electro-Opt. Syst. Design/Internat. Laser Conf., Anaheim, California.*

Deen, L. M., Walkup, J. F., and Hagler, M. O. (1975). *Appl. Opt.* **14,** 2438–2446.

Goodman, J. W. (1979). *In* "Optical Information Processing" (S. H. Lee, ed.), Topics Vol. in Appl. Phys. Springer-Verlag, New York. (to be published).

Heinz, R. A., Artman, J. O., and Lee, S. H. (1970). *Appl. Opt.* **9,** 2161–2168.

Huang, T. S. (1975). "Picture Processing and Digital Filtering." Springer-Verlag, New York.

Iwasa, S. (1976). *Appl. Opt.* **15,** 1418–1424.

Jablonowski, D. P., and Lee, S. H. (1975). *Appl. Phys.* **8,** 51–58.

Kato, H., and Goodman, J. W. (1975). *Appl. Opt.* **14,** 1813–1824.

Lee, S. H., Yao, S. K., and Milnes, A. G. (1970). *J. Opt. Soc. Amer.* **60,** 1037–1041.

Lee, S. H., Bartholomew, B., and Cederquist, J. (1976). *Proc. SPIE* **83,** 78–84.

Lohmann, A. W., and Paris, D. P. (1968). *Appl. Opt.* **7,** 651–655.

Lohmann, A. W., and Strand, T. C. (1975). *Proc. Electro-Opt. Syst. Design/Internat. Laser Conf., Anaheim, California* pp. 16–21.

Lohmann, A. W., Paris, D. P., and Werlich, H. W. (1967). *Appl. Opt.* **6,** 1139–1140.

Sawchuk, A. A. (1973). *J. Opt. Soc. Amer.* **63,** 1053–1063.

Sawchuk, A. A., and Dashiell, S. R. (1975). *Opt. Commun.* **15,** 66–70.

Sawchuk, A. A., and Peyrovian (1975). *J. Opt. Soc. Amer.* **65,** 712–715.

Strand, T. C. (1975). *Opt. Commun.* **15,** 60–65.

Strand, T. C. (1976). Ph.D. thesis, Univ. of California at San Diego.

Stroke, G. W. (1970). *Opt. Spectra* pp. 31–32 (Nov.).

Stroke, G. W., and Halioua, M. (1971). *29th Ann. Proc. Electron Microscopy Soc. Amer.*

Stroke, G. W., and Zech, R. G. (1967). *Phys. Lett.* **25A,** No. 2, 89–90 (July).

Yao, S. K., and Lee, S. H. (1971). *J. Opt. Soc. Amer.* **61,** 474–477.

10.7 MICROSCOPY

Mary E. Cox

10.7.1 Introduction

Microscopy was the original reason for the invention of holography. Gabor (1948, 1949, 1951) developed holography while trying to improve the resolution and field depth of an electron microscope. His demonstrations of the principles of holography using white light were of mixed success.

In recent years other applications of holography have become more commercially attractive. Thus the development of good quality, easily used holographic microscopes has proceeded more slowly. However, when large field volumes are to be imaged, with high resolution, it is useful to examine the desirability of using a holographic microscope.

A conventional microscope is designed to have a high transverse magnification, but at the price of a limited depth of field. A static object may be scanned, one depth layer at a time. But for a dynamic object, one which changes rapidly in time, this approach is unsatisfactory. A holographic microscope which uses a repetitively pulsed laser to record a sequence of high resolution holograms can record all the information in the volume of a dynamic object. Upon reconstruction, the object wavefronts can be examined in detail to locate the event of interest, or to follow the development of a series of events.

This section will examine in detail the current state-of-the-art in holographic microscopy. We will detail the nature of holographic magnification and scaling, looking at the effects of image magnification and aberration balancing. We will detail the techniques of microscopically augmented holography with pre- or postmagnification. The relevant equations for holographic microscopy will be discussed. The emphasis will be on their utility in determining the techniques preferred for a given application. We will emphasize the kinds of design decisions that must be made in utilizing holographic microscopy in a particular application. The emphasis in this section will be upon useful configurations, rather than upon theoretically beautiful concepts.

10.7.2 Holographic Magnification and Scaling

Holography was originally intended as a tool for the electron microscopist to form images of objects having atomic dimensions. Two factors were nec-

HANDBOOK OF OPTICAL HOLOGRAPHY
Copyright © 1979 by Academic Press, Inc.
All rights of reproduction in any form reserved.
ISBN-0-12-165350-1

10. Application Areas

essary to obtain a distortion-free optical image of a hologram recorded with an electron beam. The first was a visible light beam, whose radius of curvature was accurately scaled by the ratio of the wavelength of light to the wavelength of the electrons. This scaling of the reconstruction beam to the recording beam had to be accompanied by a corresponding enlargement of the original hologram recorded with the electron beam.

Scaling is usually accompanied by a term proportional to $\mu = \lambda_2/\lambda_1$, the ratio of the reconstruction wavelength λ_2 to the recording wavelength λ_1. Hologram magnification is designated by m, where the lateral magnification is given by the ratio of the lateral dimensions of the hologram after magnification to those before magnification, i.e., $m = x_2'/x_2 = y_2'/y_2$. When the radii of curvature of the recording and reconstructing reference waves are also changed, the total lateral magnification of the virtual image V is

$$M_{\text{lat,V}} = m \left(1 + \frac{m^2 z_1}{\mu z_c} - \frac{z_1}{z_r} \right)^{-1}, \tag{1a}$$

and the total lateral magnification of the real image R is

$$M_{\text{lat,R}} = m \left(1 - \frac{m^2 z_1}{\mu z_c} - \frac{z_1}{z_r} \right)^{-1}. \tag{1b}$$

The angular magnification, which relates the apparent size of an object when viewed through an optical instrument to its angular size when seen without that instrument, is always

$$M_{\text{ang}} = \mu/m. \tag{2}$$

The longitudinal magnification, often called depth magnification, is proportional to M_{lat}^2.

Wavelength scaling has not been widely used by holographic microscopists. With visible light only for recording and reconstruction μ lies between 0.57 and 1.75. Using coherent ultraviolet lasers to record the hologram, μ is less than 10. The limited range of values of μ, combined with the care that must be taken to avoid aberrations inherent in wavelength scaling, has made wavelength scaling a little used technique in holographic microscopy.

The enlarging of holograms is also very difficult. Typical holograms have fringe spacings of at least 1000 lines/mm. Most photographic enlargers do not have the resolution to handle these details. So magnification of the hologram is also not widely used.

Thus for most applications of holographic microscopy, one chooses $\mu = m = 1$, making $M_{\text{ang}} = 1$. To minimize aberrations, the reference wave during recording and the reconstruction wave are matched as closely as possible. Usually these are chosen as plane waves, making $z_r = z_c = \infty$, and $M_{\text{lat}} = 1$. How then does one obtain a magnified image of the object?

562

10.7.3   Microscopically Augmented Holography

The successful applications of holography to microscopy have all used a technique where the best features of conventional microscopy are augmented by holography. Knox (1960), McFee (1970), Van Ligten and Osterberg (1966), Thompson, Ward, and Zinsky (see Section 10.12), and Cox, Buckles, and Whitlow (1971) all used holography in combination with microscopy to obtain a magnified image. Those features of a conventional microscope most often used center on the good quality optics available. We will discuss in detail two forms of holographic microscopy which use a conventional microscope to form images with good resolution and large volume recording.

10.7.3.1   Premagnification

When high resolution of a small field of view is required, premagnification with a conventional microscope is often used (Van Ligten and Osterberg, 1966). A magnified real image of the object serves as the object for the hologram (Fig. 1). In this configuration the conventional microscope forms a magnified real image of the object, which serves as the object for the hologram. The reference beam bypasses the microscope. The reference beam is usually a plane wave, and its angle of incidence on the film may be varied. Upon

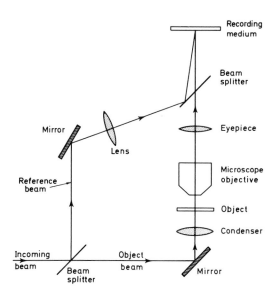

Fig. 1 Holographic microscope with premagnification using a conventional microscope. The reference beam angle may be varied to meet the film resolution requirements.

10. Application Areas

reconstruction, the object wavefronts converging to a real image or diverging from a virtual image may be studied depending on the reconstruction geometry.

Film resolution is no problem in this configuration, since the image from the microscope may be sufficiently magnified so as to overcome any limitations due to the film resolution, provided the reference beam is properly chosen. However, the object volume recorded is limited by the depth of field of the conventional microscope. The hologram cannot record more depth or lateral area than the microscope can image. In many applications this is not a serious flaw. When the object to be studied is a thin specimen, or the event sought lies in a very narrow depth range, this technique allows a variety of useful results. Holographic interferometry can be easily accomplished to locate and study objects changing with time. The growth of crystals or polymers can be detailed by taking two successive holograms and superposing them during reconstruction.

10.7.3.2 Postmagnification

When high resolution of a large field of view is required, postmagnification with a conventional microscope is often used (Leith and Upatnicks, 1965; Knox, 1966) (Fig. 2). The hologram is recorded with a plane reference wave in the near or far field of the object. Upon reconstruction the object field producing the real image can be examined with a conventional microscope.

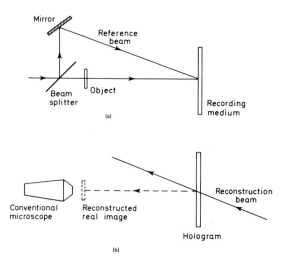

Fig. 2 Holographic (a) recording and (b) reconstruction for use with postmagnification using a conventional microscope. The reference beam angle and optical path differences may be varied to meet film resolution and laser coherence requirements.

When film resolution is a difficulty with this configuration, some initial premagnification may be done between the object and the hologram (Kogelnik, 1965; Toth and Collins, 1968). This may be done with good quality optics, if they are available, or with optics of less than optimum quality. In the latter case, these optics may be reinserted in the reconstruction stage to deaberrate the image. (Briones *et al.*, 1978; Heflinger *et al.*, 1978).

The object field has been reduced by this process to the original object size. This process is most useful if the original object is inaccessible to a conventional microscope, such as the capillaries located below the fingernail. To examine the blood flow in these capillaries in detail at high magnification, the physical fingernail cannot be traversed by the conventional microscope. However, the object field of the fingernail can be traversed. This technique allows a large field depth to be recorded, but often does not allow the maximum resolution. When the object to be studied is thick, or the event sought lies in a large depth layer, this technique supplies useful results.

10.7.4 Equations for Holographic Microscopy

Whatever technique one chooses to produce a hologram of a microscopic object, several considerations must be studied to use the chosen technique most effectively. The general equations for holography must be evaluated. The recording medium to be used must be carefully chosen with attention to speed, resolution, and processing. In this section the detailed equations of holography will be examined with special attention to their utility in holographic microscopy.

10.7.4.1 Off-Axis Holography

All successful applications of holography to microscopy have utilized an off-axis plane reference beam (Leith and Upatnicks, 1962, 1963, 1964). The use of this configuration minimizes aberrations (Meier, 1965) and allows the reconstructing wave to be made identical with the reference wave, whether one is studying the reconstructed real or virtual image. If one can choose the angle between the reference and object wave such that the fringe spacing in the interference pattern is well below the film's resolution maximum, a good quality hologram will result (Fig. 3). For a film capable of resolving a maximum of 1000 line pairs/mm, the distance between adjacent fringes d must be larger than 1 μm. If $\Omega_0 = 30°$, which means that objects of mean size of 1.22 μm are defracting the light, then

$$\sin \Omega_R = \sin \Omega_0 - (\lambda/d). \tag{3}$$

If $d = 2$ μm, and $\lambda = 0.5$ μm $= 500$ nm, then $\sin \Omega_R = \sin 30° - (0.5 \ \mu\text{m}/2 \ \mu\text{m})$ or $\Omega_R = 14.5°$. In the general case, $0 \leq \Omega_R \leq \Omega_0$.

10. Application Areas

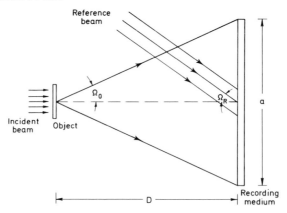

Fig. 3 Off-axis hologram geometry. Ω_0 is the largest angle between the object field and the normal to the film plane.

The choice of reference angle is also dictated by the solid angle subtended by the object at the hologram plane. Upon reconstruction the image should lie outside the reconstruction beam. This requires an analysis of the full spatial frequency spectrum of the object and is treated in Section 2.4.

In a conventional microscope the area of the object viewed in the eyepiece decreases as the third power of the numerical aperture (NA $= n \sin \theta$) of the objective. The depth of field decreases as the square of the NA. Thus the volume of the object viewed goes as

$$V \sim 1/NA^5 \qquad \text{for} \quad NA > 0.20. \tag{4}$$

Doubling the NA of the objective while keeping the eyepiece the same reduces the object volume viewed by $\frac{1}{32}$, while decreasing the minimum object resolved by $\frac{1}{2}$.

In a holographic microscope (with or without premagnification) the object space which is imaged is a much more complicated function of the NA of the system. Assuming that the reference beam angle is chosen for all object points in such a way as to not exceed the resolving power of the recording medium, then all object points of size δ or greater will be imaged in a volume V, having circular cross-section area A and depth D if the width of the film is at least

$$W = 2 \sqrt{\frac{A}{\pi} + \frac{2D \, NA}{\sqrt{1 - NA^2}}} \tag{5}$$

where

$$NA = \sin \theta = 0.61 \lambda/\delta. \tag{6}$$

The medium surrounding the object is assumed to be air.

Off-axis holograms place more stringent requirements on the coherence properties of the illuminating light source. The total requirement on the coherence length of the light source is

$$\Delta L_H \geq (a/2)\lambda\xi_r + (D/2)\lambda^2\xi_{max}^2 \qquad (7)$$

when ξ_r is the reference wave spatial frequency $= \sin \Omega_R$, a the length of the hologram plate, D the object to hologram separation, ξ_{max} the maximum object field spatial frequency, and λ the vacuum wavelength of the light used. For most applications, the first term dominates the coherence length requirements. For our example, if the object to hologram distance is 2.0 cm, $a = 35$ mm, and $\xi_{max} = 0.82 \times 10^{-6}$ m^{-1}, then

$$\Delta L_H \geq \frac{35 \times 10^{-3}\text{ m}}{2}(0.5 \times 10^{-6}\text{ m})(\sin 14.5°)$$

$$+ \frac{2 \times 10^{-2}\text{ m}}{2}(0.5 \times 10^{-6}\text{ m})^2(0.82 \times 10^{-6}\text{ m}^{-1})^2$$

$$\geq 2 \times 10^{-9}\text{ m}.$$

Without special attention, pulsed solid lasers generally have the lowest coherence length and require the most careful assessing (see Section 8.1). Most cw lasers will have more than enough coherence length.

10.7.4.2   Fresnel, Fraunhofer, and Fourier-Transform Holograms†

Most appiications of holography to microscopy record the hologram in the Fraunhofer plane; that is, the hologram is formed in a plane where superposition of plane waves originating from the various object points occurs. This is often called the far field of the object. When one is dealing with objects of microscopic size, the "far field" is usually a few millimeters from the object. There are occasions in microscopy when it is more convenient to use the Fresnel plane or the Fourier-transform plane.

When the hologram is formed in a plane where the superposition of spherical waves originating from the various object points occurs, whether the reference wave is a plane or a spherical wave, the "near field" or Fresnel plane analysis must be used (see Sections 2.2 and 4.1). One special case of a Fresnel hologram in microscopy occurs when the magnified image occurs in the plane of the film (Cox and Vahala, 1978). This image hologram technique minimizes the spatial coherence requirements of the hologram illuminating source. Image holograms can be brightly illuminated with extended sources. However, the image reso-

† See Chapter 4.

lution will degrade in planes away from the hologram plane. Color dispersion and blurring will degrade the image (see Chapter 6).

Fourier-transform holograms record the interference of two waves whose complex amplitudes at the hologram are the Fourier transforms of both the subject and reference source. The reference source must lie in the same plane as the subject. Thus the subject must effectively be a planar object, or at least its thicknesses must be small compared to the subject-to-lens distance. A lens is used to form the Fourier transform of both the subject and reference source point. The hologram is formed in the back focal plane of the lens.

For microscopy several useful properties of Fourier-transform holograms are important. The reconstructed image remains stationary when the hologram is translated. Holograms recorded on a reel of film project stationary images while the film is moving. A photograph of the output of a Fourier-transform hologram shows twin real images inverted about a center point. Quasi-Fourier-transform holograms can be formed without lenses if the point reference source is located in the same plane as the subject (see Section 4.3).

Other hologram geometries (lenseless or using lenses) can be useful in particular applications of holographic microscopy. An analysis of constraints on object motion and environment may dictate one geometry. We now consider these more detailed design considerations.

10.7.5 Design Decisions

The goal in the design of a holographic microscope is to produce the highest quality reconstructed image subject to constraints. Most holographic microscopes are "do-it-yourself" projects. Thus each design can take into account the application for which it is constructed. It is important to know who is to operate the final device. And the interaction between the operator and designer will be critical in achieving the design goals.

10.7.5.1 Object Properties

The size of the minimum resolvable microstructure within a macroscopic object is the most critical design parameter in a holographic microscope. Since objects smaller than the wavelength of the illuminating light cannot be resolved, objects much smaller than 1 μm cannot be resolved using visible light. The location and size of the film is determined by the minimum resolvable object size. The film resolution will be determined by the object size and the reference beam angle.

Object motion, or lack of motion, will determine the nature of the light source. Continuous illumination with mechanical shuttering is suitable for objects that move less than about $\lambda/20$ during the exposure. For rapid motion, pulsed light sources must be used. The maximum object speed for a given wavelength λ and exposure time Δt is of the order of $\lambda/8\ \Delta t$. Object motion may introduce another constraint involving the peak power versus energy in a light pulse of short duration. All photosensitive media have a certain range of exposure, energy per unit area, over which they will produce good quality holograms. As the total time gets shorter, the power must get larger for the same exposure (see Chapter 8). This may require large peak powers to be incident on the object and other optical devices. This can cause damage, especially if any significant portion of this power is absorbed by the object and/or optical devices.

The physical location and surroundings of the object may cause some additional problems in the design of a holographic microscope. Most objects are not in air, nor are they mounted on a slide. Many objects are in fluids, as for example, crystals growing from a melt. Some objects are in vacuum or high-pressure chambers. When the object is located in a remote or physically inaccessible place, design considerations should be taken so as to get the object field to the film plane with as few aberrations as possible. Optical devices may have to be placed in that environment. The effects of the environment must be taken into account, especially if the environment is turbulent or scatters an appreciable amount of light.

The location of the microstructure or event of interest within the macroscopic object is relevant. A fixed specimen presents few problems even to ordinary microscopy. But if random objects and/or events occur within a large sample volume, it is important that they be recorded on the hologram for a posteriori examination. Large film size, short distances from object to film, and low premagnification will all result in a larger recorded object volume. However, the holographically produced image must be studied a posteriori to locate the object and/or event originally sought.

10.7.5.2   Wavelength and Coherence of Source

Most good quality holographic microscopes have used cw or pulsed lasers to record and reconstruct the holograms. The availability of more lasers with various wavelengths and coherence properties is applauded by holographic microscopists. This wide choice also creates a problem, in that obtaining all the properties one wishes in a light source can be costly.

The first choice must be wavelength. To study red blood cells, to say the obvious, a laser with only red output should not be chosen. Be certain the object doesn't absorb a large amount of the incident light at the wavelength chosen. The wavelength should also be one that maximizes the film response. All photosensitive media properties are wavelength dependent, especially exposure. If any auxilliary optical devices are to be used in the holographic microscope, they should be antireflection coated at the chosen wavelength.

The coherence properties of the light source are critical. The best holograms

10. Application Areas

are recorded with light having a high degree of both spatial and temporal coherence. However, high spatial coherence can introduce interference between object fields originating from two (or more) widely spatially separated object points. And high temporal coherence can increase speckle.

Without special attention pulsed lasers have less spatial and temporal coherence than do most cw lasers. For most holographic microscopes amplitude division of wavefronts is more useful in obtaining the object and reference beams, provided the path lengths from the beamsplitter can be adjusted to be less than the coherence length of the light source. Since the hologram must have the largest fringe contrast achievable, the complex degree of coherence must be maximized, without introducing extraneous sources of noise.

10.7.5.3 Microscope Techniques

If the best quality hologram has been recorded, processed, and reconstructed, we now wish to discuss ways to study the reconstructed real image using standard microscope techniques. If a lens has been used during the recording of the hologram and it is reinserted into the reconstruction process in the same location it had during recording, all undiffracted light is brought to a focus (Fig. 4). Dark field illumination can be obtained by placing an

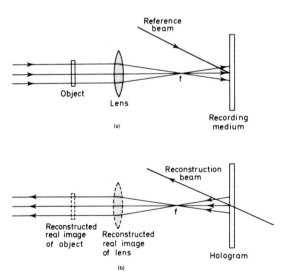

Fig. 4 Hologram (a) recording and (b) reconstruction for use when a small amount of premagnification is desired. When the lens is reinserted at the proper location, the reconstructed real image of the object is deaberrated. Also, stops can be placed at f during reconstruction to produce phase contrast, dark or bright field, or polarizing viewing.

opaque spot at the focus of the lens. Phase contrast can be obtained by placing a phase changing spot at the focus of the lens.

Interference microscopy can be done several ways. In some, two holograms are superposed and reconstructed simultaneously. Interference due to changes in the object field can be seen if one hologram is of the unoccupied object location. Interference fringes due to growth of some objects can be seen if two holograms of the object, taken at two different times, are superposed.

Under certain very restricted circumstances, abnormal structures can be located if one hologram of a normal structure is superposed with a hologram of the abnormal structure. Time averaged interferometry (see Section 10.4) can be used if the object is undergoing relatively slow motion, or if some type of periodic motion is occuring.

None of the preceding techniques destroys or alters the original hologram. Thus several different techniques can be used sequentially with the same hologram. This is very useful in that a hologram can be studied for different events using different techniques. The hologram can be stored and reused without having to repeat the original experiment.

10.7.5.4 Final Design Considerations

The last steps in designing a holographic microscope concern secondary items in that they are centered on utility and cost. After the decision has been made to construct a holographic microscope for a particular application, and the above-mentioned constraints have been incorporated into the design, considerations of how it is to be used and by whom enter into the design. At this point the designer and user should be in close communication.

The physical location of the holographic microscope within a laboratory should be carefully discussed. Darkening of the surroundings may interfere with safety. Locations and distribution of laser light beams and high voltage lines should be discussed. Safety interlocks on the lab door must be considered. Proper ventilation must be provided, especially if volatile chemicals are to be in proximity.

The training of a technician who will be primarily responsible for the use of the holographic microscope is not as simple as the training of someone to use a conventional microscope. The design must allow for maximum utilization by trained personnel without undue complications. This includes maximizing the setup and stability of the holographic microscope itself. Since the holographic microscope is a tool for use in studying microscopic phenomena, it must not get in the way of the original experiment.

The final design consideration is cost, the cost of construction and installation, and the cost of operation. Where possible, off-the-shelf subassemblies should be used. Presumably, these subassemblies have been tested for reliability and reproducibility. Overspecification of subassemblies must be avoided. The operating costs include disposable supplies and technician time.

10.7.6 Summary

Holography was originally conceived to improve the images in microscopy. To date very few successful holographic microscopes have been constructed and used. Problems involving resolution and image quality are primarily the reason for this lack of general successful application. In the future, with more cooperation between optical scientists and biomedical personnel, the holographic microscope should become an accepted optical device. Better and more versatile recording media, a better variety of light sources, and a more reasonable approach to biomedical problems should all make holographic microscopy a useful and widely accepted tool.

REFERENCES

Cox, M. E., Buckles, R. G., and Whitlow, D. (1971). *Appl. Opt.* **10**, 128.
Cox, M. E., and Vahala, K. J. (1978). *Appl. Opt.* **17**, 1455.
Briones, R. A., Heflinger, L. O., and Wuerker, R. F. (1978). *Appl. Opt.* **17**, 944.
Gabor, D. (1948). *Nature* **161**, 777.
Gabor, D. (1949). *Proc. Roy. Soc.* **A197**, 454.
Gabor, D. (1951). *Proc. Roy. Soc.* **B64**, 449.
Heflinger, L. O., Stewart, G. L., and Booth, C. R. (1978). *Appl. Opt.* **17**, 951.
Knox, C. (1966). *Science* **153**, 989.
Kogelnik, H. (1965). *Bell Syst. Tech. J.* p. 2451.
Leith, E. N., and Upatnieks, J. (1962). *J. Opt. Soc. Amer.* **52**, 1123.
Leith, E. N., and Upatnieks, J. (1963). *J. Opt. Soc. Amer.* **53**, 1377.
Leith, E. N., and Upatnieks, J. (1964). *J. Opt. Soc. Amer.* **54**, 1295.
Leith, E. N., and Upatnieks, J. (1965). *J. Opt. Soc. Amer.* **55**, 569.
McFee, R. H. (1970). *Appl. Opt.* **9**, 1834.
Meier, R. W. (1965). *J. Opt. Soc. Amer.* **55**, 987.
Toth, L., and Collins, S. A. (1968). *Appl. Phys. Lett.* **13**, 7.
Van Ligten, R. F., and Osterberg, H. (1966). *Nature* **211**, 282.

10.8 OPTICALLY RECORDED HOLOGRAPHIC OPTICAL ELEMENTS

Donald H. Close

10.8.1 Introduction

Our purpose here is to describe the use of optically recorded holograms as lens (or mirror) elements. Our objective is to convey a good understanding of the properties of such optical elements and of the type of work and level of effort involved in putting them into practical use. In Section 10.8.2, we qualitatively compare the imaging characteristics of holographic optical elements (HOEs) with those of conventional lenses and mirrors. This leads directly to the question of optical efficiency, which is surveyed in Section 10.8.3. If one wishes to utilize a HOE in an optical system, whether because of or in spite of the imaging characteristics and efficiency properties, one must consider the design and fabrication issues discussed in Sections 10.8.4 and 10.8.5, respectively. In order to give a realistic feeling for the application of HOEs, some specific examples of applications are discussed in Section 10.8.6.

Optics people have always understood holograms in terms of the lens analogy. In this analogy, the recording of the holographic fringe pattern forms an "off-axis Gabor zone plate." Since the imaging characteristics of zone plates have been known and studied for many years, the interpretation of a hologram as a superposition of zone plates provides a direct and useful understanding of the imaging properties of holograms. In these terms, the HOE is simply a hologram recorded with a point object. When used as a HOE, this point-object hologram can be "played back" with a superposition of many reference or reconstruction point sources, which together form the object to be imaged by the HOE. The superposition of the many reconstructed or image points constitutes the image formed by the HOE.

It is clear from this analogy that to cover any finite field of view the HOE cannot be "played back" strictly in the configuration in which it was recorded. Therefore, the HOE image cannot be free of aberrations. In fact, the quantitative understanding of HOE aberrations and the ability to minimize or cancel these aberrations in the HOE design process are what make possible the utilization of HOEs in practical optical systems.

Another basic insight into the performance of HOEs also follows from the

HANDBOOK OF OPTICAL HOLOGRAPHY
Copyright © 1979 by Academic Press, Inc.
All rights of reproduction in any form reserved.
ISBN-0-12-165350-1

zone plate interpretation: The HOE has multiple diffraction orders, with strong variations of optical power in the different orders. Furthermore, the optical power in each order varies rapidly with wavelength, and the amount of light diffracted into any order is highly dependent on the particular way in which the zone plate structure is fabricated. Again, the quantitative understanding of these efficiency characteristics and the ability to control them are necessary for successful application of HOEs.

The notation that we use conforms to that used in the basic references in this field. In particular, Champagne (1967) is a key reference for analysis of aberrations. Latta (1971a) made numerical studies of these aberrations and also did much of the original work in establishing raytracing for analysis and design of HOEs (Latta, 1971b). The analysis of HOEs with high optical efficiency has been based on the thick hologram theory of Kogelnik (1969). An excellent general reference is the book "Optical Holography" by Collier *et al.* (1971).

10.8.2 Imaging Properties

The most important imaging property of HOEs is that they can be designed to have any particular imaging geometry, independent of the orientation and curvature of the HOE substrate. This means that the directions of the entering and exiting rays are not controlled by the substrate normal or the refractive index. From another point of view, the HOE can be designed to transform any single, entering wavefront into any other, single, exiting wavefront, independent of a particular substrate's characteristics. The design must *consider* the substrate characteristics, and the performance at other wavelengths, angles of incidence, etc. *will* depend on the substrate. In fact, a good rule is that the more the imaging properties are required to deviate from those that would result from the laws of refraction and reflection, the stronger the dispersion, aberrations, and efficiency variations will be as the entering wave differs from the particular design assumptions. A "corollary" of this rule is that one should resort to the use of a HOE only when it is impossible to use conventional lenses and mirrors. In other words, the applications of HOEs tend to occur in specific instances where their peculiar characteristics happen to meet a specific need.

Since the HOE is a single, wavefront-transforming surface, it must have unit angular magnification. As with conventional lenses and mirrors, one must have multiple elements to build a telescope or other optical instrument. The attractive concept of "making a hologram" of an optical instrument and having that hologram perform the function of the instrument is simply not possible and is not an appropriate way to think of HOE imaging properties. We consider the HOE as a single optical surface whose "transfer function" depends on numerous parameters that can be controlled in the design process. The purpose of this section is to provide a qualitative and quantitative understanding of what parameters produce what kinds of transfer functions.

We can consider the HOE as a recording of an optical fringe pattern such that at each point in the recording material the fringe surface represents a mirror surface that reflects the design entering ray into the design exiting ray. This concept is valid only for the particular pair of conjugate waves for which the HOE is designed. It is useful because it serves to define the surface grating, which does determine the general imaging geometry of the HOE. This surface grating is the locus of points where the "mirror" fringe surfaces intersect the surface of the hologram recording material. To be specific, it is the surface of the recording material from which the transformed, or diffracted, wave *exits*. This surface grating, whether from a surface or a volume hologram, completely determines the imaging geometry, i.e., the image location, aberrations, magnification, etc., for whatever wavefront is transformed by the HOE. (Fortunately, other factors enter into determining the *efficiency* of the HOE, i.e., the amplitude of the transformed wavefront.)

10.8.2.1   The Grating Equation

Because the imaging geometry with HOEs is relatively arbitrary, it is best to use vector notation to describe it. There are four rays to be accounted for at any point of interest on the surface grating. These are the entering ray $\mathbf{C}$, the exiting ray $\mathbf{I}$, and the two rays that define the HOE design, or construction, $\mathbf{O}$ and $\mathbf{R}$. These rays are defined by unit vectors that lie along the various rays. The design rays $\mathbf{O}$ and $\mathbf{R}$ are the rays for which the "mirror fringes" above are a valid concept. For optically recorded HOEs, $\mathbf{O}$ and $\mathbf{R}$ are the object and reference rays used in the recording process. These four unit vectors and the unit vector $\mathbf{S}$ normal to the surface at the point of interest are related by the grating equation, which we write in two useful forms:

$$\mathbf{I} = \mathbf{C} - \mu(\mathbf{R} - \mathbf{O}) + \Gamma\mathbf{S}, \qquad \mu = q\lambda_c n_0/\lambda_0 n_c, \tag{1}$$

and

$$(n_c/\lambda_c)(\mathbf{I} - \mathbf{C}) \times \mathbf{S} = (qn_0/\lambda_0)(\mathbf{O} - \mathbf{R}) \times \mathbf{S}, \tag{2}$$

where Γ is determined to make $\mathbf{I}$ a unit vector, q is the integral diffraction order, λ_0 and λ_c are the wavelengths during recording and use, and n_0 and n_c are the corresponding refractive indices. These unit vectors are defined in the volume of the recording material, and they must be refracted out of the material to obtain the external ray directions.

For efficiency calculations with volume holograms, it is also necessary to have the volume grating parameters. These are the perpendicular spacing between fringe planes,

$$\Lambda = \lambda_0/n_0 \,|\, \mathbf{R} - \mathbf{O} \,|, \tag{3}$$

10. Application Areas

and the unit vector normal to the fringe planes,

$$\mathbf{N} = (\mathbf{R} - \mathbf{O})/ \mid \mathbf{R} - \mathbf{O} \mid, \tag{4}$$

If the recording material is deformed between recording and use, the fringe spacing and orientation differ from those given in Eqs. (3) and (4), and they must be redefined in terms of the deformation.

The algorithm for calculating the image or diffracted ray direction at a particular point on the HOE is therefore as follows: Trace a ray through the system until it is incident on the HOE. Refract this ray into the hologram material to obtain the unit vector $\mathbf{C}$. From the design, determine the two construction rays at the same point, $\mathbf{O}$ and $\mathbf{R}$. Use Eq. (1) to calculate the diffracted ray $\mathbf{I}$ and refract it out of the hologram material to obtain the desired image ray. If the parameter Γ is complex, then the wave in this region is evanescent.

The computation of diffracted ray directions can be more or less complicated, depending on the procedure used to design the HOE. If the HOE is defined by two simple point sources for construction, then the computation is simple. In many practical cases, however, there are optical elements in the construction beams, and thus the computation is more complex.

10.8.2.1 Consequences of the Grating Equation

Four fundamental imaging properties follow from the fact that HOE imaging is controlled by the surface grating. These are a strong variation of optical power with wavelength, strong dispersion, an independence of the HOE diffraction efficiency from its image geometry, and a duality between reflection and transmission elements. It is also true that the diffraction efficiency often changes rapidly with wavelength or angle of incidence, a characteristic that is discussed in Section 10.8.3.

The strong variation of optical power and dispersion follow from the fact that the reconstruction wavelength λ_c appears as a linear factor in the grating equation (1). The variation of optical power is shown by Champagne's expression for the equivalent focal length f of a point-source hologram recorded with object and reference point source distance of R_o and R_R,

$$\frac{1}{f} = \frac{q\lambda_c}{\lambda_o}\left(\frac{1}{R_o} - \frac{1}{R_R}\right). \tag{5}$$

The dispersion is shown by Champagne's equation for the angle of the diffracted ray, in terms of the hologram geometry (see Champagne (1967) definitions):

$$\sin \alpha_I = \frac{q\lambda_c}{\lambda_o}(\sin \alpha_o - \sin \alpha_R) + \sin \alpha_c, \tag{6}$$

The fact that the HOE image geometry is independent of the image amplitude is a very interesting and useful consequence of the grating equation. Equations (2) and (4) together show that the direction of the diffracted ray depends only on the component of the fringe normal **N** that is tangent to the substrate surface. This is illustrated in Fig. 1, which shows cross sections of three fringe systems that have the same surface component. Since the surface grating period d_s is the same for all three fringe systems, the direction of diffracted rays will be the same for all three HOEs, and therefore they will have the same image geometry, i.e., location, aberrations, etc. Since the amplitude of the diffracted rays depends not only on the three-dimensional orientation of the fringes, but also on the physical nature of the fringes (i.e., index or density variation, etc.), this amplitude will certainly be different for the three cases shown.

There are two practical consequences of this independence of image geometry and amplitude. First, it allows the HOE imaging geometry to be designed without one's being concerned about the physical recording process.

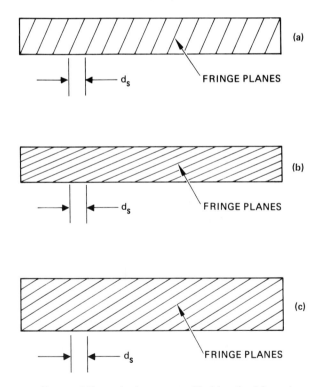

Fig. 1   Cross sections of three holograms with identical imaging geometries: (a) a transmission element, (b) a reflection element, and (c) the same reflection element, but expanded in thickness.

10. Application Areas

Second, it allows the peak efficiency of a HOE recorded at, say 488 nm, to be tuned for use at 546 nm (for example), simply by causing the recording material to change thickness by an appropriate amount. This is illustrated in Fig. 1b and 1c, where the expansion can be seen to tilt the fringes to correspond to a longer wavelength. This color shifting is very useful in practice, although again rather specific imaging conditions are involved, and the phenomenon is useful only for "thick" holograms. For reflection HOEs, the expansion factor required is the ratio of playback to construction wavelengths, or 1.119 for the example given above.

Finally, the duality between reflection and transmission elements can be stated as follows: The imaging geometry of two elements, one recorded with a given pair of object and reference waves, and the other recorded with either of those waves reflected in the substrate, will be the same. This follows from the fact that the surface grating will not change if the perpendicular component of the fringe vector changes sign. One of these elements will be a reflection hologram and the other will be a transmission hologram. Of course, the amplitudes of the diffracted waves of various orders will be very different, so the two elements in practice will perform quite differently. This duality can be useful in the HOE design process.

10.8.3 Efficiency Properties

The efficiency properties, i.e., the amplitude of the diffracted wave of interest, are determined by the physical nature of the recorded fringe pattern. These characteristics are discussed in Section 8.3, and only a brief discussion will be added here.

For HOEs, we consider only phase holograms. This is appropriate because only phase holograms have the required high diffraction efficiency and/or low loss. Two types of phase hologram are of interest: thick or volume phase holograms (both reflection and transmission) and thin or surface phase holograms. The volume holograms are recorded as a modulation of the refractive index in the bulk of the recording material. The surface holograms are recorded as a surface relief pattern on the HOE surface.

The diffraction efficiency of volume, phase HOEs is best represented by the coupled wave theory of Kogelnik (1969). Efficiency can be high for both reflection and transmission elements. Reflection elements are characterized by a high reflectivity over a narrow band of wavelengths, with the wavelength of peak reflectivity depending on the angle of incidence and the amount of expansion of the recording material between recording and use. The "Q" of the reflection HOE is approximately equal to the number of fringes recorded in the thickness of the recording material:

$$\Delta\lambda_c/\lambda_c \simeq \Lambda/T, \tag{7}$$

where T is the thickness of the material and Λ is the fringe spacing in the material. The peak efficiency is given by

$$\eta_0 = \tanh^2 \left(\frac{\pi \, \Delta n T}{\lambda_c \cos \theta_0} \right), \tag{8}$$

where $2 \, \Delta n$ is the peak-to-valley modulation of the refractive index.

Transmission phase holograms are characterized by a narrow angular range of high efficiency at a fixed wavelength and a high efficiency at some angle for a broad range of wavelengths. The angular "Q" of the transmission element is also equal to the ratio of material thickness to fringe spacing:

$$\Delta\theta \simeq \Lambda/T. \tag{9}$$

The peak efficiency of the transmission element is given by

$$\eta_0 = \sin^2 \left(\frac{\pi \, \Delta n T}{\lambda_c \cos \theta_0} \right), \tag{10}$$

where the parameters are the same as those for the reflection element. The geometry of thick reflection and transmission holograms is shown in Fig. 2.

The diffraction efficiency of surface relief holograms depends strongly on the shape of the groove and on its depth, relative to the wavelength of use. To the extent that we can model the thin phase grating as introducing a phase modualtion of amplitude 2ϕ on the incident wave, we can calculate the efficiency for special cases. Expanding the modulated wave in a Fourier series gives an efficiency of

$$\eta = 4 \sin^2 \phi / \pi^2 q^2, \qquad q \quad \text{odd}, \tag{11}$$

for square grooves and

$$\eta = J_q^2(\phi) \tag{12}$$

for sinusoidal grooves (Collier *et al.*, 1971, pp. 223–226), where q is the diffraction order and the Js are Bessel functions. Equations (11) and (12) give maximum efficiencies of 30 to 40% for $\phi = \pi$.

For high efficiency from surface reief gratings, the substrate is "blazed" and coated with an opaque layer of a reflecting metal, such as aluminum or gold. In this case, the diffraction efficiency must be calculated by a numerical solution of the boundary value problem. Computer codes for this solution have been written, and the results agree well with experimental measurements (Kalhor and Neureuther, 1971; Loewen *et al.*, 1977). In specific cases, this type of surface grating can have nearly 100% efficiency.

10.8.4 Design Issues

The most important design issue is to define when a HOE should be used in the particular optical system of concern. We certainly do not recommend that the optical designer attempt to fit a HOE into every design. On the

10. Application Areas

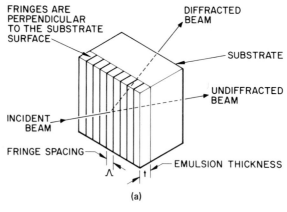

(a)

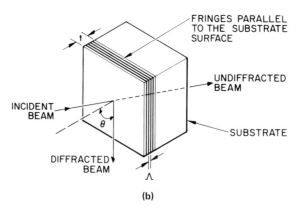

(b)

Fig. 2 Idealized geometry of thick holograms, showing parameters that determine the spectral and angular bandwidth of the diffraction efficiency: (a) a transmission element and (b) a reflection element.

contrary, we believe that a HOE should be considered only when some important function is impossible or very difficult or costly to implement with conventional lenses and mirrors. When HOE characteristics happen to match a particular need, the HOE can provide an elegant solution. Our purpose in this section is to describe the characteristics that we have found to be useful and also to discuss those which we have found to be troublesome. We also discuss some general ways of predicting HOE performance, and the tools required to perform a detailed design.

In general, the HOE must be considered as only one type of element, to be used in combination with others as needed. Our experience shows that the HOE construction beams must be considered as the products of separate optical systems. This means that the HOE is a common element in three

580

optical systems, and the parameters of all three systems must be included in the design.

10.8.4.1   Comparison with Conventional Elements

HOEs and conventional elements have one important characteristic in common: One individual element can be designed to have zero or small aberrations for one pair of conjugate points. When used at other conjugates, aberrations appear. To avoid the effects of the aberrations and still have useful performance across some extended field of view or pupil, additional elements are added to the system. The parameters available by adding elements allow the designer to reduce the overall aberrations of the system. In this sense, the design problem is one of determining what elements to add where in the system such that the available parameters do control the aberrations that must be removed.

HOEs are about as well behaved as off-axis, curved mirrors or tilted, decentered lenses. In other words, they usually add a relatively large amount of aberration to the system, typically astigmatism and coma. In addition, the HOE probably adds a large amount of dispersion, forcing use of narrow-band light or drastic measures (another HOE) to obtain a reasonable image. What the HOE offers as compensation for these problems is the ability to provide unusual geometrical configurations or special spectral characteristics. The large aberration level and the dispersion have not been useful features in our work, though they certainly can be in other cases, notably in spectroscopic applications.

A particular characteristic of HOEs that has no counterpart with conventional elements is the possible utility of elements in the construction beam optical systems. Such elements represent an ''overhead'' cost that is distributed across all HOEs made with that design. These elements provide additional parameters to control aberrations, without adding more elements to each system that is built.

10.8.4.2   Some Useful Characteristics

It seems useful at this point to give some examples of applications where HOEs are indeed useful. Some of the other applications discussed in this chapter are pertinent: providing an arbitrary, very precise wavefront transformation (Section 10.5); providing low-scattering, high-dispersion gratings (Section 10.9); and providing a one-to-many wavefront transformation (Section 10.11). Applications that are perhaps more in the spirit of this section are providing an unusual geometry (the holographic visor helmet-mounted display); providing a large element with low weight (the holographic night vision goggle); providing high reflectivity at a narrow spectral band while maintaining high overall transmission (the holographic head-up display); and providing an additional optical function without adding additional surfaces (beam sampling). These examples are discussed in Section 10.8.6.

10.8.4.3 Preliminary Design

One of the things that makes optical design somewhat of an art is that the designer has to know approximately where to start looking for a suitable solution. Experience has built up a set of starting points for conventional optics, which are called "design forms." Furthermore, aberration theory, at least for rotationally symmetric systems, is quite advanced, and the effects of actions such as moving the aperture stop can easily be calculated.

For HOEs, the situation is not so simple. First, the system tends to be more complex, probably lacking symmetry and having numerous parameters, such as elements in the construction beams. Second, optical efficiency must be taken into account as a first-order effect, since it often is a forcing function for other aspects of the design. Third, present practice is not supported by a large body of past experience that guides the design.

Having noted these difficulties, we can point to several techniques that can be used to establish some design guidelines. The first of these is the basic theory of the aberrations of single elements, such as is given by Champagne (1967). Since the system probably will have only one HOE, this theory is very useful. The basic drawback here is that most aberration theory for holograms considers only point source construction beams and planar substrates. The second valuable aid is a simple way of getting some first order or paraxial results. One example of this is the geometrical optics theory of Arsenault (1975). Another is the thin lens analogy of Sweatt (1978), which allows one to raytrace HOE systems without considering the construction beams explicitly. Finally, there are some general theoretical results that guide the choice of HOE geometry. Good examples of these are discussed in the publications of Welford and other workers at Imperial College (Welford, 1975; Smith, 1977).

10.8.4.4 Raytracing

The ultimate tool for designing HOE optical systems is computerized raytracing. This allows rapid, quantitative exploration of parameters. It is also necessary for completing the design, i.e., for the tasks of optimizing parameters and calculating tolerances on the parameters.

The geometry of HOE optical systems is such that it is appropriate to treat all rays as general skew rays. This approach is also needed for the tilted and decentered elements that often appear in HOE systems. An excellent foundation for this type of raytracing is given by Spencer and Murty (1962). The only difficult aspect to raytracing in HOE optical systems is the procedure of calculating the **O** and **R** rays at arbitrary points on the HOE, when there are optical elements in the HOE construction beams. Some kind of iterative technique is required since it is impossible to know how to launch a ray to hit the desired point on the HOE. Once this capability is acquired, the process is quite general and flexible. The only other capability that seems to be essential is to be able to automatically optimize parameters in the HOE construction beams, with merit functions defined in terms of the overall system performance.

10.8.5   Fabrication Issues

Fabrication of HOEs consists of the preparation of the exposure apparatus and the actual hologram recording in a suitable recording material. The exposure apparatus must provide the design geometry to within some given tolerance. It must also provide interferometric stability during the exposure. If the design includes optics in the HOE construction beams, these must be fabricated and aligned. Once the apparatus is prepared, the HOEs can be fabricated as rapidly as the recording material and substrates can be prepared, exposed, and processed. Achieving good cosmetic quality requires clean conditions and uniform processing, such as are standard in most optical fabrication.

The crucial issue for HOE fabrication is the recording material. For our purposes, we have found only two suitable materials: dichromated gelatin for the volume, phase holograms and conventional photoresist for the surface relief holograms. While there are certainly problems of process control, we have been able to obtain good results with these materials. Since recording media are discussed in Section 8.3, we will not discuss them further here.

With the complex geometry that seems to appear in HOE optical systems, testing the HOEs becomes a problem. This is because the HOE has large aberrations, which are canceled by large aberrations in other elements of the system. Therefore, we produce an element that by itself has poor image quality, and the only way to test it is to fabricate and align the entire system. What is needed is a separate element to cancel the aberrations of the HOE, at least at one point in the field of view. A good candidate seems to be a computer-generated hologram, which can be generated quite easily with the same raytracing program that is used to optimize the HOE design.

10.8.6.   Examples

In this section we briefly discuss four examples of optical systems in which HOEs have been utilized. These are not meant to be exhaustive, but they do indicate the sort of specific advantages that can be obtained from the use of HOEs.

10.8.6.1 Unusual Geometry

This optical system provided a helmet-mounted display that imaged a cathode ray tube surface to infinity, without obstructing the wearer's view. A reflection HOE sandwiched in the helmet visor and conforming to the shape of the visor directed light from a conventional optical system on the side of the helmet into the wearer's eye. To accomplish this, the chief ray was reflected from an angle of incidence of 47° into an angle of 13° on the same side of the normal to the surface. Optical power in the HOE imaged the system pupil onto the eye pupil of the wearer, thus providing a bright image. This geometry could not have been achieved with conventional optics.

10.8.6.2 Large Element with Low Weight

This optical system used two reflection HOEs and additional, conventional elements to image the exit face of an image intensifier tube to infinity. The optical system was mounted in a goggle arrangement to provide the user with an intensified view of his surroundings. The elements were large enough to permit an extended field of view and yet permit the user to wear normal eyeglasses. Even with this size and located away from the user's head, the HOEs did not burden the user with a large torque which would tend to bow his head. Again, optical power in the HOEs provided a pupil-forming system, and conventional elements in the HOE construction beams controlled some of the aberrations of the skewed optical system.

10.8.6.3 Narrow Spectral Reflectivity

In this optical system, the HOE was the combining element in an aircraft head-up display system. The advantage here was the ability to provide a high reflectivity over a narrow spectral band. The HOE had a peak reflectivity of about 80% and a full spectral width at the 10% reflectivity points of about 20 nm. The peak reflectivity was tuned to 543 nm, the strong, green emission line of the high-brightness, "P-43" phosphor. This system provided a bright display without reflecting enough of the light from the outside environment to induce color distortion. For example, even though the HOE reflected green light, leaves still appeared green and the sky still appeared blue when viewed through the combiner.

10.8.6.4 Additional Function

In this optical system the HOE was a surface grating, etched into the metal coating of a curved mirror. It had a diffraction efficiency of <1%, and provided a sample of the beam being reflected from the mirror. This sample came off the mirror at an angle away from the reflected beam, and the optical power of the HOE, added to the optical power of the substrate, caused the sample to come to a focus at a convenient location. By correcting the aberrations of the wavelength shift between construction and use and those caused by the skewed geometry, the aberrations as well as the direction of the main beam could be observed. To provide these same functions with conventional optics would have required a substantially more complex optical system; with the HOE, no additions at all were required.

REFERENCES

Arsenault, H. H. (1975). *J. Opt. Soc. Amer.* **65**, 903–908.

Champagne, E. B. (1967). *J. Opt. Soc. Amer.* **57**, 51–55.

Collier, R. J., Burckhardt, C. B., and Lin, L. H. (1971). "Optical Holography." Academic Press, New York.

Kalhor, H. A., and Neureuther, A. R. (1971). *J. Opt. Soc. Amer.* **61**, 43–48.

Kogelnik, H. (1969). *Bell Syst. Tech. J.* **48**, 2909–2947.

Latta, J. N. (1971a). *Appl. Opt.* **10**, 599–608, 609–618.

Latta, J. N. (1971b). *Appl. Opt.* **10**, 2698–2710.

Loewen, E. G., Neviere, M., and Maystre, D. (1977). *Appl. Opt.* **16**, 2711–2721.

Smith, R. W. (1977). *Opt. Commun.* **21**, 106–109, and earlier references.

Spencer, G. H., and Murty, M. V. R. K. (1962). *J. Opt. Soc. Amer.* **52**, 672–678.

Sweatt, W. C. (1978). *Appl. Opt.* **17**, 1220–1227.

Welford, W. T. (1975). *Opt. Commun.* **15**, 46–49, and earlier references.

10.9 SPECTROSCOPY

H. J. Caulfield

10.9.1 Definition

For reasons which should become apparent, I offer the broadest definition I know of "holographic spectroscopy." It is the use of the interference pattern between two beams of light derived from the same source to record or modify the spectrum of that source.

10.9.2 Suitable Interferometers

Suitable interferometers cause there to be two wavefronts so disposed that along some plane the phase difference between them varies linearly with position. Figure 1 shows that two plane waves incident at $+\theta$ and $-\theta$ with respect to the x axis have a phase difference of

$$\Delta\phi = (4\pi x \sin\theta)/\lambda, \tag{1}$$

where λ is the wavelength and $x = 0$ at the point of equal path lengths in the two arms of the interferometer. Spectroscopists prefer to use the wavenumber

$$\sigma = 1/\lambda \tag{2}$$

rather than the wavelength. Thus

$$\Delta\phi = 4\pi\sigma x \sin\theta. \tag{3}$$

Figure 2 shows that two point sources whose rays makes angles $+\theta$ and $-\theta$ with the perpendicular bisector of the line between the two sources have a phase difference pattern of almost the same form. Using the point separation distance $2a$ and the pattern distance S defined in Figure 2, we have

$$\Delta\phi = 2\pi(l_2 - l_1)\sigma \qquad \mathrm{mod}(2\pi)$$

$$= 2\pi\{[(a + x)^2 + S^2]^{1/2} - [(a - x)^2 + S^2]^{1/2}\}\sigma \qquad \mathrm{mod}(2\pi)$$

$$\cong 2\pi S\left\{\left[1 + \left(\frac{a + x}{S}\right)^2\right]^{1/2} - \left[1 + \left(\frac{a - x}{S}\right)^2\right]^{1/2}\right\}\sigma \qquad \mathrm{mod}(2\pi). \tag{4}$$

HANDBOOK OF OPTICAL HOLOGRAPHY
Copyright © 1979 by Academic Press, Inc.
All rights of reproduction in any form reserved.
ISBN-0-12-165350-1

10. Application Areas

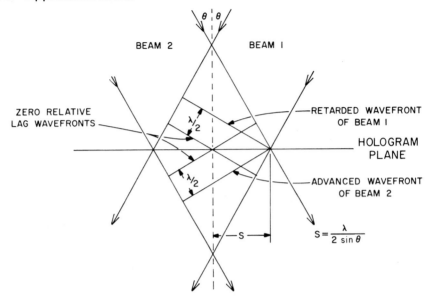

Fig. 1 Two plane waves incident at $+\theta$ and $-\theta$ with respect to the normal to the hologram plane produce a $\lambda/2$ delay over a distance S along that plane.

For $[(a + x)/S]^2 \ll 1$,

$$\Delta\phi \cong 2\pi S\left\{\left[1 + \frac{1}{2}\left(\frac{a + x}{S}\right)^2\right] - \left[1 + \frac{1}{2}\left(\frac{a - x}{S}\right)^2\right]\right\}\sigma$$

$$= \pi S\left[\frac{(a + x)^2}{S^2} - \frac{(a - x)^2}{S^2}\right]\sigma = \frac{\pi\sigma}{S}[4ax]. \tag{5}$$

But

$$a/S = \tan\theta \cong \sin\theta, \tag{6}$$

so

$$\Delta\phi \cong 4\pi\sigma x \sin\theta. \tag{7}$$

A third type of arrangement utilizes neither plane waves nor spherical waves. In this type, two identical extended sources are allowed to interfere. Figure 3 shows the arrangement. All corresponding points in the two sources have equal spacings $(2a)$ and roughly equal separations S. Thus each pair leads to a phase pattern of the same periodicity. But, more important, the net interference pattern is

$$\Delta\phi = 4\pi\sigma x \sin\theta, \tag{8}$$

where θ is measured from the center of mass of the source.

588

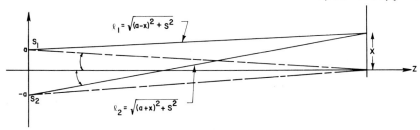

Fig. 2   Diagram for calculating the interference pattern between spherical wave sources S_1 and S_2; $\Delta\phi(x) = 2\pi(l_2 - l_1)\sigma \bmod(2\pi)$.

Some two-plane-wave interferometers are shown in Fig. 4, two-spherical-wave interferometers in Fig. 5, and source-doubled interferometers in Fig. 6. The two wavefronts can be written as

$$A(\sigma)\exp[i\phi(\sigma, x)/2], \tag{9a}$$

and

$$A(\sigma)\exp[-i\phi(\sigma, x)/2]. \tag{9b}$$

The net field is

$$F(\sigma, x) = A(\sigma)\{\exp[i\phi(\sigma, x)/2] + \exp[-i\phi(\sigma, x)/2]\}$$

$$= 2A(\sigma)\cos[\phi(\sigma, x)/2]. \tag{10}$$

For a spectrum such that the amplitude of light at wavenumbers between σ and $\sigma + d\sigma$ is $S(\sigma)\, d\sigma$, the spatial amplitude pattern is

$$P(x) = 2\int_{\sigma=0}^{\infty} S(\sigma)\cos[\phi(\sigma, x)/2]\, d\sigma$$

$$= \int_{\sigma=0}^{\infty} 2S(\sigma)\cos[2\pi\sigma x \sin\theta]\, d\sigma. \tag{11}$$

Thus $P(x)$ is the cosine Fourier transform of $S(\sigma)$.

Fig. 3   A "source-doubled" interferometer.

10. Application Areas

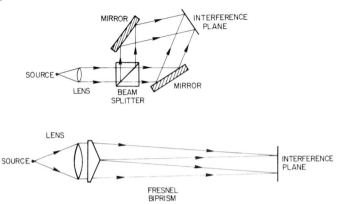

Fig. 4 Some two-plane-wave interferometers suitable for "point" sources.

Note that we have, as yet, said nothing about the dimension (call it y) normal to the beamsplitting plane. For the two-plane-wave case, the pattern is independent of y. For the two-spherical-wave case, the pattern varies only slightly with y (because of S variations).

In the absence of source doubling, the source size is restricted by Van Cittert–Zernike considerations. In particular, we can calculate (Caulfield, 1976) the resolving power

$$R = \sigma/\Delta\sigma, \tag{12}$$

where σ is the wavenumber and $\Delta\sigma$ the wavenumber resolution of the hologram. The resolving power and the source solid angle Ω are related by

$$R\Omega \leq 2\pi. \tag{13}$$

For the source-doubled interferometers that restriction does not hold.

It is also possible to show (Caulfield, 1976) that

$$R \cong N, \tag{14}$$

where N is the number of fringes in the utilized portion of the interference pattern. Also (Caulfield, 1976)

$$\Delta\sigma \cong (\sin\theta)/2L, \tag{15}$$

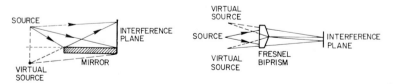

Fig. 5 Some two-spherical-wave interferometers suitable for "point" sources.

590

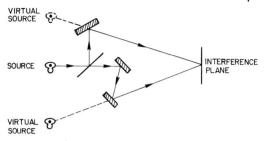

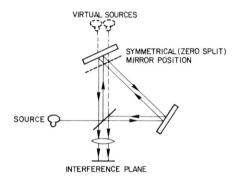

Fig. 6   Some source-doubling interferometers suitable for extended sources.

where L is the physical length of the utilized portion of the interference pattern. Note that $\Delta\sigma$ is independent of σ.

10.9.3   Recording the Hologram

A photograph of $P(x)$ is called a "spectral hologram." For $|\gamma| = 2$, the amplitude transmission of the resulting hologram is

$$H(x) = [|P(x)|^2]^{-\gamma/2} = |P(x)|$$

$$= 2 \left| \int_0^\infty S(\sigma) \cos[2\pi\sigma x \sin\theta] \, d\sigma \right|$$

$$= 2 \int_0^\infty S(\sigma)[1 + \cos(4\pi\sigma x \sin\theta)] \, d\sigma. \qquad (16)$$

For $|\gamma| \neq 2$, the hologram can give spurious "information." For these details see Caulfield (1976).

10.9.4 Spectrum Recovery

The hologram contains the spectrum in encoded form. Decoding it to measure $S(\sigma)$ can be done in many possible ways. We will now deal with some of these.

In the coherent optical Fourier transform of the light transmitted through the hologram, the amplitude

$$S_0 = 2 \int_0^\infty S(\sigma) \, d\sigma \tag{17}$$

appears at the origin and the spectrum $S(\xi)$ appears on both sides of the origin, where

$$\xi = (2\lambda_L f \sin \theta)\sigma, \tag{18}$$

and λ_L is the laser wavelength and f the focal length of the lens.

Of course the spectrum can be obtained by digital Fourier transformation. When the hologram is recorded on film, this requires microdensitometry, but if the hologram is recorded on a digital vidicon, a charge coupled device, etc., digital analysis is natural and useful.

Spectral decoding can be done in many ways. Consider a hologram made with plane waves at angles $+\theta$ and $-\theta$ with respect to the normal. Let us illuminate the hologram at an angle θ with a wavenumber σ_1. The hologram diffracts that light into many directions. The amplitude diffracted at $-\theta$ is proportional to $S(\sigma_1)$. By focusing and pinhole filtering, we can concentrate on that direction alone. By changing σ_1 (e.g., using a tunable laser), we can map out $S(\sigma)$.

Other interrogation methods include tilting the hologram while keeping the interrogation wavelength and the observation angle constant (Caulfield, 1976).

10.9.5 Spectral Filtering

If we form a spectral hologram at angles $+\theta$ and $-\theta$ and spectrum $S_H(\sigma)$ and illuminate the hologram at $+\theta$ with a spectrum $S_I(\sigma)$, then the light diffracted to $-\theta$ has a spectrum

$$S_0(\sigma) = S_H(\sigma)S_I(\sigma).$$

Thus the hologram is a spectral filter. Because we can create holograms artificially (by computer), we could make

$$S_H(\sigma) = S_0(\sigma)/S_S(\sigma),$$

where $S_S(\sigma)$ is the spectral sensitivity of a detector and $S_0(\sigma)$ is some reference spectrum. The detected signal would be

$$D_{0,I} \propto \int_0^\infty S_0(\sigma)S_S(\sigma)\,d\sigma = \int_0^\infty S_0(\sigma)S_I(\sigma)\,d\sigma.$$

Thus $D_{0,I}$ measures the cross correlation between S_0 and S_I. This is ideal for pattern recognition.

A different (and superior) way of doing spectral filtering or spectral correlation was introduced by Caulfield (1977). Here both the transmitted and the diffracted beams are monitored. The basic idea is shown schematically in Fig. 7. One interferometer produces a spatial display of the $P(x)$ pattern. A second interferometer exactly cancels the effect of the first. Thus the output spectrum and the input spectrum are identical. However, a mask $H(x)$ in the Fourier transform plane converts an input spectrum $S_I(\sigma)$ to an output spectrum

$$S_P(\sigma) = S_I(\sigma)S_H(\sigma)$$

as before.

10.9.6 Uses, Advantages, and Modifications

For recording a spectrum, holographic spectroscopy has some profound advantages over other approaches. The most cited advantage is recording speed. By focusing the light down in the y direction using a cylindrical lens and by matching the film resolution to the needed resolution, we can achieve very high speed recording of continuous events. The current record in that regard is 4×10^5 spectra per second (Tsuno and Takahashi, 1970). It is possible in principle to record spectra continuously on fast moving film (Caulfield, 1976). For pulsed sources the recording time is the pulse duration δt. The spectral frequency $\delta \nu$ resolution is limited by $\delta \nu\, \delta t \geq 1$. But $\nu = c\sigma$, so the wavenumber resolution is $\delta \sigma \geq (c\, \delta t)^{-1}$.

Other advantages include the minimization of stray light effects, background

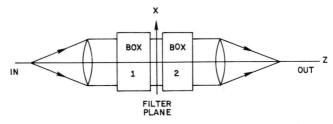

Fig. 7 A generalized holographic system contains two interferometers or "boxes," one to produce a spatially varying time delay and the second to recombine all rays with a total of zero time delay for the whole system. A hologram in the x plane acts as a spectral filter. The analogy with coherent optical image processing is clear.

suppression (by operating in portions of the Fourier-transform domain where the background is known to be small), high resolution, and (under some circumstances) cost.

For correlation and filtering the advantages are many and profound. The ability to create almost arbitrary spectral characteristics is very important. The ability to correlate with many filters in parallel using y-direction multiplexing is almost unique. With the various real time spatial light modulators, the filters become programmable. Using y-adjacent filters we can subtract two spectra.

10.9.7 Some Relatives

Diffraction gratings are close relatives of the spectral hologram. Indeed, a spectral hologram is an incoherent sum of holographically recorded gratings for each wavenumber.

Fourier-transform spectrometers are also close relatives of the spectral hologram. Fourier-transform holograms produce the spectral Fourier transform in time rather than in space.

Other spectrometers form spatial Fourier transforms and interrogate various wavelengths by oscillating a properly spaced ruling in that plane by one grating period and detecting the modulated portion of the total light (Esplin, 1978).

REFERENCES

Caulfield, H. J. (1976). Holographic spectroscopy. *In* "Advances in Holography" (N. Farhat, ed.), Vol. 2. Marcel Dekker, New York.

Caulfield, H. J. (1977). *Opt. Commun.* **23,** 344.

Esplin, R. W. (1978). *Opt. Eng.* **17,** 73.

Tsuno, T., and Takahashi, R. (1970). High speed photography of spectra, *Proc. Int. Congr. High-Speed Photography, 9th.* Soc. Motion Picture and Television, New York.

10.10 HOLOGRAPHIC CONTOURING METHODS

J. R. Varner

10.10.1 Introduction

This section gives brief descriptions of three different methods of using holography to obtain contour maps. The three methods are two-frequency holographic contouring (Varner, 1971), two-refractive-index holographic contouring (Varner, 1971), and sandwich-hologram contouring (Abramson, 1976). Some of this material has been reviewed by the author in an article (Varner, 1974) comparing the usefulness of holographic and moiré surface contouring. The relative merits of the methods will not be discussed except for the method using sandwich holograms. Only holographic contouring methods will be described here; however, sometimes a nonholographic contouring technique is more useful than a holographic technique for some applications.

A contour map is defined here as a two-dimensional image of a three-dimensional object with a superposed set of lines showing the intersections of the object with a set of equidistant planes perpendicular to the line of sight. A special case of each contouring method will be described which yields contour maps as already defined and, at the same time, is easy to explain. The three references given should help the reader who is interested in greater depth and generality.

10.10.2 Holographic Contouring Using Two Frequencies of Light

Consider the simple holographic setup sketched in Fig. 1. The object illumination consists of two coincident plane waves of frequency f_1 and f_2 (or wavelength λ_1 and λ_2) which can be considered to emanate from plane S, reflect off beamsplitter B, and then illuminate the object. Imagine a plane C which is perpendicular to the illumination beam and intersects the object. The intersection of plane C with the object is a contour line which is recorded in the following way. The dashed line represents two coincident rays, one of λ_1 and one of λ_2, traveling exactly the same path from source plane S to an object point (also intersected by plane C) and back to the recording medium; the chromatic shear due to the finite thickness of the beamsplitter B is generally

HANDBOOK OF OPTICAL HOLOGRAPHY
Copyright © 1979 by Academic Press, Inc.
All rights of reproduction in any form reserved.
ISBN-0-12-165350-1

10. Application Areas

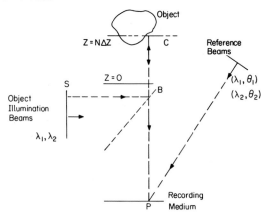

Fig. 1 Two-frequency holographic contouring.

negligible. At point P on the recording medium, the object light of each wavelength interferes with its corresponding reference beam and the intensities of these two interference patterns are recorded.

For simplicity we assume a linear relationship between the amplitude transmittance T of the recording medium and its exposure E. Thus

$$T = a_0 - a_1 E, \qquad (1)$$

which becomes

$$T = a_0 - a_1 | U_{o1} + U_{o2} + U_{r1} + U_{r2} |^2 t, \qquad (2)$$

where U_{o1}, U_{r1} represent the amplitudes of the optical electric field and t is the exposure time. The spatially varying part of E is of concern here and is of the form

$$E' = \cos(\omega x) + \cos(\omega x + \Delta \phi), \qquad (3)$$

where ω is the spatial carrier frequency and $\Delta \phi$ contains the contouring information in the form

$$\Delta \phi = (2 \pi N / \lambda_1) 2z - (2 \pi N / \lambda_2) 2z = 2 \pi N 2z (\Delta \lambda / \lambda_1 \lambda_2). \qquad (4)$$

If plane C intersects the object in a set of points P where $\Delta \phi = N 2 \pi$ (N any integer), the modulation due to E' is maximum. However, if

$$\Delta \phi = (2N - 1) \pi, \qquad (5)$$

the modulation due to E' goes to zero. Thus, upon reconstruction, bright diffraction takes place for the first case and zero diffraction for the latter. The sensitivity of contouring (the separation between bright contour lines) is given by

$$\Delta z = \lambda_1 \lambda_2 / 2n \, \Delta \lambda \qquad (\lambda^2 / 2 \, \Delta \lambda \text{ when } \Delta \lambda \text{ is small and } n = 1). \qquad (6)$$

596

Conceptually then, this explains how two-wavelength holography is used to measure contour information. Now let us look at the method that the author has found most useful (Fig. 2). Here the hologram records a real image of the object as formed by the telescopic system shown; the use of the telescope to record an image-plane hologram of the object minimizes the chromatic image decorrelation occurring upon reconstruction. The use of telecentric viewing to record contour maps applies, with only obvious differences, to all three holographic contouring systems.

Aperture A in the telescope plays an interesting role in determining the appearance of the contour map. Since aperture A is centered on the optical axis of the telescope, it passes only paraxial light reflected from the object. Only light exactly on axis gives unambiguous contour information in high contrast interference fringes. However, if A is too small, the image becomes blurred and speckled so that contour fringes and image detail both become hard to discern. Thus contour fringe contrast and image clarity are each increased by decreasing the other [see Varner (1971) for a detailed discussion].

10.10.3 Two-Refractive-Index Holographic Contouring

The schematic drawing in Fig. 3 represents a simple holographic system for obtaining a contour map of an object. The object is placed in the immersion

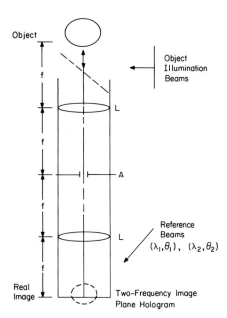

Fig. 2 The use of telescopic viewing to obtain contour maps.

10. Application Areas

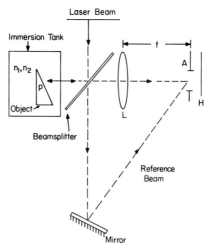

Fig. 3 A schematic drawing of two-refractive-index holographic contouring.

tank which is filled with a transparent medium of refractive index n_1. The object is then illuminated by collimated, coherent light reflected from a beam-splitter and passing through a transparent wall of the immersion tank. Again, telescopic viewing is used to control and filter the light reaching the hologram recording plane. A single holographic exposure is made using a reference beam obtained as shown (or in any convenient manner). The medium in the immersion tank is now changed to have a refractive index n_2, and, no other change being made, the hologram is exposed a second time.

When the refractive index is changed from n_1 to n_2, the optical phase change recorded for point P of the object is

$$\Delta\phi = (2\pi/\lambda)\,|\,n_2 - n_1\,|\,2z, \tag{7}$$

where z is the distance from the tank wall to P. Now if $z = N\,\Delta z$ causes ϕ to change by $N2\pi$, we have defined a set of contour planes separated by

$$\Delta z = \lambda/2\,|\,n_2 - n_1\,|. \tag{8}$$

But since $\lambda/n_1 = \lambda_1$ and $\lambda/n_2 = \lambda_2$, we find

$$z = \lambda_1\lambda_2/2\,|\,\lambda_2 - \lambda_1\,| \tag{9}$$

which is the same result we found for two-frequency contouring. As in two-frequency contouring, telescopic viewing is needed. Further, the best results are obtained by placing the object close to the transparent wall of the tank which corresponds to putting the hologram plane at the telescopic image plane in two-frequency contouring.

598

10.10.4 Sandwich-Holography Contouring

Sandwich holography has been used by Abramson (1976) to obtain contour maps as shown in Fig. 4. The technique depends upon modulating one set of interference fringes with another set. Abramson refers to these sets of interference fringes as the illumination and observation fringe patterns. By Abramson's procedure, a hologram H_1 is made using illumination beam P_1. H_1 is removed and a second hologram is made at a slightly different distance from the object using illumination beam P_2. When H_1 and H_2 are replaced in exactly their original positions, the object is seen as if illuminated by P_1 and P_2 simultaneously which gives the set of illumination fringes. A rigid sandwich is made of H_1 and H_2 in any convenient manner. When the sandwich is rotated an angle θ, the two image waves are sheared relative to one another forming the set of observation fringes.

In the region where these two fringe patterns overlap, they modulate one another as shown in Fig. 4. Any point on the object which looked dark (bright) before the sandwich was rotated will look bright (dark) after rotation if intersected by a dark observation fringe. Points intersected by bright observation fringes keep their original relative brightness after rotation since the bright observation fringe indicates a phase shift equal to an integer multiple of 360°. The net result of this intermodulation is the set of contouring fringes depicted by the dark diamonds in Fig. 4.

The spacing between adjacent bright fringes is given by

$$y = 2xd/(d^2 + x^2 - 2xd \cos B)^{1/2}, \tag{10}$$

where x is the spacing between a bright and dark observation fringe, d the spacing between a bright and dark illumination fringe, and B the angle between the illumination beam and the line of sight. Abramson describes the modifications of his system necessary to achieve plane contour surfaces. These are simply to use plane wave illumination and some approximation to telescopic viewing.

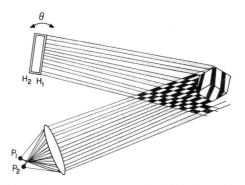

Fig. 4 Contouring by sandwich holography.

10. Application Areas

The results obtained (Abramson, 1976) by sandwich holography are contour maps of excellent visibility, and the procedure is more flexible (in this author's view) than any other holographic method. The ability to rotate the contouring planes through wide angles by manipulating the sandwich hologram is certainly not easily done by other holographic methods; as Abramson points out, this allows one to discover quickly whether a slope is toward or away from the hologram and to readily align the contour planes with some reference surface.

The angle between the line of sight and the contour planes is given by

$$\epsilon = \sin^{-1}[x \sin B/(d^2 + x^2 - 2dx \cos B)^{1/2}]. \tag{11}$$

Typically the desirable result is ϵ equal to 90°, which yields

$$y = d/\sin B, \qquad x = d/\cos B, \qquad B = \pi/2. \tag{12}$$

Abramson mentions the problem of decorrelation which occurs during the rotation of the sandwich hologram. Since the object to be contoured is diffuse (by virtual necessity), the reflected wave pattern has high spatial frequency content or a short correlation length. Thus increasing the shear by rotation can cause the contrast of the observation fringes to drop rapidly. Relatively large rotation will be necessary to obtain contour sensitivity on the order of 10 μm. The contour fringe contrast could be too low to be useful in some cases when large sandwich rotations are necessary to obtain ϵ equal to $\pi/2$.

No specific comparison of this method to nonholographic methods will be made here. In general, holographic methods require more expensive equipment and complicated procedures than nonholographic methods; sandwich holography seems to be particularly bad in this respect. However, the flexibility of sandwich holography probably compensates well for its added complexity. More definite and specific comments will have to wait for more detailed reports in the literature.

REFERENCES

Abramson, N. (1976). *Appl. Opt.* **15**, 200.

Varner, J. R. (1971). Multiple-Frequency Holographic Contouring, Ph.D. Dissertation. Univ. of Michigan, Ann Arbor.

Varner, J. R. (1974). *In* "Holographic Nondestructive Testing" (R. Erf, ed.), pp. 105–107. Academic Press, New York.

10.11 MULTIPLE IMAGE GENERATION

H. J. Caulfield

10.11.1 Introduction

There are a variety of reasons for wanting to convert a single image into an array of identical images. Likewise there are a variety of methods for doing so. The four basic types of multiple image forming systems are

(1) step and repeat cameras,
(2) incoherent optical methods,
(3) coherent nonholographic techniques, and
(4) coherent holographic methods.

Although this last method is the sole topic in this section, a brief review of the other, older approaches will serve to establish criteria by which success in multiple imaging may be judged.

Step and repeat cameras are the instruments of choice for most multiple imaging systems today. Very simply, these are not multiple image forming devices but single image forming devices which form their images sequentially in well-controlled, easily programmable places. Their advantages are availability now, easy programmability, identical quality of each image, and versatility (they can do contact printing or projection printing). The primary disadvantage appears to be lack of speed. It takes N times longer to record N images than it does to record one. This is the motivation for multiple imaging. The basic intent is clear: send one image into a ''black box'' and receive an array of identical images out.

Incoherent multiple imaging methods are of the general type shown in Fig. 1. With a point source for illumination, we obtain an image of the periodic mask in the output plane. With the input image as the source, the output is an array of copies of the input image—one for each point in the periodic mask. Readers will recognize that the word ''incoherent'' is actually misleading, because spatial coherence across the periodic mask is required. This method is discussed in detail by Thompson (1976) and references therein.

The explicitly coherent version of this system has been discussed by Thompson (1971) and later (in a slightly different form) by Kalestynski (1975). The

HANDBOOK OF OPTICAL HOLOGRAPHY
Copyright © 1979 by Academic Press, Inc.
All rights of reproduction in any form reserved.
ISBN-0-12-165350-1

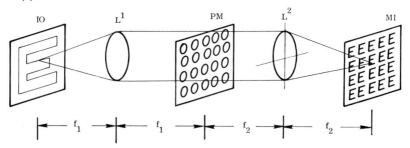

Fig. 1 Light from an incoherently radiating object IO incident on a periodic mask PM produces a multiple image MI in this configuration.

periodic structure samples the Fraunhofer diffraction pattern of the input image, so the image is the input image convolved with a periodic function.

There are two main difficulties with these two methods. First, there is a tradeoff between brightness and resolution. For good resolution we want very small pinholes. For high transmission we want big pinholes. Second, it is very difficult to obtain N equally bright images. These are the problems holographic multiple imaging must solve.

10.11.2 Some Problems Treated Elsewhere

The multiple imaging hologram as invented independently by Lu (1968) and by Groh (1968) involves construction of a specialized holographic optical element. Because these and their aberrations are discussed in Sections 2.4 and 10.11.7, we only call attention to that subject here. Likewise the image quality depends strongly on alignment during reconstruction (Section 10.11.6). Finally, that ubiquitous enemy of all coherent imaging—speckle—is a real problem (see Section 9.2).

10.11.3 Fourier-Transform Holographic Multiple Imaging

We use an input image of amplitude transmission $f(x, y)$. Its optical Fourier transform is $F(u, v)$. We make a hologram of transmission

$$H(u, v) = \sum_{m,n} \exp[-2\pi i(nx_0 u + my_0 v)],$$

where n and m are integers. The wavefront emerging from the hologram is

$$W(u, v) = F(u, v)H(u, v).$$

Optical Fourier transformation produces the image

$$w(x, y) = f(x, y) * \sum_{m,n} \delta(x - nx_0, y - my_0)$$

$$= \sum_{m,n} f(x - nx_0, y - my_0).$$

Actually, of course, we obtain $w(-x, -y)$, but this is of no consequence in most cases and can be "corrected" in any case by starting with $f(-x, -y)$ rather than $f(x, y)$ as an image. This is the method described by Lu (1968). The geometry is shown in Fig. 2.

Let us examine the question of how to record the hologram. The hologram represents plane waves traveling in many directions (one for each of the N multiple images). As a plane wave object beam and a plane wave reference beam are the only way to obtain a beam balance ratio of $K = 1$ across the hologram, that combination leads to the highest possible diffraction efficiency. Furthermore, each plane wave utilizes the full aperture of the hologram uniformly. The three basic problems are

(1)   How do we avoid cross talk?
(2)   How do we achieve maximum diffraction efficiency?
(3)   How do we achieve equal diffraction efficiency for each plane wave?

Cross talk is a phenomenon associated with nonlinear recording (see Section 8.3) whereby it may be that light which should appear at (nx_0, ny_0) appears instead at other location(s). In the perfectly periodic case postulated above (for simplicity only), those locations are also likely to be of the form $(n'x_0,$

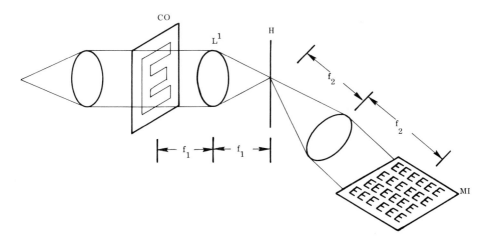

Fig. 2   Light from a coherently radiating object CO incident on a special hologram H produces a multiple image MI in this configuration.

10. Application Areas

$m'y_0$). To illustrate this consider that we want to record the hologram of N object wavefronts $O_1, O_2, \ldots, O_N$ using a single reference wavefront R. The exposure is proportional to

$$E = |O_1 + O_2 + \cdots + O_N + R|^2$$

$$= |O_1|^2 + |O_2|^2 + \cdots + |O_N|^2 + R^2$$

$$+ RO_1{}^* + RO_2{}^* + \cdots + RO_N{}^*$$

$$+ R^*O_1 + R^*O_2 + \cdots + R^*O_N$$

$$+ CT,$$

where CT is the cross talk information,

$$CT = \qquad \cdots \quad + O_1{}^*O_2 + O_1{}^*O_3 + \cdots + O_1{}^*O_N$$

$$+ O_2{}^*O_1 + \qquad \cdots \qquad + O_2{}^*O_3 + \cdots + O_2{}^*O_N$$

$$+ O_3{}^*O_1 + O_3{}^*O_2 + \qquad \cdots \qquad + \cdots + O_3{}^*O_N$$

$$\vdots \qquad \vdots \qquad \vdots \qquad \vdots \qquad \vdots$$

$$+ O_N{}^*O_1 + O_N{}^*O_2 + O_N{}^*O_3 + \cdots + \cdots .$$

For the special case of a one-dimensional array

$$O_1 = \exp[-2\pi i(x_0 u)],$$

$$O_2 = \exp[-2\pi i(2x_0 u)],$$

$$\vdots$$

$$O_N = \exp[-2\pi i(Nx_0 u)],$$

we have

$$CT = \cdots + \exp[-2\pi i(x_0 u)] + \exp[-2\pi i(2x_0 u)] + \cdots$$

$$+ \exp\{-2\pi i[(N-1)x_0 u]\} + \exp[2\pi i(x_0 u)] + \cdots$$

$$+ \exp[-2\pi i(x_0 u)] + \cdots + \exp\{-2\pi i[(N-2)x_0 u]\}$$

$$+ \exp[2\pi i(2x_0 u)] + \exp[2\pi i(x_0 u)] + \cdots + \cdots + \exp\{-2\pi i[(N-3)x_0 u]\}$$

$$\vdots \qquad \qquad \vdots \qquad \vdots \quad \vdots \qquad \qquad \vdots$$

$$+ \exp\{2\pi i[(N-1)x_0 u]\} + \exp\{2\pi i[(N-2)x_0 u]\}$$

$$+ \exp\{2\pi i[(N-3)x_0 u]\} + \cdots .$$

We can then group the terms. There are one each of $\exp[2\pi i(N-1)x_0 u]$ and $\exp[-2\pi i(N-1)x_0 u]$, two each of $\exp[2\pi i(N-2)x_0 u]$ and $\exp[-2\pi i(N-2)x_0 u], \ldots$, and $N-1$ each of $\exp[2\pi i x_0 u]$ and $\exp[-2\pi i x_0 u]$. Illuminating such a hologram with R^*, we would obtain the expected terms

604

$(R)^2O_1 + (R)^2O_2 + \cdots + (R)^2O_N$ and also terms $R^* \exp[2\pi i(N - 1)x_0u]$ $+ R^* \exp[-2\pi i(N - 1)x_0u] + 2R^* \exp[2\pi i(N - 2)x_0u] + 2R^* \exp[-2\pi i(N - 2)x_0u] + 2R^* \exp[-2\pi i(N - 2)x_0u] + \cdots + (N - 1)R^* \exp[2\pi ix_0u]$ $+ (N - 1)R^* \exp[-2\pi ix_0u]$. These terms thus appear around the reconstructing beam and need not overlap the object beam. However, if the hologram is recorded nonlinearly ($\gamma \neq 2$), some coupling occurs between these beams and the object beam. It is as though the reconstructing beam for a linearly recorded hologram were a multiple beam. This causes the input image to be convolved not with an array of delta functions but with the array of delta functions each convolved with a periodic function peaked at zero but extended out to $\pm(N - 1)x_0$ from zero. Clearly, cross talk is rendered harmless by linear recording ($|\gamma| = 2$). Often, though, linear recording is difficult or impossible. The problem then is how to prevent the cross talk altogether. There is only one way to do this. We can use one-beam-at-a-time multiplexing (see Section 5.2 for a discussion of multiplexing). As shown in Section 5.2 the diffraction efficiency of each beam is $1/N^2$ the diffraction efficiency achievable with an $N = 1$ hologram and $1/N$ of the efficiency achievable with an N-point object. On the other hand, cross talk is eliminated altogether and uniformity is readily achievable. With cross talk and $|\gamma| \neq 2$, uniformity is virtually impossible to achieve.

The throughput-hole-size tradeoff characteristic of nonholographic methods vanishes for holography. The points in the array can be diffraction limited or smaller (as viewed by the Fourier transforming lens) without sacrificing diffraction efficiency of the hologram.

10.11.4 Lensless Fourier-Transform Holographic Multiple Imaging

Suppose we use no lenses in either recording the hologram or projecting the multiple images but instead use the hologram itself to perform those functions. Figure 3a shows how we can record such a hologram, and Fig. 3b shows how to use it to project multiple images. This is the method proposed by Groh (1968).

In recording such a hologram, we should keep the reference point and object points an equal distance from the hologram to cancel the quadratic components of their wavefronts and allow uniform utilization of the hologram (see Section 4.3). Again the points may be recorded sequentially or simultaneously with the same attendant advantages and disadvantages to each approach.

10.11.5 Multiple Image Holograms

Kalestynski (1976) has suggested a variant on Groh's method which could just as well be a variant of Lu's. The idea is to record a hologram of the image

10. Application Areas

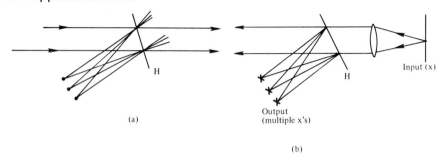

(a)

Input (x)

Output
(multiple x's)

(b)

Fig. 3 (a) We can use multiple points to construct a hologram. (b) On illuminating that hologram with light from an object (an *x* in this case), one image for each original object point is produced.

to be replicated using multiple reference beams. Then one would use a single reference beam in reconstruction to produce multiple output images. This author sees no advantage to making such a specialized multiple image hologram, when for the same effort we could make a hologram which would produce multiple images of any input image.

10.11.6   General Comments

It is clear that holographic multiple imaging has distinct advantages over step and repeat cameras in speed. It is not clear, however, that there is a net advantage. The signal strength and the signal-to-noise ratio in each of N images will vary as $1/N$ or $1/N^2$ depending on how the hologram is recorded. This limits the value of N since we want a good signal-to-noise ratio for each image. The best holograms reported to date give good images with $N \cong 1000$. The image quality varies from image to image by virtue of the off-axis holographic aberrations (e.g., astigmatism). All of the problems of coherent imaging are present. These problems may or may not be negligible. On the other hand, holography has the flexibility to do some things uniquely well. For instance, alignment of images in sequential imaging operations can be monitored easily by having one output image set aside for monitoring. When the image is properly aligned there, all of the multiple images are properly aligned.

Holographic multiple imaging is also advantageous in comparison with nonholographic multiple imaging in that it avoids the throughput-resolution trade-off and allows better image uniformity.

10.11.7   Applications

To our knowledge four general applications have been suggested for multiple image holograms. They can be used

606

(1) for image recording, e.g., for photolithography in the semiconductor field,

(2) to allow parallel optical processing to occur on a single input image,

(3) to assist in performing piecewise space-variant image processing, and

(4) (in either direction) as coupling elements for fiber optic systems.

REFERENCES

Groh, G. (1968). *Appl. Opt.* **7,** 1693.
Kalestynski, A. (1976). *Appl. Opt.* **15,** 853.
Kalestynski, A. (1975). *Appl. Opt.* **14,** 2343.
Lu, S. (1968). *Proc. IEEE* **56,** 116.
Thompson, B. J. (1971). *Laser Appl.* **1,** 33.
Thompson, B. J. (1976). *Appl. Opt.* **15,** 312.

10.12 PARTICLE SIZE MEASUREMENTS

Brian J. Thompson

10.12.1 Need for a Technique

Particle size measurement has long been a difficult technological area and one of wide diversity and importance. The techniques available are many and varied as indeed are the number of purely optical methods. Holography can fill a specific need in this field without, of course, being in any sense universally applicable. The particular strength that holography has is that dynamic situations can be handled and an image can be formed from the hologram for detailed study. Consider a field of moving particles that exist in a volume; it is an impossible task to photograph the volume if the dimensions of the individual particles are much smaller than the dimensions of the volume. For example, if the particles are 10 μm in diameter and they exist throughout a volume of 1 cm^3, an imaging system that can resolve 10 μm has a depth of field much less than 1 cm! However, it is possible to record the Fraunhofer hologram of the particles in such a volume and subsequently form an image of the particle field from the hologram. Thus a transient event is captured in the hologram to produce, at a subsequent time, a stationary image.

10.12.2 Applications

The method of Fraunhofer holography discussed in Section 4.2 was developed for the application of particle size analysis starting in 1963 (Thompson, 1963; Parrent and Thompson, 1964). Since then a significant number of papers have been written on this subject, and recent reviews give a good guide to the earlier literature [see, e.g., Thompson (1974) and Trolinger (1975)]. The method was initially used for the measurement of naturally occurring fog particles (Silverman *et al.*, 1964; Thompson *et al.*, 1966), but has since been applied to a wide range of particulate measurement and evaluation problems.

Figure 1 shows the typical configuration of such a system. The illumination is provided by a pulsed ruby laser; this is necessary, since at the resolution of a few microns and at average velocity of 100 cm/sec, an exposure time of 10^{-6} sec is required. Naturally, higher velocities require even shorter exposure times. The Q-switched ruby laser beam passes through a spatial filter and is

HANDBOOK OF OPTICAL HOLOGRAPHY
Copyright © 1979 by Academic Press, Inc.
All rights of reproduction in any form reserved.
ISBN-0-12-165350-1

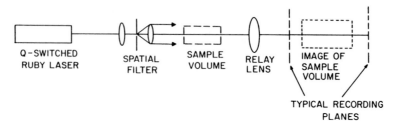

Fig. 1 Schematic diagram of a typical in-line Fraunhofer holographic system for particle size analysis.

then collimated (it must be stressed that collimation is not essential) and illuminates the sample volume. The actual volume that can be used depends upon the resolution required but is typically several cubic centimeters for particles from 2 μm and above. Before the hologram is recorded, it is often advantageous to provide some magnification of the hologram to ease the resolution requirements of the recording material. The sample itself is also imaged and typical recording planes are indicated. The details of the optical layout of these systems depends upon the particular application and the nature of the problem.

The original application to naturally occurring fog obviously led to a wider application of the method to a variety of particles and aerosol sizing problems.

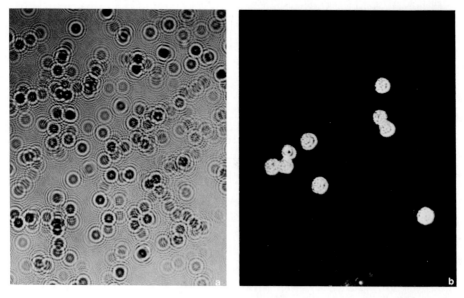

Fig. 2 (a) Hologram of a field of small pollen particles (~20 μm in diameter). (b) Image formed from a portion of that hologram shown in (a). (After Tyler, 1978.)

610

Energetic aerosols were investigated (Thompson *et al.,* 1967) and so were rocket injectors (Trolinger *et al.,* 1969). Other examples of applications include cloud chamber studies, marine plankton imaging, rocket engine exhaust studies, two-phase flow diagnostics, glass fiber measurements, dust erosion studies, snowflake and ice crystal imaging, and electron beam imaging. It is not appropriate to discuss all these areas in the present contest.

The development of these various applications has also produced additional basic studies in the holographic process itself and hence its range of applicability; for example, the extension to large and small particles, the study of the effects of index variations.

Figure 2a shows a recent example of the hologram of a field of small pollen particles that are about 20 μm in diameter. The image produced from a section of this hologram is shown in Fig. 2b. Good edge resolution is apparent at about 1 μm.

This area of application is still very active, using both in-line and off-axis methods and it is anticipated that further important results will be forthcoming.

REFERENCES

Parrent, G. B., and Thompson, B. J. (1964). *Optica Acta* **11**, 183.
Silverman, B. A., Thompson, B. J., and Ward, J. H. (1964). *J. Appl. Met.* **3**, 792.
Thompson, B. J. (1963). *J. Soc. Photo-Opt. Instr. Eng.* **2**, 43.
Thompson, B. J. (1974). *J. Phys. E* **7**, 781.
Thompson, B. J., Parrent, G. B., Ward, J. H., and Justh, B. (1966). *J. Appl. Met.* **5**, 343.
Thompson, B. J., Ward, J. H., and Zinky, W. R. (1967). *Appl. Opt.* **6**, 519.
Trolinger, J. (1975). *Opt. Eng.* **14**, 383.
Trolinger, J., Belz, R. A., and Farmer, W. M. (1969). *Appl. Opt.* **8**, 957.
Tyler, G. A. (1978). Ph.D. Thesis, Univ. of Rochester, Rochester, New York.

10.13 HOLOGRAPHIC PORTRAITURE

Walter Koechner

The development of multistage, long coherence length ruby lasers has made it possible to record human subjects (Siebert, 1968; Wuerker and Heflinger, 1971; Ansley and Siebert, 1970; McClung *et al.*, 1970; Ansley, 1970; Rundle and Higgins, 1975; Zech and Siebert, 1968; Gregor and Davies, 1969; Gregor, 1971). Because of the short pulse duration of a *Q*-switched solid state laser, mechanical instabilities and object movements are eliminated.

10.13.1 Laser Source

Q-switched ruby lasers or *Q*-switched and frequency doubled Nd:YAG lasers can be used as light sources for holographic portraiture. The salient features of these systems have been described in Section 8.1. The *Q*-switched ruby laser is by far the most commonly used laser in holographic protraiture because ruby, as compared to frequency doubled Nd:YAG, is capable of much higher energy outputs.

The distinctive features of the laser systems employed in holographic portraiture are a combination of high energy output and coherence length. A hologram of a single person requires a minimum energy of 250 mJ and a coherence length of 1 m. Holographic group portraits are usually made with 4 to 10 J of energy and a coherence length of 5 to 10 m (McClung *et al.*, 1970; Ansley, 1970; Rundle and Higgins, 1975).

Only oscillator–amplifier systems have sufficient energy and coherence length for this type of application. For portraits of a single person one amplifier is commonly used, whereas group portraits require an oscillator followed by two amplifiers. The oscillator can be *Q*-switched with a Pockels cell, Kerr cell, or with a bleacheable dye since precise timing is not required in this application.

10.13.2 Experimental Arrangements

Because of the short exposure time, mechanical stability requirements are eliminated and it becomes relatively easy to make high-quality transmission or reflection holograms.

HANDBOOK OF OPTICAL HOLOGRAPHY
Copyright © 1979 by Academic Press, Inc.

10. Application Areas

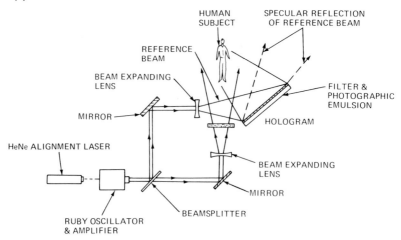

Fig. 1 Arrangement for forming a transmission hologram of a human subject employing a Q-switched ruby laser. (After Ansley, 1979.)

Figure 1 shows the arrangement for forming a transmission hologram of a human subject. The most important consideration in taking a hologram of a person is to avoid eye injuries of the subject from the laser radiation. In Fig. 1 the object beam is expanded by a negative lens and passed through a diffusing screen. If the elements are chosen properly (see Section 10.13.3 for design details), the scattered light from the diffuser does not present a radiation hazard to the model. Equally important is the optical layout of the reference beam. It is imperative that the portion (about 10%) of the reference beam which is reflected by the photographic plate is directed away from the human subject as shown in Fig. 1.

The subject is typically 1 to 2 m away from the photographic plate. The optical paths of the object and reference beam should be matched at the position of the human subject.

Figure 2 shows a similar experimental setup for recording a transmission hologram with the exception that two diffusion screens are employed for the object beam in order to provide a more uniform illumination.

The holograms are usually made on 4 × 5 or 8 × 10 in. Agfa 10E75 or 8E75 plates. The photographic plate has to be protected by a cutoff filter from fogging caused by flashlamp light or room light (e.g., Schott glass # RG-665). If, in addition, a focal plane shutter is mounted in front of the filter–photographic plate assembly, holograms can be made under daylight or normal room light conditions. The opening of the mechanical shutter has to be synchronized with the laser. Solenoid-operated mechanical shutters with free apertures of up to 6 in. are commercially available with minimum opening times of 0.4 to 0.6 sec.

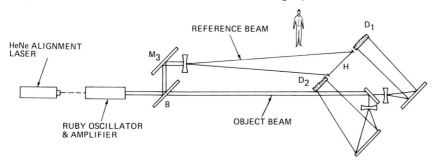

Fig. 2 Schematic diagram for making a transmission hologram using two object beams. H, holographic plate; D_1, D_2, glass plate diffusers. (After Siebert, 1968.)

Figure 3 shows an experimental arrangement for recording reflection holograms of humans. In this case the image from the hologram can be viewed with white light. The main difference as compared to Figs. 1 and 2 is the arrangement of the reference beam which enters the photographic plate from the back (see Section 5.1). In this case it is particularly important to adjust the angle such that the model is not subjected to radiation from the reference beam.

For the arrangements shown in Figs. 1 to 3, only single element glass lenses that are antireflection-coated should be used. At the high power levels of Q-switched ruby lasers, care must be taken that reflections from curved surfaces are not transmitted back into the laser. The negative lens should preferably be a plano-concave element with the concave surface facing away from the laser

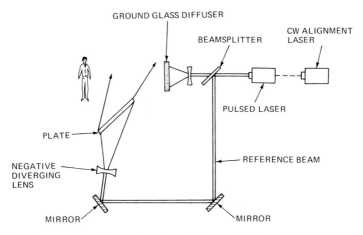

Fig. 3 Experimental configuration for making reflection (white-light) holograms of humans. (After Ansley, 1979.)

and the flat surface slightly tilted so that the back reflection misses the output port of the laser.

The high pulse power also causes problems whenever the beam is focused to a point because the focused high power ionizes air and damages any other material in its path. This means that ordinary spatial filters cannot be used to eliminate diffraction effects. Without spatial filters, the microscope objective lenses commonly used to expand the laser beam in low power holography must be replaced by diverging lenses instead.

Mirrors are essential to holographic systems. Because aluminum mirrors absorb 10% of ruby radiation, only the diverged beam should be reflected from aluminum-coated mirrors. The raw, nondiverged beam should be reflected only from mirrors capable of withstanding energies similar to those in the ruby oscillator. Beamsplitters also need to be reflection- and antireflection-coated with dielectric materials.

Neutral-density filters that are commonly used with cw lasers will be destroyed at high pulsed-laser energies. A set of glass filters is needed in which the absorbing material is diffused through the glass substrate.

10.13.3   Eye Safety Considerations

If one assumes that the human subject will not wear safety glasses, the illumination energy entering the eyes must be at a safe level.

10.13.3.1   Maximum Permissible Energy Levels

For a Q-switched ruby laser the maximum safe energy level on the retina is (ACGIH, 1972; ANSI, 1973)

$$I_{Rmax} = 0.07 \quad J/cm^2. \tag{1}$$

This peak value of energy density that can be tolerated by the retina has to be related to the energy density on the cornea of the eye.

Assume a large parallel beam incident on the cornea. The minimum spot size on the retina due to aberrations is 10 μm. If the eye is adapted to night conditions, the pupil has a diameter of about 7 mm. In this worst case situation, the focusing power of the eye increases the energy density of a parallel beam striking the cornea by a factor $(7 \text{ mm})^2/(10 \text{ } \mu m)^2 \approx 5 \times 10^5$. Dividing the retinal maximum safe energy level by this factor, one obtains $I_{Cmax} \approx 1 \times 10^{-7} \text{ J/cm}^2$ as the maximum permissible energy density on the cornea (ACGIH, 1972).

Assuming a safety factor of 10, one obtains a maximum permissible exposure level at the cornea for direct illumination or specular reflection from a Q-switched ruby laser of

$$I_{Cmax} = 1 \times 10^{-8} \quad J/cm^2. \tag{2}$$

For a daylight adapted pupil (3-mm diameter) the safe energy density can be increased by a factor of 5.

As shown in Figs. 1 to 3, the beam illuminating the subject is expanded and passed through a diffuser. In this case we must relate the energy density on the retina to the energy density at the diffuse reflecting surface. As shown in Fig. 4, an area A_D of the diffuser will be imaged to a smaller area A_R on the retina. Simple ray geometry gives the ratio of the areas as $A_D/A_R = (r/f_E)^2$.

Now let the surface energy density at the screen be I_D, and let us suppose that the screen area A_D diffuses the energy $I_D A_D$ incident on it into a solid and Ω_D. Diffusing light from a ground or sandblasted glass plate does not even remotely approach an ideal Lambertian diffuser. The light is much more peaked around small angles of the incident beam direction. Typically, for angles larger than $\pm 20°$ from the incident beam direction, the radiance of the scattered light is below 10% of the peak value (Middleton, 1960; Levi, 1968). If the area of the pupil of the eye is $A_c = (\pi/4)d_c^2$, the solid angle it subtends at the screen is A_c/r^2, and the fraction of the energy it receives is $A_c/r^2\Omega_D$. Consequently, the energy density at the retina is given by $I_R = I_D A_c/\Omega_D f_E^2$.

Note that the relations I_R and I_D are independent of r. Inserting nominal values $d_c = 7$ mm, $f = 15$ mm, $\Omega_D = \pi/10$ sr, and using $I_{Rmax} = 0.07$ J/cm², we obtain $I_{Dmax} = 0.13$ J/cm². Allowing for a safety margin, the energy density of the laser beam incident on the diffusing screen must be kept below (ANSI, 1973; Sliney and Freasier, 1973)

$$I_{Dmax} = 0.07 \text{ J/cm}^2 \tag{3}$$

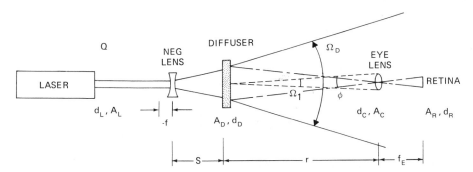

Fig. 4 Eye safety considerations: Q, energy output of laser; d_L, A_L, diameter and cross-sectional area of laser beam; d_D, A_D, laser beam diameter and area at the diffuser; d_c, A_c, diameter and area of the eye pupil; d_R, A_R, diameter and area of the image of the laser beam on the retina; ϕ, viewing angle subtended by the extended source; Ω_D, scatter angle of laser radiation; Ω_1, angle subtended by the eye pupil as seen from the diffuser; f_E, focal length of eye lens; r, range from the diffusing screen to the viewer; s, distance of the negative lens from the diffuser.

10.13.3.2 Object Beam Safety Precautions

The area of illumination on the ground glass screen is calculated according to

$$A_D = Q/I_{Dmax}, \qquad (4)$$

where Q is the laser output energy. The laser beam is expanded by a negative lens of focal length $(-f)$ located a distance s in front of the diffuser. The two parameters $(-f)$ and s can be calculated from simple ray geometry to give an energy density of I_{Dmax} at the screen for a laser output energy density $I_L = Q/A_L$. From Fig. 4 it follows $d_L/d_D = (-f)/(s - f)$ or

$$s = [(I_L/I_{Dmax})^{1/2} - 1](-f). \qquad (5)$$

Example Assume that a 1-cm-diameter output beam with an energy of 1.5 J ($I_L = 2$ J/cm²) is expanded by a negative lens of $f = -5$ cm. According to Eq. (5), a distance of $s = 21.5$ cm is required between the lens and the diffuser. The beam is expanded to an area of $A_D = 21.5$ cm² at the screen surface.

10.13.3.3 Reference Beam Safety Precautions

Direct radiation or reflection from the reference beam entering the eyes of the human subject have to be avoided under all circumstances. For example, the reference beam reflected from the photographic plate can cause eye damage. For Agfa Scientia 8E75 emulsions, the illumination at the photographic plate should be approximately 15μJ/cm². Assuming a 10% reflection from the photographic plate, the intensity reaching the eye is still 150 times greater than the permissible exposure level given by Eq. (2).

The danger is even greater from a reflection or white-light hologram because the reference beam illuminates the photographic plate from the side opposite to the subject. Only approximately 10% of the energy is absorbed by the emulsion. Precautions must obviously be taken to avoid injury to the subject. These include bringing the reference beam in at a steep angle so that the light will be reflected or transmitted away from the subject.

10.13.3.4 Avoidance of Stray Reflections

Physical barriers and shields should be erected to prevent the subject from accidentally viewing specular reflections from the reference or object beam. It is highly recommended that one determine the illumination patterns of the reference and object beam with a He–Ne alignment laser propagating in the same direction as the pulse laser beam.

10.13.4   Reconstruction†

Holographic portraits are viewed in the usual way with the expanded beam from a He–Ne or argon laser or filtered arc lamps. Reflection holograms can be reconstructed with incoherent light sources. Photographs of the holographic reconstructions are usually taken with single reflex cameras employing 35- or 50-mm focal length lenses and $f/2$ to $f/5.6$ settings. The aperture stop is a compromise between maximum depth of focus and minimum speckle. Increasing the f-number increases both the depth of focus and the speckle pattern on the holographic image.

REFERENCES

ACGIH (1972). American Conference of Governmental Industrial Hygienists, P.O. Box 1937, Cincinnati, Ohio 45201.

ANSI (1973). American National Standard for the Safe Use of Lasers, Z136.1-1973, Am. Nat. Std. Inst., New York, New York.

Ansley, D. A. (1970). *Appl. Opt.* **9**, 815.

Ansley, D. A., and Siebert, L. D. (1970). *Ann. NY Acad. Sci.* **168**, 475 (February).

Gregor, E. (1971). *Proc. SPIE*, p. 93 (April).

Gregor, E., and Davies, J. H. (1969). *Electro-Opt. Syst. Design*, p. 48 (July–August).

Levi, L. (1968). "Applied Optics: A Guide to Optical Systems Design," Vol. 1. Wiley, New York.

McClung, F. J., Jacobson, A. D., and Close, D. H. (1970). *Appl. Opt.* **9**, 103.

Middleton, W. E. K. (1960). *J. Opt. Soc. Amer.* **50**, 747.

Myers, G. E. (1973). *Electro-Opt. Syst. Design*, p. 30 (July–August).

Rundle, W. J., and Higgins, T. V. (1975). *Tech. Conf., Soc. Motion Picture Television Engrs. (SMPTE), 117th, September, Los Angeles, California.*

Siebert, L. D. (1968). *Proc. IEEE* **56**, 1242.

Sliney, D. H., and Freasier, B. C. (1973). *Appl. Opt.* **12**, 1.

Wuerker, R. F., and Heflinger, L. O. (1971). *SPIE J.* **9**, 122 (May).

Zech, R. G., and Siebert, L. D. (1968). *Appl. Phys. Lett.* **13**, 417.

† See Chapter 6.

10.14 PHOTOGRAMMETRY

N. Balasubramanian

10.14.1 Introduction

The manual of photogrammetry defines photogrammetry as the "science or art of obtaining reliable measurements by means of photography." Photogrammetrists rely on overlapping photographs for the extraction and analysis of three-dimensional information. The relative geometrical orientation of the overlapping photographs permits the reconstruction of a hypothetical three-dimensional stereomodel which is then used to measure the size, shape, and location of the object scene. On the basis of this definition one can envisage the application of photogrammetric principles whenever it is possible to produce photograms affording the adequate information. In the past, a major share of the photogrammetric work has been limited to topographic map plotting of the earth's surface because of the complexity of the photogrammetric operations. However, the application of photogrammetry to other disciplines such as engineering and biomedicine has given rise to the wide-spread acceptance of photogrammetric techniques. All the nontopographic applications are termed close range photogrammetry.

Several applications of holography to photogrammetric operations have been reported in the literature. Some of the major applications are

(1)   applications of optical data processing to achieve automation of the stereo compilation process,

(2)   holographic recording and mensuration for close range photogrammetry,

(3)   the synthesis of holographic stereomodels from stereophotographs for subsequent mensuration and display, and

(4)   the development of direct contouring techniques for obtaining metric information from objects.

The application of optical data processing techniques to achieve automation of stereocompilation has resulted in numerous different concepts and systems. A good review of these techniques is given by Balasubramanian and Leighty (1974). No effort will be made here to cover other applications such as direct holographic memories for the storage, retrieval, and display of aerial imagery.

10.   Application Areas

Greater emphasis is given here to the application of holography in close range photogrammetry and for the synthesis of holographic stereomodels.

10.14.2   Close Range Photogrammetry

It is useful first to outline the basic differences between a holographic virtual image and a photogrammetric stereomodel. In stereophotogrammetry a subjective three-dimensional model is formed by the intersection of the two conjugate bundles of rays emanating from the conjugate images of a relatively oriented stereopair of photographs. This subjective model is what is referred to as the stereomodel. Even this subjective model perceived by the eye represents only a simple fixed perspective of the object. In contrast, the virtual image of a hologram represents a true three-dimensional image and contains all the monocular parallaxes that were available in the original object. The range of perspectives that can be perceived is limited only by the aperture of the hologram over which the scattered optical wavefield corresponding to the original object has been recorded.

The methodology of extracting reliable metric information from photogrammetric stereomodels has been well established, and the same approach has been used to obtain the metric information from the holographic virtual image. Considerable work has been done with regard to the geometric fidelity of the reconstructed image and the results are well documented elsewhere (Glaser and Mikhail, 1970; Mikhail *et al.,* 1971a).

The geometric fidelity is greatly dependent on the similarity of the recording and reconstruction geometries of the hologram. A self-illuminated dot attached to an *XYZ* coordinate measuring device and placed at the virtual image space of the hologram is used as the measuring mark for performing the mensuration on the holographic virtual image. One such particular system is shown in Fig. 1. The self-illuminated dot represents the tip of an optical fiber attached to the measuring system of a wild A7 autograph. During measurement, the floating dot is brought into coincidence with the image point of interest and the coordinates of the floating dot give the coordinate of the image point. By maintaining the apparent contact of the floating dot during translation in a single plane, it is possible to generate contours and profiles that can be plotted directly. An example of such a contour, a map of a toothless dental casting, is shown in Fig. 2a.

The pointing precision, which in turn determines the accuracy and resolution of the contours, depends upon such factors as the contrast between the image and the measuring mark, image discernibility (ability to locate the top of the surface), magnification, and operator comfort during viewing. Also, the speckle present on the virtual image directly determines the resolution that is obtainable during mapping. It is very difficult to state a single number that determines the accuracy of the pointing operation in holograms. Because of

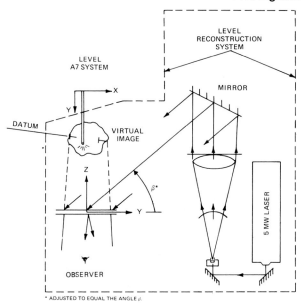

Fig. 1 Orienting the virtual image for mensuration (plan view).

the subjective nature, the accuracy varies from operator to operator. From the experience of several investigators, it is clear that the standard deviation of pointing accuracies are

$$\sigma_{xy} = 0.025 \quad \text{mm}, \qquad \sigma_z = 0.075 \quad \text{mm},$$

where σ_{xy} represents the standard deviation in the plane parallel to the hologram and σ_z represents the standard deviation perpendicular to the hologram (in the depth direction).

Direct holography offers unique advantages for close range applications. The depth of field of the reconstructed virtual image depends only on the characteristics of the coherent illumination used, and it can be adjusted to suit the task under consideration. In stereophotography, compromise on the resolution should be made in order to obtain large depth of field. The multiplicity of the perspectives permits easier pointing, increases the precision, and makes it less tedious. Even a person with monocular vision can perform the pointing, while such an effort would be impossible in stereophotogrammetry. An example is shown in Fig. 2b. Holography has its own limitations. In close range photogrammetry using stereophotography, there is no limitation on the size of the object to be recorded. However, the geometrical and physical considerations in conjunction with the requirement for coherent illumination limits the size of the objects that can be successfully recorded. In performing mensuration of the holographic virtual image, the scale is one-to-one and it is not

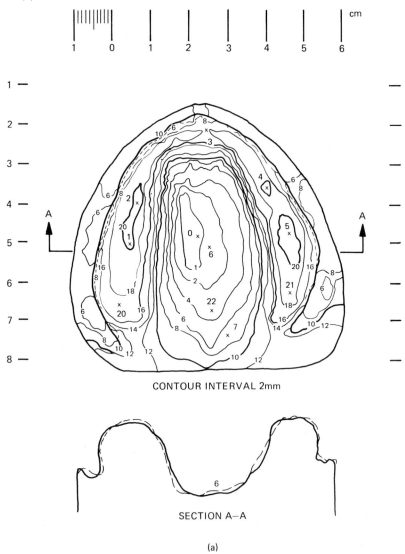

CONTOUR INTERVAL 2mm

SECTION A—A

(a)

Fig. 2 (a) Topography of a toothless dental casting, (b) monocular contouring; —— right to left, - - - left to right.

possible to obtain magnification without significantly distorting the reconstructed image. In this regard close range stereophotogrammetry offers definite advantages over direct holography. However, the ability to record and measure in three dimensions at a one-to-one scale should offer new possibilities and make holography a valuable supplement to close range photogrammetry. For

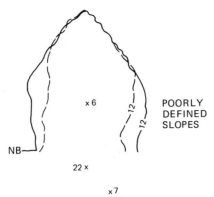

x 6

POORLY
DEFINED
SLOPES

NB

22 x

x 7

12mm CONTOUR USING ONLY ONE EYE

(b)

a good review of this subject area, see Kurtz *et al.* (1971) and Mikhail *et al.* (1971b).

10.14.3   Contour Generation with Holographic Interferometry

Several methods of generating contours representing the topography of the object using holographic interferometry have been proposed and demonstrated. The contour pattern, in the form of interferometric fringes, is produced so as to localize on or very close to the surface of the object as seen in the virtual image. The pattern is similar to the topographic contour map and should provide metric information about the object. There are three general methods of producing contours using holographic interferometry:

(1)   the method of multiple wavelength (or multiple frequency) (Heflinger and Wuerker, 1969; Hildebrand and Haines, 1967; Zelenka and Varner, 1968; Varner, 1971).

(2)   the method of multiple source (Hildebrand and Haines, 1967; Zelenka and Varner, 1968; Varner, 1971), and

(3)   the method of multiple indices.

No effort is made here to describe these various methods of generating contours or to point out their relative advantages or disadvantages. These matters are dealt with in Section 10.10. The evaluation presented here brings out the photogrammetric considerations that are necessary before the holographic contouring can become a useful tool in mensuration for close range applications.

The contours must be equal elevation contours and should represent orthoscopic projections. In all the holographic contouring methods, the contour fringes localize on or near the surface of the object (or its virtual image). When a picture of the object (or its virtual image) with the contour fringes superimposed on it is taken with a camera, as is well known in photogrammetry, one gets not an orthoscopic contour map but only a perspective contour map. Each of the contour fringes have different scale magnification and hence must be corrected individually. Hence a modification of the methods of extracting the contour information is necessary before the holographic contouring techniques can become a reliable tool for obtaining metric information. Also, the only way to tell if a contour is nearer or farther from the observer than an adjacent one is from the general appearance of the three-dimensional object being mapped. In this regard the actual three dimensionality of the virtual image helps in the process of discrimination. However, once the two-dimensional record in the form of a photograph is made, this advantage is lost. In spite of the problems of the holographic contouring methods already presented, they provide a simple means of obtaining semiquantitative metric information about the object for close range applications.

10.14.4 Holographic Stereomodels

In Section 10.14.3 it has been shown that direct holography is a valuable supplement to photogrammetry, and photogrammetric pointing techniques could be used to extract quantitative information from the holographic virtual image. If the object is either too small or too large for successful mapping, some form of scaling must be necessary to make it suitable for handling the information. Particularly if the object size is large, making coherent illumination of the object impossible, some form of intermediate recording of data must be necessary. The intermediate record could then be converted into a holographic virtual image exhibiting (from the subjective visual point of view) the relief information originally present in the object scene. Several methods of synthesizing three-dimensional holographic virtual images from several regular two-dimensional photographic records of the object scene have been proposed in the last few years. This class of holograms can be classified as composite holograms. Collier *et al.* (1971) have defined a composite hologram as "a collection of small holograms arranged in a plane with each hologram close to and often contiguous with its neighbor." The wavefronts recorded by the individual holograms are not necessarily continuous or coherent with one another. However, when the entire hologram is reconstructed, all individual reconstructions cooperate to produce a subjective three-dimensional image. A good review of these techniques has been given by Varner (1971). For additional information on composite holograms, see Section 5.5. The techniques reported have generally been proposed as novel means of storing and viewing stereoimages or as methods for reducing the information content so that three-dimensional images could be transmitted electronically. The exception to this is the holographic stereomodel, which has been developed with the subsequent processing of the model synthesized as a definite requirement.

A review of the optical considerations in recording and reconstruction of the various types of holographic stereomodels is now presented. The specific advantages and disadvantages are outlined along with their applications in topographic mapping.

The holographic stereomodel concept involves recording on a single holo-gram the overlapping images of two photographs of a common area having a proper relative orientation for subsequent viewing. Either the overlapped im-ages can be projected onto a rear projection screen and the projections re-corded holographically, or the overlapped image can be projected directly on the holographic recording photographic plate. The first case is referred to as the Fresnel holographic stereomodel, and the second method is called the focused-image holographic stereomodel.

The optical system used for recording the Fresnel holographic stereomodel is shown in Fig. 3. The stereo transparencies are projected onto the rear projection screen using coherent illumination derived from a laser. First the double projection system is relatively oriented to remove differences in scale and angular tilt at the overlapped image plane. While recording, only one of the projection systems at a time is illuminated. The entire plate records each of the projected transparencies with a different reference beam for the two exposures to allow later image separation during reconstruction. The optical system used for reconstructing the holographic stereomodel is shown in Fig. 4. Here the polarization of one of the reconstruction beams is rotated by 90° using a half wave plate. Orthogonally oriented polarizers are then used for viewing the reconstructed holographic stereomodel. The model can be viewed either orthoscopically or pseudoscopically by rotating the polarizers. Photo-grammetric analysis of the holographic stereomodel has shown that even though it is possible to perceive impressive three-dimensional display without critical alignment of the reconstructing beams, in order to obtain an undistorted model it is imperative that the reconstructing beams be critically aligned to duplicate the original reference beams. Also, it is necessary that the position of the eyes coincide with the perspective centers accurately to obtain undis-torted reconstruction of the stereomodel (Gifford and Mikhail, 1973).

The optical arrangement for recording the focused-image holographic ster-eomodel is shown in Fig. 5. This arrangement is identical to the Fresnel holographic stereomodel except for the fact that the rear projection screen has been replaced by the holographic recording plate. Even though the focused image stereomodel can be reconstructed for subsequent viewing using a system similar to the Fresnel holographic stereomodel, it is convenient to use white light or an extended incoherent source for viewing comfort. Image separation

627

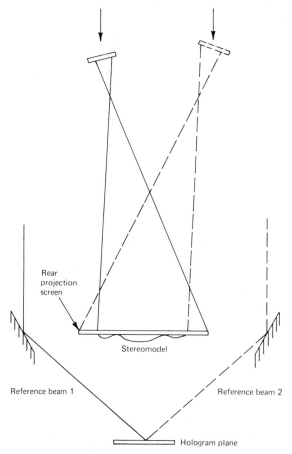

Fig. 3 Recording Fresnel holographic stereomodel.

can be obtained either by polarization or by angular discrimination of the reconstructed beams.

A focused-image holographic stereomodel can be thought of as a superposition of two carrier frequency photographs on a single emulsion. The two carrier frequencies determine the angular relationship between the reconstructed wavefronts. In a focused-image holographic stereomodel, the two carrier frequencies are essentially two off-axis zone plates that are displaced with respect to each other, thus permitting the separation of two images. When illuminated, the two off-axis zone plates produce two images of the source, the separation being identical to the eye separation. This simplified model for the recording and reconstruction of the focused-image holographic stereomodel permits easier means of producing such holographic stereomodels. If the stereopair of photographs are available in rectified form with equal scale magni-

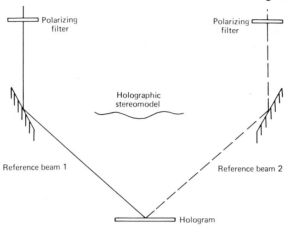

Fig. 4 Reconstruction of the holographic stereomodel.

fication, then the two photographs can be directly contact-printed onto the holographic recording plate using coherent carrier frequency illumination. This approach not only simplifies the process of recording the holographic stereo-model, but also provides advantages in terms of less geometrical distortion and greater resolution while increasing efficiency.

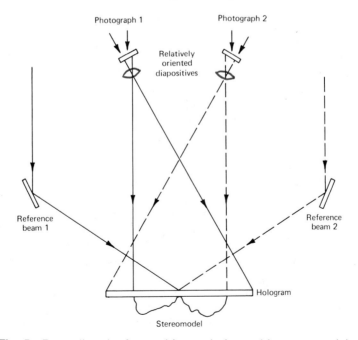

Fig. 5 Recording the focused-image holographic stereomodel.

629

10. Application Areas

Once the stereomodel has been recorded, mensuration can be performed using a self-illuminated measuring mark placed in the virtual image space, as is the case of direct holographic applications to close range photogrammetry. Thus holographic stereomodels provide simple, efficient, and permanent storage of relatively oriented stereoviews suitable for many topographic mapping applications. Fig. 6 shows a topographic map generated from one of the early holographic stereomodels.

The advantages of the holographic stereomodels have been outlined extensively in the literature. The major advantages are

(1) the relative orientation of the stereophotographs are permanently "locked in" and could be reconstructed at relative ease when required, and

(2) one has the ability to generate holographic stereomodels either from transparencies or from rectified prints.

While in the case of the Fresnel holographic stereomodel, the size of the model can be larger than the holographic recording plate; in the case of the focused-image holographic stereomodel, the size of the holographic recording plate limits the size of the model. The focused-image holographic stereomodel, however, offers many advantages as a display medium that could be used either as a training aid or a photointerpretor for qualitative analysis. Also the speckle, which is characteristic of any diffuse coherent illumination, is absent in the case of focused-image holographic stereomodels. It is also possible to superimpose a three-dimensional grid on the virtual image so as to permit a qualitative but a rapid extraction data from the holographic stereomodel (Mikhail *et al.*, 1971b). It is clear from the preceding arguments that holographic stereomodels could become a good supplement to the conventional photogrammetric operations.

10.14.5 Conclusions

Recent investigations on the applications of holography to mapping and photogrammetry indicate that there exists areas in which holography seems to offer real and practical solutions. Speckle introduced by the diffuse nature of the object still restricts the resolution and forms a barrier to the wide acceptance of the technique by the mapping community. In spite of the annoyance caused by the speckle, the holographic image has been shown to be adequate for many mensuration purposes. By interposing some optics between the virtual image and the observer (such as magnifiers), it is possible not only to reduce the size of the speckle but also magnify the original image so that the smallest points of interest exceed the main size of the speckle. In the case of the focused-image holographic stereomodel, no speckle is present since spatially incoherent light is used during reconstruction.

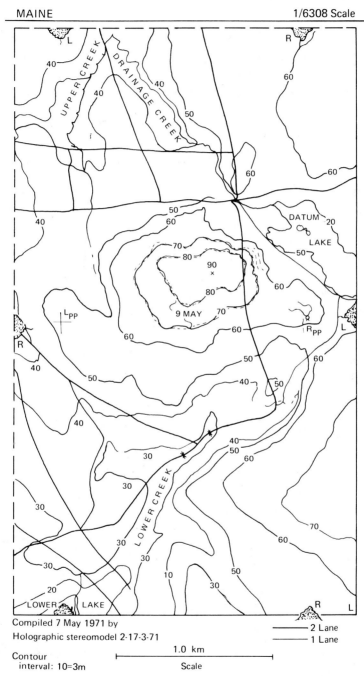

Fig. 6   Map compiled from holographic stereomodel.

10. Application Areas

REFERENCES

Balasubramanian, N., and Leighty, R. D. (eds.) (1974). *Proc. SPIE* **45**, (March).

Collier, R. J., Burckhardt, C. B., and Lin, L. H. (1971). "Optical Holography." Academic Press, New York.

Gifford, D. L., and Mikhail, E. M. (1973). Final Tech. Rep., USAETL, ETL-CR-73-14 (July).

Glaser, G. H., and Mikhail, E. M. (1970). Study of Potential Application of Holographic Techniques to Mapping. Purdue Univ., Interim Tech. Rep. ETL-CR-70-8 (December) (AD718084).

Heflinger, L. O., and Wuerker, R. F. (1969). *Appl. Phys. Lett.* **15**, 28–30.

Hildebrand, B. P., and Haines, K. A. (1967). *J. Opt. Soc. Amer.* **57**, 155–162.

Kurtz, M. K., Balasubramanian, N., Mikhail, E. M., and Stevenson, W. H. (1971). Study of Potential Application of Holographic Techniques to Mapping. Final Tech. Rep., Project DAAK-02-69-C-0563, for US Army Engineer Topographic Lab., ETL-CR-71-17 (October).

Mikhail, E. M., Glaser, G. H., and Kurtz, M. K., Jr. (1971a). Holograms for Mensuration of Close-Range Objects, *Symp. Close Range Photogrammetry, Amer. Soc. Photogrammetry, January 26–29.*

Mikhail, E. M., Kurtz, M. K., and Stevenson, W. H. (1971b). Metric Characteristics of Holographic Imagery. Paper presented at SPIE Seminar in Depth, *Quantitative Imagery in the Bio'Medical Sciences, Houston, Texas, May 10–12.*

Varner, J. R. (1971). Holographic Contouring. Presented at SPIE Seminar in Depth, *Holography, 1971, Boston, Massachusetts, April 14–15.*

Zelenka, J. S., and Varner, J. R. (1968). *Appl. Opt.* **7**, 2107–2110.

Index

Index

Index

636

Index